Marine Biochemistry

This book provides the latest comprehensive methods for isolation and other novel techniques for marine product development. Furthermore, this book offers knowledge on the biological, medical, and industrial applications of marine-derived medicinal food substances.

There has been a tremendous increase in the products derived from marine organisms for commercial application in industries every year. Functional foods of medicinal value are particularly in demand as new technology allows the stabilization of natural ingredients and their availability in pure forms to solve various human diseases. Marine flora and fauna have essential elements and trace minerals that nurture various hormones produced in the endocrine system to regulate the respective metabolisms, thereby providing a safe and healthy life to humans.

The overall presentation and clear demarcation of the contents by worldwide contributions is a novel entry point into the market of medicinal foods from the sea. The exploration of marine habitats for novel materials are discussed throughout the book.

The exploration and exploitation of the biochemistry of sea flora and fauna are limited, and this book extends the research possibilities into numerous marine habitats.

Various approaches for extracting and applying the flora and fauna are discussed. This book will be of value to researchers, marine biotechnologists, and medical practitioners, due to the vast information, as well as industrial and medical applications of marine substances all in one place.

Marine Biochemistry
Isolations and Techniques

Edited by
Se-Kwon Kim

CRC Press is an imprint of the
Taylor & Francis Group, an **informa** business

First edition published 2023
by CRC Press
6000 Broken Sound Parkway NW, Suite 300, Boca Raton, FL 33487–2742

and by CRC Press
4 Park Square, Milton Park, Abingdon, Oxon, OX14 4RN

CRC Press is an imprint of Taylor & Francis Group, LLC

ISBN: 978-1-032-30030-6 (hbk)
ISBN: 978-1-032-30212-6 (pbk)
ISBN: 978-1-003-30390-9 (ebk)

DOI: 10.1201/9781003303909

Typeset in Times
by Apex CoVantage, LLC

Contents

Preface

About 70% of the earth's surface is covered by water. The sea contains fish, seaweed, microbes, and plants. These species have traditionally been used in the development of medical applications. Bioactive metabolites found in marine-derived species can be developed into pharmaceuticals and nutraceuticals. The fact that these species live in harsh environments explains why they have unique metabolites. These species, on the other hand, are underutilized and unexplored in terms of biological applications. As a result, the current book provides an overview of these species, including their biochemistry, culture, and application, to better understand their use.

Natural products from the sea have a lot of potential and bioactivity for treating diseases: marine organic compounds, proteins, carbohydrates, lipids, and macromolecules aid in developing human-made products. However, the production of these compounds in the species is critical. Thus, understanding the biochemical perspectives of chemical production processes is vital. The current book investigates the marine biochemistry of the species in the production of marine metabolites. Furthermore, marine natural products and their chemistry are discussed, from their origins to pharmaceuticals and various natural products used to treat diseases.

The book discusses molecular identification and the phylogenetic relationship of flying fishes. Furthermore, the biochemical responses of fish to environmental stresses are discussed, and the use of fisheries byproducts for functional food is proposed. One section of the book discusses biochemistry aspects of marine microbiology. There are several microbial enzymes and derivatives, exopolysaccharides, protease inhibitors, and bioactive compounds presented.

The book *Marine Biochemistry* will be helpful to novice as well as graduate students and marine biologists, biotechnology, and industrialists. I am grateful to all the people who contributed to this book.

Se-Kwon Kim

Acknowledgments

Dr. Jae-Chul Kim is the chairman, president, and founder of the Dongwon Group. The company was established in 1969 to explore and utilize the oceans and marine resources. Chairman Kim is the pioneer of deep-sea fishing boats in South Korea. He started tuna fishing in 1958 as the first mate of Korea's first deep-sea fishing vessel.

After graduating from the National Fisheries University of Busan, he jumped into tuna fishing, and the university was renamed Pukyong National University. Furthermore, he became the captain of a fleet of deep-sea fishing boats and a new tuna fishing method. He developed and succeeded in catching large amounts of tuna in the South Pacific and Indian Oceans.

He founded Dongwon Industries in 1969 at the age of 35 and started tuna fishing as the first president. After that, he built a tuna processing plant and started to produce canned food. Based on the lessons he learned at sea as a child, he immersed himself in business management full-time.

Dongwon Group has successfully expanded its business from the fishing industry as the primary industry to the manufacturing industry as the secondary industry and financial services as the tertiary industry. Currently, Mr. Kim runs 30 affiliated companies.

To contribute to social welfare, he also established the Dongwon Educational Foundation. He devoted himself, his heart, and his soul to nurturing competent people who are the backbone of our society. He has provided scholarships to numerous college students and grants for R&D funds for researchers. He recognized the great value and potential of underutilized marine resources from his youth. He dedicated himself publishing technical books emphasizing the scientific importance of marine life and related research.

With his help, this book has been published, providing readers with the information on how the scientific values of marine life can enhance human health and well-being.

I want to thank him sincerely for his support in publishing this book.

Se-Kwon Kim
Distinguished Professor
Deptartment of Marine Science and Convergence Engineering
Hanyang University ERICA, South Korea

Editor

Se-Kwon Kim, PhD, is a Distinguished Professor at Hanyang University and Kolmar Korea Company. He worked as a Distinguished Professor in the Department of Marine Bio Convergence Science and Technology and Director of Marine Bioprocess Research Center (MBPRC) at Pukyong National University, Busan, South Korea.

He earned an MSc and a PhD at Pukyong National University and conducted his postdoctoral studies at the Laboratory of Biochemical Engineering, University of Illinois, Urbana–Champaign, Illinois, USA. Later, he became a visiting scientist at the Memorial University of Newfoundland and the University of British Colombia in Canada.

Dr. Kim served as president of the Korean Society of Chitin and Chitosan from 1986 to 1990 and the Korean Society of Marine Biotechnology from 2006 to 2007. To the credit for his research, he won the Best Paper Award from the American Oil Chemists' Society in 2002. Dr. Kim was also the chairman for Seventh Asia-Pacific Chitin and Chitosan Symposium, which was held in South Korea in 2006. He was the chief editor in the Korean Society of Fisheries and Aquatic Science from 2008 to 2009. In addition, he is the board member of the International Society of Marine Biotechnology Associations and the International Society of Nutraceuticals and Functional Food.

His major research interests are investigation and development of bioactive substances from marine resources. His immense experience in marine bioprocessing and the mass-production technologies for marine bio-industry is the key asset of holding majorly funded marine bio projects in Korea. Furthermore, he expended his research fields up to the development of bioactive materials from marine organisms for their applications in oriental medicine, cosmeceuticals and nutraceuticals. To this date, he has authored more than 750 research papers, 70 books, and 120 patents.

Contributors

Imran Ahmad
Food Agriculture and Bio-Innovation Laboratory
Chaplin School of Hospitality (F&B Science Program)
Florida International University
Miami, Florida, USA

Shiek SSJ Ahmed
Drug Discovery Lab
Faculty of Allied Health Sciences
Chettinad Academy of Research and Education
Kelambakkam, India

Sadar Aslam
Centre of Excellence in Marine Biology
University of Karachi
Karachi, Pakistan

V. Baskaran
Council of Scientific and Industrial Research—Central Food Technological Research Institute (CSIR-CFTRI)
Mysore, India

Wafa Boulajfene
Unité de Recherche de Biologie Intégrative et Ecologie Évolutive et Fonctionnelle des Milieux Aquatiques
Département de Biologie
Faculté des Sciences de Tunis
Tunis, Tunisia

Ekowati Chasanah
Center for Bioindustrial Technology
National Agency for Research and Innovation (BRIN)
Tangerang, Indonesia

Jeyapragash Danaraj
Department of Biotechnology
Karpagam Academy of Higher Education (Deemed to be University)
Coimbatore, Tamil Nadu, India

B. Deivasigamani
Centre of Advanced Study (CAS) in Marine Biology
Faculty of Marine Sciences
Annamalai University
Parangipettai, India

Harika Atmaca Ilhan
Celal Bayar University
Faculty of Science and Letters
Department of Biology
Section of Molecular Biology
Şehit Prof. Dr. İlhan Varank Kampüsü
Manisa, Turkey

Hari Eko Irianto
Research and Development Center for Marine and Fisheries Product Processing and Biotechnology
Jakarta, Indonesia

Ross A. Jeffree
Jeffree Conservation and Research
Alfords Point, New South Wales, Australia

Maushmi S. Kumar
Shobhaben Pratapbhai Patel School of Pharmacy and Technological Management
SVKM's NMIMS
Mumbai, India

Madan Kumar P
Department of Biochemistry
Council of Scientific and Industrial Research
Central Food Technological Research Institute (CSIR-CFTRI)
Mysore, India

Toshiki Nakano
Marine Biochemistry Laboratory
Graduate School of Agricultural Science
Tohoku University
Aobaku, Sendai, Japan

J. Paniagua-Michel
Department of Marine Biotechnology
Centro de Investigación Científica y de Educación Superior de Ensenada
Ensenada, México

Vandana B. Patravale
Professor of Pharmaceutics
Department of Pharmaceutical Sciences and Technology
Institute of Chemical Technology (Deemed to be University)
Mumbai, India

André Luiz Meleiro Porto
Universidade de São Paulo
Instituto de Química de São Carlos
São Carlos, Brazil

K.V. Bhaskara Rao
Department of Bio-Medical Sciences
School of Biosciences and Technology
Vellore Institute of Technology
Vellore, India

Prashakha J. Shukla
Life Science Department
C.U. Shah University
Surendranagar, India

Pitchiah Sivaperumal
Department of Pharmacology
Saveetha Dental College and Hospitals
Saveetha Institute of Medical and Technical Sciences (SIMATS)
Chennai, India

Chee Kong Yap
Department of Biology, Faculty of Science
Universiti Putra Malaysia
Serdang, Selangor, Malaysia

1 Marine Natural Products and Chemistry

From Origin to Pharmaceutics

Jeyapragash Danaraj and Saravanakumar Ayyappan

CONTENTS

1.1 INTRODUCTION

The development of modern pharmaceuticals is considered one of the major undertakings of the 20th century, and the nature acted as a significant source of therapeutics from the early eras. However, the new beginning of science and understanding of the pathological role of several diseases facilitated the design and synthesis of drugs specific to their molecular targets and thus attracted the attention from natural products to the chemically synthetic molecules as drugs (Houghten, 1994; Ortholand and Ganesan, 2004). Depending on the synthetic drugs for faster drug discovery, they were found inappropriate for our chiral world, leading to an associated 20-year low of new chemical entities in 2001 (Class, 2002). Natural products are the often structurally complex secondary metabolites that originate from the earth's flora and fauna. These specialized products possess a well-defined spatial orientation that evolved to

DOI: 10.1201/9781003303909-1

react resourcefully with their biological targets and thus occupy biologically pertinent chemical spaces, which are the starting points for drug discovery. More than 1 million potential drugs with new chemical entities have been discovered so far, and approximately 60% of the drugs available in the market are of natural origin, with reduced side effects and better absorption (Henkel et al., 1999), and 87% of human diseases, including cancer, are treated using natural products (15MNA). Since natural bioactive molecules exhibit cytotoxic effects by attacking macromolecules, natural products, hence, act as a better choice for isolating and identifying further lead structures (Newman and Cragg, 2012). As of now, roughly half of the 20 best-selling nonprotein drugs are related to natural products. It was estimated that 2 million species of plants, animals, fungi, and microorganisms were documented among 95% of the world's biodiversity that has not been evaluated for any bioactivity even today, and therefore, a scientific challenge occurs in identifying and developing efficient, effective drugs for therapeutic applications. Most natural products available in the market have originated from terrestrial sources. However, in recent years, identifying novel sources for natural products remains challenging, and the marine environment paves a clear way for the novel chemical and biological entities with promising bioactivities in pharmaceutics. Kong et al. (2010) reported that the natural products derived from marine sources were found superior to terrestrial natural products with novelties in their chemical structures.

Marine natural products emerged in the late 19th century, and biotechnology appeared as a field that provided the direction to undertake marine sources for drug development after 1980 (Newmann and Cragg, 2016). With the great potential of marine bioactive products, there is increased attention to exploiting biodiversity for drug discovery, which is challenging. It was documented that the molecular scaffolds were compared in the natural product dictionary to those in the marine natural product dictionary that showed approximately 71% were exclusively utilized by the marine organisms. Moreover, marine organisms show a higher incidence of significant bioactivity compared with terrestrial organisms (Montaser and Hendrik, 2011), and the first marine drug successfully approved by the Food and Drug Administration (FDA) and made it to the market is Cytarabine (Ara-C) for cancer during 1969, followed by Vidarabine (Ara-A) as an antiviral agent during 1976 (Figure 1.1). Most of these drugs are in several phases of clinical trials. However, still research goes on for new therapeutic agents in order to treat a large number of diseases that have no effective therapies. Many forms of bacterial, viral, fungal infections, cancer, inflammatory diseases, and neurodegenerative diseases cannot be treated successfully yet due to the lack of bioactive natural products. Moreover, the emergence of antibacterial resistance strains has developed due to the concomitant use of antibiotics results, with increased numbers of patients in hospitalization and even dying, and hence, heavy investment is needed to develop the new antibiotics that will be needed in the next millennium. This chapter deals with the significance of marine natural products for therapeutics and the sources for new biological activities and chemical structural diversity.

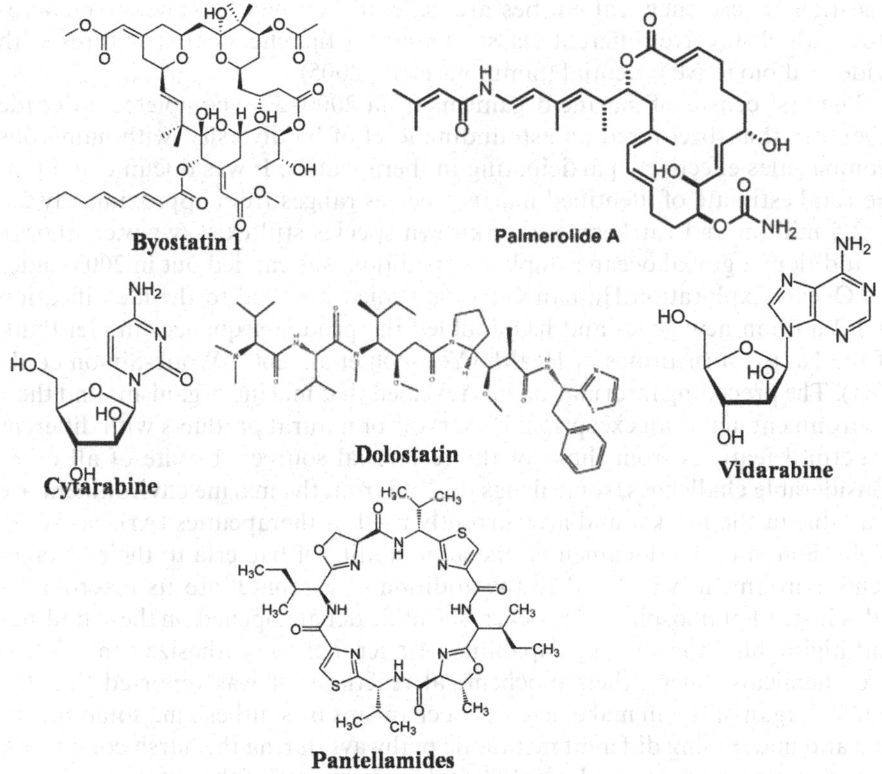

FIGURE 1.1 Chemical entities derived from marine organisms.

1.2 MARINE ENVIRONMENT: INCREDIBLE BIOLOGICAL AND CHEMICAL DIVERSITY

The revival of natural products-initiated drug discovery is coupled with the exploration of novel natural resources such as those in the marine environment, as it covers 70% of the earth's surface and signifies the largest unexplored wealthy resource. In comparison with terrestrial organisms, marine organisms do not have an illustrious history of use in traditional medicines. Nevertheless, advances in technologies and engineering, such as scuba diving and remotely operated vessels, in the last 50 years have attracted the scientific community and paved the way for exploring the resources in the marine environment for drug discovery (Cragg and Newman, 2013). The coexistence of many organisms in the marine environment increases their competitiveness and complexity for synthesizing natural products for their survival. For example, sessile organisms, such as algae, corals, sponges, and other invertebrates, are used to competing in the marine environment, of which of which exert chemicals in order to defend themselves against predators or suppress the overgrowth of other competing species or, conversely, control the motile prey species for ingestion. These chemical entities are generally termed "secondary metabolites," which involve different classes based on the chemical structures with evidenced bioactive potential (Simmons et al., 2005).

The first census of marine organisms from 2000–2010 completed a decade inventory that discovered an astounding level of biodiversity, with numerous biomolecules effectively participating in therapeutics. It was documented that the total estimate of identified marine species ranges from approximately 2.3 to 2.5 million and furthermore, unknown species still exist (www.coml.org). In addition, a global ocean sampling expedition was carried out in 2003 under the Ocean Exploration Human Genome Project that led to the identification of 1.2 million new genes and has doubled the protein sequences in GenBank of the National Institutes of Health (Yooseph et al., 2007; Wolfe-Simon et al., 2011). The preceding information has revealed that marine organisms and their environment act as an exceptional reservoir of natural products with different structural features from those of the terrestrial sources. Inspite of all other considerable challenges, some drugs derived from the marine environment are available in the market and are currently used in therapeutics (Arizza, 2013). Wolfe-Simon (2011) documented the adaptability of bacteria to their arsenic-rich environment, which led to the addition of arsenate into its macromolecules instead of phosphate. However, scientific debate opened on these findings and highlighted the striking capability of microbes to synthesize some defensive chemicals during their biochemical reactions. It was reported that the marine organisms will make defense mechanisms by synthesizing some chemicals and undergoing different metabolic pathways during the harsh conditions. For example, the deep-sea hydrothermal vents represent the extreme environments for organisms to survive, which surprisingly enclose high densities of biologically different communities with significant metabolism (Thornburg et al., 2010). Loichelins A-F, a novel peptide siderophore derived from the marine bacterium *Halomonas* sp. LOB-5, existS in the basalt weathering of the Loihi seamount east of Hawaii (Homann et al., 2009). Similarly, palmerolide A is a potent anticancer compound isolated from a cold-adapted circumpolar tunicate from Antarctica (Diyabalanage et al., 2006). However, symbiosis is an important phenomenon in the marine environment that brings nutritional scarcity and, in turn, the need for chemical defense mechanism in organisms devoid physical defensive techniques. A profound example is the symbiotic association between *Prochloron* spp. bacteria and their ascidian as animal hosts. During this symbiotic association, the bacteria provide photosynthate and defensive chemicals to the host, and in return, they will obtain the waste nitrogen (Schmidt, 2008). It was reported that patellamides, secondary metabolites that are cytotoxic in nature, were initially isolated from *Lissoclinum patella*, a marine tunicate (Ireland et al., 1982); however, the associated bacterium proved to be a true producer during symbiotic association (Schmidt et al., 2005). Concurrently, Davidson et al. (2001) reported that the symbiotic association of bacteria *Cadidatus Endobugula sertula* with marine bryozoans *Bugula neritina* results in the production of bryostatin, which exhibits cytotoxic activity. Among the marine organisms, marine sponges and macroalgae are the most prolific sources for diversified chemical compounds that also exhibit higher bioactive potential. Of the more than 5000 natural compounds derived from marine sources, marine sponges contribute about 30% of organisms with bioactive potential (Ireland et al., 1993).

Other marine sources that contribute to the natural bioactive compounds are algae, ascidians, bryozoans, cnidarians, and mollusks. In addition, different strains of phytoplankton, specifically diatoms, exhibit antimicrobial activity; however, the level of inhibition was found to be low compared to the compounds derived from other marine sources (Viso et al., 1987). From previous findings, it shows a clear picture that the marine environment represents a unique resource with massive biological diversity that potentially may lead to unique biological active chemical diversity that can be translated into biomedicine for therapeutics.

1.3 CHARACTERIZATION OF MARINE NATURAL PRODUCTS AND ITS CHEMISTRY

The marine environment is a vast area of the planet and plays a fundamental role in its physics, chemistry, and biology by providing an interconnection between different natural systems with wide range of precious ecosystem services (Halpern et al., 2012; Botana and Alfonso, 2015). These include marine natural products that have bioactive potential in therapeutics. However, the major problem for better understanding of marine metabolites chemistry and their composition is sampling difficulties. In order to characterize the bioactive metabolites, a sufficient number of samples are needed due to the instrumental resolution and bioassay approaches adopted during analysis. Modern analytical techniques, such as mass spectrometry (MS) and nuclear magnetic resonance (NMR) spectroscopy, require a small concentration of compounds in order to elucidate the structure (Kim et al., 2016). The MS technique has been used for the past two to three decades to identify the compounds present in the marine community. Furthermore, gas chromatography and liquid chromatography, coupled with MS, are considered hyphenated techniques that have higher resolutions and selectivity.

In MS, different detectors, such as photodiode array detector, electrospray ionization, and quadruple detector, are used for separating molecules with different masses and chemical structures. It was reported that a compound hierridin B were isolated from a marine cyanobacterium *Cyanobium* spp. strain and was identified and quantified with the help of MS analysis (Freitas et al., 2016). It is well known that NMR spectroscopy is most commonly used technique for the structural characterization of molecules, and the compound acremines P was isolated from a marine-derived strain of *Acremonium persicinum* that was structurally elucidated using NMR spectroscopy (Garson et al., 2017). However, NMR spectroscopy is considered less sensitive compared to MS, hindering the metabolite identification if the sample concentration is at trace levels or low amounts. For analyzing samples at low concentrations, liquid chromatography–high-resolution MS (LC-HRMS) is used, which is extremely sensitive; hence, it can be used to identify certain classes of compounds that cannot be detected by MS. This is due to the lack of ion formation, or the ion formation may get suppressed, or the ions may not be able to elute from the column to be detected (Khalifa et al., 2019). On the other hand, NMR spectroscopy has no separation process; hence, it provides a snapshot of the sample metabolomic data. NMR spectroscopy is less sensitive, more robust, and more reproducible than MS, which has universal features in metabolite detection. In recent years, the "omics" approach has played a vital role in the identification and detection of metabolome from biological entities using metabolomics tools. XCMS Online and MetaboAnalyst tools were used to identify the entire metabolite content in the biological sample. In addition, these tools help identify the metabolic pathways and gene-to-metabolomic networks that occur in marine organisms. Jeyapragash et al. (2020) have documented the comparative metabolomics analysis of seagrasses and have identified the variations that occurred in the metabolic pathways of organisms. Furthermore, this tool can be used to understand the adaptation, response, and tolerance mechanisms of marine organisms during their stress mechanisms (Jeyapragash et al., 2021).

1.4 DISCOVERY AND ORIGIN OF MARINE PHARMACEUTICS: CURRENT STATUS

During the evolution process, marine organisms have become exceptional at accumulating metabolic content by synthesizing specific and potent activities (Murray et al., 2013). Their production of bioactive compounds predominantly occurs in sessile organisms, such as algae, sponges, cnidarians, tunicates, and bryozoans, during the defensive mechanisms (Botana and Alfonso, 2015). Furthermore, the bioactive compounds released into the water rapidly dilute, and they necessitate being highly potent without losing their efficacy (Haefner, 2003); therefore, the number of natural products and novel chemical structures that exist in the oceans that have economic and health benefits was widely accepted for pharmaceutics. Government and non-governmental organizations and industrial and academic laboratories have given themselves to science for the discovery of marine-derived pharmaceuticals since the mid-1970s. Due to the availability of funding, marine-based drug discovery came into the limelight for cancer research. As a result, several marine-derived compounds are in the clinical trials, and some of the drugs been approved and are available in the market for cancer treatment, some of which include bryostatin, discodermolide, ecteinascidin, halichondrin, and others.

Bryostatin is a polyketide isolated from the bryozoan *Buguna neritina*, which is effective in both anticancer and immune-modulating activity (Suffness et al., 1989). Its mechanism of action is via the activation of protein kinase C-mediated cell-signal transduction pathways. The drug is under the clinical trial of phase II for chronic lymphocytic leukemia. Non-Hodgkin's lymphoma and multiple myeloma via a cooperative research and development agreement between the National Cancer Institute (NCI) and Bristol Myers Squibb. Ecteinascidin-743, an alkaloid isolated from the ascidian *Ecteinascidia turbinate*, is under the phase

I clinical trial for ovarian cancer. Discodermolide, a polyketide isolated from marine sponges of the class *Discodermia*, acts as an immunosuppressive and anticancer agent that inhibits cell proliferation by intrusion of the cells' microtubule network (Hart, 1997). It is also effective against breast cancer and has been licensed by the Harbor Branch Oceanographic Institution and Novartis Pharmaceutical Corporation. Another promising compound from the marine sponge is halichondrin B, isolated from marine sponge from New Zealand, *Lissodendoryx* sp. Apart from the previously mentioned drugs, many secondary metabolites, such as alkaloids, terpenoids, polyketides, peptides, steroids, sugars, and shikimic acid derivatives, were isolated from marine sponges and still research goes on in order to identify their biological activities, including anti-microbial, anti-tumor, anti-diabetic, anti-inflammatory, anti-malarial, anti-fouling, and anti-protozoal, with higher industrial and therapeutic potential (Mayer et al., 2017; Agrawal et al., 2018).

Although marine biodiversity exhibits unique chemical structures with bioactive potential, the bulk supply of chemicals for continuous clinical trials is a major problem as it depends on the large-scale collection of source organisms from the marine environment that act as suitable drug candidates. Clinical trials could be achieved with enormous amounts of drugs extracted from the source organisms. This could be achieved either by culturing the organisms or through the synthesis of compounds by precursor and elicitor feeding using economically feasible, large-scale industrial processes. The biodiscovery of marine drugs and its vision toward therapeutics had their beginning in the 1950s with the scientific achievement of Bergmann, who isolated and identified two nucleosides, spongouridine and spongothymidine, from the Caribbean sponge *Cryptotethya crypta*, which attracted researchers in the scientific community and led to the synthesis of its analogs, Vidarabine® Ara-A and Cytarabine® Ara-C, as antiviral and antitumor drugs, respectively, that was first available in the market (Botana and Alfonso, 2015). It was reported that the 30,000 new compounds of marine origin were isolated and that more than 300 patents were approved (Blunt et al., 2014).

Despite the noteworthy bioactive potential of marine natural products as a source of new drugs for therapeutics, their role has undergone numerous changes that led to a decline in pharmaceutical research and development activities in the mid-1990s. However, the area of marine drug discovery has received a new beginning in the last decade, with vast increased numbers of new compounds isolated as compared to the previous decades (Molinski et al., 2009). This renaissance might be due to the technological development in analytical and high throughput screening. In recent years, advancements in the "omics" approach and combinatorial chemistry have contributed to the medical field, being powerful tools for identifying new chemical entities with therapeutic potential for treating diseases (Bucar et al., 2013). Huge efforts have been made over the last 30 years with productive and promising results on marine products with bioactive potential; more than 18,000 new compounds from marine sources were identified in addition to the nine marine drugs currently available in the market. It includes vidarabine (Vira-A®), cytarabine

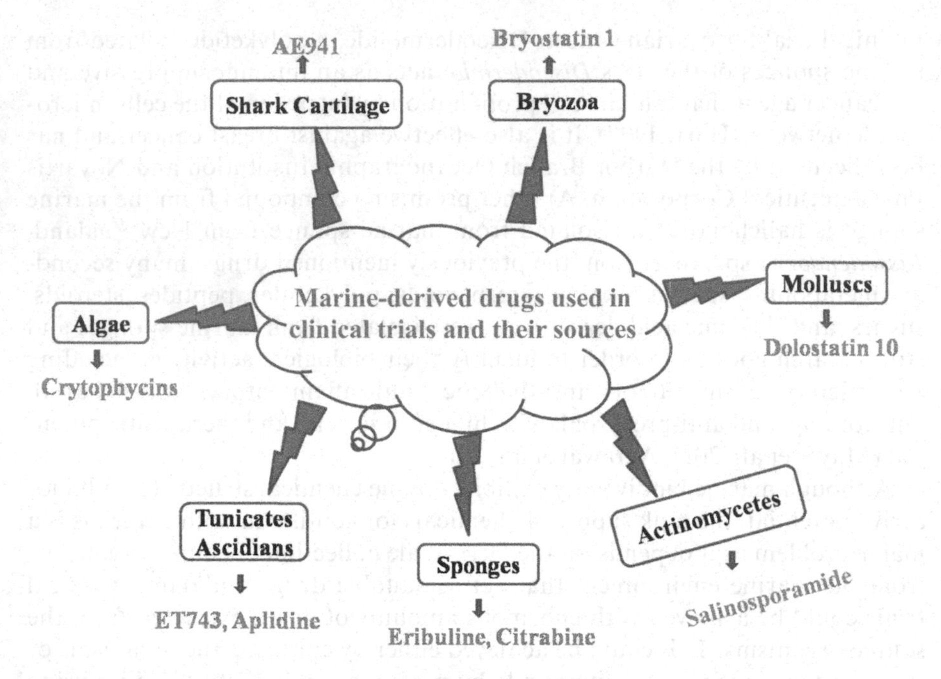

FIGURE 1.2 Marine-derived drugs used in clinical trials and their sources.

(Cytosar-U®), ziconotide (Prialt®), brentuximab vedotin (Adcetris®), eribulin mesylate (Halaven®), omega-3-acid ethyl esters (Lovaza®), nelarabine (Arranon®), trabectedin (Yondelis®), and Fludarabine Phosphate (Fludara®), and these drugs were approved by the FDA and European Agency for the Evaluation of Medicinal Products (EMEA). One over-the-counter drug, iota-carrageenan (Carragelose®), has also been approved by EMEA and is the first product isolated from the marine algae that acts as an antiviral agent. As research is still ongoing, the number of newly approved drugs from marine sources will increase. It was reported that 28 marine-derived drugs are currently in clinical trials, of which six molecules are under Phase III trials, 14 molecules are in Phase II trials and 8 molecules in Phase I (FDA, 2018; Mayer, 2018). Of the marine drugs studied for therapeutics, most of them are effective against cancer, and the compounds under clinical trial have shown better activity in cancer therapies (Figure 1.2).

1.5 FROM MARINE ORGANISMS TO NOVEL PHARMACEUTICS

1.5.1 Marine Microbes

Microorganisms have played a vital role in therapeutics since the discovery of penicillin in the late 1920s, and terrestrially, mainly microorganisms derived from the soil have been the most important resources for discovery of new drugs. It was documented that there are more than 22,000 known secondary metabolites from microorganisms, of which 70% are contributed by actinomycetes, 20% by fungi, 7% by *Bacillus* sp. and 1%–2% by other bacteria (Huang et al., 2016). It should be noted that 10% of all currently known natural bioactive products from marine source, some of which are antibiotics of microbial origin, such as penicillin, cephalosporin, streptomycin and vancomycin, are effective in reducing bacterial infections; actinomycin and mitomycin, as anticancer drugs; and cyclosporine, as immunosuppressant drugs. In addition, in the synthesis of natural products, microorganisms help accumulate enormous amounts of the drugs, or it can be metabolically engineered to produce a larger concentration by precursor and elicitor feeding. Apart from the bioactive potential of terrestrial microbes, most of the microbial biodiversity on the earth is found in the ocean, and many classes of microorganisms make symbiotic associations with the other organisms in the marine environment. Due to the unique adaptation and defensive mechanisms of the immensely diverse microorganisms in the marine environment, researchers and the pharmaceutical sector have been attracted to undertaking research on bioactive compounds for therapeutics. Since microorganisms are easy to handle and manipulate for higher production, they turned into a major focus.

It was reported that 5% of the marine bacteria present could be cultured during the 1950s, and this led to the prevalent belief that the microorganisms derived from the marine source were unculturable. This resulted in great hesitation regarding aggressive marine-oriented programs; however, as time went on, technology was developed that helped successfully cultivate marine microorganism cultures in the laboratory (Davidson, 1995). Later, the microbial biodiversity in the marine environment attracted attention among the scientific community worldwide, and researchers tended to focus on coastal sediment and organisms associated with marine invertebrates for isolating and identifying novel secondary metabolites. Trischman et al. (1994) reported that marine microbes derived from the coastal sediments, grown under saline conditions, have yielded new antibiotics, antitumor, and anti-inflammatory compounds. Furthermore, the endophytic fungi associated with marine plants and other invertebrates were found to synthesize bioactive compounds that exhibit pharmaceutical potential, and therefore, research has focused on fungi for the marine-derived drugs (Belinsky et al., 1998). In addition, the current focus is on isolating microbes from geothermal vents, which is of considerable interest as a higher biomedical resource that has shown the ability to inhibit colon tumor cell growth and prevent the replication of HIV, AIDS, and viruses (Gustafson et al., 1989).

1.5.1.1 Marine Bacteria

Bioactive compounds derived from marine *Pseudomonas* are pyrroles, pseudopeptides, pyrrolidinedione, quinoline, benzaldehyde, quinolone, phloroglucinol, phenanthrene, phenazine, zafrin, moiramides, bushrinm and andrimid (Isnansetyo and Kamei, 2009). Some of these compounds exhibit antimicrobial potential, including dibutyl phthalate and di-(2-ethylhexyl) phthalate, which have been reported to inhibit cathepsin B. In addition, bryostatins, sarcodictyin, discodermolide, and eleutherobin are the most effective cancer-treating

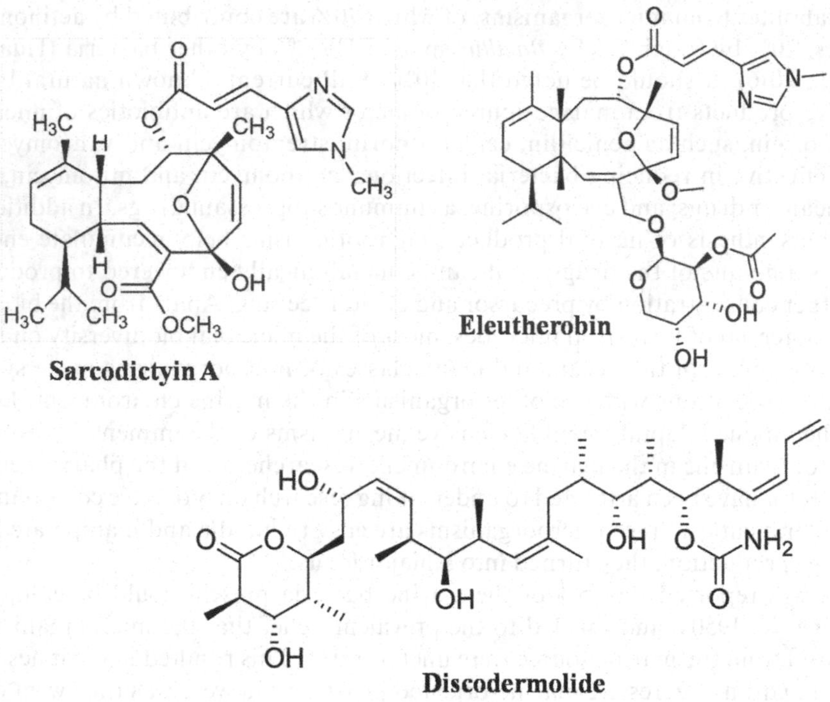

Sarcodictyin A

Eleutherobin

Discodermolide

FIGURE 1.3 Drugs derived from marine bacteria.

drugs produced mainly from bacteria (Malakar and Ahamad, 2013; Figure 1.3). Carte (1996) reported that *Lactobacilli* and *Noctiluca scintillans* exhibited potential chemoprotective against colon cancer and melanoma cancer under *in vivo* conditions. Lactobacilli as a dietary component have been found to reduce the activities of azoreductase, nitroreductase and beta-glucuronidase enzyme activity in the diet of rats. Since it reduces the functions of intestinal enzymes, it could lessen the development of colon cancer (Mitall and Garg, 1995). Sagar et al. (2013) reported that the *Halomonas* spp. strain GWS-BW-H8hM derived from a marine source suppressed the growth of HM02 (gastric adenocarcinoma), HepG2 (hepatocellular carcinoma) and MCF7 cell lines to induce apoptosis via cell cycle as compared to the actinomycin D. Ruiz-Ruiz et al. (2007) documented that the exopolysaccharides and sulfated polysaccharides derived from *H. stenophila* in the hypersaline environment exhibited the pro-apoptotic effects on leukemia cells. Two more active extracts obtained from the isolates of *Sulfitobacter pontiacus* (P1–17B (1E)) and *Halomonas axialensis* (P5–16B (5E)) reduced the HeLa and DU145 cell growth by 50–70%.

1.5.1.2 Marine Actinomycetes

Marine actinomycetes are undoubtedly the largest secondary metabolite producers among the marine microorganisms that account for 45% of the total

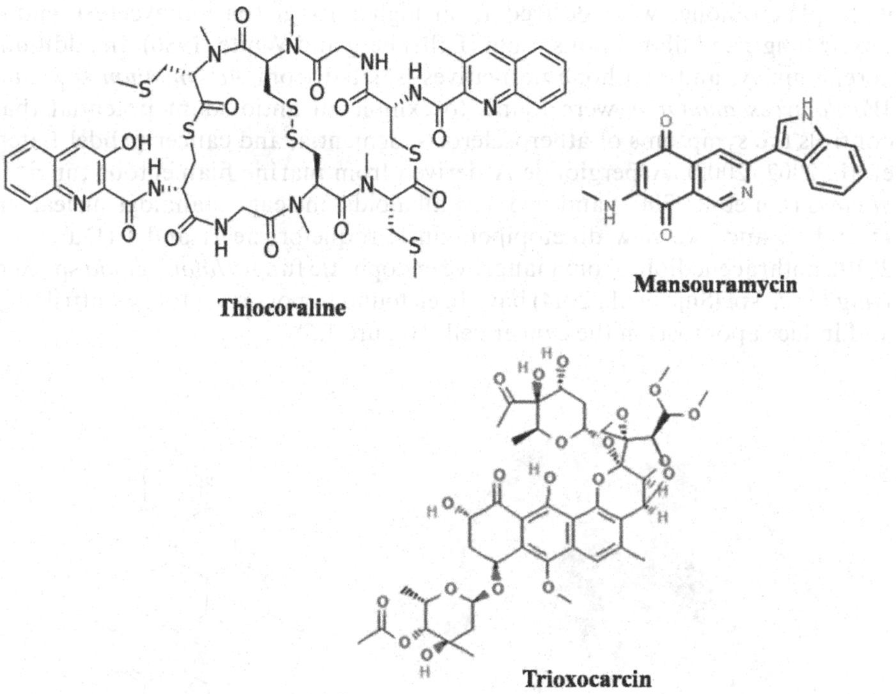

Thiocoraline

Mansouramycin

Trioxocarcin

FIGURE 1.4 Marine drugs derived from actinomycetes.

known antimicrobial metabolites. Of these compounds, 75% were derived from *Streptomyces* spp., while 25% are derived from the rare actinomycetes (Olano et al., 2009) and are considered to be a good candidate for the bioactive compounds that inhibits the growth and acts as antitumor agents (Hong et al., 2009). Some of the actinomycete-derived compounds include lucentamycins A–D isolated from *Nocardiopsis lucentensis* (Strain CNR-712), thiocoraline from *Micromonospora marina*, and trioxocarcins A–C from *Streptomyces* species. Choi et al. (2007) reported that lucentamycins A and B were found to possess cytotoxic activity against HCT-116 human colon cancer, whereas thiocoraline exhibited cytotoxic action by arresting the G1 phase of the cell cycle. Maskey et al. (2004) reported that trioxacarcins A–C showed higher antitumor activity against lung cell lines. Other cytotoxic compounds and anticancer agents derived from *Streptomyces* sp. are mansouramycins A–D (Hawas et al., 2009), macrodiolide tartrolon D (Pérez et al., 2009), and salinosporamide A derived from marine actinomycete (Fenical et al., 2009; Figure 1.4).

1.5.1.3 Marine Fungi

Marine-derived fungi correspond to a rich and bioactive source of novel anticancer agents (Newman and Hill, 2006), and most of the biologically potential principle compounds, such as leptossphaerin, leptosphaeroide and leptosphaerodione, were derived from higher fungi (basidomycetes), endophytic fungi, and filamentous fungi (Pallenberg and White, 1986). In addition, acremonin A, and xanthone derivatives isolated from *Acremonium* spp. and *Wardomyces anomalus* were found to exhibit an antioxidant potential that controls the symptoms of atherosclerosis, dementia, and cancer (Abdel-Lateff et al., 2002, 2003). Aspergiolide A derived from marine filamentous fungi *A. glaucus* (Du et al., 2007) and two new alkaloids, meleagrin analogs meleagrin D and D, and two new diketopiperazines, roquefortine H and I (Du et al., 2010), anthracenedione from mangrove endophytic fungi *Halorosellnia* sp., and *Guignardia* sp. (Suja et al., 2014) have been found to possess cytotoxic attributes and induce apoptosis in the cancer cells (Figure 1.5).

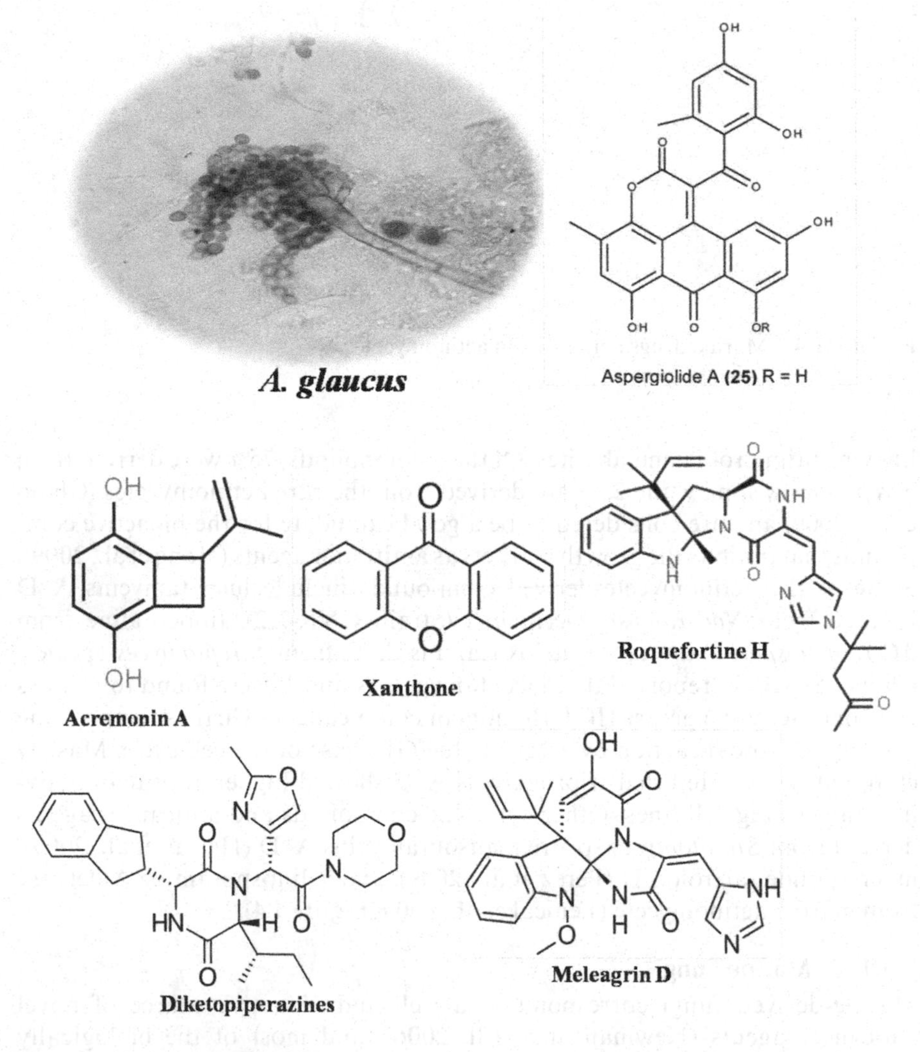

FIGURE 1.5 Marine drugs derived from fungi.

1.5.2 Marine Algae

Marine algae are one of the most prominent resources of bioactive compounds from the ocean, both ecologically and economically, included in the Asian diet for health benefits (Kim 2014). Bouga and Combet (2015) carried out an epidemiological study comparing the Japanese and Western diets and reported that the algae in the diets reduce the occurrence of chronic diseases such as coronary heart diseases, hyperlipidemia, diabetes, and cancer. Marine algae act as valuable sources of protein, fiber, polyunsaturated fatty acids, and macro and trace elements and are the predominant source of bioactive components such as polysaccharides, phytocolloids, tocopherols, terpenes, and phycocyanins that possess both nutritional and pharmaceutical potential (Alves et al., 2016). Marine algae are in the line after sponges, microbes, and phytoplankton for the marine-based drug discovery process. Many compounds derived from the algae have proved to be therapeutic agents associated with numerous health-promoting effects, including antimicrobial (Rodrigues et al., 2015), antioxidant (Pinteus et al., 2017), anti-inflammatory (de Souza et al., 2013), anticoagulant (Magalhaes et al., 2011), immunomodulatory (Pérez-Recalde et al., 2014), antidiabetic (Mayer et al., 2013), and antitumor (Alves et al., 2016). Among the compounds derived from marine algae, their structures are connected with brominated phenols, polysaccharides, and carotenoids and a large diversity of terpenoids with halogen groups (Gribble et al., 2015). Lee et al. (2012) reported that more than 3000 natural products have been isolated from marine algae for pharmaceutical potential. The compounds derived from the algae and their functions are given in Table 1.1.

1.5.3 Marine Sponges

Nearly 30% of the drugs available in the market for therapeutics are derived from marine sponges (Han et al., 2019). The initial discoveries of drugs derived from marine sponges led to the belief that it would not be last long as compared with the synthetic drugs available in the market. One of the remarkable drug discoveries from marine sponges are spongothymidine and spongouridine isolated from Caribbean sponge *Tethya crypta*. These compounds led to the development of their analog as anticancer drugs. Hirata and Uemura (1986) isolated a compound Eribulin from *Halichondria okadi* that is a synthetic halichondrin B that exhibited bioactivity against pretreated metastatic breast cancer (Figure 1.6).

1.5.4 Soft Corals

Gomaa et al. (2016) reported that one of the most widely distributed genera of soft coral is *Sarcophyton*, which composes approximately 30 species, in tropical and subtropical oceans. The species were tested for the presence of bioactive metabolites and documented that the fatty compounds such as arachidonic, eicosapentaenoic, and docosahexaenoic acids exhibited cytotoxic activity in brine shrimp.

TABLE 1.1
Some of the Compounds Derived from Marine Algae

Algae	Compound	Function	Reference
Sargassum siliquastrum	Sargachromanol E	Induce apoptosis and downregulate Bcl-xL, upregulation of Bax	Heo et al., 2011
Ecklonia cava	Dieckol	Induce the downregulation of FAK-signaling pathway results in the scavenging of free radicals and invasion of HT1080 cells	Park and Jeon, 2012
Eisenia bicyclis	Diphlorethohydroxy carmalol	Apoptosis in HL60 cells	Kang et al., 2012
Hisikia fusiformis	HFGP	Induce apoptosis in HepG2 cells	Ryu et al., 2012
Undaria pinnatifida	Fucoxanthin	Increases cisplatin treatment efficiency of HepG2 cell line	Liu et al., 2013
Laminaria digitata	Laminarin	Cause apoptosis and arrest cell cycle on human colon cancer cells	Park et al., 2013
Sargassum horneri	SHPSA	Arrest the proliferation of human colon cancer cell lines	Wang et al., 2015
Ascophyllum nodosum	Ascophylan	Increases the level of E-cathedrin and reduce N-cathedrin, which leads to inhibition of migration	Abu et al., 2015
Sargassum macrocarpum	Tuberatolide B	Reduces the cell viability of several cancer cell lines	Choi et al., 2017

Among the components of soft corals, cembranoids are the most important, which exhibited cytotoxic, ichthyotoxic, anti-inflammatory, and antagonistic activity. Crassumolide C derived from *Lobophytum crissum* inhibited the pro-inflammatory protein accumulation and cytotoxic activity (Chao and Li, 2008).

1.6 CONCLUSION

Natural products have been used as therapeutic agents for thousands of years to treat a wide spectrum of illnesses and represent an authentic starting point for drug discovery. However, there is a need for novel therapeutics because of the increased number of diseases and emerging of microbial resistance to current therapeutics. Since marine environment covers 70% of the earth's surface, its massive biodiversity seemingly produces great chemodiversity that could lead to the discovery of novel therapeutics with novel mechanisms of actions. Most of the marine organisms revealed to be interesting sources for the bioactive compounds that, in turn, have provided sustainable economic and human benefits

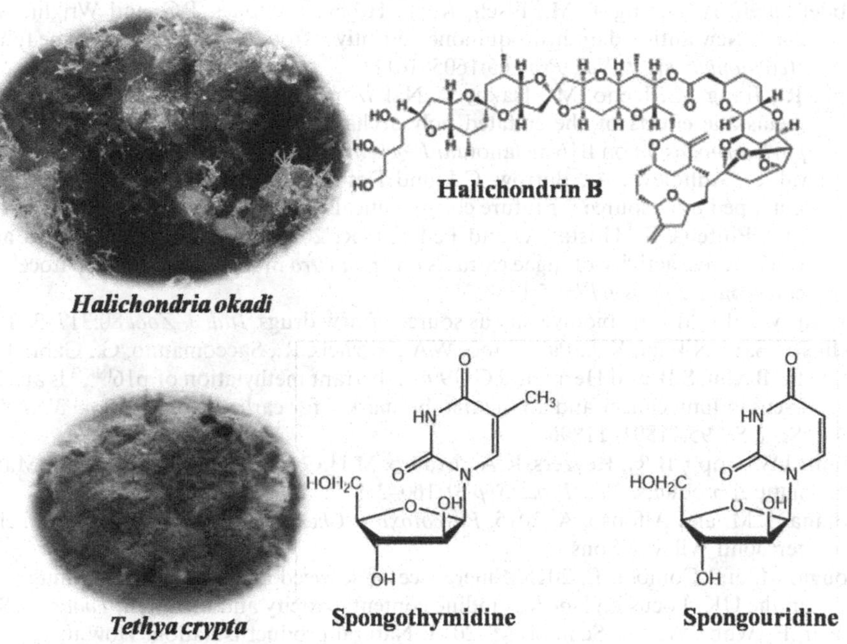

FIGURE 1.6 Drugs derived from marine sponges.

over the last several decades. Many of the compounds exhibited promising therapeutic potential for treating numerous diseases including cancer. Six among the nine marine-derived drugs currently available in the market are for cancer treatment, and some compounds with a bioactive potential against cancer treatment are under clinical trials. However, the number of compounds in the market or in clinical trial is low as compared to the total number of isolated compounds. This might be due to the diverse limits between the drug discovery and its commercialization such as difficulties in harvesting the source organisms, low product yield, insufficient investments by pharmaceutical companies, the development of environmental policies, the toxicity of the compounds, and so on. Nevertheless, these issues can be addressed by the development of advanced methodologies, such as sampling techniques, nanomole structure determination, genome sequencing, and mining. In recent years, several techniques have proved to be successful in combating several issues by high-throughput technologies to provide faster and accurate results, and in the future, the number of drugs derived from the marine source and available in the market for therapeutics will increase.

REFERENCES

Abdel-Lateff, A., Klemke, C., König, G.M. and Wright, A.D. 2003. Two new xanthone derivatives from the algicolous marine fungus Wardomyces anomalus. *J. Nat. Prod* 66:706–708.

Abdel-Lateff, A., König, G.M., Fisch, K.M., Höller, U., Jones, P.G. and Wright, A.D. 2002. New antioxidant hydroquinone derivatives from the algicolous marine fungus *Acremonium* sp. *J. Nat. Prod* 65:1605–1611.

Abu, R., Jiang, Z., Ueno, M., Isaka, S., Nakazono, S. and Okimura, T. 2015. Anti-metastatic effects of the sulfated polysaccharide ascophyllan isolated from *Ascophyllum nodosum* on B16 melanoma. *Biochem. Biophys. Res. Commun*. 458:727–732.

Agrawal, S., Adholeya, A., Barrow, C.J. and Deshmukh, S.K.2018. Marine fungi: An untapped bioresource for future cosmeceuticals. *Phytochem. Lett*. 23:15–20.

Alves, C., Pinteus, S., Horta, A. and Pedrosa, R. 2016. High cytotoxicity and anti-proliferative activity of algae extracts on an *in vitro* model of human hepatocellular carcinoma. *SpringerPlus* 5:1339.

Arizza, V. 2013. Marine biodiversity as source of new drugs. *Ital. J. Zool* 80:317–318.

Belinsky, S.A., Nikula, K.J., Palmisano, W.A., Michels, R., Saccomanno, G., Gabrielson, E., Baylin, S.B. and Herman, J.G. 1998. Aberrant methylation of $p16^{INK4a}$ is an early event in lung cancer and a potential biomarker for early diagnosis. *Proc Natl Acad Sci USA* 95:11891–11896.

Blunt, J.W., Copp, B.R., Keyzers, R.A., Munro, M.H.G. and Prinsep, M.R. 2014. Marine natural products. *Nat. Prod. Rep* 31:160–258.

Botana, L.M. and Alfonso, A. 2015. *Phycotoxins: Chemistry and Biochemistry*. Chichester: John Wiley & Sons.

Bouga, M. and Combet, E. 2015. Emergence of seaweed and seaweed-containing foods in the UK: Focus on labeling, iodine content, toxicity and nutrition. *Foods* 4:240.

Bucar, F., Wube, A. and Schmid, M. 2013. Natural product isolation: How to get from biological material to pure compounds. *Nat. Prod. Rep*. 30:525–545.

Carte, B.K. 1996. Biomedical potential of marine natural products. Marine organisms are yielding novel molecules for use in basic research and medical applications. *Bioscience*, 46(4):271–286.

Chao, H.Y. and Li, F.C. 2008. Effect of level of fibre on performance and digestion traits in growing rabbits. *Anim. Feed Sci. Technol*. 144(3–4):279–291.

Choi, Y., Kim, J., Lee, K., Choi, Y.-J., Ye, B.-R. and Kim, M.-S. 2017. Tuberatolide B suppresses cancer progression by promoting ROS-mediated inhibition of STAT3 signaling. *Mar. Drugs* 15:55.

Choi, Y.J., Kwon, E.J., Park, J.S., Kang, H.S., Kim, Y.S. and Yoo, M.A. 2007. Transcriptional regulation of the Drosophila caudal homeobox gene by bHLH-PAS proteins. *Biochim. Biophys. Acta*. 1769(1):41–48.

Class, S. 2002. Pharma overview. *Chem. Eng. News* 80:39–49.

Cragg, G.M. and Newman, D.J. 2013. Natural products: A continuing source of novel drug leads. *Biochim. Biophys. Acta* 1830:3670–3695.

Davidson, B.S. 1995. New dimensions in natural products research: Cultured marine microorganisms. *Curr. Opin. Biotechnol* 6:284–291.

Davidson, S.K., Allen, S.W., Lim, G.E., Anderson, C.M. and Haygood, M.G. 2001. Evidence for the biosynthesis of bryostatins by the bacterial symbiont "*Candidatus Endobugula sertula*" of the bryozoan Bugula neritina. *Appl. Environ. Microbiol*. 67:4531–4537.

de Souza Ferro, J.N., da Silva, J.P., Conserva, L.M. and Barreto, E. 2013. Leaf extract from *Clusia nemorosa* induces an antinociceptive effect in mice via a mechanism that is adrenergic systems dependent. *Chin. J. Nat. Med*. 11(4):385–390.

Diyabalanage, T., Amsler, C.D., McClintock, J.B.and Baker, B.J. 2006. Palmerolide A, a cytotoxic macrolide from the Antarctic tunicate *Synoicum adareanum*. *J. Am. Chem. Soc* 128:5630–5631.

Du, B., Zhong, X., Liao, X., Xu, W., Zhou, X. and Xu, S. 2010. A new antitumor arabinopyranoside from *Laurencia majuscula* induces G2/M cell cycle arrest. *Phytother. Res* 24:1447–1450.

Du, L., Zhu, T., Fang, Y., Liu, H., Gu, Q. and Zhu, W. 2007. Aspergiolide A, a novel anthraquinone derivative with naphtho [1, 2, 3-de] chromene-2, 7-dione skeleton isolated from a marine-derived fungus *Aspergillus glaucus*. Tetrahedron 63:1085–1088.

FDA. 2018. *ClinicalTrials.gov* [Online]. Available online at: https://clinicaltrials.gov/

Fenical, W., Jensen, P.R., Palladino, M.A., Lam, K.S., Lloyd, G.K. and Potts, B.C. 2009. Discovery and development of the anticancer agent salinosporamide A (NPI-0052). *Bioorg. Med. Chem.* 17:2175–2180.

Freitas, S., Martins, R., Costa, M., Leão, P., Vitorino, R., Vasconcelos, V. and Urbatzka, R. 2016. Hierridin B isolated from a marine cyanobacterium alters VDAC1, mitochondrial activity, and cell cycle genes on HT-29 colon adenocarcinoma cells. *Mar. Drugs* 14:158.

Garson, M., Hehre, W. and Pierens, G. 2017. Revision of the structure of acremine P from a marine-derived strain of Acremonium persicinum. *Molecules* 22:521.

Gomaa, N., Glogauer, M., Tenenbaum, H., Siddiqi, A. and Quiñonez, C. 2016. Social-biological interactions in oral disease: A 'cells to society' view. *PLoS ONE* 11(1):e0146218.

Gribble, G. 2015. Biological activity of recently discovered halogenated marine natural products. *Mar. Drugs* 13:4044.

Gustafson, K., Roman, M. and Fenical, W. 1989. The macrolactins, a novel class of anti-viral and cytotoxic macrolides from a deep sea marine bacterium. *J. Amer. Chem. Soc* 111:7519–7524.

Haefner, B. 2003. Drugs from the deep: Marine natural products as drug candidates. *Drug Discov. Today* 8:536–544.

Halpern, B.S., Longo, C., Hardy, D., McLeod, K.L., Samhouri, J.F. and Katona, S.K. 2012. An index to assess the health and benefits of the global ocean. *Nature* 488:615–620.

Han, B.N., Hong, L.L., Gu, B.B., Sun, Y.T., Wang, J., Liu, J.T. and Lin, H.W. 2019. Natural products from sponges. In *Symbiotic Microbiomes of Coral Reefs Sponges and Corals*. Dordrecht, The Netherlands: Springer, pp. 329–463.

Hart, S. 1997. Cone snail toxins take off. Potent neurotoxins stop fish in their tracks and may provide new pain therapies. *Bioscience* 47(3):131–134.

Hawas, U.W., Shaaban, M., Shaaban, K.A., Speitling, M., Maier, A., Kelter, G., Fiebig, H.H., Meiners, M., Helmke, E. and Laatsch, H. 2009. Mansouramycins A—D, cytotoxic isoquinolinequinones from a marine Streptomycete. *J. Nat. Prod.* 72:2120–2124.

Henkel, T., Brunne, R.M., Müller, H. and Reichel, F. 1999. Statistical investigation into the structural complementarity of natural products and synthetic compounds. *Angew. Chem. Int. Ed.* 38:643–647.

Heo, S.J., Kim, K.N., Yoon, W.J., Oh, C., Choi, Y.U. and Affan, A. 2011. Chromene induces apoptosis via caspase-3 activation in human leukemia HL-60 cells. *Food Chem. Toxicol* 49:1998–2004.

Hirata, Y. and Uemura, D. 1986. Halichondrins—antitumor polyether macrolides from a marine sponge. *Pure Appl. Chem* 58:701–710.

Homann, V.V., Sandy, M., Tincu, J.A., Templeton, A.S., Tebo, B.M. and Butler, A. 2009. Loihichelins A—F, a suite of amphiphilic siderophores produced by the marine bacterium Halomonas LOB-5. *J. Nat. Prod.* 72:884–888.

Hong, J., Zhang, J., Liu, Z., Qin, S., Wu, J. and Shi, Y. 2009. Solution structure of S. cerevisiae PDCD5-like protein and its promoting role in H(2)O(2)-induced apoptosis in yeast. *Biochemistry* 48(29):6824–6834.

Houghten, R.A. 1994. Combinatorial libraries. Finding the needle in the haystack. *Curr. Biol* 4:564–567.

Huang, C., Leung, R.K.K., Guo, M., Tuo, L., Guo, L., Yew, W.W., Lou, I., Lee, S.M.Y. and Sun, C. 2016. Genome-guided investigation of antibiotic substances produced by Allosalinactinospora lopnorensis CA15-2 T from Lop Nor region, China. *Sci. Rep* 6:20667.

Ireland, C.M., Copp, B.R., Foster, M.P., McDonald, L.A., Radisky, D.C. and Swersey, J.C. 1993. *Marine Biotechnology, Vol. 1: Pharmaceutical and Bioactive Natural Products*. D.H. Attaway and O.R. Zaborsky, eds. New York, NY: Plenum Press, pp. 1–43.

Ireland, C.M., Durso, A.R., Newman, R.A. and Hacker, M.P. 1982. Antineoplastic cyclic peptides from the marine tunicate Lissoclinum patella. *J. Org. Chem.* 47:1807–1811.

Isnansetyo, A. and Kamei, Y. 2009. Bioactive substances produced by marine isolates of Pseudomonas. *J. Ind. Microbiol. Biotechnol* 36:1239–1248.

Jeyapragash, D., Saravanakumar, A., Packiavathy, I.A.S.V. and Jeba Sweetly, D. 2021. Chlorophyll fluorescence, dark respiration and metabolomic analysis of *Halodule pinifolia* reveal potential heat responsive metabolites and biochemical pathways under ocean warming. *Mar. Environ. Res.* 164:105248.

Jeyapragash, D., Yousva, M. and Saravanakumar, A. 2020. Comparative metabolomics analysis of wild and suspension cultured cells of *H. pinifolia* (Miki) Hartog of Cymodoceaceae family. *Aquat. Bot.* 167:103278.

Kang, S.M., Kim, A.D., Heo, S.J., Kim, K.N., Lee, S.H. and Ko, S.C. 2012. Induction of apoptosis by diphlorethohydroxycarmalol isolated from brown alga. *Ishige okamurae. J. Funct. Foods* 4:433–439.

Khalifa, S.A.M., Nizar, E., Mohamed, A.F., Lei, C., Aamer, S., Mohamed-Elamir, F.H., Moustafa, S.M., Aida Abd, Saleh M.A., Syed, G.M., Fang-Rong, C., Arihiro, I., Kiyotake, S., Muaaz, A., Ulf, G. and Hesham, R.E. 2019. Marine natural products: A source of novel anticancer drugs. *Mar. Drugs* 17:491.

Kim, S.K. 2014. *Seafood Processing By-Products -Trends and Applications*. New York, NY: Springer-Verlag.

Kim, W., Kim, Y., Kim, J., Nam, B.-H., Kim, D.-G., An, C., Lee, J., Kim, P., Lee, H. and Oh, J.S. 2016. Liquid chromatography-mass spectrometry-based rapid secondary-metabolite profiling of marine Pseudoalteromonas sp. M2. *Mar. Drugs* 14:24.

Kong, D.X., Jiang, Y.Y. and Zhang, H.Y. 2010. Marine natural products as sources of novel scaffolds: Achievement and concern. *Drug Discov. Today* 15:884–886.

Lee, J., Kim, S., Jung, W.-S., Song, D.-G., Um, B.-H. and Son, J.K. 2012. Phlorofucofuroeckol-A, a potent inhibitor of aldo-keto reductase family 1 member B10, from the edible brown alga *Eisenia bicyclis. J. Korean Soc. Appl. Biol. Chem.* 55:721–727.

Liu, C.-L, Lim, Y.P. and Hu, M.L. 2013. Fucoxanthin enhances cisplatin-induced cytotoxicity via NFκB-mediated pathway and downregulates DNA repair gene expression in human hepatoma HepG2 cells. *Mar. Drugs* 11:50.

Magalhaes, K.D., Costa, L.S., Fidelis, G.P., Oliveira, R.M., Nobre, L.T.D.B., Dantas-Santos, N., Camara, R.B.G., Albuquerque, I.R.L., Cordeiro, S.L. and Sabry, D.A. 2011. Anticoagulant, antioxidant and antitumor activities of heterofucans from the seaweed Dictyopteris delicatula. *Int. J. Mol. Sci.* 12:3352–3365.

Malakar, J. and Nayak, A.K. 2013. Floating bioadhesive matrix tablets of ondansetron HCl: Optimization of hydrophilic polymer-blends. *Asian J. Pharm.* 7:174–183.

Maskey, R.P., Helmke, E., Kayser, O., Fiebig, H.H., Maier, A., Busche, A. and Laatsch, H. 2004. Anti-cancer and antibacterial trioxacarcins with high anti-malaria activity from a marine streptomycete and their absolute stereochemistry. *J. Antibiot. (Tokyo)*. 57:771–779.

Mayer, A., Rodríguez, A., Taglialatela-Scafati, O. and Fusetani, N. 2017. Marine pharmacology in 2012–2013: Marine compounds with antibacterial, antidiabetic, antifungal, anti-inflammatory, antiprotozoal, antituberculosis, and antiviral activities; affecting the immune and nervous systems, and other miscellaneous mechanisms of action. *Mar. Drugs* 15:273.

Mayer, A., Rodríguez, A., Taglialatela-Scafati, O. and Fusetani, N. 2013. Marine pharmacology in 2009–2011: Marine compounds with antibacterial, antidiabetic, antifungal, anti-inflammatory, antiprotozoal, antituberculosis, and antiviral activities; affecting the immune and nervous systems, and other miscellaneous mechanisms of action. *Mar. Drugs* 11:2510.

Mayer, A.M.S. 2018. *Marine Pharmaceuticals: The Clinical Pipeline*. [Online]. Available online at: http://marinepharmacology.midwestern.edu/clinPipeline.html.

Mitall, B.K. and Garg, S.K. 1995. Anticarcinogenic, hypocholesterolemic, and antagonistic activities of Lactobacillus acidophilus. *Crit. Rev. Microbiol* 21:175–214.

Molinski, T.F., Dalisay, D.S., Lievens, S.L. and Saludes, J.P. 2009. Drug development from marine natural products. *Nat. Rev. Drug. Discov*. 8:69–85.

Montaser, R. and Hendrik, L. 2011. Marine natural products: A new wave of drugs? *Future Med. Chem*. 3(12): 1475–1489.

Murray, P.M., Moane, S., Collins, C., Beletskaya, T., Thomas, O.P. and Duarte, A.W.F. 2013. Sustainable production of biologically active molecules of marine based origin. *New Biotechnol*. 30:839–850.

Newman, D.J. and Cragg, G.M. 2012. Natural products as sources of new drugs over the 30 years from 1981 to 2010. *J. Nat. Prod* 75:311–335.

Newman, D.J. and Hill, R.T. 2006. New drugs from marine microbes: The tide is turning. *J. Ind. Microbiol. Biotechnol*. 33:539–544.

Newman, D.J. and Cragg, G.M. 2016. Natural Products as Sources of New Drugs from 1981 to 2014. *J. Nat. Prod.* 79(3):629–61.

Olano, C., Méndez, C. and Salas, J. 2009. Antitumor compounds from marine actinomycetes. *Mar. Drugs* 7:210–248.

Ortholand, J.Y. and Ganesan, A. 2004. Natural products and combinatorial chemistry: Back to the future. *Curr. Opin. Chem. Biol*. 8:271–280.

Pallenberg, A.J. and White, J.D. 1986. The synthesis and absolute configuration of (+)-leptosphaerin. *Tetrahedron Lett*. 27:5591–5594.

Park, S. and Jeon, Y. 2012. Dieckol from *Ecklonia cava* suppresses the migration and invasion of HT1080 cells by inhibiting the focal adhesion kinase pathway downstream of Rac1-ROS signaling. *Mol. Cell*. 33:141–149.

Pérez, M., Crespo, C., Schleissner, C., Rodríguez, P., Zúñiga, P. and Reyes, F. 2009. Tartrolon D, a cytotoxic macrodiolide from the marine-derived actinomycete Streptomyces sp. MDG-04-17-069. *J. Nat. Prod*. 72:2192–2194.

Pérez-Recalde, M., Matulewicz, M.C., Pujol, C.A. and Carlucci, M.J. 2014. *In vitro* and *in vivo* immunomodulatory activity of sulfated polysaccharides from red seaweed *Nemalion helminthoides*. *Int. J. Biol. Macromol*. 63:38–42.

Pinteus, S., Silva, J., Alves, C., Horta, A., Fino, N. and Rodrigues, A.I. 2017. Cytoprotective effect of seaweeds with high antioxidant activity from the Peniche coast (Portugal). *Food Chem*. 218:591–599.

Rodrigues, D., Alves, C., Horta, A., Pinteus, S., Silva, J. and Culioli, G. 2015. Antitumor and antimicrobial potential of bromoditerpenes isolated from the red alga, *Sphaerococcus coronopifolius*. *Mar. Drugs* 13:713–726.

Ruiz-Ruiz, C., Srivastava, G.K., Carranza, D., Mata, J.A., Llamas, I., Santamaría, M., Quesada, E. and Molina, I.J. 2007. An exopolysaccharide produced by the novel halophilic bacterium Halomonas stenophila strain B100 selectively induces apoptosis in human T leukaemia cells. *Appl. Microbiol. Biotechnol*. 89:345–355.

Ryu, J., Hwang, H.J., Kim, I.H. and Nam, T.J. 2012. Mechanism of inhibition of HepG2 cell proliferation by a glycoprotein from *Hizikia fusiformis*. *Korean. J. Fish. Aquat. Sci.* 45:553–560.

Sagar, S., Esau, L., Hikmawan, T., Antunes, A., Holtermann, K., Stingl, U., Bajic, V.B. and Kaur, M. 2013. Cytotoxic and apoptotic evaluations of marine bacteria isolated from brine-seawater interface of the Red Sea. *BMC Complement. Altern. Med.* 13:29. https://doi.org/10.1186/1472-6882-13-29

Schmidt, E.W. 2008. Trading molecules and tracking targets in symbiotic interactions. *Nat. Chem. Biol.* 4:466–473.

Schmidt, E.W., Nelson, J.T., Rasko, D.A., Sudek, S., Eisen, J.A., Haygood, M.G. and Ravel, J. 2005. Patellamide A and C biosynthesis by a microcin-like pathway in Prochloron didemni, the cyanobacterial symbiont of Lissoclinum patella. *Proc. Natl. Acad. Sci* 102:7315–7320.

Simmons, T.L., Andrianasolo, E., McPhail, K., Flatt, P. and Gerwick, W.H. 2005. Marine natural products as anticancer drugs. *Mol. Cancer Ther*. 4:333–342.

Suffness, M., Newman, D.J. and K. Snader. 1989. *Bioorganic Marine Chemistry*. P.J. Scheuer, ed. New York, NY: Springer-Verlag, 3:131–168.

Suja, M., Vasuki, S. and Sajitha, N. 2014. Anticancer activity of compounds isolated from marine endophytic fungus Aspergillus terreus. *World J. Pharm. Pharm. Sci.* 3:661–672.

Thornburg, C.C., Zabriskie, T.M. and McPhail, K.L. 2010. Deep-sea hydrothermal vents: Potential hot spots for natural products discovery? *J. Nat. Prod.* 73:489–499.

Trischman, J.A., Jensen, P.R. and Fenical, W.W. 1994. Halobacillin, a cytotoxic cyclic acylpeptide of the iturin class produced by a marine bacillus. *Tetrahedron Lett*. 35:5571–5574.

Viso, A.C., Pesando, D. and Baby, C. 1987. Antibacterial and antifungal properties of some marine diatoms in culture. *Bot. Mar*. 30:41–45.

Wolfe-Simon, F., Blum, J.S. and Kulp, T.R. 2011. A bacterium that can grow by using arsenic instead of phosphorus. *Science* 332:1163–1166.

Yooseph, F., Sutton, G. and Rusch, D.B. 2007. The sorcerer II global ocean sampling expedition: Expanding the universe of protein families. *PLoS Biol*. 5:e16.

2 Application of Synchrotron Radiation Technology in Marine Biochemistry and Food Science Studies

Toshiki Nakano and Masafumi Hidaka

CONTENTS

2.1 INTRODUCTION

Due to a global decline in ocean fishery stocks, the production of farmed fish has increased in recent years. Approximately half of the world's fishery production is currently based on aquaculture. This is because aquaculture has the potential to reduce fishing pressure on threatened stocks, thereby reducing the negative effects on ecosystems and biodiversity (Pauly et al. 2002; Naylor et al. 2000; Naylor et al. 2009; Garlock et al. 2019). Accordingly, aquaculture is an important development that supports global food production. In 2018, global

DOI: 10.1201/9781003303909-2

aquaculture production was estimated to comprise approximately 82 million tons of aquatic animals (US$250 billion). At the time, aquatic animal farming was dominated by finfish (54 million tons; FAO 2020).

Because of its ability to sustainably enhance food supply, aquaculture is likely to become a major source of aquatic dietary proteins by 2050. However, in this instance, the balance between farmed and wild-caught fish, as well as the total supply of fish available for consumption by humans, will depend on future aquaculture practices. For this reason, there is a need to develop methods capable of monitoring and evaluating the impact of aquaculture on the environment and the corresponding risks (Stentiford et al. 2020; FAO 2020; Naylor et al. 2009; Naylor et al. 2000; Garlock et al. 2019). Additionally, there is also a need to evaluate the amount of wild-fish stock currently available, as well as the life history of important edible fish species in nature and the quality of post-harvested marine products (seafood) to ensure food security. In the case of Japan, a country surrounded by the sea, the effective and sustainable use of marine products is very important. Therefore, to secure sufficient quantities of marine products, technologies will need to be improved for resource conservation, aquaculture, and post-harvest (Nakano and Wiegertjes 2020; Nakano 2007, 2020, 2021b, 2021a; Nakano, Kanmuri, et al. 1999). Natural resources are often affected by climate and oceanographic conditions, while farmed fish are affected by rearing conditions and stress. Therefore, understanding the history of target species will help improve productivity, conserve resources, and maintain the health of farmed fish (Nakano et al. 2013; Nakano et al. 2014; Nakano 2020; Nakano and Wiegertjes 2020; Nakano 2007, 2016, 2021a, 2021b; Nakano et al. 2020; Wu et al. 2015; Nakano et al. 2011; Basu et al. 2001; Iwama, Afonso, and Vijayan 2006; Nakano, Kanmuri, et al. 1999; Nakano, Miura, et al. 1999; Nakano, Tosa, and Takeuchi 1995; Nakano et al. 2004).

Fish are usually distributed under refrigerated or frozen conditions after harvesting, with some being processed for the production of a variety of seafood products, including surimi (fish paste), kamaboko (salt-ground and steamed fish-paste products), and dried and salted fish. Therefore, the study of the biochemical and food-chemical properties of fish and seafood is important. In particular, the relationship between the quality of food and its microstructure has recently been reported to play an important role in understanding the properties of food products (Schoeman et al. 2016; Nakano 2019). As a result, nondestructive, noninvasive, and rapid observation techniques for fish and seafood products have been researched and developed.

Synchrotron radiation (SR), which is the electromagnetic radiation generated by charged particles accelerated in a magnetic field, can be used to observe materials and their functions at the nanoscale. SR is a powerful and advanced light source with a wide wavelength, ranging from infrared (IR) to X-ray. SR light is characterized by high brilliance (many orders of magnitude more than conventional light sources), high directivity, nonthermal radiation sources, and pulsed light emission (pulse durations less than 1 nanosecond) (Xu et al. 2018; Wang et al. 2010; Petroff 2007; Pérez and De Sanctis 2017; Li et al. 2015; Grabowski et al. 2021; Einfeld 2007; Bilderback, Elleaume, and Weckert 2005; PhoSIC 2019; Watanabe 2016; Hu et al. 2020).

SR light sources can be divided into three categories according to their energy: low energy, intermediate energy, and high energy. The SR beam is known to interact with substances in three processes: absorption, scattering, and emission. According to these processes, there are three major detection methods: (1) absorption (e.g., X-ray absorption fine structure [XAFS], Fourier transformed infrared spectroscopy [FTIR], soft/hard X-ray microscopy (scanning transmission X-ray microscopy [STXM]), and X-ray computed tomography [XRCT]); (2) scattering (e.g., X-ray diffraction [XRD], protein X-ray crystal diffraction [PX], and small-angle X-ray scattering [SAXS]); (3) detection of the emission of secondary particles (e.g. X-ray photoelectron spectroscopy [XPS] and X-ray fluorescence spectrometry [XRFS]; (Xu et al. 2018; Li et al. 2015; Hu et al. 2020; Westneat, Socha, and Lee 2008; Salditt and Töpperwien 2020; Kreft et al. 2017; Schoeman et al. 2016; Landis and Keane 2010). These experimental methods can be summarized as follows: (1) spectroscopy, (2) scattering, and (3) imaging methods (Xu et al. 2018).

In recent decades, SR-based analytical techniques have been developed and applied in the fields of material science, chemistry, physics, environmental sciences, nanosciences, biology, and structural biology (Xu et al. 2018; Li et al. 2015). There exist more than 50 facilities for SR light sources around the world. Most of facilities are third-generation types of SR light sources (Xu et al. 2018; Grabowski et al. 2021; Einfeld 2007; Bilderback, Elleaume, and Weckert 2005; Liu, Neuenschwander, and Rodrigues 2019; Hasnain and Catlow 2019; Watanabe 2016; Grochulski et al. 2017). Unfortunately, to date, information on the use of SR technology in marine biochemistry is limited. However, its application in marine biochemistry is expected to provide innovative and valuable insights within fisheries science research areas, including ecology and food science.

In this review, we describe the analytical methods of SR-based techniques and present the applications of SR techniques in the study of marine biochemistry and food science.

2.2 ANALYTICAL METHODS OF SR-BASED TECHNOLOGY

When X-rays are irradiated on material, they cause phenomena including absorption, scattering, and emission (Figure 2.1). Based on these analysis methods according to these phenomena, herein, we describe XRF, XRCT, and SAXS, which are expected to have useful applications in marine biochemistry and food analysis (Westneat, Socha, and Lee 2008; Salditt and Töpperwien 2020; Schoeman et al. 2016; Landis and Keane 2010; Hu et al. 2020; Kreft et al. 2017; Li et al. 2015; Xu et al. 2018).

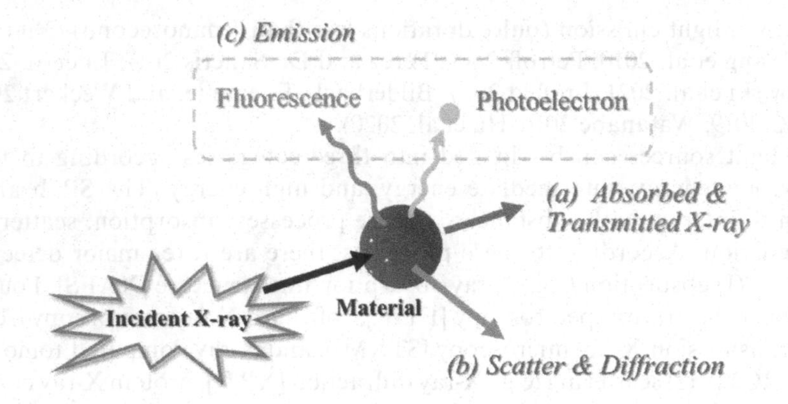

FIGURE 2.1 Schematic view of X-ray (a) absorption, (b) scattering, and (c) emission.

2.2.1 XRF

When an electron in the inner core level is excited by X-rays, the inner shell level becomes vacant. When an electron in the outer shell transitions into a vacant orbit, X-rays are emitted—this is the X-ray fluorescence. The wavelength of the fluorescent X-rays corresponds to the energy chain between the inner and outer shells. Therefore, an elemental analysis of a sample can be performed by examining the energy of the fluorescent X-rays. Depending on the elemental composition of the sample, X-rays of various energies are generated. The XRF spectrum is obtained by measuring these X-rays with a detector and plotting the energy of the X-rays on the horizontal axis and their intensity on the vertical axis. From the spectrum, the content of each element can be evaluated, and it can be used to identify the place of origin of agricultural products.

2.2.2 XRCT

XRCT is a method of three-dimensional imaging in which X-ray transmission images are acquired of a material from various angles and reconstructed from the resulting two-dimensional projection image, using the difference in X-ray transmission at each point in the material as contrast.

2.2.3 SAXS

SAXS is an analysis method that obtains structural information, such as the size and shape distribution of a material, by analyzing the scattering vector of X-rays scattered according to the structure of the material when X-rays are irradiated on it.

2.3 APPLICATIONS OF SR TECHNIQUES FOR ECOLOGICAL, ENVIRONMENTAL, AND BIOLOGICAL STUDY

2.3.1 Life History of Fish: Fish Otoliths

Fish do not have middle or outer ears, and only the inner ear is formed. The inner ear of teleost fish contains calcium carbonate ($CaCO_3$) small ear stones, called otoliths, surrounded by a gelatinous membrane. Otoliths are a pair of small hard tissues (<20 mm in diameter) whose main functions are hearing and maintaining balance. Otoliths can be classified into three types, namely, gravel stone, flat stone, and stellate stone, according to their shape. Otoliths generally refer to the largest flat stone. In the otolith, a nucleus (otolith nucleus) appears at the early stage of embryonic development, and Ca and other elements are supplied to the surface of the nucleus, which is deposited and grows concentrically. Although the shape is discoid in the early stages of development, it subsequently changes to a predominantly flattened form, characteristic of the species, depending on the rate and direction of otolith growth (Thomas and Swearer 2019; Katayama 2021; Avigliano and Volpedo 2016; Hüssy et al. 2020; Campana and Thorrold 2001; Campana 1999; Otake 2010; Katayama 2018; Cook et al. 2018). Examples of typical otoliths are shown in Figure 2.2. The shape of the otolith is known to be species-specific, and growth continues in proportion to body size during the life span of fish (Cook et al. 2018; Katayama 2021). The observation of otoliths under a light microscope reveals fine concentric rings. This ring pattern is called a circadian ring because it is formed

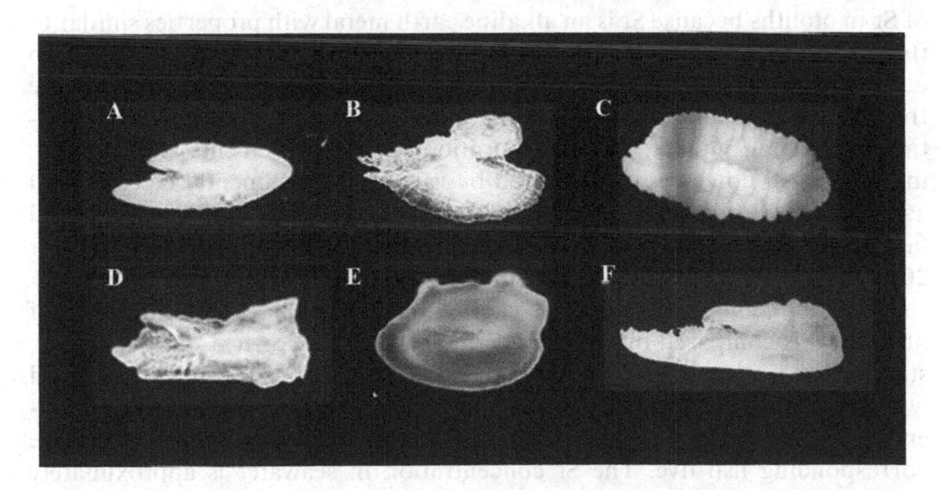

FIGURE 2.2 Typical fish otoliths: (A) sardine (*Engraulis japonicus*); (B) salmon (*Oncorhynchus keta*); (C) cod (*Gadus macrocephalus*); (D) mackerel (*Scomber australasicus*); (E) flounder (*Verasper variegates*); (F) tuna (*Thunnus thynnus*).

Credit: Dr. S. Katayama, Tohoku University.

once a day and is used to estimate the age in days and the date of hatching. In addition, the ring has two types of bands: a wide opaque band and a narrow transparent band. Opaque bands of the ring patterns are formed in summer and transparent bands are formed in winter; thus, the age of fish is known by the opaque bands and the transparent bands as annual rings. Therefore, otolith microstructure data are important for understanding the ecological and oceanographic processes underlying ecosystem dynamics (Thomas and Swearer 2019; Katayama 2021; Campana and Thorrold 2001; Otake 2010; Sponaugle 2010; Katayama 2018; Cook et al. 2018).

$CaCO_3$ crystals include aragonite, calcite, and vaterite. More than 95% of otoliths are composed of aragonite crystals of $CaCO_3$. In addition to Ca, over 30 elements, including sodium (Na), potassium (K), strontium (Sr), zinc (Zn), phosphorus (P), manganese (Mn), magnesium (Mg), silicon (Si), and iron (Fe), are present in otoliths at extremely low concentrations. However, the mechanism of trace element accumulation in otoliths remains unclear. Organic substances, such as glycoproteins, are present in otoliths (Thomas and Swearer 2019; Katayama 2021; Campana and Thorrold 2001; Otake 2010; Campana 1999). The construction of $CaCO_3$ in otoliths is regulated by enzymes. Therefore, the formation of otoliths occurs via biomineralization and may play an important role in governing otolith chemical patterns and element incorporation (Hüssy et al. 2020; Cook et al. 2018).

The otolith is a noncellular tissue characterized by a slower metabolism than other hard tissues (e.g., bones and scales), in which, once deposited, elements remain stable throughout the life cycle. Therefore, otoliths can be regarded as fish flight recorders. Many ecological studies in fish have focused on the level of Sr in otoliths because Sr is an alkaline earth metal with properties similar to those of Ca. Therefore, Sr is more likely to be incorporated into otoliths than other elements, and its concentration in otoliths is high. In addition, because the Sr concentration differs depending on the water, it reflects the concentration in the environment. The incorporation of specific elements, including Sr, in otoliths is known to be influenced by various environmental factors, such as water temperature, salinity, concentration, and stress, as well as biological factors, such as growth, ontogeny, and nutritional state (Thomas and Swearer 2019; Hüssy et al. 2020; Dimaria, Miller, and Hurst 2010).

When an ecological analysis is carried out using the Sr concentration of otoliths, it is compared by a relative value, namely, the Sr-to-Ca ratio (Sr:Ca), standardized using Ca. In aragonite crystals, this ratio is inversely correlated with the temperature at which the crystals are formed. By applying this property to otoliths, it is possible to estimate the water temperature at which the corresponding fish live. The Sr concentration in seawater is approximately 8 ppm, while that in fresh water is 1/500–1/100 of that in seawater. This difference is also reflected in the Sr concentration in otoliths. In addition to Sr, the incorporation of specific trace elements is known to be affected by environmental conditions, including temperature, salinity, pH, and environmental metal concentration (Thomas and Swearer 2019; Hüssy et al. 2020; Katayama 2021). Therefore, the life history of fish can be analyzed based on the distribution and concentration of trace elements, including Sr, Li, Ba, Mn, Mg, and Zn-to-Ca ratios, in otoliths (Thomas and Swearer 2019; Hüssy et al. 2020; Avigliano and Volpedo 2016; Katayama 2021).

In addition, Zn:Ca incorporation in the otoliths of salmonid fish may be a characteristic of phylogenic control and could be used in phylogenic research (Limburg, Gillanders, and Elfman 2010). In addition to Sr:Ca ratios, stable isotopes, such as $\delta^{18}O$, $\delta^{13}C$, and $^{87}Sr/^{86}Sr$ ratios, in the otoliths of several fish species have also been recognized as good indicators of the ecological life history of fish (Katayama 2021; Otake 2010; Cook et al. 2018). $\delta^{18}O$ can be calculated using the following equation: $[(^{18}O{:}^{16}O \text{ ratios of specimen}/^{18}O{:}^{16}O \text{ ratios of standard}) - 1] \times 1000$.

According to conventional methods for the analysis of trace elements in otoliths, the whole or part of otoliths can be collected using a precision drill, dissolved with acid, and analyzed using inductively coupled plasma mass spectrometry (ICP-MS) to measure trace elements. However, it is difficult to obtain a detailed history of the environment and growth of each individual, that is, information on the elements corresponding to the small ring crest of otoliths. To this end, laser ablation ICP-MS (LA-ICP-MS), in which a sample is irradiated with a laser beam and vaporized and introduced directly into ICP-MS, has been developed, allowing for elemental analysis at arbitrary sites on sample surfaces. Unfortunately, these methods are destructive and can only analyze part of the otolith. Additionally, these methods do not have sufficient sensitivity to detect low levels of trace elements in otoliths (Thomas and Swearer 2019; Limburg, Huang, and Bilderback 2007; Limburg and Elfman 2017; Avigliano and Volpedo 2016; Katayama 2021; Kreft et al. 2017; Carey et al. 2012).

Most elements identified in otoliths are found in trace amounts (Thomas and Swearer 2019; Limburg, Huang, and Bilderback 2007; Hüssy et al. 2020; Katayama 2021; Campana 1999). The accurate quantification of trace elements has many problems, such as matrix effects, the contamination of elements on handling, and detection limits (Limburg, Huang, and Bilderback 2007). In addition, the conventional analytical methods for otoliths are destructive and only explore limited areas of the otoliths. Accordingly, nondestructive, high-resolution methods that sample the entire surface are required (Limburg, Huang, and Bilderback 2007). Recently, SR scanning X-ray fluorescence spectrometry (SXFM), which detects X-ray fluorescence generated by irradiating a sample with a focused beam using synchrotron radiation, has been applied in the study of the distributions of trace elements on the surface of otoliths (Limburg, Huang, and Bilderback 2007; Limburg and Elfman 2017; Hüssy et al. 2020; Limburg et al. 2010; Limburg, Høie, and Dale 2010). Unlike the ICP-MS method, which detects specific isotopes and elements as targets, the SXFM method scans and detects fluorescence resulting from energy in the specimen (Limburg, Lochet, et al. 2010; Limburg, Høie, and Dale 2010; Hüssy et al. 2020; Carey et al. 2012; Kopittke et al. 2018; Kreft et al. 2017). SXFM does not require pretreatment and enables the rapid and sensitive simultaneous analysis of multiple elements on the surface of otoliths nondestructively, as well as the analysis of the chemical state and local structure of elements. SXFM also produces full-area two-dimensional plots (2D maps) of chemical variation, including elemental concentrations, throughout the otoliths (Limburg, Huang, and Bilderback 2007; Limburg and Elfman 2017; Hüssy et al. 2020; Limburg et al. 2010; Limburg, Høie, and Dale 2010). A transverse-section 2D map of a small amount of bromine (Br) in the Norwegian coastal cod *Gadus morhua* otolith was observed using SXFM (Figure 2.3; Limburg and Elfman 2017; Limburg, Høie, and Dale 2010). The distributions of both Sr and Mn in the cod otolith showed elements incorporated into the ring structure during its formation. However, the distribution of Br suggests that its uptake may be controlled by an unknown process (Limburg, Høie, and Dale 2010; Limburg and Elfman 2017). The synchrotron-based X-ray diffraction technique can directly and efficiently measure the crystallite phase and orientation at a micrometer

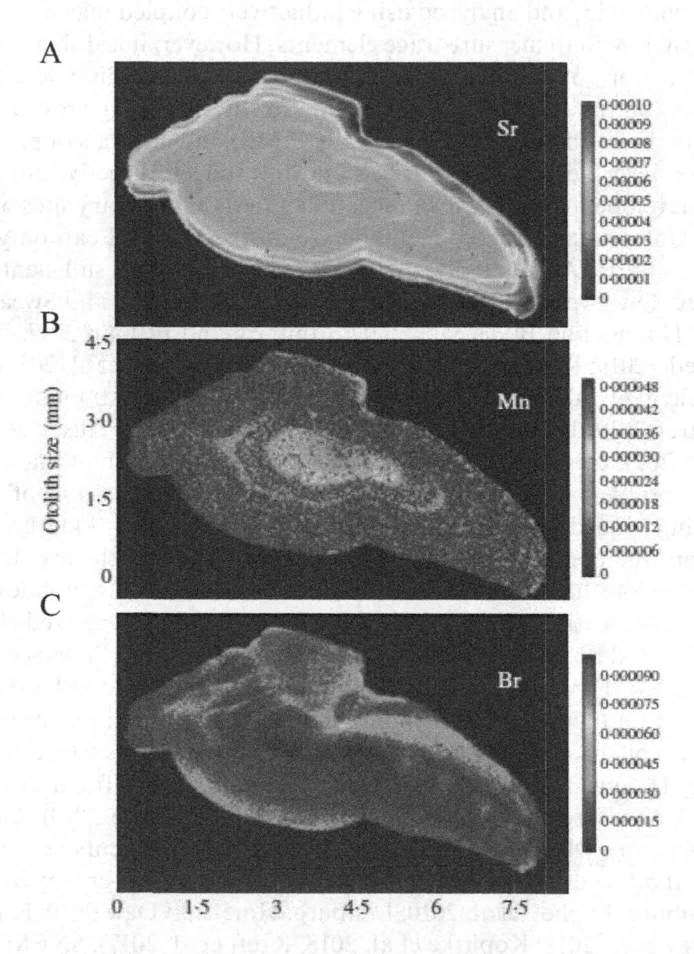

FIGURE 2.3 Transverse-section 2D maps of (a) Sr, (b) Mn, and (c) Br in Norwegian coastal cod *Gadus morhua* otolith.

Source: This figure was reprinted from Limburg and Elfman (2017) with permission.

resolution across textured polycrystalline samples, such as otoliths of small sizes (mm). The XRD method describes the biomineralogical structure and microtexture of otoliths using the XRF method (Cook et al. 2018). SR-XRCT imaging technology has also been applied to determine the nondestructive three-dimensional (3D) structure of plaice otolith and the rendering of otolith growth (Mapp et al. 2016). In addition, regarding the incorporation and existing state of Sr in otoliths, SR X-ray absorption spectroscopy, specifically extended X-ray absorption fine structure analysis, has revealed that Sr is randomly substituted for Ca in the crystal lattice of otolith aragonite (Doubleday et al. 2014).

2.3.2 SR-Based Technology in Environmental and Biological Study

In addition to ecological research, SR-based technologies have been used to study environmental and biological sciences.

SR X-ray micro-computed tomography (micro-CT) can be used to conduct quantitative morphometric analysis of fish. Micro-CT and phase-contrast micro-CT, which are nondestructive methods for high-resolution volumetric biological material analysis, can be used to visualize the organs of animals, including fish, such as zebrafish (*Danio rerio*) and medaka (*Oryzias latipes*; Weinhardt et al. 2018; Liu, Wu, and Xiao 2015; Epple and Neues 2010; Cotti et al. 2020; Chen et al. 2020; Zan et al. 2021; Westneat, Socha, and Lee 2008; Salditt and Töpperwien 2020; Yu et al. 2021). Several SR-based techniques, such as SXRF, XAFS spectra, and X-ray absorption near-edge structure (XANES) spectra, have also been applied to analyze the distribution, quantification, and chemical form (speciation) of trace or heavy metals in the tissues of aquatic organisms, such as fish (Moraes et al. 2012; Korbas et al. 2012; Espinoza-Quiñones et al. 2010; Cardoso et al. 2019; Manceau et al. 2021), sea urchin (Li et al. 2015), and algae, including phytoplankton (Xin et al. 2019; Twining et al. 2015; Lamela et al. 2018) and in the sediments of ocean muds (Kashiwabara et al. 2018), as well as in marine particles, which are one of the main reservoirs for trace elements and isotopes (Lam et al. 2015). XANES spectroscopy was applied to analyze the occurring Ca-phosphorus (P)-containing compounds, such as $CaHPO_4$ and Ca-phytate in the solid waste from a recirculation aquaculture system (Prüter et al. 2020). The geographic origin and harvesting location of fish and shellfish can also be estimated using multiple trace elemental analyses (Yamashita and Yamashita 2006; Okoshi 2006).

Many heavy metals are toxic, among which mercury (Hg) is one of the most toxic elements. The toxicity of heavy metals to organisms is related to their concentration, chemical form, and distribution. Hg is known to exist in various chemical forms, including organic methylmercury (MeHg) and inorganic mercury (Hg(II)), each with different toxicities (Manceau et al. 2021; Korbas et al. 2012; Hu et al. 2020; Fujii and Shiomi 2016). Accordingly, the objective of the study of the properties of heavy metals in organisms that humans consume is to determine their effects on both the environment and public health. The chemical forms of Hg in the muscle of fish, such as blue marlin (*Makaira* sp.), have been previously determined by SXRF and XANES (Moraes et al. 2012; Manceau et al. 2021; Cardoso et al. 2019).

2.3.3 Marine Waste and Microplastics in the Environment

Plastics and synthetic organic polymers are emerging pollutants in aquatic and terrestrial environments. Because the seafloor covers approximately 70% of the earth's surface, it is considered a major sink for plastic marine waste (or marine debris and marine litter). Jambeck et al. calculated that 274 million metric tons (Mt) of plastic waste was generated in 2010, with 4.8–12.7 million Mt of plastic waste entering the ocean (Morales-Caselles et al. 2021; Jambeck et al. 2015). Similarly, plastic waste leakage for all aquatic ecosystems in 2016 was estimated to be approximately 20 Mt (Canals et al. 2020). In fact, the annual production of plastics went from 1.5 million tons in the 1950s to 360 million tons in 2018. The most common plastics in commercial products are polyethylene (PE), polypropylene (PP), polystyrene (PS), PE terephthalate (PET), and polyvinyl chloride (PVC), and PE, PP, and PS are the most widespread plastics found in the environment. Among these types of plastics, PE accounts for 90% and is the most common plastic found in the environment (Larue et al. 2021; Park and Park 2021).

Plastics found in the environment can be categorized into different classes based on their size: macroplastics (<1 m), mesoplastics (2.5 cm–5 mm), microplastics (1 μm–5 mm), and nanoplastics (<1 μm, usually 1–100 nm). Microplastics (including nanoplastics) are ubiquitous in marine and freshwater environments, sea ice, sediments, soil, atmosphere, and living organisms, such as fish and algae. Plastics usually undergo chemical and physical modifications, leading to changes in their physicochemical properties and toxicity (Larue et al. 2021; Wootton et al. 2021; Ivleva, Wiesheu, and Niessner 2017). Microplastics have the potential to absorb and carry both inorganic and organic pollutants, including toxic chemicals, from their nearby environments due to their increased surface-to-volume ratio (Chen et al. 2021). In addition, MPs contain persistent organic pollutants, such as polycyclic aromatic hydrocarbons and heavy metals from the environment. Microplastics have also been reported to act as carriers of foreign species and microorganisms (Ivleva, Wiesheu, and Niessner 2017). The impact of microplastics on aquatic ecosystems and human health has not been fully clarified; however, the negative effects of microplastics on aquatic living resources have been observed (Xu et al. 2019; Rochman et al. 2013; Limonta et al. 2019). Additionally, various popular fish species have been reported to contain notable levels of microplastics (Rochman et al. 2015). Therefore, microplastics have become a major global environmental issue and may pose a risk to public and human health.

It is difficult to identify the characteristics of microplastics, such as their size, shape, and type, using a single analytical method. Accordingly, the identification of microplastic particles has usually been conducted using a combination of physical (morphological; e.g., microscopy) and chemical (e.g., spectroscopy) analytical methods (Xu et al. 2019; Shim, Hong, and Eo 2017). Spectroscopy can provide information on the structure and chemical composition of microplastics. FTIR spectroscopy, Raman spectroscopy, and atomic force microscopy–based infrared spectroscopy are often used to analyze microplastics (Xu et al. 2019; Shim, Hong, and Eo 2017; Park and Park 2021; Larue et al. 2021; Ivleva, Wiesheu, and Niessner 2017; Chen et al. 2021; Baruah et al. 2021). In addition to spectroscopy, thermo-analytical methods, such as thermo-extraction desorption-gas chromatography-mass spectrometry and pyrolysis gas chromatography, have been improved and intensively applied for microplastic analyses. These methods measure changes in the physical and chemical properties of plastics, depending on their thermal stability. Although thermal analysis is an alternative method for the chemical identification of certain polymer types, this analytical method is a destructive method, measuring microplastics by heating (Shim, Hong, and Eo 2017; Chen et al. 2021; Baruah et al. 2021; Park and Park 2021). In addition, sampling and sample preparation methods for microplastic analyses are known to affect the qualitative and quantitative results of microplastics. Furthermore, most sample preparation methods are complex and time-consuming (Wang et al. 2020; Stock et al. 2019). Therefore, nondestructive and simple SR-based spectroscopy is a useful method to determine the characteristics of microplastics in the specimens. Unfortunately, few examples of the study microplastics in environmental samples and marine products using SR-based techniques, such as SR-FTIR spectroscopy and SR X-ray diffraction, have been reported (Park and Park 2021; Kunz et al. 2016; Ghosh et al. 2017; Chen, Lee, and Walther 2020). However, SR-based techniques are expected to become a powerful tool for the analysis of microplastics.

2.4 APPLICATIONS OF SR-BASED TECHNOLOGY IN FOOD SCIENCE RESEARCH

The structure and distribution of the components, such as protein, lipid, and water, of the edible parts of fish are known to influence food qualities, including texture, stability, freshness, taste, sensory properties, and bacterial contamination (Schoeman et al. 2016; Alasalvar et al. 2011; Nakazawa and Okazaki 2017; Nakano 2019). Therefore, there is a need to develop a nondestructive and noninvasive observation method with which to evaluate and understand the structure and distribution of food components in order to produce high-quality products. Techniques used to observe the inner structure (microstructure) of food require destructive sample preparation or are limited to specific applications, such as electron microscopy and magnetic resonance imaging (MRI) (Schoeman et al. 2016). The applications of SR-based techniques in food science are summarized in Table 2.1. SR-based XRCT is an innovative radiographic imaging technology that enables both the nondestructive and the noninvasive 3D imaging of the microstructure of food at high resolutions (Schoeman et al. 2016). XRCT is known to visualize the content and distribution of muscle, bones, fat, and rind of meat (Miklos et al. 2015; Jensen et al. 2011). SAXS is used to obtain structural information, such as the size and shape distribution, of a material by analyzing the scattering vector of X-rays.

TABLE 2.1
Summary of the Applications of SR-Based Analytical Techniques in Food Science

Type of Food	Method	Target Components	Reference
Carp	XRCT	Fillet composition	Romvári et al. (2002)
Catfish	XRF	Heavy metal	Cardoso et al. (2019)
Cod	XRCT	NaCl, water, and protein in salted fillet	Håseth et al. (2009)
Marlin	XANES	Heavy metal	Manceau et al. (2021)
Salmon	XRCT	NaCl in salted and smoked fillet	Segtnan et al. (2009)
Tuna	XRCT	Frozen muscle structure	Sato et al. (2016)
Beef	XRCT (phase-constrast)	Meat structure, connective tissue, fat, water	Miklos et al. (2015)
Pork	SAXS	Meat structure, fat	Hoban et al. (2016)
Pork	XRCT (phase-constrast)	Fat, rind	Jensen et al. (2011)
Pork (meat and ham)	XRCT	NaCl, water, fat, and protein in salted fillet	Håseth et al. (2008)
Sausage	XRCT	Structure	Santos-Garcés et al. (2013)
Milk and cheese	SAXS	Fat, colloidal calcium phosphate	Nasuda et al. (2020)
Milk	SAXS	Casein micelle	Marchin et al. (2007)
Tofu (soybean curd)	XRCT	Frozen food structure	Sato et al. (2016); Kobayashi et al. (2020)
Walnut seed	XRCT	Internal parts	Rao et al. (2013)
Buckwheat	XRF	Metal	Kreft et al. (2017)
Rice	XANES, XRF	Heavy metal	Carey et al. (2012)
Fruit	XRCT	Total soluble solids	Li, Sun, and Cheng (2016)
Carrageenan	SAXS	Structure	Yang, Yang, and Yang (2018)
Chocolate	XRCT	Structure	Reinke et al. (2016)
Algae	XRD	Structure	Bianco-Stein et al. (2020)
Ice cream	XRCT	Structure	Do et al. (2020)

One example of food analysis using SAXS is the nanostructure analysis of dairy products. Nanostructure analysis of dairy products (Nasuda et al. 2020), casein (Marchin et al. 2007), and the quality of meat (Hoban et al. 2016) have been conducted using SAXS or ultrasmall-angle XS. Casein micelles, which play a major role in the aggregation process of cheese, are distributed over a relatively wide size range (20–600 nm in diameter), with an average diameter of approximately 100 nm (Marchin et al. 2007). Although transmission electron microscope (TEM), including cryo-TEM, is necessary to observe the structure of these micelles, samples must be dried and frozen, which is problematic due to the fact that they differ greatly from the original state of the food (Aoki et al. 2017). On the other hand, SAXS can be used at room temperature and pressure to obtain nano-level structural information under conditions close to those of the original food. Using SAXS measurements to analyze cheese, the differences in the chainlike structure of milk fat and the formation of particles of approximately 6–17 nm, depending on the aging period of the cheese, have been reported (Nasuda et al. 2020). XRCT and SAXS are not unique to SR. It is possible to measure XRCT and SAXS using laboratory-level equipment. We demonstrate the advantages of using SR-based X-rays to measure both SAXS and CT images.

SAXS and XRCT integrate the intensity of transmitted and scattered X-rays, respectively, which are obtained by irradiating a sample with X-rays for a certain period. In the case of laboratory X-rays, the intensity of the X-rays is relatively weak, and it takes a relatively long time to integrate them into the samples in order to obtain a signal. For this integration, a single measurement may take 10–30 min for SAXS and ≥12 h for XRCT. During this time, the samples are exposed to air. In the case of soft materials that contain moisture, such as food products, exposing the sample to air will cause it to dry and become damaged, which creates problems for both SAXS and XRCT. In the case of SAXS, data are obtained using information from the beginning and end of X-ray exposure, and the resulting information can be misleading if it changes throughout the measurement period. However, this problem is more serious in the case of XRCT: if the sample is deformed during the measurement, 3D reconstruction is not possible. One of the advantages of using synchrotron radiation for SAXS and XRCT is that the necessary data collection can be completed in a shorter time than when using a laboratory X-ray source because of the use of high-flux X-rays. In our experience of measuring foods using both laboratory and synchrotron radiation, SAXS measurements that took 20 to 30 minutes in the laboratory could be completed in less than a minute when using beam line19B2 at RIKEN/SPring-8 in Hyogo, Japan. In the case of XRCT, the same measurement only took 5 min using BL14B2 at RIKEN/SPring-8. Comparing laboratory-level and SR-based XRCT, the spatial resolution of the laboratory-level CT system is at the micron level, which is comparable to that of SR-based CT. On the other hand, the characteristics of SR, such as beam shape, high spatial coherence, and high intensity, provide advantages; in particular, monochromaticity provides a relatively large advantage. In the case of laboratory CT, the imaging contrast is expressed as CT values given in Hounsfield units (HU) and are attenuation coefficients relative to water (Håseth et al. 2008). The value of HU is not absolute and varies depending on the X-ray voltage and other conditions used in the measurement. On the other hand, the contrast of the image obtained by CT imaging using monochromatic X-rays of synchrotron radiation is expressed as the linear absorption X-ray coefficient, which is the absolute amount of X-ray absorption. Each element has a wavelength-dependent intrinsic mass absorption coefficient, such that if the wavelength (energy) of the X-rays used and the mass of the element contained in each voxel are known, the amount of absorption can be calculated.

An example of an XRCT analysis of salted salmon is in the production of salted salmon (Schoeman et al. 2016; Segtnan et al. 2009). The reduction of the concentration of salt (i.e., for health reasons and due to its negative impact on quality) requires a nondestructive analysis of salt content to monitor the salting process. Noting that the density of NaCl is approximately twice that of salmon fillets, Segtnan *et al.* investigated the possibility of detecting NaCl by XRCT (Segtnan et al. 2009). They found that the NaCl content could be predicted locally via a combination of a calibration curve for NaCl obtained by combining three different X-ray voltages (80, 110, and 130 keV) and fat prediction by near-IR interaction imaging.

2.4.1 Frozen Fish Products

Recently, researchers have investigated the quality of frozen marine products using the SR-based XRCT technique in an attempt to address the need to preserve marine products by refrigeration and freezing. Preservation methods include chilled preservation at temperatures ranging from above freezing to 10°C, superchilling (ice temperature) preservation at 0°C or below without freezing, partial freezing preservation at partial freezing temperatures range from near-freezing temperatures (around –3°C), and freezing preservation at –18°C or below. In general, when food is frozen, it is necessary to quickly pass through the maximum ice crystal formation zone, found at –5°C to –1°C. When frozen, water separates from proteins, and water molecules combine to form ice crystals. If the ice crystal size is large, the cells and tissues can be damaged, resulting in deterioration after thawing. In addition, if the temperature during cryopreservation is poorly controlled, the protein in the tissue can become denatured, with incomplete water absorption by the protein during thawing (Nakazawa and Okazaki 2017; Sato et al. 2021). All these factors can lead to a significant reduction in the quality of frozen foods and are involved in determining the quality of frozen foods, including ice crystals, proteins, lipid components, and chemical reactions. In fish muscle, which is generally composed of more than 70% water, cryopreservation controls the size of ice crystals and the proteins and lipids contained therein, which, in turn, affects the quality of the thawed fish. Therefore, in order to control ice crystal formation, it is necessary to observe and objectively evaluate the effects of crystal formation on the formation, growth, size, and location of cells and peripheral tissues.

Several methods for the observation of ice crystals in tissues have been proposed, including freeze substitution, low-temperature scanning electron microscopy, and microslicer imaging systems, a low-temperature adhesive film method (Kono 2017). However, most of these methods not only are destructive but also require the preparation of specimens, including the fixation and sectioning of tissues, and have the disadvantage that only local and 2D observations can be made. Compared with other foods, such as meat, ice crystals in fish and marine products are rarely observed (Kono 2017; Hagiwara, Kobayashi, and Kimizuka 2017; Syamaladevi et al. 2012; Sato, Kajiwara, and Sano 2016). The XRCT method, which is also used for medical examination, can be used to

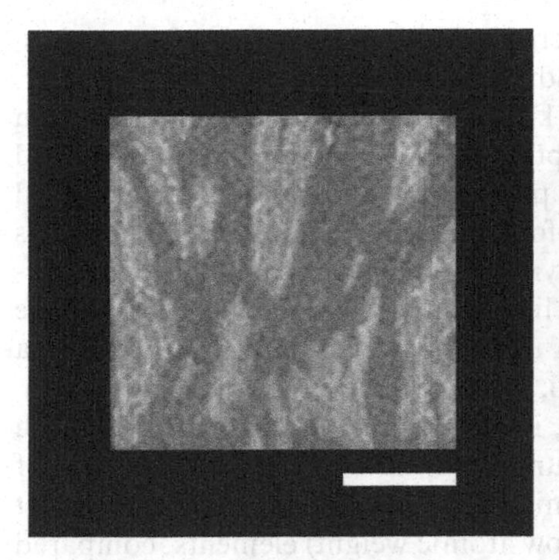

FIGURE 2.4 2D and 3D images of frozen "toro meat," a fatty muscle, from bluefin tuna *Thunnus thynnus*. Scale bar: 1000 μm.

observe the internal structure of a specimen by reconstructing the X-ray transmission image; however, in the case of frozen food, the contrast is weak because the density difference between the ice crystal and unfrozen parts is small, and the resolution is very low. The CT of frozen and fresh tuna muscle was measured using the SR-based XRCT method, which is a high-brightness X-ray light source (Sato, Kajiwara, and Sano 2016). While only frozen tuna showed a contrast between regions of high and low X-ray absorption, the amount of absorption in the low X-ray absorption region was consistent with the theoretical amount of absorption in ice. The reason for the contrast observed in the 3D imaging of the frozen tuna may be that the metal ions contained in the muscle act as a contrast agent (Sato, Kajiwara, and Sano 2016; Hagiwara, Kobayashi, and Kimizuka 2017). Very recently, we have also succeeded in performing the 3D imaging of frozen "toro meat," a fatty muscle, from bluefin tuna *Thunnus thynnus*. The toro meat specimen was imaged at beamline BL14B2 of RIKEN/SPring-8 using SR-based XRCT with a wavelength of 1 Å. XRCT was able to clearly distinguish the protein, fat, and ice of frozen muscle according to the frequency distributions of the X-ray absorption coefficient. As shown in Figure 2.4, the SR-based XRCT technique revealed that the concentrated protein (red area) floated in the matrix of ice crystals (pink area) and fat (light pink area) like a small boat.

2.5 CONCLUSION AND FUTURE DIRECTIONS

This review article provides a novel overview of the applications of SR technology in the field of marine biochemistry.

Synchrotrons are particle accelerators in which charged particles circulate along a closed path. A charged particle, an electron, in the magnetic field, is accelerated to the speed of light. Because SR is extremely bright compared to conventional X-ray generators and microscopes, it is gentle to soft materials, has a variable wavelength (energy), and a high directivity, similar to a laser. SR-based X-ray contrast resolution is known to be >1000 times higher than the density resolution of X-ray absorption. SR can measure in nanosecond order with a pulse duration of 50–100 picoseconds (Nakano 2019; Pérez and De Sanctis 2017; Rao et al. 2013). Therefore, in particular, SR-based imaging is recently attracting many scientists studying on medical and biological applications. At present, a next-generation SR light source facility, which will provide the first synchrotron beam in 2023, is under construction at the Aobayama campus of Tohoku University in Sendai, Japan (Figure 2.5; PhoSIC 2019).

This new light source, a soft X-ray, is more than 100 times brighter than a third-generation light source and runs on 5 MW, which is 1/10th that of RIKEN/SPring-8, with an ultra-low emittance and an energy of 3 GeV. A soft X-ray has a high sensitivity for light (low atomic weight) elements, compared with hard X-ray (Tanino et al. 2017). Accordingly, next-generation light sources are suited to the observation of soft materials, such as biological substances and food, for the study of their functions at the nanoscale. This light source represents a cutting-edge technology for the visualization of material function and dynamics and will help establish and accelerate collaboration between industry and academia for the development of SR-based applications using a new concept known as a "coalition concept" (PhoSIC 2019). SR-based technology has the potential to provide innovative and scientific information about ecology, environmental science, resource science, life science, and food science to establish sustainable development goals (Willis et al. 2021; Visbeck et al. 2014; Jacob-John et al. 2021; Thilsted et al. 2016; Farmery et al. 2021;

FIGURE 2.5 The estimated completed drawing of a next-generation SR light source facility "NanoTerasu" in Sendai, Japan.

Credit: Photon Science Innovation Center (PhoSIC).

Azra et al. 2021; Galgani et al. 2021; Nakano 2019). In particular, information regarding marine science obtained by SR-based research is likely to improve and enrich our quality of life in the near future. Thus, further investigation on SR-based technology is required to advance our knowledge of the relationship between the application of next-generation SR light sources and marine biochemistry.

ACKNOWLEDGMENT

T.N. is grateful to F. Galgani at Station de Corse, Institut Français de Recherche pour l'Exploitation de la Mer (IFREMER), France, R. Kobayashi at College of Bioresource Science, Nihon University, Japan, and T. Suzuki, Department of Food Science and Technology, Tokyo University of Marine Science and Technology, Japan, for their valuable discussions. The authors thank N. Yagi, M. Sato, K. Osaka, and I. Hirosawa at JASRI/SPring-8, Japan; T. Fujii, M. Harata, Y. Kanayama, S. Katayama, M. Komai, T. Nochi, Y. Ochiai, Y. Takeoka, S. Tazo, K. Watanabe, and T. Yamaguchi at the Graduate School of Agricultural Science; and J. Kawamura at the University Research Administration Center; K. Kanie and W. Yashiro at the Institute of Multidisciplinary Research for Advanced Materials, Tohoku University, Japan, for their valuable suggestions. The authors are also grateful to S. Akiyama at Maruse Akiyama Shoten Ltd., Japan; N. Asami and H. Onodera at Miyagi Food Industry Council, Japan; D. Takahashi, R. Saitou, and S. Kon at the Economic Affairs Bureau, Sendai City Hall, Japan, for their supports. Some studies performed by T.N. were financially supported by a fund from Sendai City Existing Synchrotron Radiation Facility Creation Project (Trial Use Project) from Sendai City, in part by a Grant-in-Aid for Scientific Research (KAKENHI), a fund from the Core-to-Core Program (A. Advanced Research Networks) from JSPS, Japan, and a fund from A Scheme to Revitalize Agriculture, Forestry and Fisheries in Disaster Area of Fukushima Prefecture through Deploying Highly Advanced Technology from MAFF, Japan.

REFERENCES

Alasalvar, C., F. Shahidi, K. Miyashita, and U. Wanasundara. 2011. "Seafood quality, safety, and health applications: An overview." In *Handbook of seafood quality, safety and health applications*, edited by C. Alasalvar, F. Shahidi, K. Miyashita and U. Wanasundara, 1–10. Chichester, West Sussex: John Wiley & Sons.

Aoki, T., R. Mizuno, T. Kimura, and S. Dosako. 2017. "Models of the structure of casein micelle and its changes during processing of milk." *Milk Science* 66:125–143. doi: 10.11465/milk.66.125.

Avigliano, Esteban, and Alejandra Vanina Volpedo. 2016. "A review of the application of otolith microchemistry toward the study of Latin American fishes." *Reviews in Fisheries Science & Aquaculture* 24 (4):369–384. doi: 10.1080/23308249.2016.1202189.

Azra, Mohamad Nor, Victor Tosin Okomoda, Meisam Tabatabaei, Marina Hassan, and Mhd Ikhwanuddin. 2021. "The contributions of shellfish aquaculture to global food security: Assessing its characteristics from a future food perspective." *Frontiers in Marine Science* 8. doi: 10.3389/fmars.2021.654897.

Baruah, A., A. Sharma, S. Sharma, and R. Nagraik. 2021. "An insight into different microplastic detection methods." *International Journal of Environmental Science and Technology*. doi: 10.1007/s13762-021-03384-1.

Basu, N., T. Nakano, E. G. Grau, and G. K. Iwama. 2001. "The effects of cortisol on heat shock protein 70 levels in two fish species." *General and Comparative Endocrinology* 124 (1):97–105.

Bianco-Stein, Nuphar, Iryna Polishchuk, Gabriel Seiden, Julie Villanova, Alexander Rack, Paul Zaslansky, and Boaz Pokroy. 2020. "Helical microstructures of the mineralized coralline red algae determine their mechanical properties." *Advanced Science* 7 (11):2000108.

Bilderback, Donald H., Pascal Elleaume, and Edgar Weckert. 2005. "Review of third and next generation synchrotron light sources." *Journal of Physics B: Atomic, Molecular and Optical Physics* 38 (9):S773-S797. doi: 10.1088/0953-4075/38/9/022.

Campana, S. E. 1999. "Chemistry and composition of fish otoliths: Pathways, mechanisms and applications." *Marine Ecology Progress Series* 188:263–297.

Campana, S. E., and S. R. Thorrold. 2001. "Otoliths, increments, and elements: Keys to a comprehensive understanding of fish populations?" *Canadian Journal of Fisheries and Aquatic Sciences* 58:30–38.

Canals, Miquel, Christopher K. Pham, Melanie Bergmann, Lars Gutow, Georg Hanke, Erik Van Sebille, Michela Angiolillo, Lene Buhl-Mortensen, Alessando Cau, Christos Ioakeimidis, Ulrike Kammann, Lonny Lundsten, George Papatheodorou, Autun Purser, Anna Sanchez-Vidal, Marcus Schulz, Matteo Vinci, Sanae Chiba, Francois Galgani, Daniel Langenkämper, Tiia Möller, Tim W. Nattkemper, Marta Ruiz, Sanna Suikkanen, Lucy Woodall, Elias Fakiris, Maria Eugenia Molina Jack, and Alessandra Giorgetti. 2020. "The quest for seafloor macrolitter: A critical review of background knowledge, current methods and future prospects." *Environmental Research Letters*. doi: 10.1088/1748-9326/abc6d4.

Cardoso, Márcia, Renata De Faria Barbosa, Gislene Torrente-Vilara, Gabriela Guanaz, Edgar Francisco Oliveira De Jesus, Eliane Teixeira Mársico, Roberta De Oliveira Resende Ribeiro, and Felipe Gusmão. 2019. "Multielemental composition and consumption risk characterization of three commercial marine fish species." *Environmental Pollution* 252:1026–1034. doi: 10.1016/j.envpol.2019.06.039.

Carey, Anne-Marie, Enzo Lombi, Erica Donner, Martin D. De Jonge, Tracy Punshon, Brian P. Jackson, Mary Lou Guerinot, Adam H. Price, and Andrew A. Meharg. 2012. "A review of recent developments in the speciation and location of arsenic and selenium in rice grain." *Analytical and Bioanalytical Chemistry* 402 (10):3275–3286. doi: 10.1007/s00216-011-5579-x.

Chen, Donglei, Henning Blom, Sophie Sanchez, Paul Tafforeau, Tiiu Märss, and Per E. Ahlberg. 2020. "The developmental relationship between teeth and dermal odontodes in the most primitive bony fish Lophosteus." *eLife* 9. doi: 10.7554/elife.60985.

Chen, Jennifer Yee-Shian, Yao-Chang Lee, and Bruno A. Walther. 2020. "Microplastic contamination of three commonly consumed seafood species from Taiwan: A pilot study." *Sustainability* 12 (22):9543. doi: 10.3390/su12229543.

Chen, Jiaxin, Weimu Wang, Hui Liu, Xiaohui Xu, and Jihong Xia. 2021. "A review on the occurrence, distribution, characteristics, and analysis methods of microplastic pollution in ecosystems." *Environmental Pollutants and Bioavailability* 33 (1):227–246. doi: 10.1080/26395940.2021.1960198.

Cook, Phil K., Cristian Mocuta, Élise Dufour, Marie-Angélique Languille, and Loïc Bertrand. 2018. "Full-section otolith microtexture imaged by local-probe X-ray diffraction." *Journal of Applied Crystallography* 51 (4):1182–1196. doi: 10.1107/s1600576718008610.

Cotti, Silvia, Ann Huysseune, Wolfgang Koppe, Martin Rücklin, Federica Marone, Eva M. Wölfel, Imke A. K. Fiedler, Björn Busse, Antonella Forlino, and P. Eckhard Witten. 2020. "More bone with less minerals? The effects of dietary phosphorus on the post-cranial skeleton in zebrafish." *International Journal of Molecular Sciences* 21 (15):5429. doi: 10.3390/ijms21155429.

Dimaria, R. A., J. A. Miller, and T. P. Hurst. 2010. "Temperature and growth effects on otolith elemental chemistry of larval Pacific cod, Gadus macrocephalus." *Environmental Biology of Fishes* 89 (3–4):453–462. doi: 10.1007/s10641-010-9665-2.

Do, G., S. Sase, R. Kobayashi, M. Sato, Y. Bae, T. Maeda, S. Ueno, and T. Araki. 2020. "Determining the internal structure of ice cream using cryogenic microtome imaging and X-ray computed tomography." *Japan Journal of Food Engineering* 21 (3):113–121.

Doubleday, Z. A., H. H. Harris, C. Izzo, and B. M. Gillanders. 2014. "Strontium randomly substituting for calcium in fish otolith aragonite." *Analytical Chemistry* 86 (1):865–869. doi: 10.1021/ac4034278.

Einfeld, D. 2007. "Synchrotron light sources, status and new projects." In *Brilliant light in life and material sciences*, edited by V. Tsakanov and H. Wiedemann, 3–20. Cham, Switzerland: Springer.

Epple, M., and F. Neues. 2010. "Synchrotron microcomputer tomography for the non-destructive visualization of the fish skeleton." *Journal of Applied Ichthyology* 26 (2):286–288. doi: 10.1111/j.1439-0426.2010.01422.x.

Espinoza-Quiñones, F. R., A. N. Módenes, S. M. Palácio, N. Szymanski, R. A. Welter, M. A. Rizzutto, C. E. Borba, and Alexander D. Kroumov. 2010. "Evaluation of trace element levels in muscles, liver and gonad of fish species from São Francisco River of the Paraná Brazilian state by using SR-TXRF technique." *Applied Radiation and Isotopes* 68 (12):2202–2207. doi: 10.1016/j.apradiso.2010.06.001.

FAO. 2020. "The state of world fishery and aquaculture 2020 (SOFIA 2020)." doi: 10.4060/ca9229en. Rome, Italy: Food and Agriculture Organization of the United Nations.

Farmery, A. K., K. Alexander, K. Anderson, J. L. Blanchard, C. G. Carter, K. Evans, M. Fischer, A. Fleming, S. Frusher, E. A. Fulton, B. Haas, C. K. Macleod, L. Murray, K. L. Nash, G. T. Pecl, Y. Rousseau, R. Trebilco, I. E. Van Putten, S. Mauli, L. Dutra, D. Greeno, J. Kaltavara, R. Watson, and B. Nowak. 2021. "Food for all: Designing sustainable and secure future seafood systems." *Reviews in Fish Biology and Fisheries*. doi: 10.1007/s11160-021-09663-x.

Fujii, T., and K. Shiomi. 2016. *New Food Hygiene*. Tokyo: Kouseusha Kouseikaku.

Galgani, Francois, Aleke Stoefen-O. Brien, Judith Weis, Christos Ioakeimidis, Qamar Schuyler, Iryna Makarenko, Huw Griffiths, Joan Bondareff, Dick Vethaak, Alan Deidun, Paula Sobral, Konstantinos Topouzelis, Penny Vlahos, Fernanda Lana, Martin Hassellov, Olivia Gerigny, Bera Arsonina, Archis Ambulkar, Maurizio Azzaro, and Maria João Bebianno. 2021. "Are litter, plastic and microplastic quantities increasing in the ocean?" *Microplastics and Nanoplastics* 1 (1). doi: 10.1186/s43591-020-00002-8.

Garlock, Taryn, Frank Asche, James Anderson, Trond Bjørndal, Ganesh Kumar, Kai Lorenzen, Andrew Ropicki, Martin D. Smith, and Ragnar Tveterås. 2019. "A global blue revolution: Aquaculture growth across regions, species, and countries." *Reviews in Fisheries Science & Aquaculture* 28 (1):107–116. doi: 10.1080/23308249.2019.1678111.

Ghosh, Pradipta, Steven Van Petegem, Helena Van Swygenhoven, and Atul H. Chokshi. 2017. "An in-situ synchrotron study on microplastic flow of electrodeposited nanocrystalline nickel." *Materials Science and Engineering: A* 701:101–110. doi: 10.1016/j.msea.2017.06.075.

Grabowski, Marek, David R. Cooper, Dariusz Brzezinski, Joanna M. Macnar, Ivan G. Shabalin, Marcin Cymborowski, Zbyszek Otwinowski, and Wladek Minor. 2021. "Synchrotron radiation as a tool for macromolecular X-Ray Crystallography: A XXI century perspective." *Nuclear Instruments and Methods in Physics Research Section B: Beam Interactions with Materials and Atoms* 489:30–40. doi: 10.1016/j.nimb.2020.12.016.

Grochulski, Pawel, Michel Fodje, Shaun Labiuk, Tomasz W. Wysokinski, George Belev, Malgorzata Korbas, and Scott M. Rosendahl. 2017. "Review of Canadian Light Source facilities for biological applications." *Nuclear Instruments and Methods in Physics Research Section B: Beam Interactions with Materials and Atoms* 411:17–21. doi: 10.1016/j.nimb.2017.01.065.

Hagiwara, T., R. Kobayashi, and N. Kimizuka. 2017. "Possible technologies for improving the quality of frozen marine products." In *Innovative technologies for improving the quality of marine products in frozen state,* edited by E. Okazaki, K. Konno and T. Suzuki, 90–111. Tokyo: Kouseisha Kouseikaku.

Håseth, T. T., M. Høy, B. Egelandsdal, and O. Sørheim. 2009. "Nondestructive analysis of salt, water, and protein in dried salted cod using computed tomography." *Journal of Food Science* 74 (3):E147–E153.

Håseth, T. T., M. Høy, J. Kongsro, A. Kohler, O. Sørheim, and B. Egelandsdal. 2008. "Determination of sodium chloride in pork meat by computed tomography at different voltages." *Journal of Food Science* 73 (7):E333–E339. doi: 10.1111/j.1750-3841.2008.00883.x.

Hasnain, S. Samar, and C. Richard A. Catlow. 2019. "Synchrotron science in the UK: NINA, the SRS and Diamond." *Philosophical Transactions of the Royal Society A: Mathematical, Physical and Engineering Sciences* 377 (2147):20190147. doi: 10.1098/rsta.2019.0147.

Hoban, J. M., D. L. Hopkins, N. Kirby, D. Collins, F. R. Dunshea, M. G. Kerr, K. Bailes, J. J. Cottrell, B. W. B. Holman, W. Brown, and E. N. Ponnampalam. 2016. "Application of small angle X-ray scattering synchrotron technology for measuring ovine meat quality." *Meat Science* 117:122–129. doi: 10.1016/j.meatsci.2016.03.005.

Hu, Huaiqiang, Jiating Zhao, Liming Wang, Lihai Shang, Liwei Cui, Yuxi Gao, Bai Li, and Yu-Feng Li. 2020. "Synchrotron-based techniques for studying the environmental health effects of heavy metals: Current status and future perspectives." *TrAC Trends in Analytical Chemistry* 122:115721. doi: 10.1016/j.trac.2019.115721.

Hüssy, Karin, Karin E. Limburg, Hélène De Pontual, Oliver R. B. Thomas, Philip K. Cook, Yvette Heimbrand, Martina Blass, and Anna M. Sturrock. 2020. "Trace element patterns in otoliths: The role of biomineralization." *Reviews in Fisheries Science & Aquaculture*:1–33. doi: 10.1080/23308249.2020.1760204.

Ivleva, Natalia P., Alexandra C. Wiesheu, and Reinhard Niessner. 2017. "Microplastic in Aquatic Ecosystems." *Angewandte Chemie International Edition* 56 (7):1720–1739. doi: 10.1002/anie.201606957.

Iwama, G. K., L. O. B. Afonso, and M. M. Vijayan. 2006. "Stress in fishes." In *The physiology of fishes (3rd Ed.)*, edited by D. H. Evans and J. B. Claiborne, 319–342. Boca Raton, FL: CRC Press.

Jacob-John, Jubin, Clare D'Souza, Tim Marjoribanks, and Stephen Singaraju. 2021. "Synergistic interactions of SDGs in food supply chains: A review of responsible consumption and production." *Sustainability* 13 (16). doi: 10.3390/su13168809.

Jambeck, J. R., R. Geyer, C. Wilcox, T. R. Siegler, M. Perryman, A. Andrady, R. Narayan, and K. L. Law. 2015. "Plastic waste inputs from land into the ocean." *Science* 347:768–771.

Jensen, Torben H., Arvid Böttiger, Martin Bech, Irene Zanette, Timm Weitkamp, Simon Rutishauser, Christian David, Elena Reznikova, Jürgen Mohr, Lars Bager

Christensen, Eli V. Olsen, Robert Feidenhans'L, and Franz Pfeiffer. 2011. "X-ray phase-contrast tomography of porcine fat and rind." *Meat Science* 88 (3):379–383. doi: 10.1016/j.meatsci.2011.01.013.

Kashiwabara, Teruhiko, Ryuichi Toda, Kentaro Nakamura, Kazutaka Yasukawa, Koichiro Fujinaga, Sayuri Kubo, Tatsuo Nozaki, Yoshio Takahashi, Katsuhiko Suzuki, and Yasuhiro Kato. 2018. "Synchrotron X-ray spectroscopic perspective on the formation mechanism of REY-rich muds in the Pacific Ocean." *Geochimica et Cosmochimica Acta* 240:274–292. doi: 10.1016/j.gca.2018.08.013.

Katayama, S. 2018. "A description of four types of otolith opaque zone." *Fisheries science* 84 (5):735–745.

Katayama, S. 2021. *Fish ecology revealed by eloquent otoliths*. Tokyo: Kouseisha Kouseikaku.

Kobayashi, R., T. Ishiguro, A. Ozeki, K. Kawai, and T. Suzuki. 2020. "Property changes of frozen soybean curd during frozen storage in 'Kori-tofu' manufacturing process." *Food Hydrocolloids* 104:105714.

Kono, S. 2017. "Quick observation method of ice crystal in frozen fish." In *Innovative technologies for improving the quality of marine products in frozen state*, edited by E. Okazaki, K. Konno and T. Suzuki, 112–121. Tokyo: Kouseisha Kouseikaku.

Kopittke, Peter M., Tracy Punshon, David J. Paterson, Ryan V. Tappero, Peng Wang, F. Pax C. Blamey, Antony Van Der Ent, and Enzo Lombi. 2018. "Synchrotron-based X-ray fluorescence microscopy as a technique for imaging of elements in plants." *Plant Physiology* 178 (2):507–523. doi: 10.1104/pp.18.00759.

Korbas, Malgorzata, Tracy C. Macdonald, Ingrid J. Pickering, Graham N. George, and Patrick H. Krone. 2012. "Chemical form matters: Differential accumulation of mercury following inorganic and organic mercury exposures in zebrafish larvae." *ACS Chemical Biology* 7 (2):411–420. doi: 10.1021/cb200287c.

Kreft, Ivan, et al. 2017. "New insights into structures and composition of plant food materials." *Journal of Microbiology, Biotechnology and Food Sciences* 7 (1):57–61. doi: 10.15414/jmbfs.2017.7.1.57-61.

Kunz, Alexander, Bruno A. Walther, Ludvig Löwemark, and Yao-Chang Lee. 2016. "Distribution and quantity of microplastic on sandy beaches along the northern coast of Taiwan." *Marine Pollution Bulletin* 111 (1–2):126–135. doi: 10.1016/j.marpolbul.2016.07.022.

Lam, Phoebe J., Benjamin S. Twining, Catherine Jeandel, Alakendra Roychoudhury, Joseph A. Resing, Peter H. Santschi, and Robert F. Anderson. 2015. "Methods for analyzing the concentration and speciation of major and trace elements in marine particles." *Progress in Oceanography* 133:32–42. doi: 10.1016/j.pocean.2015.01.005.

Lamela, Paula A., Roberto D. Pérez, Carlos A. Pérez, and Guillermina A. Bongiovanni. 2018. "Use of synchrotron radiation X-ray fluorescence and X-ray absorption spectroscopy to investigate bioaccumulation, molecular target, and biotransformation of volcanic elements." *X-Ray Spectrometry* 47 (4):305–319. doi: 10.1002/xrs.2843.

Landis, Eric N., and Denis T. Keane. 2010. "X-ray microtomography." *Materials Characterization* 61 (12):1305–1316. doi: 10.1016/j.matchar.2010.09.012.

Larue, Camille, Géraldine Sarret, Hiram Castillo-Michel, and Ana Elena Pradas Del Real. 2021. "A critical review on the impacts of nanoplastics and microplastics on aquatic and terrestrial photosynthetic organisms." *Small* 17 (20):2005834. doi: 10.1002/smll.202005834.

Li, Jiang-Lin, Da-Wen Sun, and Jun-Hu Cheng. 2016. "Recent advances in nondestructive analytical techniques for determining the total soluble solids in fruits: A review." *Comprehensive Reviews in Food Science and Food Safety* 15 (5):897–911.

Li, Yu-Feng, Jiating Zhao, Ying Qu, Yuxi Gao, Zhenghang Guo, Zuoliang Liu, Yuliang Zhao, and Chunying Chen. 2015. "Synchrotron radiation techniques for nanotoxicology." *Nanomedicine: Nanotechnology, Biology and Medicine* 11 (6):1531–1549. doi: 10.1016/j.nano.2015.04.008.

Limburg, K. E., and M. Elfman. 2017. "Insights from two-dimensional mapping of otolith chemistry." *Journal of Fish Biology* 90 (2):480–491. doi: 10.1111/jfb.13048.

Limburg, K. E., Bronwyn Gillanders, and Mikael Elfman. 2010. "Patterns and magnitude of Zn:Ca in otoliths support the recent phylogenetic typology of Salmoniformes and their sister groups." *Canadian Journal of Fisheries and Aquatic Sciences* 67 (4):597–604. doi: 10.1139/f10-014.

Limburg, K. E., Hans Høie, and Darren S. Dale. 2010. "Bromine patterns in Norwegian coastal Cod otoliths—a possible marker for distinguishing stocks?" *Environmental Biology of Fishes* 89 (3–4):427–432. doi: 10.1007/s10641-010-9660-7.

Limburg, K. E., Rong Huang, and Donald H. Bilderback. 2007. "Fish otolith trace element maps: New approaches with synchrotron microbeam x-ray fluorescence." *X-Ray Spectrometry* 36 (5):336–342. doi: 10.1002/xrs.980.

Limburg, K. E., Aude Lochet, Debra Driscoll, Darren S. Dale, and Rong Huang. 2010. "Selenium detected in fish otoliths: A novel tracer for a polluted lake?" *Environmental Biology of Fishes* 89 (3–4):433–440. doi: 10.1007/s10641-010-9671-4.

Limonta, Giacomo, Annalaura Mancia, Assja Benkhalqui, Cristiano Bertolucci, Luigi Abelli, Maria Cristina Fossi, and Cristina Panti. 2019. "Microplastics induce transcriptional changes, immune response and behavioral alterations in adult zebrafish." *Scientific Reports* 9 (1). doi: 10.1038/s41598-019-52292-5.

Liu, Huiqiang, Xizeng Wu, and Tiqiao Xiao. 2015. "Technical Note: Synchrotron-based high-energy x-ray phase sensitive microtomography for biomedical research." *Medical Physics* 42 (10):5595–5603. doi: 10.1118/1.4929551.

Liu, L., R. T. Neuenschwander, and A. R. D. Rodrigues. 2019. "Synchrotron radiation sources in Brazil." *Philosophical Transactions of the Royal Society A: Mathematical, Physical and Engineering Sciences* 377 (2147):20180235. doi: 10.1098/rsta.2018.0235.

Manceau, A., S. Azemard, L. Hédouin, E. Vassileva, D. Lecchini, C. Fauvelot, P. Swarzenski, P. Glatzel, P. Bustamante, and M. Metian. 2021. "The chemical forms of mercury in blue marlin billfish: Implications for human exposure." *Environmental Science and Technology Letters* 8 (5):405–411.

Mapp, J. J. I., M. H. Fisher, R. C. Atwood, G. D. Bell, M. K. Greco, S. Songer, and E. Hunter. 2016. "Three-dimensional rendering of otolith growth using phase contrast synchrotron tomography." *Journal of Fish Biology* 88 (5):2075–2080. doi: 10.1111/jfb.12949.

Marchin, Stéphane, Jean-Luc Putaux, Frédéric Pignon, and Joëlle Léonil. 2007. "Effects of the environmental factors on the casein micelle structure studied by cryo transmission electron microscopy and small-angle x-ray scattering/ultrasmall-angle x-ray scattering." *The Journal of Chemical Physics* 126 (4):045101. doi: 10.1063/1.2409933.

Miklos, R., M. S. Nielsen, H. Einarsdottir, R. Feidenhans'l, and R. Lametsch. 2015. "Novel X-ray phase-contrast tomography method for quantitative studies of heat induced structural changes in meat." *Meat Science* 100:217–221. doi: 10.1016/j.meatsci.2014.10.009.

Moraes, Paula M., Felipe A. Santos, Cilene C. F. Padilha, José C. S. Vieira, Luiz F. Zara, and Pedro De M. Padilha. 2012. "A preliminary and qualitative metallomics study of mercury in the muscle of fish from Amazonas, Brazil." *Biological Trace Element Research* 150 (1–3):195–199. doi: 10.1007/s12011-012-9502-x.

Morales-Caselles, Carmen, Josué Viejo, Elisa Martí, Daniel González-Fernández, Hannah Pragnell-Raasch, J. Ignacio González-Gordillo, Enrique Montero, Gonzalo M. Arroyo, Georg Hanke, Vanessa S. Salvo, Oihane C. Basurko, Nicholas Mallos,

Laurent Lebreton, Fidel Echevarría, Tim Van Emmerik, Carlos M. Duarte, José A. Gálvez, Erik Van Sebille, François Galgani, Carlos M. García, Peter S. Ross, Ana Bartual, Christos Ioakeimidis, Gorka Markalain, Atsuhiko Isobe, and Andrés Cózar. 2021. "An inshore—offshore sorting system revealed from global classification of ocean litter." *Nature Sustainability* 4 (6):484–493. doi: 10.1038/s41893-021-00720-8.

Nakano, T. 2007. "Microorganisms." In *Dietary supplements for the health and quality of cultured fish*, edited by H. Nakagawa, M. Sato and D. M. Gatlin III, 86–108. Oxfordshire, UK: CAB International (CABI).

Nakano, T. 2016. "Studies on stress and stress tolerance mechanisms in fish." *Nippon Suisan Gakkaishi* 82 (3):278–281. doi: 10.2331/suisan.WA2290.

Nakano, T. 2019. "Possibility of utilization of synchrotron radiation in fisheries science." In *Guidance on the use of synchrotron radiation*, edited by Tohoku synchrotron radiation facility promotion office, 42–48. Tokyo: AGNE Gijutsu Center.

Nakano, T. 2020. "Stress in fish and application of carotenoid for aquafeed as an anti-stress supplement." In *Encyclopedia of Marine Biotechnology*, edited by Se-K Kim, 2999–3019. Hoboken, USA: John Wiley & Sons Publications.

Nakano, T. 2021a. "Elucidation of environmental stress in fish and its application for farming healthy fish." *La mer* 59 (1–2):39–45.

Nakano, T. 2021b. "Stress and prevention against stress with supplements in fish." *Yoshoku Business (Aquaculture Business)* 58 (9):20–24.

Nakano, T., L. O. Afonso, B. R. Beckman, G. K. Iwama, and R. H. Devlin. 2013. "Acute physiological stress down-regulates mRNA expressions of growth-related genes in coho salmon." *PLoS ONE* 8 (8):e71421. doi: 10.1371/journal.pone.0071421.

Nakano, T., M. Kameda, Y. Shoji, S. Hayashi, T. Yamaguchi, and M. Sato. 2014. "Effect of severe environmental thermal stress on redox state in salmon." *Redox Biology* 2 (1):772–776. doi: 10.1016/j.redox.2014.05.007.

Nakano, T., T. Kanmuri, M. Sato, and M. Takeuchi. 1999. "Effect of astaxanthin rich red yeast (*Phaffia rhodozyma*) on oxidative stress in rainbow trout." *Biochimica et Biophysica Acta* 1426 (1):119–125.

Nakano, T., Y. Miura, M. Wazawa, M. Sato, and M. Takeuchi. 1999. "Red yeast *Phaffia rhodozyma* reduces susceptibility of liver homogenate to lipid peroxidation in rainbow trout." *Fisheries Science* 65 (6):961–962.

Nakano, T., K. Osatomi, N. Miura, Y. Aikawa-Fukuda, K. Kanai, A. Yoshida, H. Shirakawa, A. Yamauchi, T. Yamaguchi, and Y. Ochiai. 2020. "Effect of bacterial infection on the expression of stress proteins and antioxidative enzymes in Japanese flounder." In *Evolution of marine coastal ecosystems under the pressure of global changes*, edited by H. J. Ceccaldi, Y. Henocque, T. Komatsu, P. Prouzet, B. Sautour, and J. Yoshida, 111–127. Cham, Switzerland: Springer-Nature Switzerland AG.

Nakano, T., Y. Shoji, H. Shirakawa, Y. Suda, T. Yamaguchi, M. Sato, and R. H. Devlin. 2011. "Daily expression patterns of growth-related genes in growth hormone transgenic coho salmon, *Oncorhynchus kisutch*." *La mer* 49:111–117.

Nakano, T., M. Tosa, and M. Takeuchi. 1995. "Improvement of biochemical features in fish health by red yeast and synthetic astaxanthin." *Journal of Agricultural and Food Chemistry* 43 (6):1570–1573.

Nakano, T., M. Wazawa, T. Yamaguchi, M. Sato, and G. K. Iwama. 2004. "Positive biological actions of astaxanthin in rainbow trout." *Marine Biotechnology* 6 (Suppl.):S100–S105.

Nakano, T., and G. Wiegertjes. 2020. "Properties of carotenoids in fish fitness: A review." *Marine Drugs* 18 (11):10.3390/md18110568. doi: 10.3390/md18110568.

Nakazawa, N., and E. Okazaki. 2017. "Effect of freezing and storage conditions on the quality of seafood." In *Innovative technologies for improving the quality of marine products in frozen state*, edited by E. Okazaki, K. Konno and T. Suzuki, 36–59. Tokyo: Kouseisha Kouseikaku.

Nasuda, Yuko, Masato Ohnuma, Michihiro Furusaka, Kaoru Hara, and Toshinori Ishida. 2020. "Nanostructure analysis of dairy products." *Nippon Shokuhin Kagaku Kogaku Kaishi* 67 (6):186–192. doi: 10.3136/nskkk.67.186.

Naylor, R. L., R. J. Goldburg, J. H. Primavera, N. Kautsky, M. C. M. Beveridge, J. Clay, C. Folke, J. Lubchenco, H. Mooney, and M. Troell. 2000. "Effect of aquaculture on world fish supplies." *Nature* 405 (6790):1017–1024.

Naylor, R. L., R. W. Hardy, D. P. Bureau, A. Chiu, M. Elliott, A. P. Farrell, I. Forster, D. M. Gatlin, R. J. Goldburg, K. Hua, and P. D. Nichols. 2009. "Feeding aquaculture in an era of finite resources." *Proceedings of the National Academy of Sciences* 106 (36):15103–15110.

Okoshi, K. 2006. "Determination of geographic origin of the oyster Crassostrea gigas using multi-elemental analyses: Application for oyster culture." In *Technologies for species and origin identification of fisheries products*, edited by Y. Fukuda, S. Watabe and K. Nakamura, 128–138. Tokyo: Kouseisha Kouseikaku.

Otake, T. 2010. "Analysis of otolith." In *The fundamentals of fish ecology*, edited by K. Tsukamoto, 100–109. Tokyo: Kouseisha Kouseikaku.

Park, Hanbai, and Beomseok Park. 2021. "Review of microplastic distribution, toxicity, analysis methods, and removal technologies." *Water* 13 (19):2736. doi: 10.3390/w13192736.

Pauly, D., V. Christensen, S. Guenette, T. J. Pitcher, U. R. Sumaila, C. J. Walters, R. Watson, and D. Zeller. 2002. "Towards sustainability in world fisheries." *Nature* 418 (6898):689–695.

Pérez, Serge, and Daniele De Sanctis. 2017. "Glycoscience@Synchrotron: Synchrotron radiation applied to structural glycoscience." *Beilstein Journal of Organic Chemistry* 13:1145–1167. doi: 10.3762/bjoc.13.114.

Petroff, Yves. 2007. "The future of synchrotron radiation." *Journal of Electron Spectroscopy and Related Phenomena* 156–158:10–19. doi: 10.1016/j.elspec.2006.11.039.

PhoSIC, Photon science innovation center. 2019. *The next generation synchrotron radiation facility*. Edited by Tohoku University. Sendai, Japan.

Prüter, Julia, Sebastian Marcus Strauch, Lisa Carolina Wenzel, Wantana Klysubun, Harry Wilhelm Palm, and Peter Leinweber. 2020. "Organic matter composition and phosphorus speciation of solid waste from an African catfish recirculating aquaculture system." *Agriculture* 10 (10):466. doi: 10.3390/agriculture10100466.

Rao, Donepudi V., M. Bhaskaraiah, Roberto Cesareo, Antonio Brunetti, Tako Akatsuka, Tetsuya Yuasa, Zhong Zhong, Tohoru Takeda, and Giovanni E. Gigante. 2013. "Synchrotron-based non-destructive diffraction-enhanced imaging systems to image walnut at 20 keV." *Journal of Food Measurement and Characterization* 7 (1):13–21. doi: 10.1007/s11694-012-9134-z.

Reinke, Svenja K., Fabian Wilde, Sergii Kozhar, Felix Beckmann, Josélio Vieira, Stefan Heinrich, and Stefan Palzer. 2016. "Synchrotron X-ray microtomography reveals interior microstructure of multicomponent food materials such as chocolate." *Journal of Food Engineering* 174:37–46.

Rochman, Chelsea M., Eunha Hoh, Tomofumi Kurobe, and Swee J. Teh. 2013. "Ingested plastic transfers hazardous chemicals to fish and induces hepatic stress." *Scientific Reports* 3 (1). doi: 10.1038/srep03263.

Rochman, Chelsea M., Akbar Tahir, Susan L. Williams, Dolores V. Baxa, Rosalyn Lam, Jeffrey T. Miller, Foo-Ching Teh, Shinta Werorilangi, and Swee J. Teh. 2015. "Anthropogenic debris in seafood: Plastic debris and fibers from textiles in fish and bivalves sold for human consumption." *Scientific Reports* 5 (1):14340. doi: 10.1038/srep14340.

Romvári, R., Cs Hancz, Zs Petrási, T. Molnár, and P. Horn. 2002. "Non-invasive measurement of fillet composition of four freshwater fish species by computer tomography." *Aquaculture International* 10 (3):231–240.

Salditt, Tim, and Mareike Töpperwien. 2020. "Holographic imaging and tomography of biological cells and tissues." In *Nanoscale photonic imaging, topics in applied physics 134*, edited by Tim Salditt, A. Egner, and D. R. Luke, 339–376. Cham, Switzerland: Springer International Publishing.

Santos-Garcés, E., J. Laverse, P. Gou, E. Fulladosa, P. Frisullo, and M. A. Del Nobile. 2013. "Feasibility of X-ray microcomputed tomography for microstructure analysis and its relationship with hardness in non-acid lean fermented sausages." *Meat Science* 93 (3):639–644.

Sato, Masugu, Kentaro Kajiwara, and Norimichi Sano. 2016. "Non-destructive three-dimensional observation of structure of ice grains in frozen food by X-ray computed tomography using synchrotron radiation." *Japan Journal of Food Engineering* 17 (3):83–88. doi: 10.11301/jsfe.17.83.

Sato, Minoru, Kouichi Watanabe, Akiko Yamauchi, Toshiyasu Yamaguchi, Toshiki Nakano, Yoshihiro Ochiai, Satoshi Katayama, Yukari Niwa, and Yusuke Yamada. 2021. "Tissue observation and quality evaluation of olive flounder thawed by 100 MHz electromagnetic radiation." *Nippon Suisan Gakkaishi* 87 (3):275–280. doi: 10.2331/suisan.20-00049.

Schoeman, Letitia, Paul Williams, Anton Du Plessis, and Marena Manley. 2016. "X-ray micro-computed tomography (µCT) for non-destructive characterisation of food microstructure." *Trends in Food Science & Technology* 47:10–24. doi: 10.1016/j.tifs.2015.10.016.

Segtnan, V. H., M. Hoy, O. Sorheim, A. Kohler, F. Lundby, P. Wold, and R. Ofstad. 2009. "Noncontact salt and fat distributional analysis in salted and smoked salmon fillets using X-ray computed tomography and NIR interactance imaging." *Journal of Agricultural and Food Chemistry* 57:1705–1710.

Shim, Won Joon, Sang Hee Hong, and Soeun Eo Eo. 2017. "Identification methods in microplastic analysis: A review." *Analytical Methods* 9 (9):1384–1391. doi: 10.1039/c6ay02558g.

Sponaugle, Su. 2010. "Otolith microstructure reveals ecological and oceanographic processes important to ecosystem-based management." *Environmental Biology of Fishes* 89 (3–4):221–238. doi: 10.1007/s10641-010-9676-z.

Stentiford, G. D., I. J. Bateman, S. J. Hinchliffe, D. Bass, R. Hartnell, E. M. Santos, M. J. Devlin, S. W. Feist, N. G. H. Taylor, D. W. Verner-Jeffreys, R. van Aerle, E. J. Peeler, W. A. Higman, L. Smith, R. Baines, D. C. Behringer, I. Katsiadaki, H. E. Froehlich, and C. R. Tyler. 2020. "Sustainable aquaculture through the One Health lens." *Nature Food* 1:468–474.

Stock, Friederike, Christian Kochleus, Beate Bänsch-Baltruschat, Nicole Brennholt, and Georg Reifferscheid. 2019. "Sampling techniques and preparation methods for microplastic analyses in the aquatic environment—A review." *TrAC Trends in Analytical Chemistry* 113:84–92. doi: 10.1016/j.trac.2019.01.014.

Syamaladevi, Roopesh M., Kalehiwot N. Manahiloh, Balasingam Muhunthan, and Shyam S. Sablani. 2012. "Understanding the influence of state/phase transitions on ice recrystallization in Atlantic salmon (Salmo salar) during frozen storage." *Food Biophysics* 7:57–71.

Thilsted, Shakuntala Haraksingh, Andrew Thorne-Lyman, Patrick Webb, Jessica Rose Bogard, Rohana Subasinghe, Michael John Phillips, and Edward Hugh Allison. 2016. "Sustaining healthy diets: The role of capture fisheries and aquaculture for improving nutrition in the post-2015 era." *Food Policy* 61:126–131. doi: 10.1016/j.foodpol.2016.02.005.

Thomas, Oliver R. B., and Stephen E. Swearer. 2019. "Otolith biochemistry—a review." *Reviews in Fisheries Science & Aquaculture* 27 (4):458–489. doi: 10.1080/23308249.2019.1627285.

Twining, Benjamin S., Sara Rauschenberg, Peter L. Morton, and Stefan Vogt. 2015. "Metal contents of phytoplankton and labile particulate material in the North Atlantic Ocean." *Progress in Oceanography* 137:261–283. doi: 10.1016/j.pocean.2015.07.001.

Visbeck, Martin, Ulrike Kronfeld-Goharani, Barbara Neumann, Wilfried Rickels, Jörn Schmidt, Erik Van Doorn, Nele Matz-Lück, and Alexander Proelss. 2014. "A sustainable development goal for the ocean and coasts: Global ocean challenges benefit from regional initiatives supporting globally coordinated solutions." *Marine Policy* 49:87–89. doi: 10.1016/j.marpol.2014.02.010.

Wang, Bing, Zhe Wang, Weiyue Feng, Meng Wang, Zhongbo Hu, Zhifang Chai, and Yuliang Zhao. 2010. "New methods for nanotoxicology: Synchrotron radiation-based techniques." *Analytical and Bioanalytical Chemistry* 398 (2):667–676. doi: 10.1007/s00216-010-3752-2.

Wang, Zhong-Min, Mahima Parashar, Sutapa Ghosal, and Jeff Wagner. 2020. "A new method for microplastic extraction from fish guts assisted by chemical dissolution." *Analytical Methods* 12 (45):5450–5457. doi: 10.1039/d0ay01277g.

Watanabe, M. 2016. "The characteristics of synchrotron radiation and its importance." In *A guide to synchrotron radiation science*, edited by M. Watanabe, S. Sato, I. Munro, G. S. Lodha, 1.1–1.22. New Delhi: Narosa Publishing House Pvt. Ltd.

Weinhardt, Venera, Roman Shkarin, Tobias Wernet, Joachim Wittbrodt, Tilo Baumbach, and Felix Loosli. 2018. "Quantitative morphometric analysis of adult teleost fish by X-ray computed tomography." *Scientific Reports* 8 (1). doi: 10.1038/s41598-018-34848-z.

Westneat, Mark W., John J. Socha, and Wah-Keat Lee. 2008. "Advances in biological structure, function, and physiology using synchrotron X-ray imaging." *Annual Review of Physiology* 70 (1):119–142. doi: 10.1146/annurev.physiol.70.113006.100434.

Willis, Kathryn A., Catarina Serra-Gonçalves, Kelsey Richardson, Qamar A. Schuyler, Halfdan Pedersen, Kelli Anderson, Jonathan S. Stark, Joanna Vince, Britta D. Hardesty, Chris Wilcox, Barbara F. Nowak, Jennifer L. Lavers, Jayson M. Semmens, Dean Greeno, Catriona Macleod, Nunnoq P. O. Frederiksen, and Peter S. Puskic. 2021. "Cleaner seas: Reducing marine pollution." *Reviews in Fish Biology and Fisheries*. doi: 10.1007/s11160-021-09674-8.

Wootton, Nina, Patrick Reis-Santos, and Bronwyn M. Gillanders. 2021. "Microplastic in fish — A global synthesis." *Reviews in Fish Biology and Fisheries* 31 (4):753–771. doi: 10.1007/s11160-021-09684-6.

Wu, H., A. Aoki, T. Arimoto, T. Nakano, H. Ohnuki, M. Murata, H. Ren, and H. Endo. 2015. "Fish stress become visible: A new attempt to use biosensor for real-time monitoring fish stress." *Biosensors and Bioelectronics* 67:503–510. doi: 10.1016/j.bios.2014.09.015.

Xin, Xiaying, Gordon Huang, Chunjiang An, and Renfei Feng. 2019. "Interactive toxicity of triclosan and nano-TiO2 to green alga Eremosphaera viridisin Lake Erie: A new perspective based on fourier transform infrared spectromicroscopy and synchrotron-based X-ray fluorescence imaging." *Environmental Science & Technology* 53 (16):9884–9894. doi: 10.1021/acs.est.9b03117.

Xu, Jun-Li, Kevin V. Thomas, Zisheng Luo, and Aoife A. Gowen. 2019. "FTIR and Raman imaging for microplastics analysis: State of the art, challenges and prospects." *TrAC Trends in Analytical Chemistry* 119:115629. doi: 10.1016/j.trac.2019.115629.

Xu, W., Y. Liu, A. Marcelli, P. P. Shang, and W. S. Liu. 2018. "The complexity of thermoelectric materials: Why we need powerful and brilliant synchrotron radiation sources?" *Materials Today Physics* 6:68–82. doi: 10.1016/j.mtphys.2018.09.002.

Yamashita, Y., and M. Yamashita. 2006. "Estimation of fishing location by multiple trace elemental analysis." In *Technologies for species and origin identification of fisheries products*, edited by Y. Fukuda, S. Watabe, and K. Nakamura, 121–127. Tokyo: Kouseisha Kouseikaku.

Yang, Zhi, Huijuan Yang, and Hongshun Yang. 2018. "Characterisation of rheology and microstructures of κ-carrageenan in ethanol-water mixtures." *Food Research International* 107:738–746.

Yu, Fucheng, Ke Li, Feixiang Wang, Haipeng Zhang, Xiaolu Ju, Mingwei Xu, Guohao Du, Biao Deng, Honglan Xie, and Tiqiao Xiao. 2021. "Double-exposure method for speckle-tracking x-ray phase-contrast microtomography." *Journal of Applied Physics* 129 (7):073101. doi: 10.1063/5.0043053.

Zan, Guibin, Sheraz Gul, Jin Zhang, Wei Zhao, Sylvia Lewis, David J. Vine, Yijin Liu, Piero Pianetta, and Wenbing Yun. 2021. "High-resolution multicontrast tomography with an X-ray microarray anode—structured target source." *Proceedings of the National Academy of Sciences* 118 (25):e2103126118. doi: 10.1073/pnas.2103126118.

3 Biodiscovery of Marine Microbial Enzymes in Indonesia

Ekowati Chasanah, Pujo Yuwono,
Dewi Seswita Zilda, and Siswa Setyahadi

CONTENTS

3.1 INTRODUCTION

Enzymes are biocatalysts widely used in various industries from feed, food, detergent, leather, textiles, laundry, pharmaceuticals, cosmetics, and fine chemicals which we used daily. Enzymes work efficiently and specifically in mild process conditions so that they can reduce environmental pollution and minimize the cost of production. More than 80% of the global market of enzymes are for industrial application (Kumar, Sangwan, Singh, & Gill, 2014). The demand for enzymes is increasing, along with people's awareness of healthy living through healthy, safe food and other daily needs products produced in a more environmentally friendly manner.

In 2019, the value of the global enzyme market was at US$5.6, while in 2020–2027, the compound annual growth rate (CAGR) was expected to grow up to 6.4% (US$6 billion). The demand for some enzymes is predicted to rise, especially for the end-use industries, such as biofuel, home cleaning, animal feed, and food and beverage. The demand for carbohydrase and protease is also anticipated to increase in food and beverage applications, particularly in China, India, and Japan. Another factor which affects the large-scale demand is the increase in industrialisation due to the advancement in the nutraceutical sector (Market analysis Report, 2020). In 2020, the value of food enzymes was

DOI: 10.1201/9781003303909-3

US$1715.3 million, and by the end of 2026, it is predicted to reach US$2348.2 million. The food enzymes market is segmented into carbohydrase, protease and lipase. Among the global industrial enzyme producers, Novozymes is the largest player of the food enzymes producers, that is, protease, followed by DuPont as transglutaminase producers and AB Enzyme for α-amylase (Market analysis Report, 2020). However, China is also becoming an enzyme producer, and it was reported that 100 companies were involved in enzyme production in China in 2010, in addition to small to medium-sized producers of various enzymes (Kumar et al., 2014).

Even though animals and plants can and have been used as enzyme sources, microbes have served as the largest producer of commercial enzymes. Globally, microbes account for a two-thirds share of commercial protease producers, and some special characteristics of microbial enzymes, such as being small in size, dense and spherical in structure, make microbial enzymes are preferable for industrial application (Razzaq et al., 2019). Compared to those from plant and animals, microbes are relatively inexpensive to produce, and their catalytic activities are more predictable and controllable so that they can be produced within a short period and be sustainable independent of season (Cheng, Ismail, Kamaruding, Saidin, & Danish-daniel, 2020).

It was reported that during 2019, Indonesia imported 5464 tons of enzyme valued at US$45,380 from China, Finland and Denmark. Therefore, it is an urgent need for Indonesia to fulfill local industrial needs by producing local enzymes while also supporting global needs for unique enzymes. As 75% of its territory is marine and geographically lies on the equator, Indonesia is blessed with various unique marine environments, making it rich in biodiversity and chemicals as well, including microbes producing enzymes. Therefore, enzymes biodiscovery from unique marine locations such as around marine volcanoes, deepsea and other unique places will produce unique enzymes as well. This chapter reports a review on the present status of marine microbial enzyme research from Indonesia to support the food industry and pharmaceuticals, namely, protease, and marine carbohydrase, namely, chitinase, chitosanase, agarase, alginate lyase, and carrageenase.

3.2 BIODISCOVERY OF MARINE MICROBIAL ENZYMES

3.2.1 Protease Enzyme

Protease, or proteolytic enzyme, is widely used in various industries, accounting for about 60% of the total global enzymes sales. The value of the global proteases market is US$1645 million in 2020 and, by the end of 2026, is expected to reach US$2084.7 million, growing between 2021 and 2026 at a compound annual growth rate (CAGR) of 3.4% (https://www.marketstudyreport.com/reports/global-proteases-market-researchreport-2020). Microbial proteases that can tolerate harsh conditions, work best at a wide range of pH and use specific substrates are a needed priority for industries (Razzaq, 2019). Protease has been applied widely in food and feed industries and has been used in the detergent, waste management, leather and cosmetics industries. Alkaline protease has been an effective additive in laundry detergent formulations, and alkaline protease has been reported to degrade fibrin (protease fibrinolytic enzymes) could be applied as an anticancer drug and in thrombolytic therapy in the future (Razzaq et al., 2019).

Proteases are categorized as hydrolases, enzymes of class 3, subclass 3.4, peptide hydrolases or peptidase according to the Nomenclature Committee of the International Union of Biochemistry and Molecular Biology (Mamo & Assefa, 2018). They can be grouped as exopeptidases and endopeptidases; exopeptidase hydrolyzes the peptide bond proximal to the amino or carboxy terminal of the substrate, whereas endopeptidases cut the peptide bonds from the termini of the substrate. Based on the catalysis mechanism, that is, the hydrolysis of amide bonds in peptide substrates, proteases are classified as serine proteinases (EC 3.4.21), cysteine proteinases (EC 3.4.22), aspartyl proteinases (EC3.4.23), metalloproteinases (EC 3.4.24) and threonine peptidases (EC 3.4.25) (Kieliszek, Pobiega, Piwowarek, & Kot, 2021).

One of the serine proteases is the keratinase enzyme (E.C. 3.4.99.11), which is able to hydrolyze the disulfide hydrogen bonds in the keratin proteins. Keratinolytic proteases are also generally from metalloprotease regardless of the microbial source (Nnolim, Udenigwe, Okoh, & Nwodo, 2020). Microbial keratinases are continuously gaining an important role due to their effective bioconversion of hard keratin–rich wastes, that is, feathers from the poultry industry, as cost-effective substrates, and they are utilized to produce high-value products such as amino acids and bioactive peptides. In the feed industry, keratinolytic peptidase enzymes, that is, a protease that hydrolyzes keratin in fibrous animal protein such as feather, horns, hair, and nails as natural waste, could be used as a supplement in the animal feed and leather industries, in detergent formulation, in cosmetics and in organic fertilizer production (Kalaikumari et al., 2018; Nnolim et al., 2020). The biodiscovery of the keratinase enzyme from marine microbes has been reported (Lintang, Suhartono, Hwang, & Pyun, 2007; Rahayu, Bata, & Hadi, 2014). Thermostable keratinolytic protease was secreted by the L-23 isolate from a coastal hot spring in North Sulawesi, Indonesia. The enzyme was maximally produced when the L-23 isolate was cultivated using a selective medium containing 1% chicken feather powder at 70°C and a pH 7, and the keratinolytic protease worked best at 65°C and pH 7. The pure enzyme had an estimated molecular weight of 47 and 64 kDa. Keratinase has also been reported to be produced intracellularly and extracellularly by *Bacillus* sp. and *Bacillus licheniformis* MB-2 isolated from Indonesia. The keratinase was used to produce chicken feather meal. A substitution feeding trial showed that keratin meal from chicken feather can be used to substitute as much as 53% without significant difference ($P > 0.05$) on growth, feed consumption and conversion of the growing layer (Rahayu et al., 2014). The thermostable keratinolytic enzyme was also produced by *Brevibacillus thermoruber* LII, which is optimally active at temperatures between 45–55°C and a pH of 6–7. The *Brevibacillus thermoruber* LII bacteria were able to degrade untreated chicken feathers after 24 h incubation in a liquid medium (Zilda et al., 2012). In addition to the feed industry, the application of keratinase was also reported in cosmetics products as a bioactive compound in the preparation of topical products used as hair removals (Nnolim et al., 2020), as well as for dehairing sheepskin. In a concentration of 5% or 86.75 U/g keratinase (w/dwt. skin), the enzyme was able to remove the hair of sheepskin very efficiently similar to that of the chemical methods used traditionally (Kalaikumari et al., 2018).

Proteolytic enzymes have been used as meat tenderizers, and under controlled conditions and the right concentration of the proteolytic enzyme, the toughness of the meat will be reduced and the eating quality of meat enhanced (Madhusankha & Thilakarathna, 2021). Protease enzymes also have an important role in the dairy industry, namely, cheese, yogurt, kefir, and other so-called fermented dairy products. It was reported that some LAB (lactic acid bacteria) produce microbial proteases that may replace chymosin in cheese production; bakeries, that is, in the production of bread and pastries; brewing; food additives, such as protein hydrolyzates; and the feed industry (Kieliszek et al., 2021). Aspartic proteases have been reported to be applied in the food and beverage industries, such as in cheese making for the milk-clotting process, in fruit juices and alcoholic drink products to make clear of for hydrolyzing protein turbidity complex, and in the bread industry by modifying wheat gluten (Mamo & Assefa, 2018). The biodiscovery of protease to support the food industry to improve meat quality or as a tenderizer, and other applications are widely open along with the development of the food industry for both local and global needs.

In industrial processes involving polymers, thermostable proteases that are able to work at high temperatures are advantageous and are usually more preferred since the solubility of the reactants is increased at high temperatures. Using thermostable enzymes will also minimize contamination by mesophilic microbes. A number of thermostable proteases have been reported from Indonesia. Thermostable proteases had been isolated from the Sabang seabed, Weh Island, from a sand sample from approximately 15 m below the sea surface. The best isolate, PLS A, is moderately halophilic and had an optimal cultivation in a thermus medium for protease production at 65°C and a pH 7 for 18 h incubation (Iqbalsyah, Malahayati, Atikah, & Febriani, 2019). Thermostable protease has also been produced from *Brevibacillus thermorubber* LII; the pure protease enzyme was reported as homo-hexameric with a molecular weight of 215 kDa. The protease not only worked best at 60°C and a pH of 8 but also was active in an acidic buffer (up to pH 4), totally inhibited by PMSF (phenylmethanesulfonyl fluoride) and EDTA (ethylenediaminetetraacetic acid), indicating the enzyme was a neutral metallo-serine protease (Zilda, 2014). The *Brevibacillus thermorubber* have been reported to produce a thermostable protease that is active on keratin and fibroin degradation as well (Zilda, 2021). Thermostable protease, which worked best at 45°C and pH 7.0 was also reported from *Bacillus licheniformis* HSA3–1a from a hot spring of South Sulawesi (Rutu, Natsir, & Arfah, 2015). Thermostable protease also has been reported to be produced by bacteria from marine hydrothermal vent in the South Kawio Islands of South Sulawesi. The microbe has a maximum growth rate constant of μmax = 0.21 h^{-1}, and the optimum temperature for enzyme production was at 65°C and a pH between 6.6–7.5, 32 h. The protease enzyme work best at 70°C and a pH of 8 (Muharram & Aryantha, 2012).

Collagenase enzymes are metalloprotease enzymes that hydrolyze a collagen that is the most abundant protein in marine animals; therefore, collagenase is important in the nitrogen recycling process in the ocean. Collagenase is also important in preparing peptides for collagen hydrolyzates that have significant value due to their important functions, namely, (1) as nutrition by providing amino acids and small peptides, (2) as basic functions by improving connective tissues such as skin and bone, (3) as biological functions by possessing bioactive benefits such as being antioxidants and antihypertensives and (4) as having other useful functions, including antifreeze, metal chelation, and edible film formation (Chen, Li, & Huang, 2020). The important role as a meat tenderizer is also recognized (Madhusankha & Thilakarathna, 2021). The biodiscovery of collagenase in Indonesia has been reported, including collagenase from *Bacillus subtilis* ATCC 6633, which are active at the optimum condition of 50°C and at an optimum pH of collagenase 7–9 (Rochima et al., 2016), and from *Bacillus* sp. 6–2 isolated from fish waste of a local market in Makassar, Indonesia, which worked best at a pH of 7.0 and a temperature of 40°C (Sartika, Natsir, Dali, & Leliani, 2019). Protease has been produced by bacteria isolate HFSI-5 and -8 isolated from the gut of the sea cucumber *Holothuria scabra* from Indonesian waters. The protease has been reported to show thrombolysis activities, as good as Nattokinase, a commercial fibrinolysisi agent; therefore, these proteases might have the potential to be developed as a thrombolysis agent (Hidayati et al., 2021). Collagenase with a molecular weight of 124 and 26 kD was produced from *B. licheniformis* F11.4, which worked best at a temperature of 50°C and a pH of 7.0 (Baehaki et al., 2012). The collagenase was also reported to hydrolyze other protein substrates such as casein, gelatin, and fibrin.

Another thermostable protease, from *Bacillus* sp. BII.1 and *Bacillus* sp. BII,.2 isolated from a hot spring in Banyuwedang, Bali, Indonesia, has been used to make fish protein hydrolyzate (FPH). The first screening had no bitter taste in the FPH product. The protease then was used to produce FPH using yellowstripe scad (*Selaroides leptolepis*) (Fawzya, Nursatya, Susilowati, & Chasanah, 2020; Fawzya, Martosuyono, & Zilda, 2017) and sardinella (Chasanah, Susilowati, Yuwono, Zilda, & Fawzya, 2019). The FPH was used as an ingredient to enrich protein content improve the digestibility and protein absorption of various food products, targeting babies up to 6 months through enriching formula for high protein–complementary food using FPH (Putri et al., 2019), children, and pregnant mothers. FPH is a very strategic product in countries with problems of malnourished children, such as in Indonesia (Fawzya & Irianto, 2020). In addition, FPH can also be used as a protein supplement, as in sport energy drinks, in food products for older people and in sports nutrition, as well as in other healthy products (Fawzya & Irianto, 2020). By using local protease, protein sufficiency in such countries such as Indonesia that have rich fisheries will be easier and more economical and could be achieved by converting noneconomical fish commodities to FPH (Fawzya & Chasanah, 2020).

3.2.2 Chitinolytic Enzymes

Chitin-derivative products, namely, N-acetyl glucosamine (GlcNAc), glucosamine, and chitooligosaccharides, are produced by chitinolytic enzyme activity on a chitin polymer substrate. GlcNAc and glucosamine have been widely used in food, cosmetics and pharmaceutical industries, whereas chitooligosaccharides have been reported to enhance human health due to their bioactivity and antimicrobial, antioxidant, antitumor and anti-inflammatory properties (Dukariya & Kumar, 2020). The enzymatic production of GlcNAc from chitin was claimed to be more environmentally friendly, to produce higher yields, and to have higher bioactivities compared to those chemically processed like those used in industries today (Zhang et al., 2020). Chitinolytic enzymes, that is, chitinase, have also been applied for the bio-control of phytopathogens and insects (Mubarik, Mahagiani, Anindyaputri, Sugeng, & Rusmana, 2010). Chitinolytic enzymes have also been used to prepare fungal biomass as a medium for cultivating bacteria or microorganisms by converting the chitin in the mycelium and producing mono- and or disaccharides that can be readily used by the bacteria/microorganisms (Stumpf, Vortmann, Dirks-Hofmeister, Moerschbacher, & Philipp, 2019).

For decades, chitin has been well known as the second polymer, after cellulose, and is available in nature in an α or β form in the matrix with other materials such as protein and minerals. It can be found as a major structural component of arthropod exoskeleton and fungal cell walls, and the α-chitin form is being dominant. Commercial industries use chemical processes to obtain chitin, which is not considered environmentally friendly because they produce chemical waste. Environmentally friendly chitin production processes using lactic acid bacteria *Lactobacillus acidophilus* FNCC 116 to demineralize shrimp shells and deproteinize using *Bacillus licheniformis* F11.1 have been reported. Chitin produced by a biologically process has been reported to have a higher viscosity than that of chemically processed, which was more uncontrollable (Wahyuntari, Junianto, & Setyahadi, 2011). After chitin is obtained, derivative products from chitin that can be applied in various industries, namely, chitosan, chitooligosaccharides, chitin and chitosan, and the monomer, namely, glucosamine and N-acetyl glucosamine, can be produced using chitinolytic enzymes. The chitinolytic enzymes discussed in the following, namely, chitin deacetylase (CDA), chitinase, and chitosanase, are groups of enzymes that hydrolyze chitin and produce derivative products.

CDA (E.C. 3.5.1.4.1) is a key enzyme in replacing the unfriendly chemical process of chitosan production by hydrolyzing the acetamido group of chitin polymers. CDAs have been isolated from several microorganisms, first from *Mucor rouxii* (Araki & Ito, 1975); however, CDA research is not reported on as much as chitinase and chitosanase are, which might be due to a problem in the CDA assay. Besides CDA, chitin oligosaccharide deacetylase (COD; EC 3.5.1.105) is also able to produce β-N-acetyl-D-glucosaminyl-(1,4)-D-glucosamine (GlcNAc-GlcN) from $(GlcNAc)_2$ (Hirano et al., 2015). Enzyme assays for deacetylation activity of CDAs and COD are based on monitoring acetate release by ultraviolet (UV) absorbance changes, radiolabeled substrates and coupled enzymatic assays or formation of free amino groups with chromogenic or fluorogenic reagents, such as fluorescamine, o-phthalaldehyde, or ninhydrin (Pascual & Planas, 2018). COD and chitinase (glycosyl hydrolase [GH] or GH family 18) had been reported to be produced by marine bacteria, *Vibrio parahaemolyticus* KN1699 isolated from a Yatsu dry beach (Narashino, Chiba Prefecture, Japan) (Hirano et al., 2015). *Vibrio parahaemolyticus* is a member of *Vibrionacea*, a family of gram-negative and facultative anaerobes bacteria under the phylum Proteobacteria, inhabitants of fresh- or saltwater/marine environments. This result gives insight into marine environments, including Indonesia's marine environments, being a rich source of CDA or COD enzymes.

Chitinase (EC:3.2.1.14) and chitosanase (EC 3.2.1.132) have similar actions, hydrolyzing β-glycosidic bonds of chitin and chitosan polymers into their monomer form, namely, N- acetylglucosamine and N-glucosamine, respectively. The difference between both of them is that chitinase can hydrolyze β 1–4 glycosidic linkage in 100% acetylated chitin while it cannot hydrolyzed 100% deacetylated chitin/chitosan and vice versa; chitosanase is not able to hydrolyzed 100% acetylated chitin but is able to hydrolyze 100% deacetylated chitin/chitosan.

Chitosanases are capable of hydrolyzing β 1–4 glycosidic linkage of commercial chitosan which have 30–60% acetyl groups, and chitinase, lysozyme, and cellulase enzymes are also capable of hydrolyzing commercial chitosan with 20–35% acetyl groups (Tremblay, Yamaguchi, Fukamizo, & Brzezinski, 2001). In fish and mollusks, chitinases present in fish gut take a role in digesting chitin present in their prey, while in insects and crustaceans, chitinases are used in molting process, catalyzing exoskeletal molting. Chitinases in plant and seaweed are acting as self-defense proteins against fungi, and in bacteria, chitinolytic enzymes, are used in producing simple molecules from chitin for their nutritional needs as previously mentioned.

Chitosanases are important enzymes in the production of COS (chitooligosaccharide), and based on their amino acid sequence, chitosanases are classified into six GH families, and based on their cleavage specificity, they are grouped into four classes, namely, class I, which can cleave D–D (D = deacetylated, A = acetylated) and A–D; class II chitosanase (GH 8), which is capable of only cleaving the D–D; class III chitosanases, which cleave D–A in addition to cleaving D–D; and class IV GH, which is able to cleave all bonds except A–A. However, Weikert, Niehues, Moerschbacher, Cord-Landwehr, and Hellmann (2017) proposed the classification should be revisited. The bioactivity of COS as hydrolysis products is not only dependent on the degree of polymerization (DP) and the fractionation of acetylation (FA) but also on the pattern of acetylation (PA), and so far, the classification reported is not closely correlated to the sequence specificity since the products produced are not similar or different.

Research on chitinolytic enzymes in Indonesia has been dominated by screening the activity of chitinase and chitosanase, followed by lab-scale production, optimization and characterization, and application study of the enzymes but very limited study on CDA or COD. Table 3.1 shows research on chitinase and chitosanase conducted in Indonesia.

TABLE 3.1
Microbial Chitinase and Chitosanase from Indonesian Marine and Fisheries Products

No	Microbes (isolated from)	Chtinase/chitosanase; optimum T, H	Production Medium	References
1	*Serratia marcescens* PT-6 (shrimp pond sediment)	Chitinase; pH 6, 37°C	1.5 % colloidal chitin, 0.5 % starch, and 0.1 % yeast extrac	Agustiar et al. (2019)
2	*Aeromonas media* KLU 11.16 (shrimp waste processing)	Chitosanase; optimum pH 6 and temperature of 30°C. Activator: 1 mM Fe3+, Inhibitor: detergent (1 mM Triton X-100 and SDS). This enzyme digested 85% DA better than 100% DA chitosan	Minimal Synthetic Medium (MSM) supplemented with 0.5% colloidal chitin as the sole carbon source	Chasanah E., Patantis, Zilda, Ali, & Risjani (2011) Chasanah, Ali, & Ilmi (2012)
3	*Bacillus cereus* (KKT 1, KKT 14, and KKT 19), *Bacillus thuringiensis* (KKT 6), *Enterobacter cloacae* (LCK 20), *Pseudomonas stutzeri* (LCK 17), and *Stenotrophomonas maltophilia* (THK1) (an Indonesian traditional fermented fish product: rusip)	Screening in choloidal chitin agar	K2HPO4 0.1% (w/v), MgSO4.7H2O 0.01% (w/v), NaCl 0.1% (w/v), (NH4)2SO4 0.7% (w/v), yeast extract 0.05% (w/v), colloidal chitin 2% (w/v), agar 1% (w/v)	Puspita et al. (2017)
4	*Bacillus licheniformis* MB-2 (from Tompaso geothermal springs, North Sulawesi)	1. Chitinase; 67 kDa; T: 70°C: temperature and pH of the enzyme were and 6.0	0.5% colloidal chitin, 0.7% (NH4)2SO4, 0.1% K2HPO4, 0.1% NaCl, 0.01% MgSO4? 7H2O, 0.05% yeast extract and 1.5% agar (pH 7.0) and incubated at 55?C on a rotary shaker at 180 rev min)1	Toharisman, Suhartono, Spindler-Barth, Hwang, & Pyun (2005)

No	Microbes (isolated from)	Chtinase/chitosanase; optimum T, H	Production Medium	References
		2. Chitosanase; 75 kDa. Worked best at 70°C and pH 6.0 and 7.0.	0.24% chitosan, 0.25% casiton, 1% MgSO4, 1.4% K2HPO4, 0.02% CaCl2·2H2O, 0.002% FeSO4·7H2O (w/v)	Chasanah, Hariyadi, Witarto, Hwang, & Suhartono (2006)
5	*Stenotrophomonas maltophilia* 2123 (shrimp processing waste)	Chitosanase; pH 6, 50°C	Minimal Synthetic Medium (MSM) supplemented with 0.5% colloidal chitin as the sole carbon source	Fawzya, Pratitis, & Chasanah (2009)
6	34bs bacterial isolate from sponges	Chitosanase; pH 6–7, T 60°C; the half-life of chitosanase activity: 8.34 hours (at 37°C) and 55.12 min at 50°C	Minimal Synthetic Medium (MSM) supplemented with 0.5% colloidal chitin as the sole carbon source	Chasanah, Zilda, & Uria (2009)
7	SDI23, MDR23, SDI15, SDI13 dan BLT12 bacterial isolates from shrimp *petis*	Chitinase; pH 6, T: 45°C	K2HPO4 0.1% (b/v), MgSO4.7H2 0.01% (b/v), NaCl 0.1% (b/v), (NH4)2 SO4 0.7% (b/v), Yeast Extract 0.05% (b/v), & colloidal chitin 2%	Orinda, Puspita, Putra, & Lelana (2015)
8	Pseudomonas (the skin of blue swimmer crab (*Portunus pelagicus*)	Chitinase, screening	0.1% K2HPO4, 0.01% MgSO4.7H2O, 0.1% NaCl, 0.7% (NH4)2SO4, 0.05% yeast extract, 2% colloidal chitin, and 1% agar	Sulistijowati, Sudin, & Harmain (2021)
9	*Bacillus* sp.(from Ronto, a traditional fermented shrimp products)	Chitinase, screening	0.1% K2HPO4, 0.01% MgSO4.7H2O, 0.1% NaCl, 0.7% (NH4)2SO4, 0.05% yeast extract, 2% colloidal chitin, and 1% agar	Haryogya & Ustadi (2020)
10	*Providencia stuartii* (rotten shrimp)	Chitinase, screening	0.1% K2HPO4, 0.01% MgSO4.7H2O, 0.1% NaCl, 0.7% (NH4)2SO4, 0.05% yeast extract, 2% colloidal chitin, and 1% agar	Hardoko, Josephine, Handayani, & Halim (2020)

(*Continued*)

TABLE 3.1
(Continued)

No	Microbes (isolated from)	Chtinase/chitosanase; optimum T, H	Production Medium	References
11	*Pseudomonas stutzeri* PT5	Chitinase: pH 6 and T 37°C	The addition of 0.1% of ammonium phosphate and 0.1% of maltose increased the chitinase activity	Chalidah et al. (2018)
12	*Microminospora* T5a1 (from terasi, a fermented shrimp product)	Chitosanase; 24 hours at 37°C	0.1% K2HPO4, 0.01% MgSO4.7H2O, 0.1% NaCl, 0.7% (NH÷4)2SO4, 0.05% yeast extract and 0.5% colloidal chitin as inducer	Patantis, Zilda, Fawzya, & Chasanah (2020); Chasanah, Fawzya, Ariani, & Maruli (2013)
13	*Bacillus licheniformis* BPPTCC2 (isolated from shrimp waste)	Chitinase; pH 6.0 temperature of 50°C	Minimal media with chitin	Maggadani, Setyahadi, & Harmita (2017)

Our worked on crude chitosanase produced by *Microminospora* T5a1 and *Stenothopromonas maltophilia* showed that when the crude chitosanase were used to produce COS, the COS was able to inhibit *Aspergillus flavus* as good as that of sodium benzoate and potasium sorbate in their commercial application of 1000 ppm (Chasanah, Fawzya, Ariani, & Maruli, 2013). The COS was also able to inhibit 10% of the growth of *Staphylococcus aureus* and was toxic to He La cell line with LC_{50} of 120 ppm (Fawzya et al., 2009). While COS produced using *B.licheniformis* MB-2 was capable of inhibiting *Pseudomonas aerugenosa*, *Salmonella typhimurium*, *Escherichia coli*, *S. aureus*, *Listeria monocytogenes*, and *Bacillus cereus* at the minimum inhibition concentration value of 321 ppm for *Salmonella typhymurium* and 402 ppm for *P. aeruginosa*, *S. aureus*, and *E. coli*. Using the contact time of 24 h, all pathogenic bacteria tested was reduced by 2–5 log cycles (Chasanah, Meidina, & Suhartono, 2017). The chitinase derived from *Bacillus licheniformis* BPPTCC2 can be used to produce GlcNAc, using optimum conditions of pH 6.0 and temperature of 50°C, 0.2 U enzyme and 3% of substrate for 5 days. The yield of GlcNAc produced was 99.41% (Maggadani, Setyahadi, & Harmita, 2017).

3.2.3 Agarase, Alginate Lyase, and Carrageenase

Macroalgae or seaweed contains a high content of polysaccharides, namely, sulfated (fucoidan, agaran, carrageenan, ulvan) and nonsulfated (laminaran, alginate), both of which are soluble and rich in dietary fibers (30–75%). The biological production of derivative products from macroalgae polysaccharides, namely, oligosaccharides, are in high demand for the nutraceutical industry due to their beneficial properties, such as their antitumor, antiviral, anticoagulant and anti-inflammatory activities (Zhu & Yin, 2015; Chauhan & Arunika, 2016); therefore, the biologically process of derivative products could be used as ingredients for the health care, cosmetics, and food industries and agriculture, which have great value. A recent publication showed that a health-promoting effect, namely, short-chain fatty acids (SCFAs), is produced when oligosaccharides from laminaran, alginate and agar are used by the gut microbiome as prebiotics while oligosaccharides from fucoidan, carrageenan and ulvan need more research to confirm these positive effects (Gotteland et al., 2020). Agarase, alginate lyase and carrageenase are enzymes that hydrolyze marine polysaccharides from macroalgae. The exploration and application of these enzymes in marine countries such as Indonesia are very strategic for producing high-value derivative products from macroalgae, of which Indonesia is claimed as the biggest producer.

Agarase is a hydrolytic enzyme that is capable of hydrolyzing agar into agar-oligosaccharides. There are grouped as α-agarases and β-agarases according to the sites of hydrolysis. The α-agarases cleave the α-L-(1,3) linkages of agarose to produce oligosaccharides, namely, agarobiose, whereas the β-agarases cleave the β-D-(1,4) linkages of agarose to produce neoagarooligosaccharides. Oligosaccharide products with 3,6-anhydro-L-galactopyranose at the reducing end are produced by α-agarases while products with D-galactopyranoside residues at the reducing end are produced by β-agarases (Li, Sha, Zilda, Hu, & He, 2014) (Anggraeni & Ansorge-Schumacher, 2021). Biodiscovery research on agarase from Indonesia reported that *Bacillus* sp. BI-3 produce thermostable agarase when cultured in a medium composed of 0.3% (w/v) peptone, 0.3% (w/v) yeast extract, 0.3% (w/v) NaCl, and 2.0% (w/v) agar incubated at 55°C. The bacteria were isolated from a hot spring in Kalianda, Lampung. The pure enzyme, with a size of 58 kDa, was obtained by running the cell-free crude enzyme onto a Q-Sepharose column (2.6 × 40 cm), followed by a Sephacryl S-200 column. The agarase enzyme works best at a pH of 6.4 and a temperature of 70°C, but it is stable and active in a pH range of 5.8–8.0 at 80°C for 15 min. This enzyme is interesting because it produced neoagarobiose as the final product, which has great potential in the cosmetics industry since the reported product can perform both moisturizing and whitening effects on skin (Kobayashi, Takisada, Suzuki, Kirimura, & Usami, 1997). Another interesting agarase from Indonesia has been reported from *Microbulbifer elongatus* PORT2 that have been isolated from seawater in Batu Karas, Pangandaran, West Java, Indonesia. PORT2 recombinant agarases (which were expressed in *E. coli*) worked best at 50°C as thermostable agarases even though the bacteria *Microbulbifer elongatus* were from a mesophilic environment. The agarases' activity produced not only the saccharides neagarohexaose (NA6), neoagarotetraose (NA4) and neoagraobiose (NA2), which are typical agar-derived products, but also the modified ones from Indonesian natural agar, which is promising potential novel bioactivity (Anggraeni & Ansorge-Schumacher, 2021). Agarase has also been detected in *Vibrio alteromonas*, *Salinivibrio* and *Marinobacter* that were isolated from marine sediment of Bara Caddi, South Sulawesi, Indonesia (Zilda, Patantis, Prawira, Sibero, & Fawzya, 2021).

Alginate is a linear complex polysaccharide extracted from brown seaweed consisting of β-dmannuronic acid (M) and α-l-guluronic acid (G) units of widely varying composition and sequence. Alginate lyases can degrade alginate polymers into oligosaccharides or monosaccharides, and like other oligosaccharides, they have a great potential for applications in the food, cosmetic and pharmaceutical industries. There are two enzymes degrading alginate, namely, alginate lyase, that is, polyM lyases (EC 4.2.2.3) that work on M-rich alginates, and polyG lyases (EC 4.2.2.11), which works on G-rich alginates (Zhu & Yin, 2015). Research on alginate lyase from *Bacillus megaterium* S245 associated with brown seaweed *Sargassum crassifolium* from Indonesia has been reported (Subaryono, Ardilasari, Peranginangin, Zakaria, & Suhartono, 2016). The enzymes worked best at 45°C and a pH of 7.0, uniquely working toward the two substrates, both polymannuronate and polyguluronate. The alginate oligosaccharide (AOS) produced by the enzyme has the ability to induce cell proliferation of human lymphocytes, namely, CD 8 cells or cytotoxic T-cell and noncell CD4/CD8. AOS was produced by reacting 50 U of enzymes with alginate as a substrate; within 2 h, the reaction showed the highest index of lymphocyte proliferation (Subaryono, Perangiangin, Suhartono, & Zakaria, 2017). Another alginate lyase has been identified secreted by bacterium strain T513 isolated from rotten seaweed *Turbinaria*, which has 95% similarity with *Bacillus tequilensis* strain 10b. The enzyme active at an optimum temperature of 50°C and a pH of 8 and is stable at a pH of 4–9 and a temperature of 45°C (Zilda et al., 2019).

Seaweed-associated fungi have been isolated from brown and red seaweed from Sepanjang Beach, Gunung Kidul, Yogyakarta. Of 29 fungal isolates obtained, 11 isolates produced amylase, agarase, alginate and carrageenase enzymes. The prospective fungi isolate was identified as *Aspergillus sydowii* (Hutapea et al., 2021). From the same place, bacteria identified as *Salinicola zeshunii*, *Bacillus piscis* and *Bacillus licheniformis* associated with brown seaweed *Chaetomorpha* sp. were detected for potentially producing agarase and alginate lyase (Wijaya et al., 2021). From Pantai Panjang and Teluk Awur, Jepara, Indonesia, a fungus, identified as *Penicillium oligosporum*, has been reported to produce agarase, alginate lyase, carrageenase and amylase as well (Ayuningtyas et al., 2021).

Carrageenan is a sulfated polysaccharide that is an important component in the cell walls of red seaweeds. Carrageenan has been commercially produced and used as gelling, thickening, and stabilizing agents in the food industry and others, and their derived products, namely, oligosaccharides, have been reported to be active against tumors and viruses and are rich in antioxidant and immunomodulation activities. Therefore, carrageenan oligosaccharide has a significant potential for biomedical and physiological applications. It has various important applications, such as bioethanol production, textile industry, as a detergent additive and for isolating the protoplast of algae (Chauhan & Arunika, 2016; Gui, Gu, Zhang, & Li, 2021). The enzymes that are involved in biologically producing oligosaccarides are kappa (κ-), iota (ι-) and lambda (λ-) carrageenases. They are endohydrolaze enzymes that hydrolyze β-(1–4) linkages of carrageenans internally, yielding oligo-carrageenan products. From sediment of a hot spring of Lampung, Indonesia, a κ-carrageenase had been reported to be produced by carrageenan-degrading bacterium *Bacillus* sp. HT19. The purified enzyme works best at 60°C and a pH of 7.0. The product hydrolysate was neo-carrabiose that has been reported can improve the cucumber against cucumber mosaic virus and increased the activity of antioxidant enzymes in infected plants (Li, Pan, Xie, Zhang, & Gu, 2018). Carrageenan degrading enzymes or carrageenases was detected from bacteria associated with *Kappahycus alvarezii* namely *Labrenzia* sp., *Alteromonas* sp., *Vibrio* sp., *Celeribacter* sp., *Pseudoalteromonas* sp., *Phaeobacter* sp. and *Cobetia* sp. (Yusriyyah, Citra, Tassakka, & Latama, 2021). *Vibrio*, which are marine bacteria, are also reported to produce carrageenase. A *Vibrio* sp. strain NJ-2 isolated from rotten red algae secreted carrageenase with a molecular mass of 33 kDa and worked best at 40°C and a pH of 8.0. At optimum conditions, the enzyme showed a maximal activity of 937 U/mg and hydrolyzed the κ-carrageenan into 2–8 unit of D-oligosaccharides (Zhu & Ning, 2016).

3.3 ENZYME PRODUCTION

In laboratory studies, the production of enzymes is conducted in small volumes using Erlenmeyer flasks from 1–10 L for fermentation. In general, the bioprocess technology for microbial enzyme production is developed in three stages or scales, namely, (1) the laboratory scale, also called bench scale, where basic screening procedures are conducted; (2) the pilot plant, where the optimal step of operating conditions are obtained; and (3) plant scale, where industries apply the enzyme production process at an economic scale. Before pilot plant–scale stage, the enzyme produced in the bench or lab-scale production has to be optimized so that the enzyme can be produced in high yield, stable and economically feasible. Native enzymes are usually low in activity and yield and are not stable. Therefore, genetic modification and process engineering, including substrate modification and optimization, are important steps after finding the targeted enzymes.

For hydrolaze enzymes, in general, the medium used to screen and cultivate bacteria or other microbes producing enzymes is a minimal media added to the enzyme substrate that is usually in a polymer form. Keratinolytic enzyme, for example, as a used minimal medium supplemented with 1% w/v feather was reported to be the most efficient when compared to other media (Kalaikumari et al., 2018; Lintang et al., 2007). Since feather is the only carbon and nitrogen source in feather minimal medium, the bacteria have to produce keratinase in order to break down the feather into a simpler form so that the bacteria can access and use the carbon and nitrogen present in chicken feather. Since media are very important both technically and economically, the optimization of media is usually conducted using RSM (response surface methodology) preceded with a Plackett-Burman analysis to screen the significant variables used in the RSM. When the production was scaled up to 50-L and 500-L fermenter, the maximum keratinase activity at a pilot scale (500-L bioreactor), the production of keratinase was achieved at 1735 U/mL with a shorter time production compared to laboratory scale, and this is advantageous (Kalaikumari et al., 2018).

Our laboratory experience working with the production of proteolytic and other enzymes also started with isolating, identifying, characterizing, and optimizing the bacteria at a laboratory scale, followed by scaling up into the pilot plant scale. In this step, protease-producing bacterial isolates, namely, BII-1, BII-2 and BII-6 isolates identified as *Bacillus licheniformis*, isolates BII-3 and BII-4 identified as *Bacillus subtilis*, and isolate LII identified as *Brevibacillus thermoruber* BII-1, have been isolated from a hot spring in Banyuwedang, Bali, Indonesia (Zilda et al., 2012). The bacterial isolate showed the capability to degrade skim milk by forming a large clear zone surrounding its colony. The BII-1 protease showed an optimum temperature at 50°C and was used to scale up production at 100 L. Since the cost of media composed about 60% of total the production cost in the enzyme production, we tried using technical-grade skim milk instead of skim milk pro analysis (p.a.) used in lab work. The use of technical-grade skim milk for protease production was feasible; the protease activity produced using technical-grade skim milk was as good as that of skim milk p.a. (Fawzya, Nursatya, Susilowati, & Chasanah, 2020). The scaling up of the enzyme production using 10-L controlled fermenter was conducted using the same conditions as in the bench-scale experiments, and it was successfully producing protease with similar activity as that of the lab work. However, when the production was upscaled to pilot plant scale (100-L fermenter), it was more complicated than the bench-scale production as many factors must be adjusted as conditions, such as agitation and aeration affecting the dissolved oxygen and airflow rate in the fermentation, in turn, determine the activity of the protease enzyme produced. In our experience, the activity of protease produced in 100-L fermenter was 10–15% lower than its bench-scale counterpart. Therefore, selecting a fermenter and optimizing the fermenter's condition become important for obtaining excellent mixing and reasonably good heat and mass transfer rate so that the suitable conditions needed for the bacteria to grow and secrete the enzymes are obtained.

3.4 CONCLUDING REMARKS

As a maritime country, Indonesia has second-longest coastal line in the world; therefore, fisheries sectors, including macroalgae, are becoming national commodities. Indonesia is the second-largest cultivator of macroalgae, especially agar-producing macroalgae, in the world; however, about 60–70% of them were exported overseas as dried macroalgae, including to China and the Philippines (Anggraeni & Ansorge-Schumacher, 2021). During 2019, as much as 9,746,946 tons of dried macroalgae were exported (https://www.beritadaerah.co.id/2021/03/15/lima-provinsi-dengan-jumlah-produksi-rumput-laut terbesar/). Processing macroalgae into products, both polymers and their derivatives, as mentioned in the previous section, will bring significant added value to the country. Other marine polymers, such as chitin and chitosan, are also abundant, since Indonesia is also the main producer of shrimp. In 2020, Indonesia exported 208,000 tons, and it has targeted 727,000 tons of shrimp to be exported in 2024 in the form of ready-to-cook or ready-to-eat shrimp (Hidayatulah & Taufiqurohman, 2020). Shrimp and other crustacean shells are the main source of chitin/chitosan since shrimp shell contains about 25–40% chitin and 15–20% chitin in the shells. Therefore, the waste from shrimp shells will be enormous, and the bioconversion of this waste to chitin, chitosan, and the derivative products as COS will also generate a significant value of currency for Indonesia.

The application of these marine polymers in various food industries is enormous. These polymer-derivative products, namely, oligosaccharides, either oligosaccharide, agar, COS, AOS and carrageenan oligosaccharides, have been reported to have specific bioactivity, as mentioned in previous section, and the demand for these ingredient products in the global market is increasing steadily to support not only the food industry but also the feed, pharmaceutical, and cosmetics industries, among others.

From what we delivered, Indonesia is also rich in unique biocatalysts that can potentially be used as bioconversion agent-producing ingredients and processing aids. Until now, Indonesia is importing microbial biocatalysts and ingredients to support their industries. Microbial enzymes, especially from marine life, offer a uniqueness that industries need most due to their safety, specificity and environmentally friendly processes. It is estimated that the growth of the enzyme industry is increasing both local and globally, so it is possible to develop domestic enzyme industries in Indonesia. But the challenge is to create a strategy that strongly supports national industries by utilizing the genetic resources originating in Indonesia.

CONFLICTS OF INTEREST

The authors declare that there is no conflict of interests regarding the publication of this article.

REFERENCES

Agustiar, A. A., Faturrohmah, I., Sari, B. W., Isnaini, N. B., Puspita, I. D., Triyanto, . . . Ustadi. (2019). Increasing Chitinase Activity of Serratia marcescens PT-6 through Optimization of Medium Composition. *Squalen Bulletin of Marine and Fisheries Postharvest and Biotechnology*, *14*(3), 113–120.

Anggraeni, S. R., & Ansorge-Schumacher, M. B. (2021). Characterization and Modeling of Thermostable GH50 Agarases From Microbulbifer elongatus PORT2. *Marine Biotechnology*, *23*, 809–820.

Araki, Y., & Ito, E. (1975). A Pathway of Chitosan Formation in *Mucor rouxii*. *European Journal of Biochemistry*, *55*, 71–78.

Ayuningtyas, E., Siber, M., Hutapea, N., Frederick, E., Murwani, R., Zilda, D., . . . Radjasa, O. (2021). Screening of Extracellular Enzyme from Phaeophyceae-Associated

Fungi Screening of Extracellular Enzyme from Phaeophyceae-Associated Fungi. In *IOP Conf. Ser.: Earth Environ. Sci. 750 012005* (pp. 1–9). https://doi.org/10.1088/1755-1315/750/1/012005

Baehaki, A., Suhartono, M. T., Sukarno, Syah, D., Sitanggang, A. B., Setyahadi, S., & Meinhardt, F. (2012). Purification and Characterization of Collagenase from *Bacillus licheniformis* F11.4. *Afr. J. Microbiol. Res.*, *6*(10), 2373–2379. https://doi.org/10.5897/AJMR11.1379

Chalidah, N., Khotimah, I., Hakim, A. R., Meata, B., Puspita, I., Nugraheni, P., . . . Pudjiraharti, S. (2018). Chitinase Activity of *Pseudomonas stutzeri* PT5 in Different Fermentation Condition. In *IOP Conf. Ser.: Earth Environ. Sci. 139 012042* (pp. 1–10).

Chasanah, E., Ali, M., & Ilmi, M. (2012). Identification and Partial Characterization of Crude Extracellular Enzymes from Bacteria Isolated from Shrimp Waste Processing. *Squalen Bulletin* of *Marine and Fisheries Postharvest and Biotechnology*, *7*(1), 11–18.

Chasanah, E., Fawzya, Y. N., Ariani, F., & Maruli, K. (2013). Bioactivity of Chitooligosaccharide Produced by Chitosan using *Microminospora* T5a1 Chitosanase as Antifungal. *JPB Kelautan Dan Perikanan*, *8*(1), 65–72.

Chasanah, E., Hariyadi, P., Witarto, A. B., Hwang, J. K., & Suhartono, M. T. (2006). Biochemical Characteristics of Chitosanase From the Indonesian *Bacillus licheniformis* MB-2. *Molecular Biotechnology*, *3*, 93–102.

Chasanah, E., Meidina, M., & Suhartono, M. T. (2017). Antibacterial Potency of Chitosan Oligomer Produced by Bacillus licheniformis MB-2 Chitosanase. *Indonesian Fisheries Research Journal*, *14*(2), 91–95.

Chasanah, E., Patantis, G., Zilda, D. S., Ali, M., & Risjani, Y. (2011). Purification and Characterization of *Aeromonas media*. *Journal of Coastal Development*, *15*(1), 104–113.

Chasanah, E., Susilowati, R., Yuwono, P., Zilda, D. S., & Fawzya, Y. N. (2019). Amino Acid Profile of Biologically Processed Fish Protein Hydrolysate (FPH) Using Local Enzyme to Combat Stunting. In *IOP Conference Series: Earth and Environmental Science* (Vol. 278(1), p. 012013). IOP Publishing.

Chasanah, E., Zilda, D. S., & Uria, A. R. (2009). Screening and Characterization of Bacterial Chitosanase from Marine Environment. *Journal of Coastal Development*, *12*(2), 64–72.

Chauhan, P. S., & Arunika, C. (2016). Bacterial Carrageenases: An Overview of Production and Biotechnological Applications. *3 Biotech*, *6*(146), 1–18. https://doi.org/10.1007/s13205-0160461-3

Chen, M., Li, Y., & Huang, G. (2020). Potential Health Functions of Collagen Bioactive Peptides : A Review. *American Journal of Biochemistry and Biotechnology*, *16*(4), 507–519. https://doi.org/10.3844/ajbbsp.2020.507.519

Cheng, T. H., Ismail, N., Kamaruding, N., Saidin, J., & Danish-daniel, M. (2020). Industrial Enzymes-producing Marine Bacteria from Marine Resources. *Biotechnology Reports*, *27*(e00482).

Dukariya, G., & Kumar, A. (2020). Distribution and Biotechnological Applications of Chitinase : A Review. *International Journal of Biochemistry and Biophysics 8*, *8*(2), 17–29. https://doi.org/10.13189/ijbb.2020.080201

Fawzya, Y. N., & Irianto, H. E. (2020). Fish Protein Hydrolysates in Indonesia: Their Nutritional Values, Health Benefits, and Potential Applications. In *Marine Niche: Applications in Pharmaceutical Sciences* (pp. 283–297). Singapore: Springer.

Fawzya, Y. N., Nursatya, S. M., Susilowati, R., & Chasanah, E. (2020). Characteristics of Fish Protein Hydrolysate From Yellowstripe Scad (Selaroides leptolepis) Produced by a Local Microbial Protease. In *E3S Web of Conferences* (Vol. 147, p. 03017). EDP Sciences.

Fawzya, Y. N., Pratitis, A., & Chasanah, E. (2009). Karakterisasi enzim kitosanase dari isolat bakteri KPU 2123 dan aplikasinya untuk produksi oligomer kitosan. *Jurnal Pasacapanen dan Bioteknologi Kelautan dan Perikanan*, *4*(1), 69–78.

Gotteland, M., Riveros, K., Gasaly, N., Carcamo, C., Magne, F., Liabeuf, G., . . . Rosenfeld, S. (2020). The Pros and Cons of Using Algal Polysaccharides as Prebiotics. *Frontiers in* Nutrition, *7*(163), 1–15. https://doi.org/10.3389/fnut.2020.00163

Gui, Y., Gu, X., Zhang, Q., & Li, J. (2021). Expression and Characterization of a Thermostable Carrageenase from an Antarctic *Polaribacter* sp. NJDZ03. *Frontiers in Microbiology*, 12(March), 1–11. https://doi.org/10.3389/fmicb.2021.631039

Hardoko, Josephine C., Handayani, R., & Halim, Y. (2020). Isolation, Identification and Chitinolytic Index of Bacteria from Rotten Tiger Shrimp (*Penaeus monodon*) Shells. *AACL Bioflux*, *13*(1), 360–371.

Haryogya, A. M., & Ustadi. (2020). Isolation and Molecular Identification Chitinolytic Bacteria from Ronto. In *E3S Web of Conferences 147* (Vol. 3030, pp. 0–7).

Hidayati, N., Nurrahman, N., Fuad, H., Munandar, H., Zilda, D. S., Ernanto, A. R., . . . & Ethica, S. N. (2021, March). *Bacillus tequilensis* Isolated From Fermented Intestine of *Holothuria scabra* Produces Fibrinolytic Protease With Thrombolysis Activity. In *IOP Conference Series: Earth and Environmental Science* (Vol. 707, No. 1, p. 012008). IOP Publishing.

Hidayatulah, T., & Taufiqurohman, M. (2020). Ekspor udang melesat lampaui target 2020. *Lokadata*, February 10, 2021. https://lokadata.id/artikel/ekspor-udang-masih-bisa-tumbuh-lebihtinggi.

Hirano, T., Sugiyama, K., Sakaki, Y., Hakamata, W., Park, S., & Nishio, T. (2015). Structurebased Analysis of Domain Function of Chitin Oligosaccharide Deacetylase from *Vibrio parahaemolyticus*. *FEBS Letters*, *589*(1), 145–151. https://doi.org/10.1016/j.febslet.2014.11.039

Hutapea, N., Sibero, M., Ayuningtyas, E., Frederick, E., Wijayanti, D., Sabdono, A., . . . Murwani, R. (2021). Seaweed-Associated Fungi from Sepanjang Beach, Gunung-Kidul, Yogyakarta as Potential Source of Marine Polysaccharides-Degrading Enzymes. In *IOP* Conf. *Ser.: Earth Environ. Sci. 750 012007* (pp. 1–14). https://doi.org/10.1088/17551315/750/1/012007

Iqbalsyah, T. M., Malahayati, M., Atikah, A., & Febriani, F. (2019). Cultivation Conditions for Protease Production by A Thermo-Halostable Bacterial Isolate Pls A. *Jurnal Natural*, *19*(February), 18–23. https://doi.org/10.24815/jn.v19i1.11971

Kalaikumari, S. S., Vennila, T., Monika, V., Raj, K. C., Gunasekaran, P., & Rajendhran, J. (2018). Bioutilization of Poultry Feather for Keratinase Production and its Application in Leather Industry. *Journal of Cleaner Production*, (October). https://doi.org/10.1016/j.jclepro.2018.10.076

Kieliszek, M., Pobiega, K., Piwowarek, K., & Kot, A. M. (2021). Characteristics of the Proteolytic Enzymes Produced by Lactic Acid Bacteria. *Molecules*, *26*, 1–15.

Kobayashi, R., Takisada, M., Suzuki, T., Kirimura, K., & Usami, S. (1997). Neoagarobiose as a Novel Moisturizer with Whitening Effect. *Bioscience, Biotechnology, and Biochemistry*, *61*(I), 162–163.

Kumar, V., Sangwan, P., Singh, D., & Gill, P. K. (2014). Global Scenario of Industrial Enzyme Market. In A. K. Sharma & Vikas Beniwal (Eds.), *Industrial Enzymes: Trends, Scope and Relevance* (pp. 173–196). Nova Publisher. https://doi.org/10.13140/2.1.3599.0083

Li, J., Pan, A., Xie, M., Zhang, P., & Gu, X. (2018). Characterization of a Thermostable κ Carrageenase from a Hot Spring Bacterium and Plant Protection Activity of the Oligosaccharide Enzymolysis Product. *Science of Food and Agriculture*. https://doi.org/10.1002/jsfa.9374

Li, J., Sha, Y., Zilda, D. S., Hu, Q., & He, P. (2014). Purification and Characterization of Thermostable Agarase from *Bacillus* sp. BI-3, a Thermophilic Bacterium Isolated from Hot Spring. *Journal of Microbiology and Biotechnology*, *24*(1), 19–25.

Lintang, R., Suhartono, M. T., Hwang, J. A. E. K., & Pyun, Y. U. R. (2007). Thermostable Chicken Feather Degrading Enzymes from L-23 Isolate from Indonesia. *Microbiology Indonesia*, *1*(3), 109–113.

Madhusankha, G. D. M. P., & Thilakarathna, R. C. N. (2021). Meat Tenderization Mechanism and the Impact of Plant Exogenous Proteases: A Review. *Arabian Journal of Chemistry*, 14(2), 102967. https://doi.org/10.1016/j.arabjc.2020.102967

Maggadani, B. L., Setyahadi, S., & Harmita. (2017). Screening and Chitinolytic Activity Evaluation of Nine Indigenous Bacteria Isolate. *Journal of Pharmaceutical Sciences and Research*, *4*(1), 2407–2354.

Mamo, J., & Assefa, F. (2018). The Role of Microbial Aspartic Protease Enzyme in Food and Beverage Industries. *Hindawi Journal of Food Quality*, *ID7957269*, 1–15.

Market Analysis Report. 2020. Industrial Enzymes Market Size, Share & Trends Analysis Report By Product (Carbohydrases, Proteases, Lipases, Polymerases & Nucleases), By Source, By Application, By Region, and Segment Forecasts, 2020–2027. Grand View Research. Report ID: 978-1-68038-844-2.

Mubarik, N. R., Mahagiani, I., Anindyaputri, A., Sugeng, S., & Rusmana, I. (2010). Chitinolytic Bacteria Isolated from Chili Rhizosphere : Chitinase Characterization and Its Application as Biocontrol for Whitefly (Bemisia tabaci Genn.). *American Journal of Agricultural and Biological Sciences*, *5*(April), 430–435. https://doi.org/10.3844/ajabssp.2010.430.435

Muharram, L. H., & Aryantha, I. N. P. (2012). Karakterisasi Protease Termofilik Asal Hydrothermal Vent Laut Dalam Kepulauan Kawio Sulawesi Utara. *Journal of Science, Technology and Entrepreneur*, *1*(1), 18–29.

Nnolim, N. E., Udenigwe, C. C., Okoh, A. I., & Nwodo, U. U. (2020). Microbial Keratinase : Next Generation Green Catalyst and Prospective Applications. *Frontiers in Microbiology*, 11(December), 1–20. https://doi.org/10.3389/fmicb.2020.580164

Orinda, E., Puspita, I. D., Putra, M. P., & Lelana, I. Y. B. (2015). Chitinolytic Enzyme Activity of Isolate SDI23 from Petis and The Activity of Its Partially Purified Enzyme in Different pH and Temperature. *Journal of Fisheries Science*, *XVII*(2), 96–102.

Pascual, S., & Planas, A. (2018). Screening Assay for Directed Evolution of Chitin Deacetylases: Application to Vibrio cholerae Deacetylase Mutant Libraries for Engineered Specificity. *Analytical Chemistry*, *90*, 10654–10658. https://doi.org/10.1021/acs.analchem.8b02729

Patantis, G., Zilda, D., Fawzya, Y., & Chasanah, E. (2020). Purification of Chitosanase from *Stenotrophomonas maltophilia* KPU 2123 and *Micromonospora* sp. T5a1 for Chitooligosacharide Production. In *IOP Conf. Ser.: Earth Environ. Sci 404 012078* (pp. 1–8). https://doi.org/10.1088/1755-1315/404/1/012078

Puspita, I. D., Wardani, A. R. I., Oki, Puspitasari, A., Nugraheni, P. S. I. H., Putra, M. P., Pudjiraharti, S., & Ustadi. (2017). Occurrence of Chitinolytic Bacteria in Shrimp Rusip and Measurement of their Chitin-degrading Enzyme Activities. *Biodiversitas*, *18*(3), 1275–1281. https://doi.org/10.13057/biodiv/d180354

Putri, Y. I., Anwar, S., Afifah, D. N., Chasanah, E., Fawzya, Y. N., & Martosuyono, P. (2019). Optimasi Formula MP-ASI Bubuk Sumber Protein dengan Substitusi Hidrolisat Protein Ikan dan Tepung Kacang Hijau Menggunakan Response Surface Methodology. *Jurnal Aplikasi Teknologi Pangan*, *8*(4), 123–129. https://doi.org/10.17728/jatp.4346

Rahayu, S., Bata, M., & Hadi, W. (2014). Protein Concentrate Substitution Using Feather Meal Processes by Phisicochemistry and Fermentation. *Agripet*, *14*(1), 31–36.

Razzaq, A., Shamsi, S., Ali, A., Ali, Q., Sajjad, M., Malik, A., & Ashraf, M. (2019). Microbial Proteases Applications. *Frontiers in Bioengineering and Biotechnology, 7*(June), 1–20. https://doi.org/10.3389/fbioe.2019.00110

Rochima, E., Sekar, N., Buwono, I. D., Afrianto, E., & Pratama, I. R. (2016). Isolation and Characterization of Collagenase from *Bacillus subtilis* (Ehrenberg, 1835); ATCC 6633 for Degrading Fish Skin Collagen Waste from Cirata Reservoir, Indonesia. Aquatic *Procedia, 7*, 76–84. https://doi.org/10.1016/j.aqpro.2016.07.010

Rutu, I., Natsir, H., & Arfah, R. (2015). Production of Protease Enzyme from Bacteria in Hot Spring of South Sulawesi, *Bacillus licheniformis* HSA3–1a. *Marina Chimica Acta, 16*(1), 10–17.

Sartika, Natsir, H., Dali, S., & Leliani. (2019). Production and Characterization of Collagenase from *Bacillus* sp. 6–2 Isolated from Fish Liquid Waste. *Indonesian Chimica Acta, 12*(1), 58–66.

Stumpf, A. K., Vortmann, M., Dirks-Hofmeister, M. E., Moerschbacher, B. M., & Philipp, B. (2019). Identification of a Novel Chitinase From Aeromonas hydrophila AH-1N for the Degradation of Chitin Within Fungal Mycelium. *FEMS Microbiology Letters, 366*(1), fny294.

Subaryono, Ardilasari, Y., Peranginangin, R., Zakaria, F. R., & Suhartono, M. T. (2016). Bacillus *megaterium* S245 Shows Activities toward Polymannuronate and Polyguluronate. *Squalen Bulletin of Marine and Fisheries Postharvest and Biotechnology, 11*(2), 45–52.

Subaryono, Perangiangin, R., Suhartono, M. T., & Zakaria, F. R. (2017). Imunomodulator Activity of Alginate Oligosaccharides from Alginate *Sargassum crassifolium. JPHPI, 20*(1), 63–73.

Sulistijowati, R., Sudin, & Harmain, R. M. (2021). Chitinase Activity Potential and Identification of Chitinolytic Bacteria Isolated of Swimmer Crab' s Cell. *International Journal of Agricultural and Biological Engineering, 14*(3), 228–231. https://doi.org/10.25165/j.ijabe.20211403.5273

Toharisman, A., Suhartono, M. T., Spindler-Barth, M., Hwang, J., & Pyun, Y.-R. (2005). Purification and Characterization of a Thermostable Chitinase from *Bacillus licheniformis* MB-2. *World Journal of Microbiology & Biotechnology, 21*, 733–738. https://doi.org/10.1007/s11274-0044797-1

Tremblay, H., Yamaguchi, T., Fukamizo, T., & Brzezinski, R. (2001). Mechanism of Chitosanase-Oligosaccharide Interaction: Subsite Structure of *Streptomyces* sp. N174 Chitosanase and the Role of Asp57 Carboxylate. *Journal of Biochemistry, 130*, 679–686.

Wahyuntari, B., Junianto, & Setyahadi, S. (2011). Process Design of Microbiological Chitin Extraction Process Design of Microbiological Chitin Extraction. *Microbiology, 5*(1), 39–45. https://doi.org/10.5454/mi.5.1.7

Weikert, T., Niehues, A., Moerschbacher, B. M., Cord-landwehr, S., & Hellmann, M. J. (2017). Reassessment of Chitosanase Substrate Specificities and Classification. *Nature* Communication, 1–11. https://doi.org/10.1038/s41467-017-01667-1

Wijaya, A., Sibero, M., Zilda, D., Windiyana, A., Wijayanto, A., Fredericl, E., . . . Radjasa, O. (2021). Preliminary Screening of Carbohydrase-Producing Bacteria from *Chaetomorpha* sp. in Sepanjang Beach, Yogyakarta, Indonesia. In *IOP Conf. Ser.: Earth Environ. Sci. 750* 012027 (pp. 1–8). https://doi.org/10.1088/1755-1315/750/1/012027

Yusriyyah, A. A., Citra, A., Tassakka, M. A., & Latama, G. (2021). Identification of the Potential of Degrading Carrageenan in Red Algae *Kappaphycus alvarezii* Symbiotic Bacteria. *International Journal of Environment, Agriculture and Biotechnology, 6*(1), 81–85. https://doi.org/10.22161/ijeab

Zhang, A., Mo, X., Zhou, N., Wang, Y., Wei, G., Hao, Z., & Chen, K. (2020). Identification of Chitinolytic Enzymes in *Chitinolyticbacter meiyuanensis* and Mechanism of

Efficiently Hydrolyzing Chitin to N -Acetyl Glucosamine. *Frontiers in Microbiology*, *11*(October), 1–11. https://doi.org/10.3389/fmicb.2020.572053

Zhu, B., & Ning, L. (2016). Purification and Characterization of a New K-Carrageenase from the Marine Bacterium *Vibio* sp. NJ-2. *Journal of Microbiology and Biotechnology*, *26*(2), 255–262.

Zhu, B., & Yin, H. (2015). Alginate Lyase: Review of Major Sources and Classification, Properties, Structure-function Analysis and Applications. *Bioengineered*, *6:3*(June), 125–131.

Zilda, D. S., Patantis, G., Prawira, S. S., Sibero, M. T., & Fawzya, N. (2021). Screening and Identification of Agarase-producing Bacteria from Sediment Sample of Bara Caddi Sea, South Sulawesi. *JPB Kelautan Dan Perikanan*, *16*(1), 11–21.

Zilda, D. S., Yulianti, Y., Sholihah, R. F., Subaryono, Fawzya, N. F., & Irianto, H. E. (2019). A Novel *Bacillus* sp. Isolated from Rotten Seaweed: Identification and Characterization Alginate Lyase its Produced. *Biodiversitas*, *20*(4), 1166–1172. https://doi.org/10.13057/biodiv/d200432

Zilda, D. Z. (2014). Microbial Transglutaminase: Source, Production and Its Role to Improve Surimi Properties. *Squalen Bulletin of Marine and Fisheries Postharvest and Biotechnology*, *9*(1), 35–44.

Zilda, D. Z., Harmayani, E., Widada, J., Asmara, W., Irianto, H. E., Patantis, G., & Fawzya, Y. N. (2012). Screening of Thermostable Protease Producing Microorganisms Isolated From Indonesian Hotspring. *Squalen Bulletin of Marine and Fisheries Postharvest and Biotechnology*, *7*(3), 105–114.

4 Soluble Potentially Toxic Metals (Cu and Pb) in the Different Tissues of Marine Mussel *Perna viridis*

Health Risk Perspectives

Chee Kong Yap, Franklin Berandah Edward, Wan Mohd Syazwan, Noor Azrizal-Wahid, Wan Hee Cheng, Wen Siang Tan, Moslem Sharifinia, Alireza Riyahi Bakhtiari, Muskhazli Mustafa, Hideo Okamura, Khalid Awadh Al-Mutairi, Salman Abdo Al-Shami, Mohamad Saupi Ismail, and Amin Bintal

CONTENTS

4.1 INTRODUCTION

Marine mussels are a major internationally traded seafood commodity (Toyofuku, 2006). Besides, from the nutritional point of view, marine mussels are important food sources of essential trace metals, certain vitamins (Cheong and Lee, 1984) as well as protein. Some commercial shellfish may contain polyunsaturated *n-3* fatty acids, widely known to reduce risks of cardiovascular disease (Kromhout et al., 1985). However, due to the well-reported studies on health risks in relation to heavy metals consumption,

DOI: 10.1201/9781003303909-4

dietary exposure to heavy metals has been a major concern (Petroczi and Naughton, 2009).

Metal-contaminated seafood may cause toxicity to humans if it is excessively consumed (Kromhout et al., 1985). Therefore, the estimation of metal concentrations in edible tissues of marine mussels is crucial since bioaccumulation and biomagnification of these toxic metals increase over time and might eventually pose toxic threat to consumers. On the other hand, potentially toxic Cu and Pb could be extremely harmful as they potentially accumulated in food (marine organisms) and transferred through the food chain to human (Aschner, 2002; Cheng and Yap, 2015). From the public health point of view, elevated metals levels that exceed the permissible limit evidently have tarnished the image of the food. Heavy metals, such as Cu and Pb, are of particular interest to public health as they are commonly detected in coastal areas and originate from natural geological sources and human activities (Yap et al., 2004, 2016). Therefore, detailed studies on the safety of marine mussels as food are required to know their safety for human consumption.

This study aims to provide information on the soluble fraction (SF) and the insoluble fraction (IF) of Cu, and Pb in the edible tissues of the green-lipped mussel *Perna viridis*. Besides, human health risk assessments of Cu and Pb in the mussels based on total and SF concentrations were also studied.

4.2 MATERIALS AND METHODS

The sampling of the green-lipped mussel *P. viridis* was conducted in February and April 2006 (Table 4.1) from Pantai Lido (PLido), Kampung Pasir Puteh (KPP), and Telok Mas (TMas). The first two locations are located in the Straits of Johore, and they are potentially receiving domestic and industrial discharges from the surrounding areas. Samples from Telok Mas was collected by fishermen from the Malacca coastal water. Since the mussels from the three sites were of the same species and were collected in the same month, therefore the potential causes due to interspecific variation and seasonal changes were minimal.

Upon reaching the laboratory, about 20–25 of similar-sized mussels were dissected and pooled into different parts: crystalline style (CS), foot, gill, gonad, mantle, muscle and remainder.

The SF and IF of metals in the soft tissues of the mussels were separated using modified procedures according to Bragigand et al. (2004; Figure 4.1). This method (Bragigand et al., 2004) suggested that the soft tissues (0.5 g dry weight) to be minced and homogenized with 2 mL of Ultra-turrax in a TRIS buffer solution 20nM, NaCl 150 mM, pH 8.6, at 4°C, based on 4 mL/g fresh weight. The samples were then put through centrifugation at 25 rpm for 55 min to separate SF and IF. Soluble fraction (S1) was then analyzed using Atomic Absorption Spectrophotometer (AAS), while insoluble fraction (P1) was further digested with 2.5 mL of nitric acid (HNO_3) (1 mL of HNO_3 per 0.2 g of P1). After the acid digestion, the IF pellets were analyzed for heavy metals in based on fresh weight.

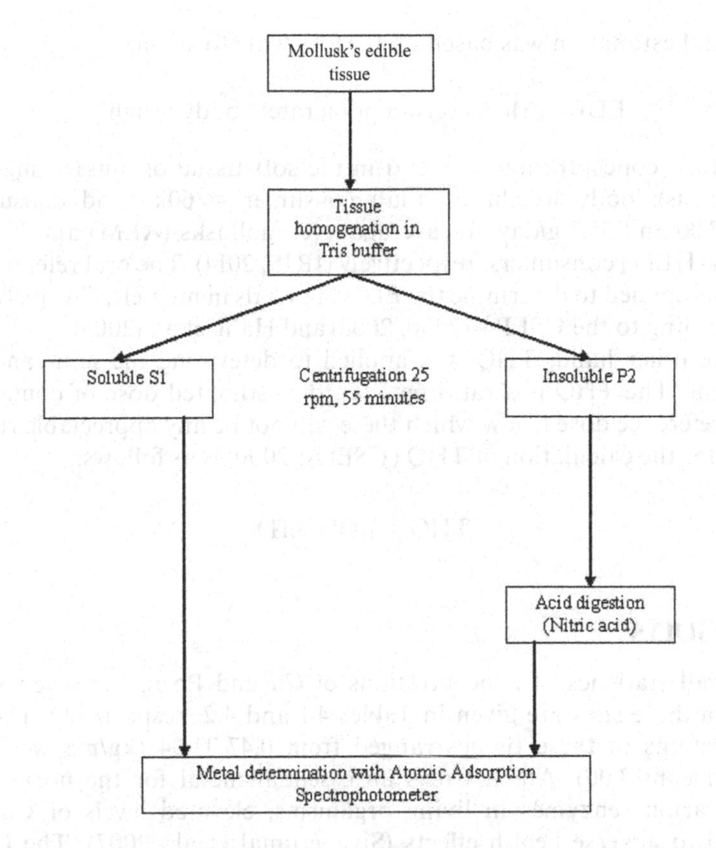

FIGURE 4.1 Estimation of the soluble fraction (SF = S1) and the insoluble fraction (IF = P2) of Cu and Pb in mussels (modified from Bragigand et al., 2004).

Approximately 0.50 g (duplicates) was weighed, placed into digesting tube and digested in 10 ml of nitric acid (HNO_3). Sample digestions were conducted in a heating block digester. For the first hour, samples were digested on a heating block at 40°C, and for the last 3 hours, the temperature was raised up to 140°C to ensure all samples were fully digested. The samples were allowed to cool and were later diluted with 40 ml distilled water. Then all samples were filtered with Whatman No.1 filter paper and analyzed for Cu and Pb using the AAS. A procedural blank was prepared in order to check for sample accuracy. The quality of the analytical procedures had been checked with quality assurance.

4.2.1 Health Risk Assessment

The estimated daily intake (EDI) and target hazard quotient (THQ) values were calculated to determine the once-or long-term potential hazardous exposure to metals through consumption of mussels (USEPA, 1989) from Peninsular Malaysia.

The EDI estimation was based on the following formula:

$$EDI = (Mc \times \text{consumption rate}) / \text{body weight} \quad (4.1)$$

where Mc = concentration of metal in the soft tissue of mussel (mg/kg wet weight basis); body weight of adult consumer = 60kg and consumption rate = 17.86 and 35.7 g/day, for average-level mollusks (ALM) and high-level mollusks (HLM) consumers, respectively (IRIS, 2014). The oral reference dose (RfD) was applied to determine the EDIs of metals in mussels; Cu: 40.0 and Pb: 3.50 according to the USEPA (2000, 2008) and Hang et al. (2009).

On the other hand, THQ was applied to determine the non-cancer risk assessment. The THQ is a ratio between the estimated dose of contaminant and the reference dose below which there will not be any appreciable risk. The formula for the calculation of THQ (USEPA, 2000) is as follows:

$$THQ = EDI / RfD \quad (4.2)$$

4.3 RESULTS

The overall statistics of concentrations of Cu and Pb in the seven soft tissues from three sites are given in Tables 4.1 and 4.2, respectively. The metal concentrations in these tissues ranged from 0.47–11.04 (kg/mg wet weight) for Cu (mean: 3.06). Albeit Cu is an essential metal for the normal function of various enzymes in living organisms, elevated levels of Cu intake may lead to adverse health effects (Sivaperumal et al., 2007). The Cu concentrations reported in this study was within the permissible limit set by US Environmental Protection Agency (USEPA, 1983) (120 mg/kg wet weight), Ministry of Agriculture, Fisheries and Food (MAFF, 2000) (20 mg/kg wet weight (w/w)), World Health Organization (WHO, 1996) (30 mg/kg w/w), and Malaysian Food Regulations (30 mg/kg w/w; MFR, 1985). The lower levels of Cu (1.32–5.37 mg/kg w/w) than the above permissible limits in the soft tissues of mangrove snails *Nerita lineata* collected from Peninsular Malaysia was also reported by Cheng and Yap (2015), and within the Cu ranges (0.48–17.5 mg/kg wet weight) of 40 *P. viridis* populations collected (2002–2009) from Peninsular Malaysia (Yap et al., 2016).

The ranges of Pb concentrations (mg/kg wet weight) in the seven soft tissues were 1.03–30.0 (mean: 6.70). Lead is a nonessential metal known to cause detrimental effects on the nervous system and normal functions of the kidney (Ylmaz et al., 2010). The Pb concentrations reported in this study were above the permissible limits set by USEPA (4.0 mg/kg wet weight; USEPA, 1983), Ministry of Agriculture, Fisheries and Food (2.0 mg/kg wet weight; MAFF, 2000), WHO (2.0 mg/kg wet weight; WHO, 1996) and FAO (0.5 mg/kg wet weight; FAO, 1983). In particular, all sites exceeded 2.00 mg/kg wet weight, set by Malaysian Food Regulations 1985 (MFR, 1985). The higher levels of Pb (0.21 to 10.5 mg/kg wet weight) than the above permissible limits in the soft tissues of mangrove snails *N. lineata* collected from Peninsular Malaysia was also reported by Cheng and Yap (2015), and also higher than the Pb ranges (0.27–10.4 mg/kg wet weight) of 40 *P. viridis* populations collected (2002–2009) from Peninsular Malaysia (Yap et al., 2016).

The mean percentages of the concentration of each metal in the SF and IF of Cu and Pb in the seven different soft tissues of *P. viridis* are also shown in Tables 4.1 and 4.2, respectively.

For Cu (Table 4.1), the SF ranged from 2.22 to 48.9% while the IF ranged from 51.1 to 97.8%. These Cu IFs were all above 50% and agreed with Bragigand et al. (2004)'s finding who reported the oysters of Restronguet Creek, Arcachon and Ronce had a higher percentage of insoluble Cu (75%, 76% and 57%, respectively) than soluble Cu.

For Pb (Table 4.2), the SF ranged from 3.69 to 47.4% while the IF ranged from 52.6 to 96.3%. As found in the soft tissues of the *P. viridis* from Telok Mas, the soluble Pb ranged between 6.22–43.7%, of the total Pb concentrations. The soluble Pb in the mantle (43.7%) of the Telok Mas mussel was found to be elevated in the SF.

The EDI and THQ values of Cu for ALM and HLM consumers in the total concentrations and SFs of Cu are presented in Table 4.3. The THQ values of Cu based on total concentrations of Cu ranged from 0.00–0.08 for ALM, and 0.01–0.16 for HLM. The THQ values of Cu based on SF of Cu ranged from 0.00–0.04 for ALM, and 0.00–0.07 for HLM. All the THQ values were below 1.0, indicating noncarcinogenic risks of Cu to the consumers based on total Cu concentrations and SF concentrations in the seven edible tissues of mussels.

The EDI and THQ values of Pb for ALM and HLM consumers in the total concentrations and SFs of Pb are presented in Table 4.4. Except for CS, the other six tissues with the THQ values of Pb based on total concentrations of Pb ranged from 0.09–0.77 for ALM. However, for HLM, there are 13 out of 21 tissues of three sites had THQ values that were higher than 1.0. The THQ values of Pb based on SF of Pb ranged from 0.01–0.27 for ALM, and 0.03–0.55 for HLM. These THQ values based on SF were all less than 1.0, denoting non-carcinogenic risks of Pb to the consumers based on Pb SF concentrations in the seven edible tissues of mussels.

4.4 DISCUSSION

High percentages of Cu and Pb were mainly found in the IFs (Cu: 51.1–97.8%; Pb: 52.6–96.3%) in the seven soft tissues of the mussels. This could be attributed to the low detoxification rate of mineral granules of the mussels from their soft tissues (Geffard et al. 2002). The elevated metal concentrations found in the IFs could be due to the metal were tightly bound to the metallothionein (Roesijadi, 1992) and another possibility was that the metallothionein not yet saturated by those metal ions. The high metals concentrations in IF may be caused by the slow depuration of metals in different tissues due to the formation of metal-thiolate complex with the cysteine residues inside the lysosomes (Yap et al., 2003) which could result in high level of metals found in the IFs. For Cu, Ettajani et al. (2000)

TABLE 4.1
Concentrations (mg/kg wet weight) of Cu in the Soluble Fraction (SF) and Insoluble Fraction (IF), and the Summation of SF and IF (SUM), and Their Percentages

		mg/kg			%			
		SF	IF	SUM	SF	IF	SF	IF
CS	PLido	4.92	5.70	10.6	46.4	53.6	29.8	70.2
	KPP	0.63	8.07	8.70	7.22	92.8		
	TMas	3.95	7.09	11.0	35.7	64.3		
Foot	PLido	0.40	1.12	1.52	26.6	73.4	24.5	75.5
	KPP	0.43	1.21	1.65	26.2	73.8		
	TMas	0.26	1.01	1.27	20.7	79.3		
Gill	PLido	0.32	0.63	0.95	33.5	66.5	30.4	69.6
	KPP	0.30	3.18	3.48	8.73	91.3		
	TMas	0.72	0.76	1.48	48.9	51.1		
Gonad	PLido	0.22	1.21	1.42	15.1	84.9	26.7	73.3
	KPP	0.68	0.91	1.58	42.7	57.3		
	TMas	0.37	1.31	1.68	22.3	77.7		
Mantle	PLido	0.21	2.45	2.66	7.81	92.2	6.88	93.12
	KPP	0.15	2.85	3.01	5.11	94.9		
	TMas	0.11	1.32	1.43	7.71	92.3		
Muscle	PLido	0.21	0.26	0.47	44.7	55.3	29.8	70.2
	KPP	0.37	2.90	3.28	11.3	88.7		
	TMas	0.21	0.42	0.62	33.3	66.7		
Remain	PLido	0.28	1.17	1.45	19.2	80.8	13.6	86.4
	KPP	0.08	3.69	3.78	2.22	97.8		
	TMas	0.43	1.79	2.22	19.4	80.6		
	Minimum	0.08	0.26	0.47	2.22	51.1		
	Maximum	4.92	8.07	11.04	48.9	97.8		
	Mean	0.73	2.34	3.06	23.1	76.9		
	Median	0.32	1.31	1.65	20.7	79.3		
	Std Error	0.27	0.48	0.68	3.23	3.23		
	Skewness	2.74	1.48	1.77	0.32	–0.32		
	Kurtosis	5.87	1.17	1.76	–1.15	–1.15		

Note: CS = crystalline style; Remain = remaining soft tissues; PLido = Pantai Lido; KPP = Kampung Pasir Puteh; TMas = Telok Mas.

reported that the distribution of this metal between the SF and IF in oysters *Crassostrea gigas* is little. Thomson et al. (1985) reported that 90% of Cu IF in *C. gigas* gills were in the form of granules.

The metals present in the SFs were more bioavailable to consumers than metals bound to IFs (Wallace et al., 2003). The bioavailability of dietary metals

TABLE 4.2
Concentrations (mg/kg wet weight) of Pb in the Soluble Fraction (SF) and Insoluble Fraction (IF), and the Summation of SF and IF (SUM) and Their Percentages

		mg/kg			%			
		SF	IF	SUM	SF	IF	SF	IF
CS	PLido	3.21	10.2	13.5	23.8	76.3	20.2	79.8
	KPP	2.57	27.4	30.0	8.57	91.4		
	TMas	0.29	0.74	1.03	28.2	71.8		
Foot	PLido	1.91	6.29	8.20	23.3	76.7	21.38	78.7
	KPP	1.57	5.21	6.79	23.2	76.8		
	TMas	0.25	1.19	1.43	17.4	82.6		
Gill	PLido	0.43	6.74	7.17	5.98	94.0	20.9	79.1
	KPP	1.65	6.53	8.18	20.2	79.8		
	TMas	0.63	1.10	1.73	36.4	63.6		
Gonad	PLido	0.57	7.18	7.75	7.31	92.7	23.6	76.4
	KPP	2.11	6.31	8.42	25.0	74.9		
	TMas	0.72	1.14	1.86	38.6	61.4		
Mantle	PLido	1.81	4.07	5.88	30.8	69.2	27.29	72.7
	KPP	0.27	6.92	7.19	3.69	96.3		
	TMas	0.50	0.55	1.05	47.4	52.6		
Muscle	PLido	1.93	2.89	4.82	39.9	60.0	20.4	79.7
	KPP	0.83	5.20	6.02	13.7	86.3		
	TMas	0.15	1.93	2.09	7.35	92.7		
Remain	PLido	2.25	4.57	6.82	33.0	66.9	24.1	75.9
	KPP	1.37	7.71	9.08	15.1	84.9		
	TMas	0.42	1.31	1.73	24.14	75.9		
	Minimum	0.15	0.55	1.03	3.69	52.6		
	Maximum	3.21	27.4	30.0	47.4	96.3		
	Mean	1.21	5.49	6.70	22.5	77.5		
	Median	0.83	5.20	6.79	23.3	76.7		
	Std Error	0.20	1.26	1.38	2.69	2.69		
	Skewness	0.55	2.72	2.46	0.20	–0.20		
	Kurtosis	-0.82	8.29	7.07	–0.85	–0.85		

Note: CS = crystalline style; Remain = remaining soft tissues; PLido = Pantai Lido; KPP = Kampung Pasir Puteh; TMas = Telok Mas.

to consumers was also determined by the physico-chemical characteristics of metal in the soft tissues of seafood (Mouneyrac et al., 1998; Bragigand et al., 2004). This is important to understand the flow of metals concentrations to the next trophic level. Therefore, it is crucial to investigate the distribution of metals accumulated in seafood into SF and IF and to model the digestibility of the

TABLE 4.3
Estimated Daily Intake (EDI) and Target Hazard Quotient (THQ) of Cu for Average-Level Mollusk (ALM) and High-Level Mollusk (HLM) Consumers in the Total Concentrations and Soluble Fractions (SF) of Cu

	Site	Total EDI ALM	Total EDI HLM	Total THQ ALM	Total THQ HLM	SF EDI ALM	SF EDI HLM	SF THQ ALM	SF THQ HLM
CS	PLido	3.16	6.32	0.08	0.16	1.47	2.93	0.04	0.07
	KPP	2.59	5.18	0.06	0.13	0.19	0.37	0.00	0.01
	TMas	3.28	6.57	0.08	0.16	1.17	2.35	0.03	0.06
Foot	PLido	0.45	0.90	0.01	0.02	0.12	0.24	0.00	0.01
	KPP	0.49	0.98	0.01	0.02	0.13	0.26	0.00	0.01
	TMas	0.38	0.76	0.01	0.02	0.08	0.16	0.00	0.00
Gill	PLido	0.28	0.57	0.01	0.01	0.09	0.19	0.00	0.00
	KPP	1.04	2.07	0.03	0.05	0.09	0.18	0.00	0.00
	TMas	0.44	0.88	0.01	0.02	0.22	0.43	0.01	0.01
Gonad	PLido	0.42	0.85	0.01	0.02	0.06	0.13	0.00	0.00
	KPP	0.47	0.94	0.01	0.02	0.20	0.40	0.01	0.01
	TMas	0.50	1.00	0.01	0.03	0.11	0.22	0.00	0.01
Mantle	PLido	0.79	1.58	0.02	0.04	0.06	0.12	0.00	0.00
	KPP	0.89	1.79	0.02	0.04	0.05	0.09	0.00	0.00
	TMas	0.43	0.85	0.01	0.02	0.03	0.07	0.00	0.00
Muscle	PLido	0.14	0.28	0.00	0.01	0.06	0.12	0.00	0.00
	KPP	0.97	1.95	0.02	0.05	0.11	0.22	0.00	0.01
	TMas	0.19	0.37	0.00	0.01	0.06	0.12	0.00	0.00
Remain	PLido	0.43	0.86	0.01	0.02	0.08	0.17	0.00	0.00
	KPP	1.12	2.25	0.03	0.06	0.02	0.05	0.00	0.00
	TMas	0.66	1.32	0.02	0.03	0.13	0.26	0.00	0.01

Note: CS = crystalline style; Remain = remaining soft tissues; PLido = Pantai Lido; KPP = Kampung Pasir Puteh; TMas = Telok Mas.

IF (Bragigand et al., 2004). This is the basis of human health risk consumption on the commercial mussels.

It was reported that the higher the level of pollution, the more significant the metals binding to cytosolic ligands is in *C. gigas* (Geffard et al., 2002). This could be the main cause of the elevated percentages of metals detected in the SFs of the foot and mantle of the mussels, where these tissues are continuously in contact with the polluted external environment. Besides, higher percentages of metal were observed in the SF than the IF of other soft tissues also suggesting the mussels could be exposed to high level of metal contamination in their living habitats. The rapid elimination of cytosolic compounds due to the short half-lives of the metal may explain the high percentages of metals in the SFs as reported by Roesijadi and Klerks (1989) in the gill of *Crassostrea virginica*. The

TABLE 4.4
Estimated Daily Intake (EDI) and Target Hazard Quotient (THQ) of Pb for Average-Level Mollusk (ALM) and High-Level Mollusk (HLM) Consumers in the Total Concentrations and Soluble Fractions (SF) of Pb

	Site	Total EDI ALM	Total EDI HLM	Total THQ ALM	Total THQ HLM	SF EDI ALM	SF EDI HLM	SF THQ ALM	SF THQ HLM
CS	PLido	4.00	8.00	**1.14**	**2.29**	0.95	1.91	0.27	0.55
	KPP	8.94	17.86	**2.55**	**5.10**	0.77	1.53	0.22	0.44
	TMas	0.31	0.61	0.09	0.18	0.09	0.17	0.02	0.05
Foot	PLido	2.44	4.88	0.70	**1.39**	0.57	1.14	0.16	0.32
	KPP	2.02	4.04	0.58	**1.15**	0.47	0.94	0.13	0.27
	TMas	0.43	0.85	0.12	0.24	0.07	0.15	0.02	0.04
Gill	PLido	2.13	4.27	0.61	**1.22**	0.13	0.26	0.04	0.07
	KPP	2.44	4.87	0.70	**1.39**	0.49	0.98	0.14	0.28
	TMas	0.51	1.03	0.15	0.29	0.19	0.37	0.05	0.11
Gonad	PLido	2.31	4.61	0.66	**1.32**	0.17	0.34	0.05	0.10
	KPP	2.51	5.01	0.72	**1.43**	0.63	1.25	0.18	0.36
	TMas	0.55	1.11	0.16	0.32	0.21	0.43	0.06	0.12
Mantle	PLido	1.75	3.50	0.50	**1.00**	0.54	1.08	0.15	0.31
	KPP	2.14	4.28	0.61	**1.22**	0.08	0.16	0.02	0.05
	TMas	0.31	0.62	0.09	0.18	0.15	0.30	0.04	0.08
Muscle	PLido	1.43	2.87	0.41	0.82	0.57	1.15	0.16	0.33
	KPP	1.79	3.58	0.51	**1.02**	0.25	0.49	0.07	0.14
	TMas	0.62	1.24	0.18	0.35	0.05	0.09	0.01	0.03
Remain	PLido	2.03	4.06	0.58	**1.16**	0.67	1.34	0.19	0.38
	KPP	2.70	5.40	0.77	**1.54**	0.41	0.81	0.12	0.23
	TMas	0.51	1.03	0.15	0.29	0.12	0.25	0.04	0.07

Note: CS = crystalline style; Remain = remaining soft tissues; PLido = Pantai Lido; KPP = Kampung Pasir Puteh; TMas = Telok Mas.

Values in bold are the highest ones.

possibly saturation of metal-binding protein inside the cell by excessive metal ions, may result the coming metal ion only adsorb onto the cell wall, hence increasing the concentrations in the SFs.

According to Wallace et al. (2003), the bioavailability of metals in the SFs of prey (cytosol and proteins) are higher than those in the IFs of predators (cell walls and exoskeleton, for crustaceans, and metal concretions/granules). Metals may be released form IFs during the digestion process in which the pHs and enzymes vary, increasing the potential of bioavailable SF (Amiard et al., 2008). Therefore, the estimation of total metals concentrations may cause an overestimation of concentrations that are bioavailable to a consumer (Wallace and Luoma, 2003).

4.5 CONCLUSION

The present study investigated the SF and IF of potentially toxic Cu and Pb in the seven different tissues of *P. viridis* collected from three sites in Peninsular Malaysia. Based on the THQ values of the Cu and Pb obtained from this study, it was suggested that the consumption of the mussel from the three sites would not pose non-carcinogenic risks to the consumers, however, excessive consumption of the mussel is not recommended to avoid metals toxic due to the presence of the metal concentrations in the SF. This study has provided a more accurate estimation of the concentrations of Cu and Pb which were bioavailable to the consumers based on the different tissues of the mussels. These new estimates potentially contribute to re-evaluation of the maximum permissible limits of food standards for metals, which is currently based on total metal concentrations which could lead to an overestimation of the metal quantities likely to be bioavailable to a consumer.

REFERENCES

Amiard, J. C., Triquet, C. A., Charbonnier, L., Mesnil., Rainbow, P. S., and W. X. Wang. 2008. Bioaccessibility of essential and non-essential metals in commercial shellfish from Western Europe and Asia. *Food and Chemical Toxicology* 46:2010–2022.

Aschner, M. 2002. Neurotoxic mechanisms of fish-borne methylmercury. *Environmental Toxicology and Pharmacology* 12:101–104.

Bragigand, V., Berthet, B., Amiard, J. C., and P. S. Rainbow. 2004. Estimates of trace metal bioavailabilty to humans ingesting contaminated oysters. *Food and Chemical Toxicology* 42:1893–1902.

Cheng, W. H., and C. K. Yap. 2015. Potential human health risks from toxic metals via mangrove snail consumption and their ecological risk assessments in the habitat sediment from Peninsular Malaysia. *Chemosphere* 135:156–165.

Cheong, L., and H. B. Lee. 1984. *Mussel farming*. Satis Extension Manual Series. No. 5. Bangkok: Southeast Asian Fisheries Development Centre.

Ettajani, H., Berthet, B., Amiard, J. C., and L. Chevolot. 2000. Determination of cadmium partitioning in microalgae and oysters: Contribution to the assessment of trophic transfer. *Archives of Environmental Contamination and Toxicology* 40:209–221.

FAO (Food Administration Organization). 1983. *Compilation of legal limits for hazardous substances in fish and fishery products*. FAO Fisheries Circular. 764. FAO, Rome, 102p. http://www.fao.org/fi/oldsite/eims_search/1_dett.asp?calling=simple_s_result&lang=fr&pub_id=65155

Geffard, A., Amiard, J. C., and C. Amiard-Triquet. 2002. Kinetic of metal elimination in oyster from a contaminated estuary. *Comparative Biochemistry and Physiology Part C* 131:281–293.

Hang, X. S., Wang, H. Y., Zhou, J. M., Ma, C. L., Du, C. W., and X. Q. Chen. 2009. Risk assessment of potentially toxic element pollution in soils and rice (*Oryza sativa*) in a typical area of the Yangtze River Delta. *Environmental Pollution* 157:2542–2549.

IRIS (Integrated Risk Information System). 2014. *US environmental protection agency*. http://cfpub.epa.gov/ncea/iris/index.cfm?fuseaction=iris.showSubstanceList (accessed February 04, 2014).

Kromhout, D., Bosschieter, E. B., and C. C. Lezenne. 1985. The inverse relationship between fish consumption and 20-year mortality from coronary heart disease. *The New England Journal of Medicine* 312:1205–1209.

MAFF (Ministry of Agriculture, Fisheries and Food). 2000. *Monitoring and surveillance of non-radioactive contaminants in the aquatic environment and activities regulating the disposal of wastes at sea, 1997*. Aquatic environment monitoring report number 52. Lowestoft: MAFF.

MFR (Malaysian Food Regulations). 1985. *Malaysian law on food and drugs*. 985. Malaysian Law, Kuala Lumpur.

Mouneyrac, C., Amiard, J. C., and C. Amiard-Triquet. 1998. Effects of natural factors (salinity and body weight) on cadmium, copper, zinc and metallothionein-like protein levels in resident populations of oysters *Crassostrea gigas* from polluted estuary. *Marine Ecology Progress Series* 162:125–135.

Petroczi, A., and D. P. Naughton. 2009. Mercury, cadmium and lead contamination in seafood: A comparative study to evaluate the usefulness of Target Hazard Quotients. *Food and Chemical Toxicology* 47:298–302.

Roesijadi, G. 1992. Metallothionein in metal regulation and toxicity in aquatic animals. *Aquatic Toxicology* 22:81–113.

Roesijadi, G., and P. L. Klerks. 1989. Kinetic analysis of cadmium binding to metallothionein and other intracellular ligands in oyster gills. *Journal of Experimental Zoology* 251:1–12.

Sivaperumal, P., Sankar, T. V., and P. G. V. Nair. 2007. Heavy metal concentrations in fish, shellfish and fish products from internal markets of India vis-a'-vis international standards. *Food Chemistry* 102:612–620.

Thomson, J. D., Pirie, B. J. S., and S. G. George. 1985. Cellular metal distribution in the pacific oyster, *Crassostrea gigas* (Thun) determined by quantitative X-ray microprobe analysis. *Journal of Experimental Marine Biology and Ecology* 85:37–45.

Toyofuku, H. 2006. Joint FAO/WHO/IOC activities to provide scientific advice on marine biotoxins (research report). *Marine Pollution Bulletin* 52:1735–1745.

USEPA (US Environmental Protection Agency). 1983. *Methods for chemical analysis of water and waste*. EPA Report 600/4–79–020. Cincinnati: Office of Water, USEPA.

USEPA (US Environmental Protection Agency). 1989. *Guidance manual for assessing human health risks from chemically contaminated, fish and shellfish*. EPA-503/8–89–002. USEPA, Washington DC.

USEPA (US Environmental Protection Agency). 2000. *Risk-based concentration table*. Philadelphia, PA: USEPA, Washington, DC.

USEPA (US Environmental Protection Agency). 2008. *USEPA, integrated risk information system*. CRC.

Wallace, W. G., Lee, B. G., and S. N. Louma. 2003. Subcellular compartmentalization of Cd and Zn in two bivalves. I. Significance of metal-sensitive fractions (MSF) and biologically detoxified metal (BDM). *Marine Ecology Progress Series* 249:183–197.

Wallace, W. G., and S. N. Luoma. 2003. Subcellular compartmentalization of Cd and Zn in two bivalves. II. Significance of trophically available metal (TAM). *Marine Ecology Progress Series* 257:125–137.

WHO (World Health Organization). 1996. Health criteria other supporting information. In *Guidelines for drinking water quality*, 2nd edn (pp. 31–388). Geneva: WHO.

Yap, C. K., Cheng, W. H., Karami, A., and A. Ismail. 2016. Health risk assessments of heavy metal exposure via consumption of marine mussels collected from anthropogenic sites. *Science of the Total Environment* 553:285–296.

Yap, C. K., Ismail, A., Omar, H., and S. G. Tan. 2003. Accumulation, depuration and distribution of cadmium and zinc in the green-lipped mussel *Perna viridis* (L) under Laboratory conditions. *Hydrobiologia,* 498:151–160.

Yap, C. K., Ismail, A., and S. G. Tan. 2004. Heavy metal (Cd, Cu, Pb and Zn) concentrations in the green-lipped mussel *Perna viridis* (Linnaeus) collected from some wild and aquacultural sites in the west coast of Peninsular Malaysia. *Food Chemistry* 84:569–575.

Ylmaz, A. B., Sangun, M. K., Yag, D., and C. Turan. 2010. Metals (major, essential to non-essential) composition of the different tissues of three demersal fish species from Iskenderun Bay, Turkey. *Food Chemistry,* 123:410–415.

5 Transformation of Natural Products by Marine-Derived Microorganisms

Thayane Melo de Queiroz and
André Luiz Meleiro Porto

CONTENTS

5.1 INTRODUCTION

In recent decades, some researchers have investigated the biotransformation of natural products and synthetic compounds by marine-derived microorganisms. These microorganisms can be isolated from algae, sponges, crustaceans, fish, and molluscs, among other organisms (Birolli et al., 2019). The enzymes present in these microorganisms may have some advantages, such as stability over a wide range of pH and temperature and tolerance to high pressure and high salt concentrations (Sakar et al., 2010; Queiroz et al., 2020). Thus, the enzymes from marine-derived microorganisms have great potential to be applied in various transformations of natural and synthetic compounds into new derivatives.

Some studies carried out in our group have reported the use of marine-derived fungi in the biotransformation of different natural products. Among the natural compounds studied were progesterone (steroid) and the terpenes *rac*-camphor, (-)-ambrox®, (-)-sclareol, and (+)-sclareolide (Martins et al., 2015; De Paula and Porto, 2020). Thus, it was proposed in this work to present the main studies described in the literature that describe the biotransformation of

DOI: 10.1201/9781003303909-5

natural products (terpenoids, flavonoids, chalcones, polyketides, and steroids) by marine-derived microorganisms.

5.2 TERPENES AND TERPENOIDS

Terpenoids are organic compounds derived from C_5 isoprene units that are normally joined together in a *head-to-tail* condensation. These compounds are found in plants and microorganisms, and they constitute an extensive class of compounds with different carbon chain sizes. The members of this class of compounds are represented by monoterpenes (C_{10}), sesquiterpenes (C_{15}), diterpenes (C_{20}), triterpenes (C_{30}), and tetraterpenes (C_{40}) (Dewick, 2009). The terpenoids have important biological properties, such as anti-inflammatory, antitumor, antimicrobial, antiparasitic, and insecticide activities, among others (De Carvalho and Da Fonseca, 2006; Bakkali et al., 2008; Schwab et al., 2013; Martins et al., 2015).

Some examples of terpenoid biotransformation that have been described in the literature will be presented below. The marine bacteria *Vibrio cholerae*, *Listonella damsela*, and *Vibrio alginolyticus*, isolated from sediments of Daya Bay (China), were used in the biotransformation of *D*-limonene. The biotransformation experiments were carried out in liquid culture medium containing peptone and yeast extract as carbon and nitrogen sources. These experiments were incubated in an orbital shaker (120 rpm, 28°C) for 6 days. Different compounds were identified by gas chromatography-mass spectrometry (GC-MS), including possible biotransformation products (Figure 5.1). Furthermore, sesquiterpenes and triterpenes were identified, which were not detected in the control experiments. Houjin and co-workers believe that the presence of *D*-limonene activates the biosynthetic pathways of other terpenoids (Houjin et al., 2006).

In another study, Li et al. (2007) described the biotransformation of the sesquiterpene cyclonerodiol by the ascomycete *Penicillium* sp. and the actinomycete

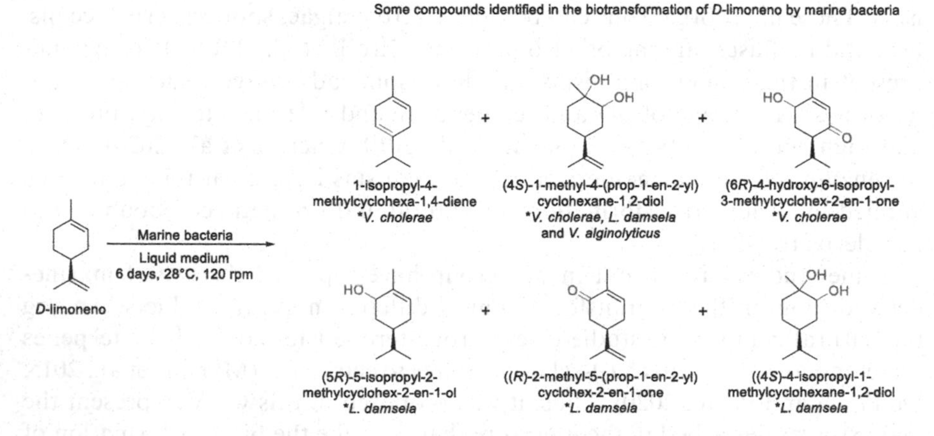

FIGURE 5.1 Biotransformation of *D*-limonene by marine bacteria (Houjin et al., 2006).

Streptomyces sp., isolated from the alga *Sargassum thunbergii* (Korea) and from the sea plant *Zostera marina* (Korea), respectively. The microbial transformation of cyclonerodiol was performed in a culture medium containing soytone (0.1%), soluble starch (1.0%), and seawater under conditions of 29°C and 130 rpm for 72 hours. The biotransformation experiments employing the fungus *Penicillium* sp. allowed for the isolation of the metabolite cyclonerodiol mannopyranoside (12.0 mg). The actinomycete *Streptomyces* sp. favoured the formation of compounds (10*Z*)-cyclonerotriol (3.4 mg) and (10*E*)-cyclonerotriol (7.6 mg; Figure 5.2). These compounds were characterised by nuclear magnetic resonance (NMR), high resolution mass spectrometry (HRMS), and Fourier-transform infrared (FTIR).

Only hydroxylated products were obtained in the biotransformation of (1*S*,2*E*,4*R*,6*R*,7*E*,11*E*)-2,7,11-cembratriene-4,6-diol by *Bacillus* sp. NC5, *Bacillus* sp. NK8, and *Bacillus* sp. NK7, isolated from the sponge *Negombata magnifica*. The bacteria *Bacillus* sp. NC5 promoted the formation of (1*S*,2*E*,4*R*,6*R*,7*Z*,11*E*)-2,7,11-cembratriene-4,6,19-triol (y = 12 mg), (1*S*,2*E*,4*R*,6*R*,7*E*, 9*S*,11*E*)-2,7,11-cembratriene-4,6,9-triol (y = 4.5 mg), (1*S*,2*E*,4*R*,6*R*,7*E*,11*Z*)-2,7,11-cembratriene-4,6,20-triol (y = 8 mg) and (1*S*,2*E*,4*R*,6*R*,7*E*,11*E*,10*R*)-2,7,11-cembratriene-4,6,10-triol (y = 21 mg). *Bacillus* sp. NK8, and *Bacillus* sp. NK7 favoured the formation of (1*S*,2*E*,4*R*,6*R*,7*E*,11*E*,10*S*)-2,7,11-cembratriene-4,6,10-triol (y = 9 mg) and (1*S*,2*E*,4*R*,6*R*, 7*E*,11*E*,13*R*)-2,7,11-cembratriene-4,6,13-triol (y = 11 mg), respectively (culture medium containing tryptic soy, 14 days) (Figure 5.3). The biotransformation products were characterised by NMR, HRMS and FTIR (Sayed et al., 2008).

In another example of hydroxylation by marine-derived microorganisms, the monoterpene geraniol was hydroxylated to 1,7-dihydroxy-3,

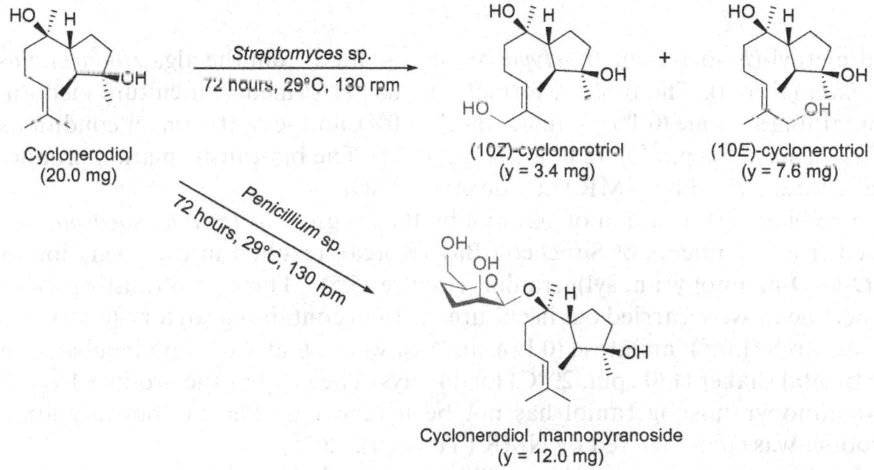

FIGURE 5.2 Microbial transformation of cyclonerodiol by the marine-derived fungi *Streptomyces* sp. and *Penicillium* sp. (Li et al., 2007).

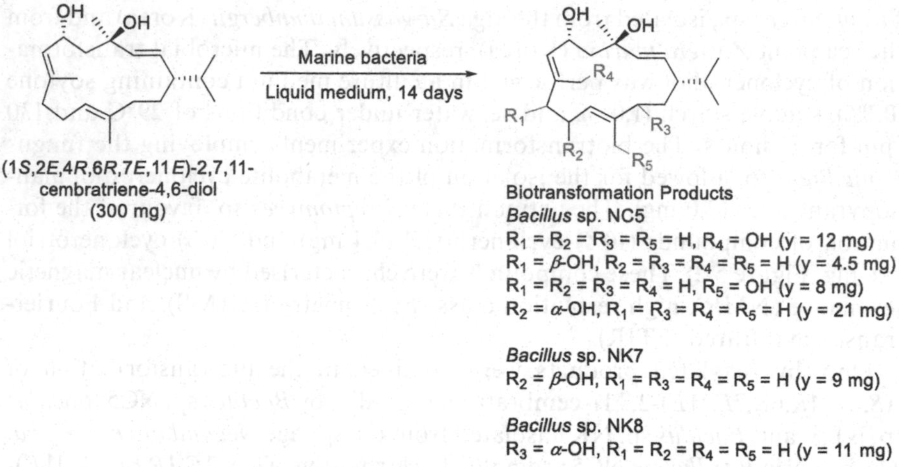

OH OH

Marine bacteria

Liquid medium, 14 days

(1*S*,2*E*,4*R*,6*R*,7*E*,11*E*)-2,7,11-cembratriene-4,6-diol (300 mg)

Biotransformation Products

Bacillus sp. NC5

$R_1 = R_2 = R_3 = R_5 = H, R_4 = OH$ (y = 12 mg)

$R_1 = \beta$-OH, $R_2 = R_3 = R_4 = R_5 = H$ (y = 4.5 mg)

$R_1 = R_2 = R_3 = R_4 = H, R_5 = OH$ (y = 8 mg)

$R_2 = \alpha$-OH, $R_1 = R_3 = R_4 = R_5 = H$ (y = 21 mg)

Bacillus sp. NK7

$R_2 = \beta$-OH, $R_1 = R_3 = R_4 = R_5 = H$ (y = 9 mg)

Bacillus sp. NK8

$R_3 = \alpha$-OH, $R_1 = R_2 = R_4 = R_5 = H$ (y = 11 mg)

FIGURE 5.3 Biohydroxylation of 1*S*,2*E*,4*R*,6*R*,7*E*,11*E*)-2,7,11-cembratriene-4,6-diol by the marine bacteria *Bacillus* sp. (Sayed et al., 2008).

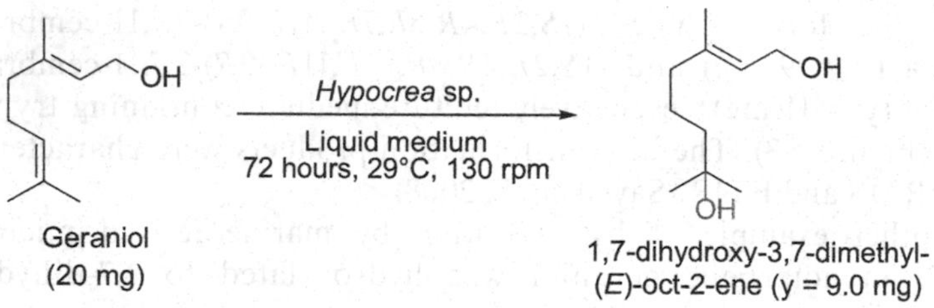

Geraniol (20 mg)

Hypocrea sp.

Liquid medium
72 hours, 29°C, 130 rpm

1,7-dihydroxy-3,7-dimethyl-(*E*)-oct-2-ene (y = 9.0 mg)

FIGURE 5.4 Biohydroxylation of geraniol by *Hypocrea* sp. (Leutou et al., 2009).

7-dimethyl-(*E*)-oct-2-ene by *Hypocrea* sp., isolated from the alga *Undaria pinnatifida* (Korea). The biotransformation was performed in a culture medium containing soytone (0.1%), soluble starch (1.0%), and seawater under conditions of 29°C and 130 rpm for 72 hours (Figure 5.4). The biotransformation product was characterised by NMR (Leutou et al., 2009).

The biotransformation of geraniol by the fungus *Thielavia hyalocarpa*, isolated from sediments of Suncheon Bay (Korea), resulted in the formation of 1-*O*-(α-*D*-mannopyranosyl)geraniol (Figure 5.5). These biotransformation experiments were carried out in culture medium containing soytone (0.1%), soluble starch (1.0%), mannose (0.1%), and sea water, and they were incubated in an orbital shaker (130 rpm, 29°C) for 14 days. The yield of the product 1-*O*-(α-*D*-mannopyranosyl)geraniol has not been reported. The biotransformation product was characterised by NMR (Yun et al., 2015).

Koshimura and co-workers (2009) reported the biotransformation of bromosesquiterpenes by the fungi *Rhinocladiella atrovirens* NRBC 32362 and *Rhinocladiella* sp. K-001, isolated from the alga *Stypopodium zonale*. The

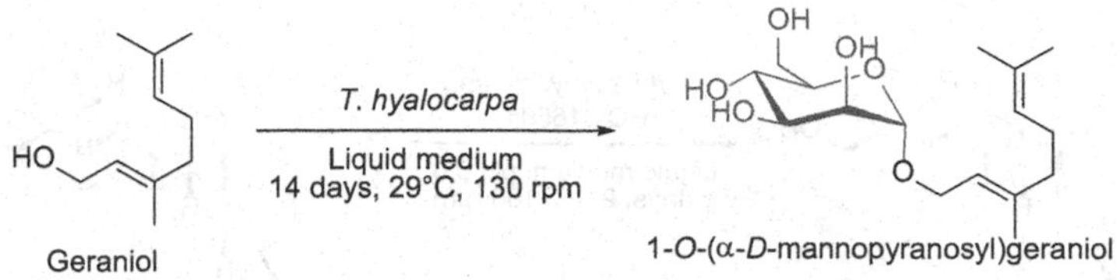

FIGURE 5.5 Biotransformation of geraniol by the fungus *T. hyalocarpa* (Yun et al., 2015).

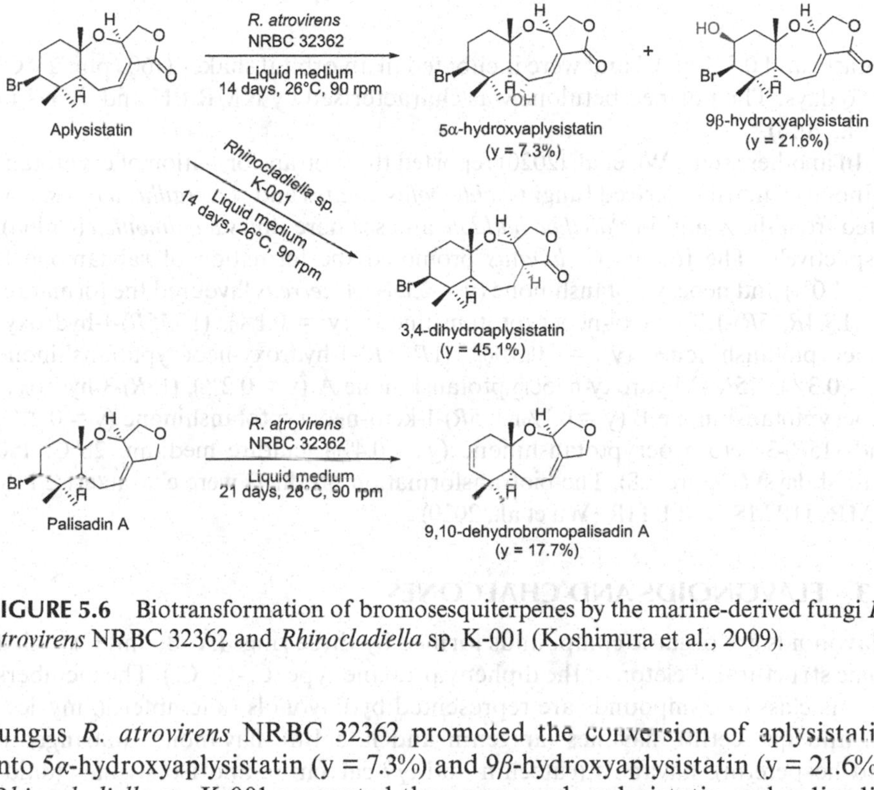

FIGURE 5.6 Biotransformation of bromosesquiterpenes by the marine-derived fungi *R. atrovirens* NRBC 32362 and *Rhinocladiella* sp. K-001 (Koshimura et al., 2009).

fungus *R. atrovirens* NRBC 32362 promoted the conversion of aplysistatin into 5α-hydroxyaplysistatin (y = 7.3%) and 9β-hydroxyaplysistatin (y = 21.6%). *Rhinocladiella* sp. K-001 converted the compounds aplysistatin and palisadin A into 3,4-dihydroaplysistatin (y = 45.1%) and 9,10-dehydrobromopalisadin A (y = 17.7%), respectively (culture medium containing peptone, yeast extract, malt extract, glucose, and seawater, 26°C, 90 rpm, 14-21 days; Figure 5.6) (Koshimura et al., 2009).

The regioselective oxidation of the triterpene betulin to betulone (y = 43.4%) was promoted by *Dothideomycete* sp. HQ 316564, which was isolated from the coral *Galaxea fascicularis* L. (China). In this study, 52 microbial strains were screened, and only the marine-derived fungus *Dothideomycete* sp. HQ 316564 catalysed the biotransformation of betulin (Figure 5.7). The biotransformation experiments were carried out in potato dextrose broth (PDB) (2% glucose, 20%

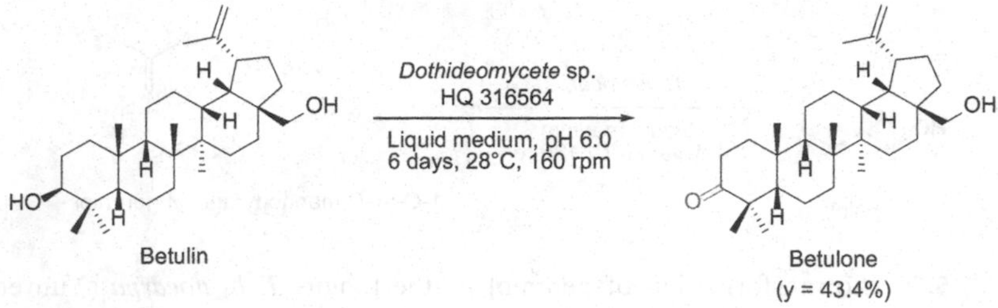

FIGURE 5.7 Regioselective oxidation of the triterpene betulin to betulone by the marine-derived fungus *Dothideomycete* sp. HQ 316564 (Liu et al., 2013).

potato) and 0.3% NaCl and were incubated in an orbital shaker (160 rpm, 28°C) for 6 days. The purified betulone was characterised by NMR (^{1}H and ^{13}C) (Liu et al., 2013).

In another study, Wu et al. (2020) reported the biotransformation of cryptotanshinone by marine-derived fungi *Cochliobolus lunatus* and *Aspergillus terreus*, isolated from the zoanthid *Palythoa haddoni* and sea hare *Aplysia pulmonica* (China), respectively. The fungus *C. lunatus* promoted the formation of salviamone B (y = 1.0%) and neocryptotanshinone (y = 2.1%). *A. terreus* favoured the formation of (1*S*,4*R*,15*R*)-1,18-epoxy-neocryptotanshinone (y = 0.1%), (1*S*,15*R*)-1-hydroxy-neocryptotanshinone (y = 0.3%), (1*R*,15*R*)-1-hydroxy-neocryptotanshinone (y = 0.3%), (15*R*)-3-hydroxy-neocryptotanshinone A (y = 0.2%), (15*R*)-3-hydroxy-neocryptotanshinone B (y = 0.1%), (15*R*)-1-keto-neocryptotanshinone (y = 0.2%), and (15*R*)-3-keto-neocryptotanshinone (y = 0.4%) (culture medium, 28°C, 180 rpm, 4 days) (Figure 5.8). The biotransformation products were characterised by NMR, HRMS, and FTIR (Wu et al., 2020).

5.3 FLAVONOIDS AND CHALCONES

Flavonoids are organic compounds formed by three rings (A, B, and C) with a basic structural skeleton of the diphenylpropane type (C_6-C_3-C_6). The members of this class of compounds are represented by flavonols (kaempferol, myricetin and quercetin), flavones (apigenin and luteolin), flavanones (naringenin and hesperetin), flavanols (catechin and *epi*-catechin), and isoflavones (daidzein and genistein), among others. Flavanones occur predominantly in citrus fruits, flavones occur in herbs, isoflavones occur in vegetables, flavanols occur in fruits, and flavonols occur in fruits and vegetables (Yao et al., 2004; Raffa et al., 2017). Studies suggest that flavonoids have important biological properties, such as anti-inflammatory, antioxidant, hepatoprotective, antithrombotic, antiviral, antibacterial, antifungal, anticarcinogenic, and vasodilating activities (Williams et al., 2004; Soobrattee et al., 2005; Haytowitz et al., 2013; Wu et al., 2013; Rosa et al., 2017).

De Lise and co-workers (2016) described the isolation, recombinant expression in *E. coli*, and partial characterisation of a α-L-rhamnosidase (α-RHA)

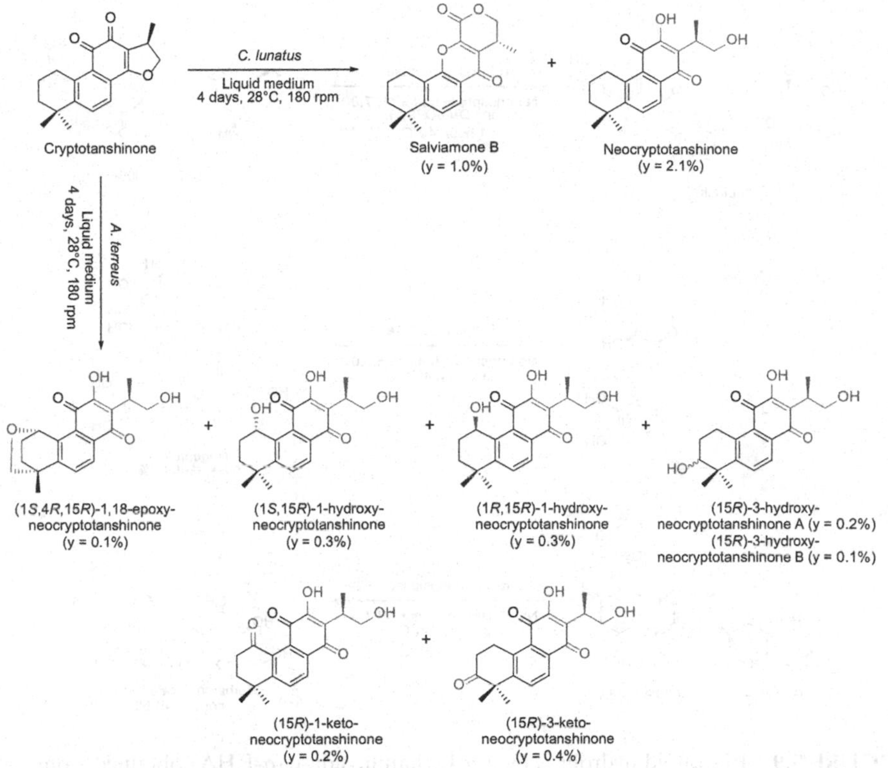

C. lunatus
Liquid medium
4 days, 28°C, 180 rpm
Cryptotanshinone
Salviamone B
(y = 1.0%)
Neocryptotanshinone
(y = 2.1%)
A. terreus
Liquid medium
4 days, 28°C, 180 rpm
(1S,4R,15R)-1,18-epoxy-neocryptotanshinone
(y = 0.1%)
(1S,15R)-1-hydroxy-neocryptotanshinone
(y = 0.3%)
(1R,15R)-1-hydroxy-neocryptotanshinone
(y = 0.3%)
(15R)-3-hydroxy-neocryptotanshinone A (y = 0.2%)
(15R)-3-hydroxy-neocryptotanshinone B (y = 0.1%)
(15R)-1-keto-neocryptotanshinone
(y = 0.2%)
(15R)-3-keto-neocryptotanshinone
(y = 0.4%)

FIGURE 5.8 Biotransformation of cryptotanshinone by the marine-derived fungi *C. lunatus* and *A. terreus* (Wu et al., 2020).

obtained from the marine bacteria *Novosphingobium* sp. PP1Y, which was isolated from surface seawater (Italy). In this study, the α-RHA enzyme was used in the hydrolysis of the flavonoids naringin, rutin, and neohesperidin dihydrochalcone into their glycosylated derivatives. The reactions were performed in Na-phosphate buffer (pH 7.0) under magnetic stirring at 40°C for 1–3 hours (Figure 5.9).

In another study, marine-derived fungi *Westerdykella* sp. CBMAI 1679, *Acremonium* sp. CBMAI 1676, *Cladosporium* sp. CBMAI 1237, *Aspergillus* sp. CBMAI 1829, *Penicillium oxalicum* CBMAI 1996, *Penicillium citrinum* CBMAI 1186, and *Mucor racemous* CBMAI 847 were employed in the bioreduction of flavanone (±)-2-phenylchroman-4-one, obtaining the compound flavan-4-ol with yields ranging from 2% to 36%. The fungi *Penicillium raistrickii* CBMAI 931, *Fusarium* sp. CBMAI 1676, and *Aspergillus sydowii* CBMAI 935 promoted the formation of 2′-hydroxy-dihydrochalcone and 2′,4-dihydroxy-dihydrochalcone from (±)-2-phenylchroman-4-one. The reactions were performed in phosphate buffer (Na_2HPO_4/KH_2PO_4 0.1 mol L^{-1}, pH 7.0) at 32°C and 130 rpm for 7

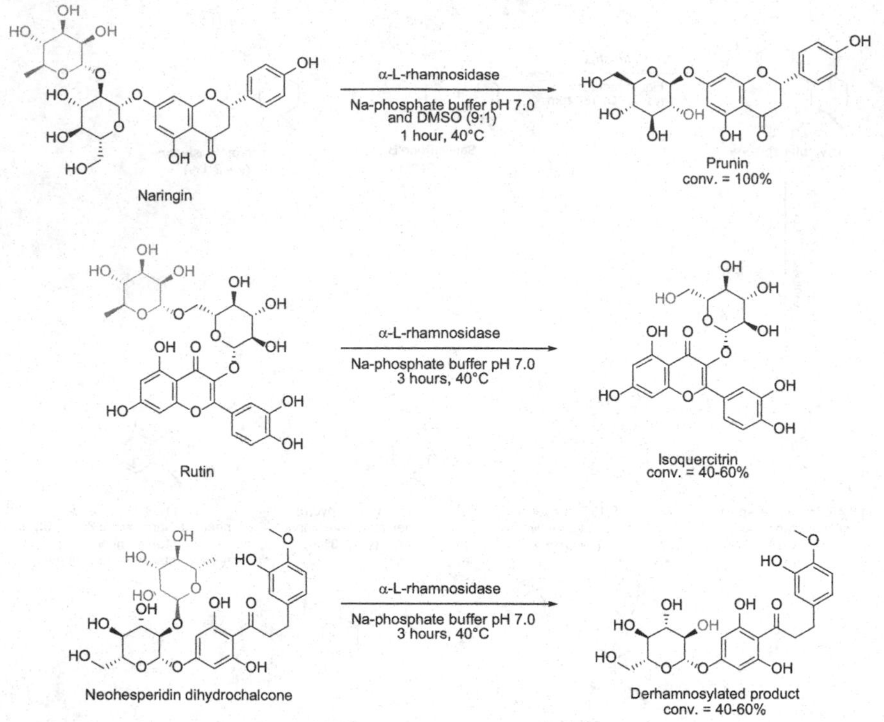

FIGURE 5.9 Flavonoid hydrolysis by a α-L-rhamnosidase (α-RHA) obtained from the marine-derived bacteria *Novosphingobium* sp. PP1Y (De Lise et al., 2016).

days (Figure 5.10). All compounds were analysed by HPLC-PDA using chiral and non-chiral column (De Matos et al., 2021a).

The chalcones are precursors in the biosynthetic pathway of several flavonoid derivatives. As well as for flavonoids, there are also studies that describe the biotransformation of chalcones by marine-derived fungi. For example, *Penicillium raistrickii* CBMAI 931, *Cladosporium* sp. CBMAI 1237, *Aspergillus sydowii* CBMAI 935, *Westerdykella* sp. CBMAI 1679, *Penicillium oxalicum* CBMAI 1996, *Penicillium citrinum* CBMAI 1186, *Mucor racemous* CBMAI 847, and *Aspergillus sclerotiorum* CBMAI 849 promoted the regioselective and chemoselective biotransformation of 2′-hydroxychalcone, forming the products 2′-hydroxy-dihydrochalcone, (±)-2-phenylchroman-4-one, and 2′,4-dihydroxy-dihydrochalcone, showing that these microorganisms have the potential to be used in several biotechnological applications. The reactions were performed in phosphate buffer (Na_2HPO_4/KH_2PO_4 0.1 mol L^{-1}, pH 8.0) at 32°C and 130 rpm for 7 days (Figure 5.11). The biotransformation products were characterised by NMR, MS and FTIR (De Matos et al., 2021b).

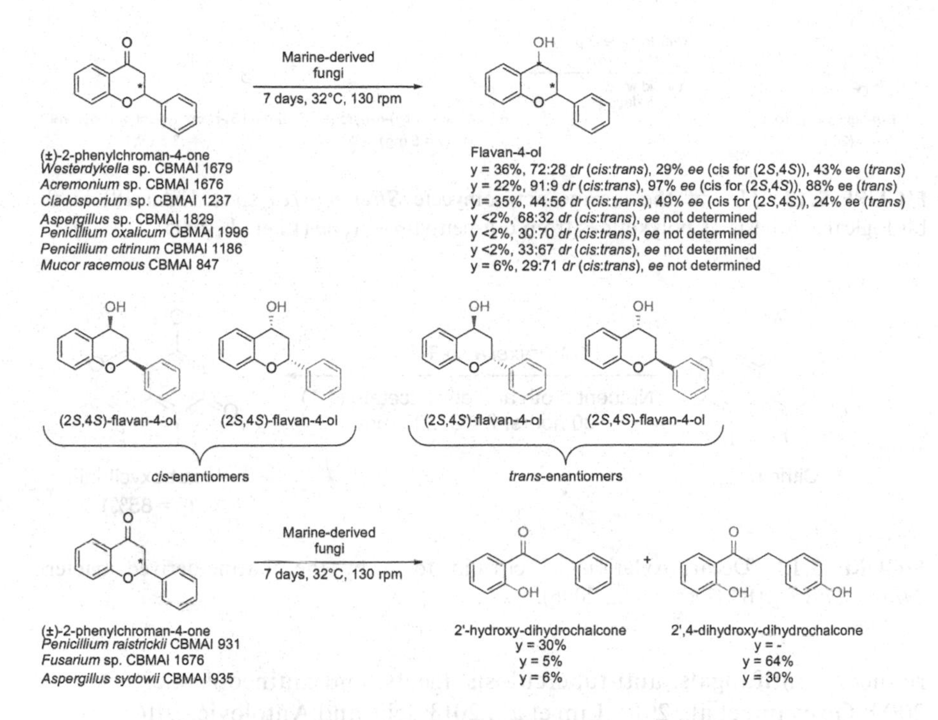

FIGURE 5.10 Microbial transformations of flavanone (±)-2-phenylchroman-4-one by marine-derived fungi (De Matos et al., 2021a).

FIGURE 5.11 Biotransformation of 2′-hydroxychalcone by marine-derived fungi (De Matos et al., 2021b).

5.4 POLYKETIDES

Polyketides are a class of natural products normally formed from the condensation of methylmalonyl-CoA, propanoyl-CoA and acetyl-CoA units by the action of polyketide synthase (PKS) enzymes. These compounds are found in plants, fungi, and bacteria (Hertweck, 2009; Dewick, 2009). The polyketides have relevant biological properties and are used as antibiotics, cholesterol

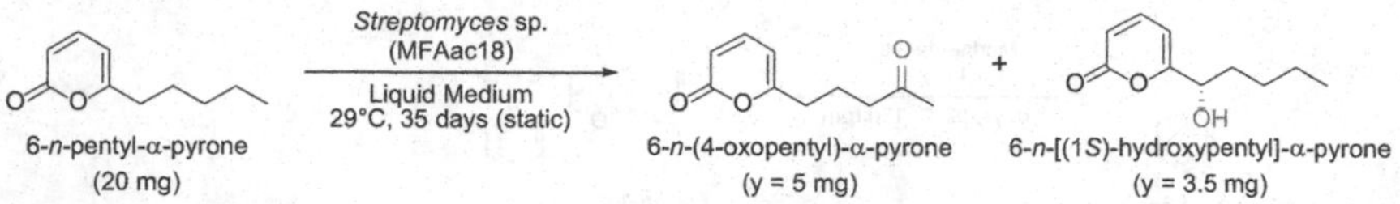

FIGURE 5.12 Use of the marine actinomycete *Streptomyces* sp. (MFAac18) in the biological synthesis of polyketides from 6-*n*-pentyl-*α*-pyrone (Li et al., 2005).

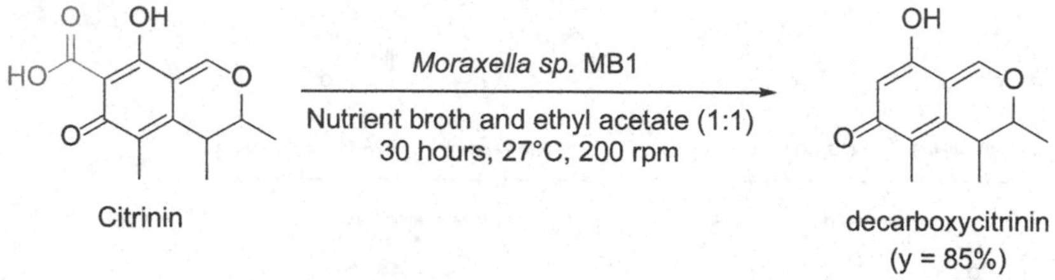

FIGURE 5.13 Decarboxylation of citrinin toxin by the marine-derived bacteria *Moraxella* sp. MB1 (Devi et al., 2006).

reducers, antifungals, anti-tuberculosis agents, and antineoplastics (Dewick, 2009; Goswami et al., 2012; Lim et al., 2013; Jelić and Antolović, 2016).

Li and co-workers (2005) reported the biological synthesis of 6-*n*-(4-oxopentyl)-*α*-pyrone (5 mg) and 6-*n*-[(1*S*)-hydroxypentyl]-*α*-pyrone (3.5 mg) polyketides from 6-*n*-pentyl-*α*-pyrone promoted by the actinomycete *Streptomyces* sp. (MFAac18), isolated from the sea plant *Zostera marina* (Korea; Figure 5.12). The reaction was carried out in culture medium (0.1% soytone, 1.0% soluble starch, and seawater) at 29°C for 35 days. (static), and the biotransformation products were characterised by NMR.

Only one product was obtained in the biotransformation of citrinin toxin by the marine-derived bacteria *Moraxella* sp. MB1, which was isolated from the alga *Elysia* sp. (India). This bacterium promoted the decarboxylation of citrinin, forming a non-toxic compound: decarboxycitrinin (85%). The biotransformation reaction was performed in a biphasic system formed by nutrient broth and ethyl acetate (1:1) under conditions of 27°C and 200 rpm for 30 hours (Figure 5.13). The biotransformation product was characterised by NMR (Devi et al., 2006).

5.5 OTHER CLASSES (STEROID, COUMARIN, QUINONE, AND LIGNAN)

There are studies that describe the biotransformation of other classes of natural products using marine organisms, such as steroids, quinones, isocoumarins, and lignan, among others. For example, De Paula and Porto (2020) described the biotransformation of progesterone by culture broth and mycelia from the marine-derived fungus *Aspergillus sydowii* CBMAI 935. The biotransformation of progesterone employing the culture broth from *A. sydowii* CBMAI 935 yielded testosterone (24%) and testololactone (36%). In the presence of mycelia, only testololactone (87%) was formed (culture medium containing 2% malt extract, pH 7.4, 32°C, 130 rpm, 7 days). The biotransformation products were characterised by NMR, MS, and FTIR (Figure 5.14).

The marine-derived bacteria *Stappia* sp. BAac008, which was isolated from alga *Enteromorph compressa* (Korea), promoted the hydroxylation of dihydroisocoumarin (-)-mellein, forming the compound (3*R*,4*S*)-4-hydroxymellein. The biotransformation reaction was carried out in culture medium (0.1% soytone, 1.0% soluble starch, and seawater) at 29°C and 130 rpm for 72 hours (Figure 5.15). The (3*R*,4*S*)-4-hydroxymellein was characterised by NMR, MS and FTIR (Zhile et al., 2010).

In another example of hydroxylation of natural products by marine-derived bacteria, the dipyrroloquinone terreusinone was converted to terreusinol by the actinomycete *Streptomyces* sp., isolated from the sea plant *Zostera marina* (Korea). The reaction was carried out in culture medium (0.1% soytone, 1.0% soluble starch, and seawater), at 29°C for 35 days (static; Figure 5.16). The biotransformation product was characterised by NMR. This study was one of the first reported in the literature describing the biotransformation of natural products by marine-derived microorganisms. Furthermore, the results showed that the genus *Streptomyces* has potential to be used as a biocatalyst in biooxidation reactions (Li et al., 2003).

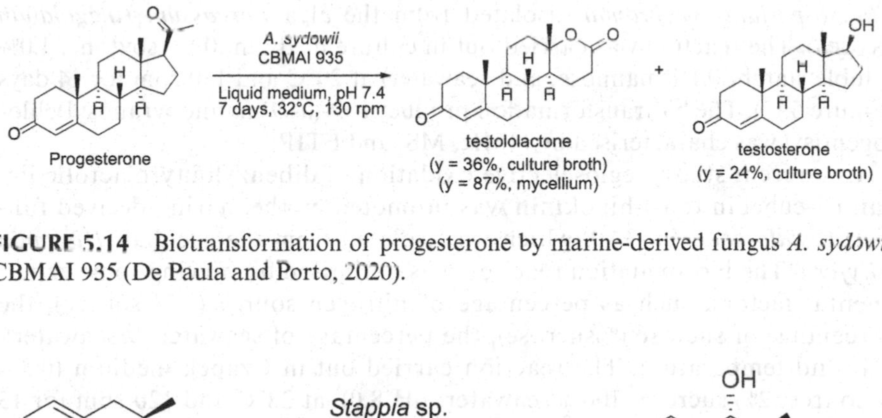

FIGURE 5.14 Biotransformation of progesterone by marine-derived fungus *A. sydowii* CBMAI 935 (De Paula and Porto, 2020).

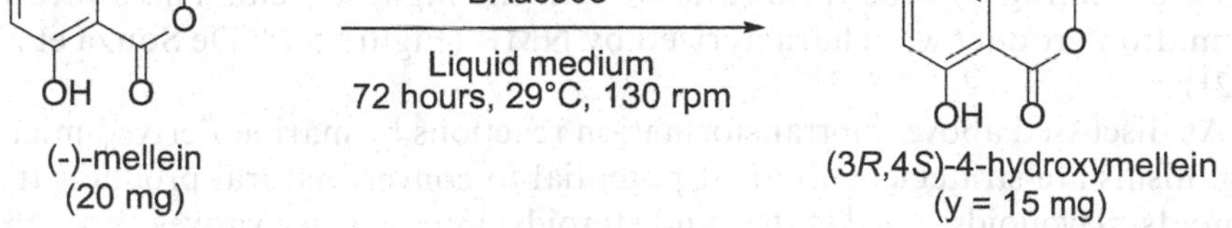

FIGURE 5.15 Biohydroxylation of (-)-mellein by marine-derived *Stappia* sp. BAac008 (Zhile et al., 2010).

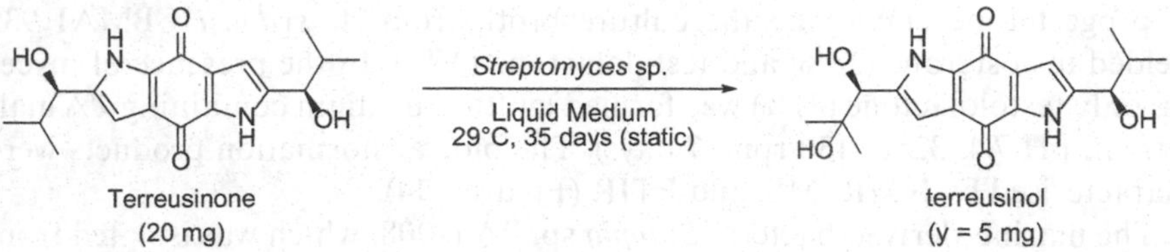

FIGURE 5.16 Biohydroxylation of terreusinone by *Streptomyces* sp. (Li et al., 2003).

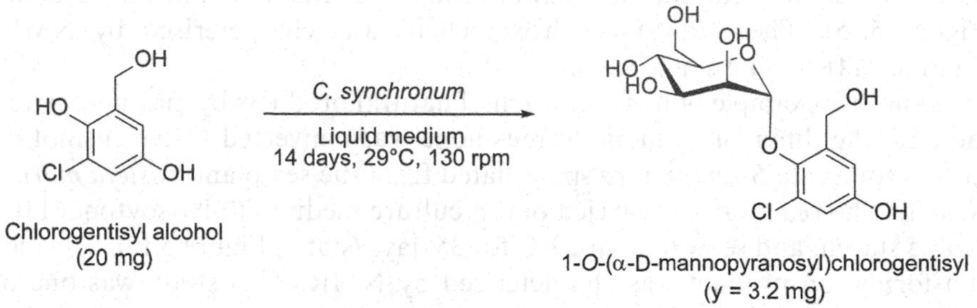

FIGURE 5.17 Microbial transformation of chlorogentisyl alcohol by the marine-derived fungus *Chrysosporium synchronum*. (Yun et al., 2011).

Yun et al. (2011) described the microbial transformation of chlorogentisyl alcohol into 1-*O*-(*α*-*D*-mannopyranosyl)chlorogentisyl alcohol by *Chrysosporium synchronum*, isolated from the alga *Sargassum ringgoldium* (Korea). The reaction was carried out in culture medium (0.1% soytone, 1.0% soluble starch, 0.1% mannose, and seawater) at 29°C and 130 rpm for 14 days (Figure 5.17). The biotransformation product 1-*O*-(α-*D*-mannopyranosyl)chlorogentisyl was characterised by NMR, MS, and FTIR.

In a recent study, regioselective oxidation of dibenzylbutyrolactolic lignan (-)-cubebin to (-)-hinokinin was promoted by the marine-derived fungus *Absidia coerulea* 3A9, which was isolated from the ascidian *Distaplia stilyfera*. The bio-oxidation reaction was optimised by varying some experimental factors, such as percentage of nitrogen source (%N source), the percentage of sucrose (%sucrose), the percentage of seawater (%seawater), pH, and temperature. The reaction carried out in Czapek medium (0.3% N source, 2% sucrose, 100% seawater, pH 8.0), at 28°C and 120 rpm for 13 days, obtaining (-)-hinokinin (27.6%) with the highest yield. The biotransformation product was characterised by NMR (Figure 5.18; De Souza et al., 2021).

As discussed above, biotransformation reactions by marine-derived microorganisms are strategic with great potential to convert natural products (terpenoids, flavonoids, polyketides, and steroids) into new derivatives through a low-cost and easy biological route.

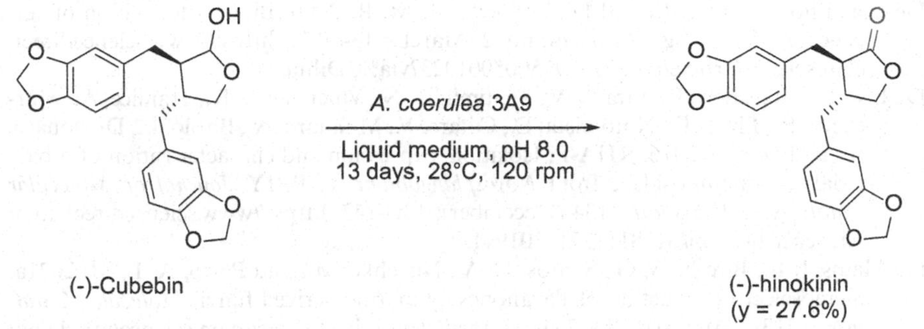

FIGURE 5.18 Oxidation of dibenzylbutyrolactolic lignan (-)-cubebin to (-)-hinokinin by marine-derived fungus *Absidia coerulea* 3A9 (De Souza et al., 2021).

5.6 CONCLUSION

From the research carried out in databases for the development of this work, it can be concluded that there is still a small number of studies in the literature describing the biotransformation of natural products by marine-derived organisms (fungi, bacteria, and algae), revealing the potential yet to be explored in this area of research. Furthermore, it can be concluded that some marine fungi and bacteria are promising sources of enzymes capable of catalysing biotransformation reactions in natural products, such as bio-oxidation, bioreduction, decarboxylation, and hydrolysis, among others.

ACKNOWLEDGMENTS

TMQ is thankful for the fellowship provided by the Coordenação de Aperfeiçoamento de Pessoal de Nível Superior (CAPES/Proc. 1732109). This study was financed by the Coordenação de Aperfeiçoamento de Pessoal de Nível Superior—Brazil (CAPES)—Finance Code 001. ALMP thanks the Fundação de Amparo à Pesquisa do Estado de São Paulo (FAPESP, Projects 2016/20155-7 and 2019/07654-2) and Conselho Nacional de Desenvolvimento Científico e Tecnológico (CNPq, Project 302528/2017-2).

REFERENCES

Bakkali, F., Averbeck, S., Averbeck, D., and Idaomar, M. 2008. Biological effects of essential oils-a review. *Food and Chemical Toxicology*, no. 2 (February): 446–475. https://nidoreessentia.com/wp-content/uploads/2018/11/Biological-effects-of-essential-oils-%E2%80%93-A-review.pdf.

Birolli, W. G., Lima, R. N., and Porto, A. L. M. 2019. Applications of marine-derived microorganisms and their enzymes in biocatalysis and biotransformation, the underexplored potentials. *Frontiers in Microbiology*, 10 (August): 1:30. https://www.frontiersin.org/articles/10.3389/fmicb.2019.01453/full

De Carvalho, C. C. C. R., and Da Fonseca, M. M. R. 2006. Biotransformation of terpenes. *Biotechnology Advances*, no. 2 (March): 134–142. https://www.sciencedirect.com/science/article/pii/S0734975005001023?via%3Dihub.

De Lise, F., Mensitieri, F., Tarallo, V., Ventimiglia, N., Vinciguerra, R., Tramice, A., Marchetti, R., Pizzo, E., Notomista, E., Cafaro, V., Molinaro, A., Birolo, L., Di Donato, A., and Izzo, V. 2016. RHA-P: Isolation, expression and characterization of a bacterial α-L-rhamnosidase from *Novosphingobium* sp. PP1Y. *Journal of Molecular Catalysis B: Enzymatic*, 134 (December): 136–147. https://www.sciencedirect.com/science/article/pii/S1381117716301941.

De Matos, I. L., Birolli, W. G., Santos, D. A., Nitschke, M., and Porto, A. L. M. 2021a. Stereoselective reduction of flavanones by marine-derived fungi. *Molecular Catalysis*, 513 (August): 111734. https://www.sciencedirect.com/science/article/pii/S2468823121003515.

De Matos, I. L., Nitschke, M., and Porto, A. L. M. 2021b. Regioselective and chemoselective biotransformation of 2′-hydroxychalcone derivates by marine-derived fungi. *Biocatalysis and Biotransformation*, no. 1 (July): 1–11. https://www.tandfonline.com/doi/epub/10.1080/10242422.2021.1956909?needAccess=true.

De Paula, S. F. C., and Porto, A. L. M. 2020. Cascate reactions of progesterone by mycelia and culture broth from marine-derived fungus *Aspergillus sydowii* CBMAI 935. *Biocatalysis and Agricultural Biotechnology*, 25 (May): 101856. https://www.sciencedirect.com/science/article/pii/S1878818119319577.

De Souza, J. M., Santos, M. F. C., Pedroso, R. C. N., Pimenta, L. P., Siqueira, K. A., Soares, M. A., Dias, G. M., Pietro, R. C. L. R., Ramos, H. P., Silva, M. L. A., Pauletti, P. M., Veneziani, R. C. S., Ambrósio, S. R., Braun, G. H., and Januário, A. H. 2021. Optimization of (-)-cubebin biotransformation to (-)-hinokinin by the marine fungus *Absidia coerulea* 3A9. *Archives of Microbiology*, no. 7 (September): 4313–4318. https://rd.springer.com/article/10.1007%2Fs00203-021-02417-0.

Devi, P., Naik, C. G., and Rodrigues, C. 2006. Biotransformation of citrinin to decarboxycitrin using an organic solvent-tolerant marine bacterium, *Moraxella* sp. MB1. *Marine Biotechnology*, no. 2 (March): 129–138. https://rd.springer.com/article/10.1007%2Fs10126-005-5021-5.

Dewick, P. M. 2009. *Medicinal natural product: A biosynthetic approach*. New York: Wiley.

Goswami, S., Vidyarthi, A. S., Bhunia, B., and Mandal, T. 2012. A review on lovastatin and its production. *Journal of Biochemical Technology*, no. 1 (August): 581–587. https://jbiochemtech.com/article/a-review-on-lovastatin-and-its-production.

Haytowitz, D. B., Bhagwat, S., and Holden, J. M. 2013. Sources of variability in the flavonoid content of foods. *Procedia Food Science*, 2 (May): 46–51. https://www.sciencedirect.com/science/article/pii/S2211601X13000096.

Hertweck, C. 2009. The biosynthetic logic of polyketide diversity. *Angewandte Chemie International Edition*, no. 26 (June): 4688–4716. https://onlinelibrary.wiley.com/doi/epdf/10.1002/anie.200806121.

Houjin, L., Wenjian, L., Chuanghua, C., Yipin, Z., and Yongcheng, L. 2006. Biotransformation of limonene by marine bacteria. *Chinese Journal of Analytical Chemistry*, 34: 946–950.

Jelić, D., and Antolović, R. 2016. From erythromycin to azithromycin and new potential ribosome-binding antimicrobials. *Antibiotics*, no. 3 (September): 1–13. https://www.mdpi.com/2079-6382/5/3/29.

Koshimura, M., Utsukihara, T., Kawamoto., M., Saito, M., Horiuchi, C. A., and Kuniyoshi, M. 2009. Biotransformation of bromosesquiterpenes by marine fungi. *Phytochemistry*, no. 17 (December): 2023–2026. https://www.sciencedirect.com/science/article/abs/pii/S0031942209003574?via%3Dihub.

Leutou, A. S., Yang, G., Nenkep, V. N., Siwe, X. N., Feng, Z., Khong, T. T., Choi, H. D., Kang, J. S., and Son, B. W. 2009. Microbial transformation of a monoterpene, geraniol, by the marine-derived fungus *Hypocrea* sp. *Journal of Microbiology and Biotechnology*, no. 10 (July): 1150–1152. https://www.koreascience.or.kr/article/JAKO200935736658147.page.

Li, X., Kim, S. K., Jung, J. H., Kang, J. S., Choi, H. D., and Son, B. W. 2005. Biological synthesis of polyketides from 6-*n*-pentyl-*α*-pyrrone by *Streptomyces* sp. *Bulletin of the Korean Chemical Society*, no. 11 (November): 1889–1890. http://koreascience.or.kr/article/JAKO200502727343332.pdf.

Li, X., Kim, Y. H., Jung, J. H., Kang, J. S., Kim, D. K., Choi, H. D., and Son, B. W. 2007. Microbial transformation of the bioactive sesquiterpene, cyclonerodiol, by the ascomycete *Penicillium* sp. and the actinomycete *Streptomyces* sp. *Enzyme and Microbial Technology*, no. 5 (April): 1188–1192. https://www.sciencedirect.com/science/article/pii/S014102290600456X?via%3Dihub.

Li, X., Lee, S. M., Choi, H. D., Kang, J. S., and Son, B. W. 2003. Microbial transformation of terreusinone, an ultraviolet-A (UV-A) protecting dipyrroloquinone, by *Streptomyces* sp. *Chemical and Pharmaceutical Bulletin*, no. 12 (December): 1458–1459. https://www.jstage.jst.go.jp/article/cpb/51/12/51_12_1458/_article.

Lim, L. E., Vilchèze, C., Ng, C., Jacobs, W. R., Ramón-García, S., and Thompson, C. J. 2013. Anthelmintic avermectins kill *Mycobacterium tuberculosis*, including multidrug-resistant clinical strains. *Antimicrobial Agents and Chemotherapy*, no. 2 (February): 1040–1046. https://journals.asm.org/doi/10.1128/AAC.01696-12.

Liu, H., Lei, X. L., Li, N., and Zong, M. H. 2013. Highly regioselective synthesis of betulone from betulin by growing cultures of marine fungus *Dothideomycete* sp. HQ 316564. *Journal of Molecular Catalysis B: Enzymatic*, 88 (April): 32–35.

Martins, M. P., Ouazzani, J., Arcile, G., Jeller, A. H., De Lima, J. P. F., Seleghim, M. H. R., Oliveira, A. L. L., Debonsi, H. M., Venâncio, T., Yokoya, N. S., Fujii, M. T., and Porto, A. L. M. 2015. Biohydroxylation of (-)-ambrox®, (-)-sclareol, and (+)-sclareolide by whole cells of brazilian marine-derived fungi. *Marine Biotechnology*, no. 2 (April): 211–218. https://rd.springer.com/article/10.1007%2Fs10126-015-9610-7.

Queiroz, T. M., Ellena, J., and Porto, A. L. M. 2020. Biotransformation of ethinylestradiol by whole cells of Brazilian marine-derived fungus *Penicillium oxalicum* CBMAI 1996. *Marine Biotechnology*, no. 5 (October): 673–682. https://rd.springer.com/article/10.1007%2Fs10126-020-09989-w.

Raffa, D., Maggio, B., Raimondi, M. V., Plescia, F., and Daidone, G. 2017. Recent discoveries of anticancer flavonoids. *European Journal of Medicinal Chemistry*, 142 (December): 213–228. https://www.sciencedirect.com/science/article/pii/S0223523417305561?via%3Dihub.

Rosa, G. P., Seca, A. M. L., Barreto, M. C., and Pinto, D. C. G. A. 2017. Chalcone: A valuable scaffold upgrading by green methods. *ACS Sustainable Chemistry & Engineering*, 9 (July): 7467–7480. https://pubs.acs.org/doi/abs/10.1021/acssuschemeng.7b01687.

Sakar, S., Pramanik, A., Mitra, A., and Mukherjee, J. 2010. Bioprocessing data for the production of marine enzymes. *Marine Drugs,* no. 4 (April): 1323–1372. https://www.ncbi.nlm.nih.gov/pmc/articles/PMC2866489/.

Sayed, K. A. E., Laphookhieo, S., Baraka, H. N., Yousaf, M., Hebert, A., Bagaley, D., Rainey, F. A., Muralidharan, A., Thomas, S., and Shah, G. V. 2008. Biocatalytic and semisynthetic optimization of the anti-invasive tobacco (1*S*,2*E*,4*R*,6*R*,7*E*,11*E*)-2,7,11-cembratriene-4,6-diol. *Bioorganic & Medicinal Chemistry*, no. 6 (March): 2886–2893. https://www.sciencedirect.com/science/article/pii/S096808960701125X?via%3Dihub.

Schwab, W., Fuchs, C., and Huang, F. C. 2013. Transformation of terpenes into fine chemicals. *European Journal of Lipid Science and Technology*, 115 (September): 3–8. https://onlinelibrary.wiley.com/doi/epdf/10.1002/ejlt.201200157.

Soobrattee, M. A., Neergheen, V. S., Luximon-Ramma, A., Aruoma, I. O., and Bahorun, T. 2005. Phenolics as potential antioxidant therapeutic agents: Mechanism and actions. *Mutation Research*, no. 1 (November): 200–213. https://daneshyari.com/article/preview/9908999.pdf.

Williams, R. J., Spencer, J. P. E., and Evans, C. R. 2004. Flavonoids: Antioxidant or signaling molecules? *Free Radical Biology and Medicine*, 36: 838–849. https://www.sciencedirect.com/science/article/pii/S0891584904000334?via%3Dihub.

Wu, J. S., Meng, Q. Y., Zhang, Y. H., Shi, X. H., Fu, X. M., Zhang, P., Li, X., Shao, C. L., and Wang, C. Y. 2020. Annular oxygenation and rearrangement products of cryptotanshinone by biotransformation with marine-derived fungi *Cochliobolus lunatus* and *Aspergillus terreus*. *Bioorganic & Medicinal Chemistry*, 103 (October): 104192. https://www.sciencedirect.com/science/article/pii/S0045206820314899?via%3Dihub.

Wu, T., He, M., Zang, X., Zhou, Y., Qiu, T., Pan, S., and Xu, X. 2013. A structure-activity relationship study of flavonoids as inhibitors of *E. coli* by membrane interaction effect. *Biochimica et Biophysica Acta*, no. 11 (November): 2751–2756. https://www.sciencedirect.com/science/article/pii/S0005273613002757.

Yao, L. H., Jiang, Y. M., Shi, J., Tomás-Barberán, F. A., Datta, N., Singanusong, R., and Chen, S. S. 2004. Flavonoids in food and their health benefits. *Plant Foods for Human Nutrition*, no. 3 (July): 113–122. https://rd.springer.com/article/10.1007%2Fs11130-004-0049-7.

Yun, K., Kondempudi, C. M., Choi, H. D., Kang, J. S., and Son, B. W. 2011. Microbial mannosidation of bioactive chlorogentisyl alcohol by the marine-derived fungus *Chrysosporium synchronum*. *Chemical and Pharmaceutical Bulletin*, no. 4 (April): 499–501. https://www.jstage.jst.go.jp/article/cpb/59/4/59_4_499/_pdf/-char/en.

Yun, K., Kondempudi, C. M., Leutou, A. S., and Son, B. W. 2015. New Production of a monoterpene glycoside, 1-O-(α-D-mannopyranosyl) geraniol, by the marine-derived fungus Thielavia hyalocarpa. *Bulletin of the Korean Chemical Society*, 36 (August): 2391–2393. https://onlinelibrary.wiley.com/doi/epdf/10.1002/bkcs.10451.

Zhile, F., Nenkep, V. N., Yun, K., Zhang, D., Choi, H. D., Kang, J. S., and Son, B. W. 2010. Biotransformation of bioactive (-)-mellein by a marine isolate of bacterium *Stappia* sp. *Journal of Microbiology and Biotechnology*, no. 6 (April): 985–987. https://www.jmb.or.kr/submission/Journal/020/JMB020-06-05_FDOC_1.pdf.

Zhou, S., Wang, F., Wong, E. T., Fonkem, E., Hsieh, T. C., Wu, J.M., and Wu, E. 2013. Salinomycin: A novel anti-cancer agent with known anti-coccidial activities. *Current Medicinal Chemistry*, no. 33 (August): 4095–4101. https://www.ncbi.nlm.nih.gov/pmc/articles/PMC4102832/.

6 Zooplankton and $^{210}Po/^{210}Pb$ in Southwest Pacific Waters

Ross A. Jeffree, Ron Szymczak, Gillian Peck, and Scott W. Fowler

CONTENTS

DOI: 10.1201/9781003303909-6

6.1 INTRODUCTION

The role of zooplankton and their particulate products in the biogeochemical cycling and vertical transport of natural radionuclides, such as ^{210}Po and ^{210}Pb, in the marine environment has been the subject of both field and experimental laboratory studies for many years (Cherry et al., 1975; Heyraud et al., 1976; Beasley et al., 1978; Fowler and Fisher, 2004; Rodriguez y Baena et al., 2007; Fowler, 2011). Such information has become increasingly important in recent years as a result of studies that document the usefulness of measuring the disequilibrium between the $^{210}Po/^{210}Pb$ pair in sinking particulate matter in order to estimate the export flux of particulate organic carbon in the upper water column (Stewart et al., 2007, 2010; Rutgers van der Loeff and Geibert, 2008; Tang et al., 2018a, 2018b).

The main objectives of this field-based study were to evaluate the potentially enhancing effects of oligotrophy on the bioaccumulation of ^{210}Po and ^{210}Pb in zooplankton and their fecal pellets and to critically evaluate a proposed model of the role of zooplankton and their fecal pellets in controlling ^{210}Po concentrations in the surface waters of low-productivity regions of the ocean (Jeffree et al., 1997). Our previous studies on the biogeochemical behavior of ^{210}Po and particle-reactive trace metals in the oligotrophic regions of French Polynesia and the Timor Sea, respectively, indicated the capacity of these systems to concentrate these elements in seawater, and in a way that was enhanced by declining biological productivity and the associated reduced organic particle fluxes in the euphotic zone (Jeffree et al., 1997; Jeffree and Szymczak, 2000). Moreover, as a result of this proposed mechanism, ^{210}Po concentrations in zooplankton from French Polynesia were enhanced up to previously unreported high levels (Jeffree et al., 1997).

Detailed knowledge of how zooplankton and the associated food chain dynamics help govern the oceanic cycling of radionuclides is accruing; nevertheless, most of our present information comes from measurements made in the more productive, temperate-zone eutrophic waters (Fowler and Fisher, 2004; Rodriguez y Baena et al., 2007; Stewart et al., 2007, 2008). However, the overall importance of ^{210}Po bioaccumulation by marine organisms in the radiological dose received by the world's population from the consumption of seafood (Aarkrog et al., 1997) warrants further study of its biogeochemical behavior in oligotrophic systems, especially in view of their predicted areal expansion due to the current and future increase in oceanic temperatures (Bopp et al., 2001; Behrenfeld et al., 2006; Bindoff et al., 2019; Cheng et al., 2019), particularly in the southern hemisphere (Sarmiento et al., 2004). In fact, geochemical mass balance studies based on ^{210}Po seawater profiles measured in various oligotrophic regions of the Atlantic and Pacific Oceans have documented large ^{210}Po deficiencies in the water column to depths of 3000 m (Nozaki et al., 1990; Kim, 2001; Chung and Wu, 2005). Physical transport and focusing of radionuclide-enriched particles, as well as ^{210}Po bioaccumulation by cyanobacteria and the subsequent food chain transfer, have been hypothesized as possible mechanisms for producing these large ^{210}Po deficits.

In this study undertaken in the southwest Pacific, we test hypotheses of the enhancing effect of oligotrophy on ^{210}Po (and its progenitor ^{210}Pb) in zooplankton and their fecal pellets and their role in determining ^{210}Po water concentrations, as previously identified in euphotic-zone waters of French Polynesia (Jeffree et al., 1997). To help refine our evaluation of the proposed model, in the present investigation, we include site-specific measurements of ^{210}Po and ^{210}Pb in both dissolved and particulate fractions of surface seawater as well as in freshly produced fecal pellets from the collected zooplankton, along with onboard measurements of fecal pellet production rates for the zooplankton sampled at each station. These parameters were not directly measured in the earlier French Polynesian study.

6.2 MATERIALS AND METHODS

6.2.1 Study Region and Sampling Methods

All samples were collected from an oligotrophic region in the southwest Pacific located between Fiji and New Caledonia within the geographical coordinates shown in Figure 6.1. Seawater and zooplankton were sampled at 13 stations during two periods of the annual cycle on three separate cruises, *viz.* March 2001, November 2001 and March 2002. Forty-liter water samples were collected at each station from 5 m below the surface using 5-L polyvinyl chloride Niskin bottles. Upon recovery, the water was filtered through either 47-mm- or 125-mm-diameter 0.4-µm Nuclepore membranes. Ten liters of filtered seawater were transferred to a glass vessel for the pre-concentration of radionuclides (^{210}Pb and ^{210}Po). The filters containing the particulate matter were placed in filter holders and stored at 4°C for later processing ashore of

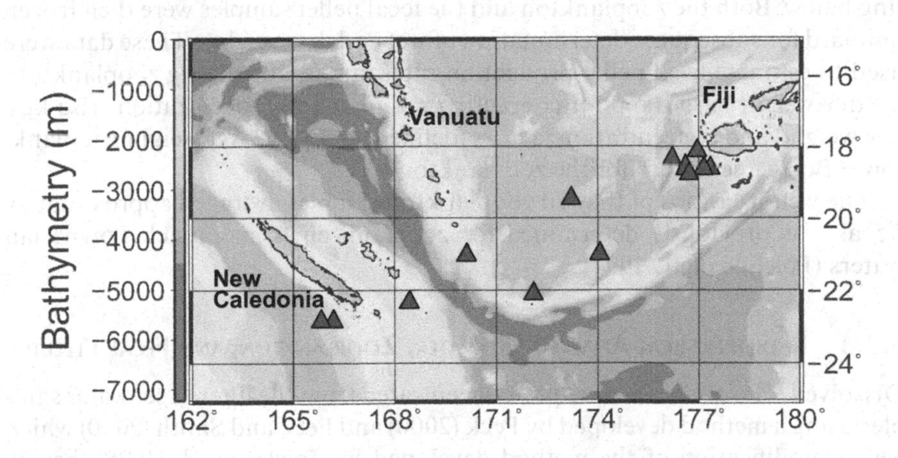

FIGURE 6.1 Geographical locations where zooplankton and water samples were obtained for this study. Lighter-shaded areas are shallow waters, and dark-shaded areas are deep waters.

a subset of those collected. Zooplankton was collected at night in the upper water column by making a series of short horizontal tows at a depth of approximately 10 m using standard netting methodologies previously described in Poletico et al. (1994). In brief, a 0.5-m diameter net with 300-µm mesh and a filtering cod end designed to collect live zooplankton was used. Towing speeds varied between 2–3 kn depending on the sea state. Following the sample collection, small subsamples were preserved in 5% formaldehyde for 24–48 hr and then stored in 70% alcohol for their subsequent taxonomic characterization. In all samples, calanoid, followed by cyclopoid copepods, were by far the numerically dominant zooplankton organisms as was previously observed in oceanic samples from French Polynesia (Jeffree et al., 1997). Some samples taken during March 2002 also showed increased abundances of foraminifera. Other pelagic groups present, in decreasing order of general abundance, were crab larvae, mysids, salps, pteropods and chaetognaths. Nauplii and copepodites were not a significant component of any sample.

6.2.2 Determination of Zooplankton Biomass and Fecal Pellet Production Rates

Zooplankton biomass was calculated from the water volume sampled, which was a function of (1) net diameter and (2) measurements from a flow meter suspended in the center of the entrance to the net.

Immediately following the net sampling, any large detritus and extraneous material were removed, and after each was cleaned, the mixed-zooplankton sample was then poured into a fecal pellet collection vessel (La Rosa, 1976) containing continually aerated, filtered seawater for a period of 7–12 hr to allow the release of fecal pellets from the live zooplankton. These pellets passed through a 200-µm separation net to be subsequently concentrated on a 20-µm mesh netting below. Both the zooplankton and the fecal pellet samples were then frozen onboard for subsequent determination of wet and dry weights. These data were used to estimate fecal pellet production rates (mg fecal pellets/g zooplankton/hr; dry weight basis), and for computing radionuclide concentrations (Bq kg^{-1} dry weight) and concentration factors (defined as Bq kg^{-1} wet weight zooplankton ÷ Bq kg^{-1} seawater) for the zooplankton.

The wet/dry ratios of thawed zooplankton samples averaged approximately 7.5 as was previously determined for zooplankton from French Polynesian waters (Poletico et al., 1994).

6.2.3 Radiochemical Analysis of Water, Zooplankton, and Fecal Pellets

Dissolved ^{210}Po and ^{210}Pb were pre-concentrated from the filtered seawater samples using a method developed by Peck (2000) and Peck and Smith (2000) which was a modification of the method developed by Towler et al. (1996), briefly summarized here. Filtered seawater (10 L) was acidified with HCl and standard additions of the yield tracers ^{209}Po (0.1 Bq) and stable Pb (3 mg) were made. After 1–2 hr, the samples were neutralized by the addition of a sodium hydroxide solution. MnO_2-coated magnetite was added to each sample and stirred vigorously for 1 hr, and the magnetite was then simply collected by means of a magnet. The loaded magnetite adsorbent was stored under distilled water in 50-mL plastic bottles and returned to the shore laboratory for further processing.

In the laboratory, the MnO_2 was dissolved in H_2O_2/HCl, a small amount of ascorbic acid was added to the solution, and ^{210}Po was spontaneously plated onto a silver disc that was then analyzed by alpha spectrometry (Smith and Hamilton, 1984). ^{210}Po was counted immediately, and the activity of the ^{210}Pb in the seawater sample was inferred from a second plating and the measurement of ^{210}Po after allowing several months for ingrowth from the ^{210}Pb. The recovery of the stable Pb yield tracer was measured by graphite furnace atomic absorption spectroscopy, and the appropriate correction was made to the ^{210}Pb activity. ^{209}Po was measured by alpha spectrometry to determine the efficiency of the ^{210}Po measurements.

Filters containing particulate ^{210}Po and ^{210}Pb were spiked with standard additions of the yield tracers ^{209}Po (0.1 Bq) and stable Pb (3 mg) and digested in gently boiling 50% HNO_3 for 3 hr. The samples were filtered and the filtrate evaporated to near dryness and brought up with dilute HCl. The ^{210}Po and ^{210}Pb were then measured as described earlier for the seawater. The fecal pellet and zooplankton samples were first weighed wet, then oven-dried to constant weight, reweighed and subsequently analyzed for ^{210}Po and ^{210}Pb in a manner similar to that for the filtered particulates. Zooplankton and fecal pellet sample masses ranged between 2.2–19.9 g dry weight and 0.03–0.3 g dry weight, respectively.

Regarding analytical quality assurance, the Australian Nuclear Science & Technology Organization (ANSTO) Radiochemical Laboratory had previously participated in a tripartite inter-comparison exercise of ^{210}Po measurements in zooplankton subsamples from French Polynesia, as reported in Jeffree et al. (1997). This inter-comparison showed only minor (about 25%) analytical differences between three laboratories (ANSTO, Australia; SCPRI, France; and LESE, French Polynesia), as compared to between-sample variation. The method employed for radiochemical analysis of the zooplankton in this study was also used in an international inter-comparison study of ^{210}Po and ^{210}Pb in seaweed, the results of which were reported in Outola et al. (2006).

6.2.4 Modeling of Zooplankton Biomass-Determined ^{210}Po and ^{210}Pb Water Concentrations in the Euphotic Zone

The objective of our modeling exercise was to determine whether a simple single-compartment model of the thermally stratified euphotic zone, from which ^{210}Po and ^{210}Pb removal is mediated by the production, sinking and export of zooplankton fecal pellets, could reasonably predict the resulting ^{210}Po and ^{210}Pb water concentrations measured in the euphotic zone. We used a similar methodology from a previous study (Jeffree et al., 1997) and the steps involved were as follows:

1. Determination of the zooplankton biomass-mediated rate of ^{210}Po and ^{210}Pb removal from the euphotic zone, based on the station-specific empirical values determined in this study.
2. Use of these data in a simple mathematical model previously developed, which did predict the empirical relationship between zooplankton biomass and ^{210}Po concentrations in zooplankton in French Polynesian waters (Jeffree et al., 1997); namely, $Cw = R/k(1 - e^{-kt})$, where Cw is the ^{210}Po concentration in water (dissolved), R is the rate of ^{210}Po input to the water column predominantly from the entry of ^{210}Pb to the sea surface by wet and dry deposition and its *in situ* decay to ^{210}Po, t is time and k is the rate constant for loss of ^{210}Po from the euphotic zone waters that depends on

   a. The measured rate of fecal pellet production per unit mass of zooplankton per 100 m^3, where ^{210}Po and ^{210}Pb removal is mediated by the production, rapid sinking and export of zooplankton fecal pellets (50–60 m/day; Small et al., 1979).
   b. The measured ^{210}Po concentration in zooplankton fecal pellets.
3. Statistical comparison of the model-derived ^{210}Po and ^{210}Pb water concentrations with the site-specific empirical values so as to assess the predictive capability of the model and, hence, the plausibility of our hypothesis about the importance of zooplankton in determining ^{210}Po and ^{210}Pb concentrations in the ambient water.

For this modeling approach, we have made the following assumptions:

1. The system is at steady state, in that the same amounts of ^{210}Po and ^{210}Pb are entering the euphotic zone as are leaving via zooplankton fecal pellets. Under these conditions, according to the equation $Cw = R/k$ $(1 - e—^{kt})$, $1 - e—^{kt} \approx 1$, and therefore, $Cw = R/k$ (Whicker and Schultz, 1982).
2. The depth of the euphotic zone is set at 100 m, into which the annual input of ^{210}Pb, and consequently ^{210}Po from ingrowth, is delivered. Whereas a depth of 200 m was used in a previous study in the ultra-oligotrophic French Polynesian waters (Jeffree et al., 1997), the choice of 100 m is more representative of the limit of photosynthetic activity in these waters, which range from oligotrophy to mesotrophy (Wang et al., 2005; Van Wambeke et al., 2018). ^{210}Pb input rates vary over time, and it was not possible to directly measure wet and dry atmospheric inputs in our immediate study area. Therefore, we use an empirically derived annual input rate of 80 Bq m^{-2} year^{-1} previously measured at Fiji (Turekian et al., 1977) just adjacent to our study region (see Figure 6.1).

6.3 RESULTS

6.3.1 ^{210}Po and ^{210}Pb Concentrations in Water Compartments, Zooplankton, and Their Fecal Pellets

Dissolved ^{210}Po concentrations in seawater ranged over a factor of 3.4 between 0.47–1.58 mBq L^{-1}, with a mean value of 1.02 mBq L^{-1}. ^{210}Pb levels varied by a factor of 3.9 between 0.81–3.19 mBq L^{-1}, with a mean value of 1.7 mBq L^{-1}, as shown in Table 6.1 and Figure 6.2. These values were very consistent with those measured approximately three decades earlier (Nozaki et al., 1976) for samples taken in a similar general geographical region (15–20° S, 165–180° E), that is, ranging from 0.63–1.17 mBq L^{-1} for ^{210}Po (mean = 0.92; n = 5) and from 1.75–2.45 mBq L^{-1} for ^{210}Pb (mean = 2.13; n = 5) and seen in Figure 6.2. A comparison of the means of the two data sets (this study and Nozaki et al., 1976) by t test for unequal sample sizes (Zar, 1996) showed that they were not significantly different ($p > 0.05$) in either ^{210}Po- or ^{210}Pb-dissolved water concentrations.

In Table 6.2 are shown the concentrations of ^{210}Po and ^{210}Pb in suspended particulate matter for a subset of six water samples and the ratios of dissolved:particulate water concentrations for each nuclide, based on these particulate values and the dissolved water concentrations given in Table 6.1. The ^{210}Po particulate water concentrations vary over a factor of five, from 0.10–0.52 mBq L^{-1}, and for ^{210}Pb they vary by over an order of magnitude between 0.02–0.25 mBq L^{-1}. Figure 6.3 compares the means (and SDs) of these ratios of dissolved:particulate phase concentrations of ^{210}Po and ^{210}Pb, which clearly demonstrates the heightened affinity by an order of magnitude of ^{210}Po for particulates relative to ^{210}Pb.

Concentrations of ^{210}Po varied between 565 and 1736 Bq kg^{-1} in mixed zooplankton and 830–3107 Bq kg^{-1} in their fecal pellets (Table 6.1). We performed a t test for dependent samples that showed that the mean ^{210}Po concentration in fecal pellets of 1940 Bq kg^{-1} was significantly higher ($p < 0.01$, n = 12) than its mean concentration of 1283 Bq kg^{-1} in the zooplankton producing them. The ^{210}Pb concentrations ranged between 47–551 Bq kg^{-1} in zooplankton and between 44 617 Bq kg^{-1} for their fecal pellets, as shown in Table 6.1. However, the mean ^{210}Pb concentration of 340 Bq kg^{-1} in fecal pellets was not significantly different from that in their zooplankton producers (250 Bq kg^{-1}; $p > 0.05$, n = 12).

A comparison by t test confirmed that the mean ^{210}Po concentration in zooplankton (mean = 1283 Bq kg^{-1}) was significantly ($p < 0.0001$, n = 12) greater than the corresponding mean ^{210}Pb concentrations (mean = 250 Bq kg^{-1}) and by a factor of about 5. This difference in concentrations between ^{210}Po and ^{210}Pb was a little more pronounced in fecal pellets, with means of 1940 Bq kg^{-1} for ^{210}Po and 340 Bq kg^{-1} for ^{210}Pb being significantly different ($p < 0.0001$, n = 12) and by a factor of 5.7.

^{210}Po concentration factors in zooplankton varied by a factor of 4.7 between $0.7–3.3 \times 10^5$ with a mean of 1.8×10^5, as shown in Table 6.1. Interestingly, Uddin et al. (2018) have recently reported a similar range of ^{210}Po concentration factors ($0.8–5.3 \times 10^5$) in six species of copepods from the northern Arabian Gulf. These

TABLE 6.1

^{210}Po and ^{210}Pb Concentrations in Seawater, Zooplankton, and Their Fecal Pellets and Radionuclide Concentration Factors (CFs) in Zooplankton Collected from the Southwestern Pacific Ocean

Sample Number	Sampling Period	Zooplankton Biomass (mgDW/m^3)	Zooplankton (Bq/kg DW)			Seawater [dissolved (mBq/L)]			Fecal Pellets (Bq/kg DW)			CF for Zooplankton (Bq/kgWW: Bq/L)	
			^{210}Po	^{210}Pb	^{210}Po/ ^{210}Pb	^{210}Po	^{210}Pb	^{210}Po/ ^{210}Pb	^{210}Po	^{210}Pb	^{210}Po/ ^{210}Pb	^{210}Po	^{210}Pb
1	Mar 2002	2.09	1736	266	6.5	0.70	1.67	0.4	2655	617	4.3	3.3×10^5	2.1×10^4
2	Mar 2002	7.12	1404	324	4.3	1.52	1.57	1.0	1389	438	3.2	1.2×10^5	2.8×10^4
3	Mar 2002	1.40	1039	187	5.6	0.47	1.72	0.3	2660	351	7.6	2.9×10^5	1.4×10^4
4	Mar 2002	2.22	998	209	4.8	0.92	3.19	0.3	2529	389	6.5	1.4×10^5	0.9×10^4
5	Mar 2002	4.68	1319	196	6.7	0.93	2.52	0.4	1784	419	4.3	1.9×10^5	1.0×10^4
6	Nov 2001	2.91	1285	302	4.2	1.23	0.81	1.5	830	115	7.2	1.4×10^5	5.0×10^4
7	Nov 2001	2.53	1348	551	2.4	0.68	0.81	0.8	1468	408	3.6	2.6×10^5	9.1×10^4
8	Nov 2001	2.66	1580	528	3	1.58	0.97	1.6	2578	390	6.6	1.3×10^5	7.3×10^4
9	Nov 2001	0.7	1239	266	4.7	0.83	2.35	0.4	1734	431	4.0	2.0×10^5	1.5×10^4
10	Nov 2001	0.14	572	90	6.4	-	-	-	-	-	-	-	-
11	Nov 2001	1.96	565	227	2.5	1.15	1.36	0.8	1241	267	4.6	0.66×10^5	2.2×10^4
12	Mar 2001	-	1452	61	23.8	-	-		1306	44	29.7	-	-
13	Mar 2001	-	1433	47	30.5	1.22	1.26	1.0	3107	209	14.9	1.6×10^5	0.5×10^4
Mean		2.6	1283	250	8.1	1.02	1.7	0.8	1940	340	8.0	1.8×10^5	3.3×10^4
(range)		(0.14–7.12)	(565–1736)	(47–551)	(2.4–30.5	(0.47–1.58)	(0.81–3.19)	(0.3–1.6)	(830–2655)	(44–617)	(3.2–29.7)	($0.7–3.3 \times 10^5$)	($0.9–9.1 \times 10^4$)
Median		2.22	1285	227	4.8	0.93	1.57	0.8	1755	390	5.6	1.6×10^5	2.1×10^4

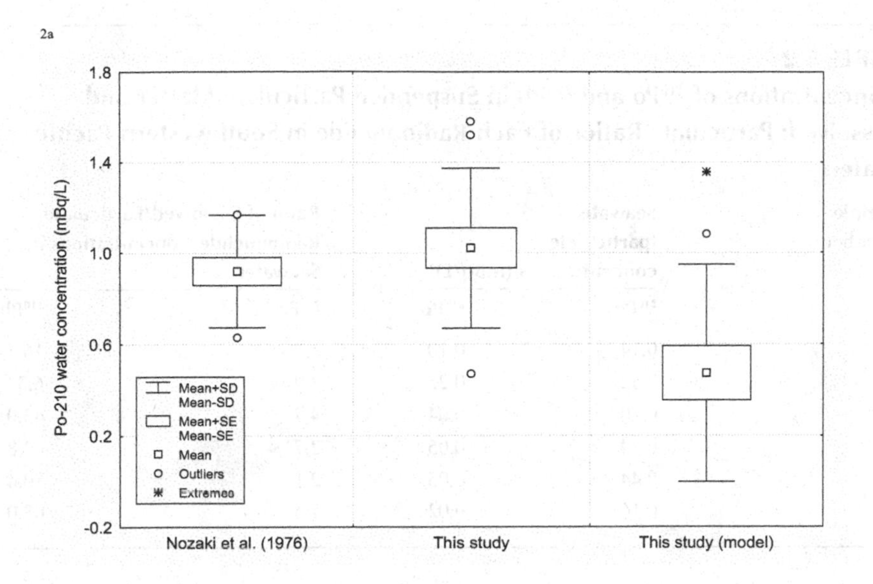

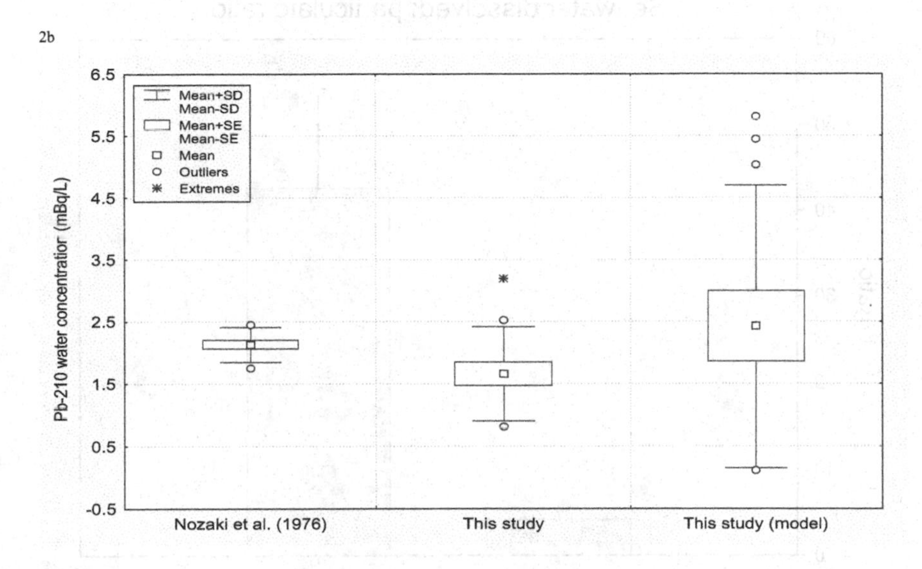

FIGURE 6.2 A comparison of mean water concentrations of (a) ^{210}Po and (b) ^{210}Pb as measured in the region shown in Figure 1 by Nozaki et al. (1976) and in this study along with the model-generated values.

values from the two data sets are elevated by approximately an order of magnitude compared to a typical concentration factor of 3×10^4, which the IAEA (2004) recommends as a measure of central tendency among the values assessed before 2004, as given in Table 6.3. Similarly, ^{210}Pb concentration factors in our study, ranging from $0.9–9.1 \times 10^4$ with a mean of 3.3×10^4, were substantially

TABLE 6.2
Concentrations of ^{210}Po and ^{210}Pb in Suspended Particulate Matter and Dissolved: Particulate Ratios of Each Radionuclide in Southwestern Pacific Waters

Sample Number	Seawater [particulate concentrations (mBq/L)]		Ratio of Dissolved/Particulate Radionuclide Concentrations in Seawater	
	^{210}Po	^{210}Pb	^{210}Po	^{210}Pb
1	0.14	0.10	5.0	16.7
2	0.52	0.25	2.9	6.3
3	0.10	0.04	4.7	43.0
4	0.34	0.05	2.7	63.8
5	0.44	0.05	2.1	50.4
13	0.16	0.02	7.6	63.0

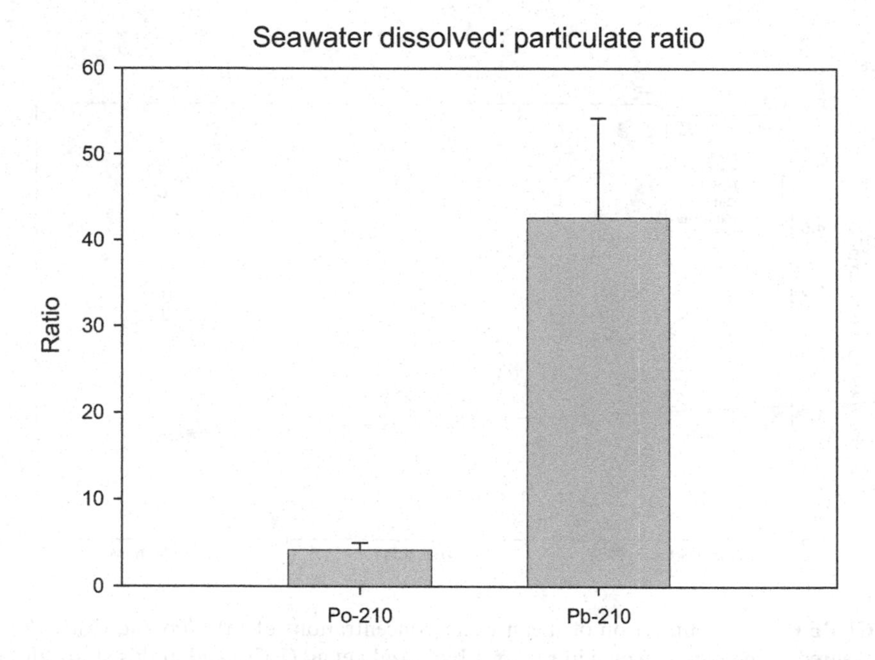

FIGURE 6.3 Mean ratios of dissolved:particulate concentrations of ^{210}Po and ^{210}Pb in seawater.

elevated compared to that (1×10^3), which was considered to be a typical value in marine waters (IAEA, 2004). Therefore, compared to typical ^{210}Po and ^{210}Pb concentration factors previously reported from other regions, our data from highly oligotrophic waters suggest that both radionuclides are considerably enhanced in zooplankton relative to their concentrations in ambient seawater.

TABLE 6.3
Comparison of Concentration Factors (CFs) for ^{210}Po and ^{210}Pb in Zooplankton from Southwestern Pacific Waters with Previously Reported Data Assessed by the IAEA (2004)

Radionuclide	Zooplankton CF	IAEA Recommended CF Value for Zooplankton (range of reported values)
^{210}Po	$0.7–3.3 \times 10^5$	3×10^4 (5×10^3–5×10^4)
^{210}Pb	$0.9–9.1 \times 10^4$	1×10^3 (5×10^2–2×10^3)

TABLE 6.4
Measurements of Rates of Fecal Pellet Production by Zooplankton (Mainly Copepods) from Southwestern Pacific Waters

Sample No.	Fecal Pellet Mass (mg dry)	Zooplankton Mass (g dry)	Period of Collection (hr)	Fecal Pellet Production Rate (mg dry fecal pellet/ g dry zooplankton/hr)
1	324	9.2	9.5	3.8
2	204	7.3	10.5	2.7
3	179	19.8	7	1.3
4	38	19.9	10.5	0.2
5	73	11.4	6.8	0.9
6	52	7.9	8	0.8
7	40	8.4	8	0.6
8	27	2.2	12	1.0
9	13	2.0	11.8	0.6
10	60	4.9	10	1.2
11	42	11.2	10.8	0.3
12	29	2.2	11.3	1.2
13	132	7.0	11.6	1.6
14	67	14.7	11	0.4

6.3.2 Fecal Pellet Production Rates

The measured fecal pellet production rates are shown in Table 6.4. The fecal pellet production rate was plotted against the total period of pellet production and collection (not shown), which varied between 6.75 and 12 hr, to show that there was no obvious pattern of reduced production rate per hour with an increasing total period of collection, a result that was confirmed by linear regression analysis ($p > 0.05$, $n = 14$). Rates varied between 0.18–3.78 (mean = 1.2) mg dry fecal pellet/g dry zooplankton/hr. To assess the validity of these estimates using production rates measured onboard, we have compared them with similar measurements of fecal pellet carbon production rate in tropical northeastern Pacific waters (0.03–0.04 μg at C/mg dry zooplankton/hr) reported by Small et al. (1983). Converting these latter rates to total dry weight values gives an average rate of 0.42 mg dry fecal pellet/g dry zooplankton/hr which is within the range of values measured in our study.

6.3.3 Comparisons of Modeled and Measured ^{210}Po and ^{210}Pb Water Concentrations

For each zooplankton sample for which biomass, fecal pellet production rate and their ^{210}Po and ^{210}Pb concentrations were measured, we have calculated an annual rate of removal of each radionuclide from a euphotic zone of one square meter surface area and extending to 100 m in depth. These values provided empirically derived, station-specific measures of loss-rate constants (k), so that according to the formula, Cw = R/k, the resultant ^{210}Po and ^{210}Pb water concentrations were predicted and then statistically compared with the measured empirical values. The calculated rates of ^{210}Po removal by zooplankton fecal pellets were around the range of 10^1–10^2 mBq g dry zooplankton^{-1} year^{-1}, which was similar to values previously determined for zooplankton from French Polynesian waters (Jeffree et al., 1997). In Figure 6.2 are shown the means and measures of variation for ^{210}Po and ^{210}Pb water concentrations, for both the modeled and the two sets of empirical data, namely, from Nozaki et al. (1976) and from this study. For both radionuclides, there is no significant difference between the modeled and each set of empirical values ($p > 0.05$), indicating a reasonably good predictive capability of the model for both ^{210}Po and ^{210}Pb. Variation about the mean is similar between our measured and modeled ^{210}Po water concentrations, but for ^{210}Pb, the modeled values showed a greater variance compared to the empirical values.

6.3.4 Relationship between Zooplankton Biomass and Their ^{210}Po Concentrations

In Figure 6.4 are shown the ^{210}Po concentrations in zooplankton plotted as a function of zooplankton biomass for 11 samples (Table 6.1) collected in the southwestern Pacific (this study) and, for comparison, the greater number of samples previously collected in French Polynesian waters (Jeffree et al., 1997). All the ^{210}Po concentrations in zooplankton from the southwestern Pacific fall within the range of those ^{210}Po values in French Polynesian zooplankton, which are also highly right-skewed in their frequency distribution of ^{210}Po concentrations (Figure 3a, Jeffree et al., 1997). A comparison of their median concentrations shows that the southwestern Pacific value of 1285 Bq kg^{-1} is comparable to the 1000 Bq kg^{-1} measured in the French Polynesian samples. Both these median values are much higher than a median value of 200 Bq kg^{-1} determined by Jeffree et al. (1997) for marine zooplankton from a broad range of other geographical locations as well as measures of central tendency in numerous

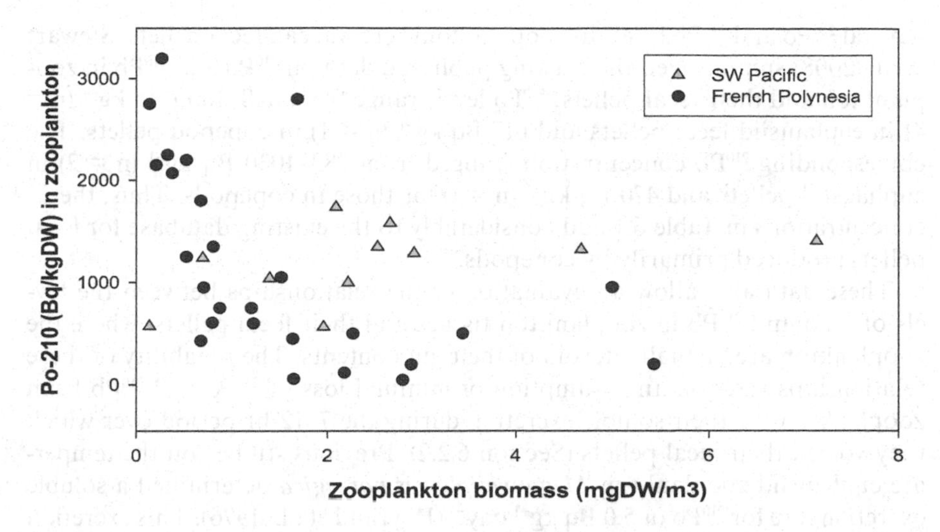

FIGURE 6.4 Plot of ^{210}Po concentrations in zooplankton as a function of their biomass for samples from the southwestern Pacific (this study) and French Polynesia (from Jeffree et al. 1997).

^{210}Po concentrations in various zooplankton species reviewed by Stewart et al. (2008). However, the highest concentration of ^{210}Po measured in southwestern Pacific zooplankton of 1736 Bq kg^{-1} is lower by nearly a factor of two than the maximum concentration for French Polynesian zooplankton (3200 Bq kg^{-1}) which show an increase in their ^{210}Po levels with declining biomasses, particularly within the range of 0.14–1 mg dry weight m^{-3}.

In our previous studies, the results of which motivated the present investigation, low and declining biomass and associated particle flux rates were associated with exponentially increasing ^{210}Po concentrations in zooplankton (Jeffree et al., 1997), and Cd, Co, Pb, Mn, Cr and Ni concentrations in the ambient seawater (Jeffree and Szymczak, 2000). In the highly oligotrophic waters of French Polynesia, the exponential increase of ^{210}Po concentrations in zooplankton becomes evident as their biomass declines from about 1–0.14 mg dry weight m^{-3}. In our southwestern Pacific study, although the overall range ran from 7.12 down to 0.14 mg m^{-3}, there were only two biomass values measured below 1 mg m^{-3} compared to 13 values used in our earlier study (Jeffree et al., 1997).

6.4 DISCUSSION

6.4.1 ^{210}Po and ^{210}Pb Concentrations in Zooplankton and Their Fecal Pellets

In contrast to considerable numbers of measurements of these radionuclides in seawater, there are few field measurements available in the literature of natural ^{210}Po and ^{210}Pb levels in zooplanktonic crustacean fecal pellets. Stewart et al. (2008) summarized the existing published data on ^{210}Po and ^{210}Pb in zooplankton and their fecal pellets. ^{210}Po levels range from 648–1000 Bq kg^{-1} (n = 4) in euphausiid fecal pellets and 617 Bq kg^{-1} (n = 1) in copepod pellets. The corresponding ^{210}Pb concentrations ranged from 383–1030 Bq kg^{-1} (n = 2) in euphausiid pellets and 470 Bq kg^{-1} (n = 1) for those in copepods. Thus, the 12 concentrations in Table 6.1 add considerably to the existing database for fecal pellets produced primarily by copepods.

These data also allow an evaluation of the relationships between the levels of ^{210}Po and ^{210}Pb in zooplankton tissues and their fecal pellets, where the zooplankton are virtually devoid of their gut contents. The reliability of these relationships rests on an assumption of minimal loss of ^{210}Po and ^{210}Pb from zooplankton via their soluble excretion during the 7–12-hr period over which they voided their fecal pellets (Section 6.2.2). Previous studies on the temperate euphausiid zooplankton *Meganyctiphanes norvegica* determined a soluble excretion rate for ^{210}Po of 5.0 Bq kg^{-1} day^{-1} (Heyraud et al., 1976). This excretion rate is nearly an order of magnitude less than the rate of loss via the fecal excretion route (Heyraud et al., 1976; Stewart et al., 2008), which is also comparable between tropical small and large zooplankton (mainly copepods) and temperate euphausiids (Small et al., 1983), thus adding support to the validity of the relationships determined in this current investigation between radionuclide levels in zooplankton and their fecal pellets.

The concentrations of ^{210}Po varied between 565–1736 Bq kg^{-1} in mixed zooplankton and 830–3107 Bq kg^{-1} in their fecal pellets. It is noteworthy that the mean value for ^{210}Po in fecal pellets from these oligotrophic waters is the highest reported concentration to date for fecal pellets and is a factor of 3 higher than corresponding levels in copepod pellets from the more mesotrophic waters of the northwestern Mediterranean Sea (Rodriguez y Baena et al., 2007). This observation suggests that ^{210}Po levels in the phytoplankton and particles ingested by these zooplankton are also significantly enhanced in these waters. Temporal variations in their concentrations may accordingly explain the range of ^{210}Po concentrations in zooplankton and their fecal pellets. It also noteworthy that the ^{210}Po:^{210}Pb ratios in zooplankton and fecal pellets are particularly elevated for samples taken in March 2001 (Table 6.1), although the seawater values are not also aberrant.

The mean ^{210}Po concentration in zooplankton was also found to be greater than the corresponding ^{210}Pb concentrations by a factor of about 5. This difference in concentrations was a little more pronounced in fecal pellets, by a factor of 5.7. This ratio is considerably enhanced compared to the ^{210}Po:^{210}Pb ratios reported for copepod fecal pellets (1.4) and euphausiid pellets (1.0–2.1) from the Mediterranean Sea (Rodriguez y Baena et al., 2007; Stewart et al., 2008). It is also higher than ratios ranging from 0.7–1.9 in salp and pteropod feces from the same waters (Krishnaswami et al., 1985; Rodriguez y Baena et al., 2007; Stewart et al., 2008) and much higher than the ratio of about 1.0 determined experimentally with Mediterranean euphausiids feeding on radio-labeled brine shrimp (Stewart et al., 2005).

From these few comparisons, it seems clear that highly oligotrophic waters can lead to enhanced ^{210}Po:^{210}Pb ratios in the feces of zooplankton inhabiting them, and it is likely that both the food source and the assimilative capacity of the zooplankton will have a strong effect on the fractionation of these two radionuclides in excreted feces.

Our measurements of zooplankton fecal pellets collected under field conditions have demonstrated their ability to highly concentrate ^{210}Po and ^{210}Pb relative to measured dissolved levels present in the water column. This is particularly the case for ^{210}Po concentrations in fecal pellets that are significantly elevated ($p < 0.01$) by a mean factor of 50% compared to their zooplankton producers. Such empirical results support the hypothesis of a zooplankton-mediated role in the determination of ^{210}Po water concentrations in the euphotic zone. They are also consistent with food and other ingested particulate matter being the major source of ^{210}Po and ^{210}Pb to zooplankton (Stewart and Fisher, 2003). In fact, enhanced concentrations of many biologically essential and non-essential trace elements in zooplankton fecal pellets, compared to corresponding levels in the organisms producing them and the prey they ingest, have been reported previously (Fowler, 1977), and it appears to be a common phenomenon for other natural radionuclides as well (Rodriguez y Baena et al., 2007; Stewart et al., 2008).

The capacity of zooplankton to appreciably influence ^{210}Po water concentrations in the euphotic zone of these oligotrophic waters was a major hypothesis of investigation in this empirical and associated modeling study. In this context, the results of our study indicate the following factors would *a priori* particularly enhance the potential of zooplankton to absorb and export ^{210}Po and also ^{210}Pb from the euphotic zone in these southwestern Pacific waters, namely,

1. The elevated accumulative capacity of zooplankton as expressed in their concentration factors for ^{210}Po and ^{210}Pb, both in absolute terms (CFs of 10^5 for ^{210}Po and 10^4–10^5 for ^{210}Pb) and relative to typical values in marine waters for zooplankton (CFs of 10^4 for ^{210}Po and 10^3 for ^{210}Pb; IAEA, 2004).
2. The further enhancement of ^{210}Po levels in fecal pellets relative to these elevated levels in zooplankton.
3. The heightened levels of ^{210}Pb in fecal pellets which were comparable to those ^{210}Pb levels determined in zooplankton.
4. The heightened ^{210}Po levels in zooplankton fecal pellets relative to those from zooplankton living in other more biologically productive marine waters.

6.4.2 Comparison of Modeled and Measured ^{210}Po and ^{210}Pb Concentrations in Seawater

The simple model we have employed has predicted a mean ^{210}Po water concentration that is about a factor of only 2 below the measured mean value. This modeled result is consistent with the hypothesis of a major role of zooplankton fecal pellets in determining ^{210}Po water concentrations, although the modeling was constrained by the availability of a single regional value of ^{210}Po atmospheric input (R), rather than site-specific values for R (Figure 6.2a, Table 6.1). Moreover, the model has been similarly able to predict mean measured ^{210}Pb water concentrations, based on its measured levels in zooplankton fecal pellets and their production rates (Figure 6.2b; Table 6.1). The use of the measured site-specific values for zooplankton in this model to predict mean ^{210}Po water concentrations gave values much closer to empirical values (by a factor of 2) than in a previous investigation. This previous investigation used the same model equation to predict ^{210}Po concentrations in French Polynesian zooplankton from predicted water concentrations (Jeffree et al., 1997), but (of necessity) based on non-site-specific, published values for (1) rates of fecal pellet production per unit biomass of copepod and (2) a ^{210}Po level in euphausiid fecal pellets. Although the ranges of modeled ^{210}Po zooplankton concentrations predicted in this previous study did encompass the measured levels, they varied by more than a factor of 4 from the empirical values.

The model used to predict ^{210}Po and ^{210}Pb water concentrations ascribed their loss from the euphotic zone as being solely due to the fecal pellets produced by zooplankton that were identified as being dominated numerically by calanoid and cyclopoid copepods. Additional export vectors for these radionuclides include sinking particulate aggregates, which are associated with phytoplankton blooms observed in this oceanic region (Caffin et al., 2018; Van Wambeke et al., 2018) and other zooplankton groups. Zooplankton taxa, such as salps and pteropods, have also been strongly implicated in the removal of ^{210}Po and ^{210}Pb from the euphotic zone due to their elevated concentrations in fecal pellets. For example, in a previous study in the Mediterranean, the fecal pellets of salps contained significantly higher levels of both ^{210}Po and ^{210}Pb compared to those of mixed-zooplankton samples, which were mainly composed of copepods (Farber Lorda et al., 2013). Both salps and pteropods were also present among the mixed zooplankton samples used in this study. However, their potential role in the removal of both radionuclides from the euphotic zone in the southwestern Pacific is likely to be minimal due to the following:

1. They were low in occurrence in all samples relative to the numerical dominance of copepods, which would indicate minor roles for them in the vertical flux of ^{210}Po and ^{210}Pb even if their fecal pellets were enhanced in both ^{210}Po and ^{210}Pb concentrations.
2. The presence of other important transfer vectors (including sinking particulate aggregates) for ^{210}Po and ^{210}Pb removal from the water column, which had not been included in our model predictions, would have resulted in modeled values for their water concentrations, which were appreciably elevated compared to the measured values. This would follow from the ^{210}Po and ^{210}Pb removal rates being underestimated by considering only zooplankton fecal pellets as their sole transport vector. The modeled mean ^{210}Po water concentration actually indicates

an enhanced role of zooplankton fecal pellets in the removal of ^{210}Po from water by a factor of 2. The modeled mean water concentration of ^{210}Pb slightly underestimates (about 40%) the role of zooplankton fecal pellets in ^{210}Pb removal from the water column.

These findings would discount any hypothesis of additional vectors of importance for the vertical transport of these radionuclides from the euphotic zone. However, the high variances around modeled mean water concentrations of ^{210}Po and particularly ^{210}Pb do detract from this conclusion.

6.4.3 Effect of Oligotrophy on ^{210}Po and ^{210}Pb Concentrations in Zooplankton

Our results for ^{210}Po and ^{210}Pb concentrations in zooplankton show a consistency with published values, in that the concentration factor for ^{210}Po is elevated relative to that for ^{210}Pb. However, they also contrast with previous reported concentration factors as values for both radionuclides are elevated by about an order of magnitude over what are considered concentration factors of central tendency in zooplankton (IAEA, 2004). Furthermore, the comparably elevated concentrations of ^{210}Po and ^{210}Pb in fecal pellets suggest that enhanced concentrations in southwestern Pacific zooplankton are a reflection of the heightened concentrations in their dietary phytoplankton and particulate material (Stewart et al., 2005).

In a previous investigation of French Polynesian zooplankton, concentration factors for ^{210}Po could not be calculated because ^{210}Po concentrations in the ambient water were not measured (Jeffree et al., 1997). In that study, the ^{210}Po concentrating mechanism was explained by low and declining biomass and associated particle flux rates that resulted in exponentially increasing ^{210}Po concentrations in zooplankton. The measurement of the elevated concentration factors for ^{210}Po in zooplankton in the present study may have identified another mechanism, or component of the same mechanism, that enhances its concentrations in these oligotrophic waters of the South Pacific.

The comparison in Figure 6.4 of the ^{210}Po concentrations in zooplankton from southwestern Pacific and French Polynesian waters as a function of their biomass did show very similar median values, although the southwestern Pacific zooplankton did not show the inverse relationship between zooplankton biomass and their ^{210}Po concentrations that had been established for French Polynesian zooplankton (Jeffree et al., 1997). However, there were only two samples taken in the southwestern Pacific within the lowest biomass range between 1–0.14 mg m^{-3}, where French Polynesian zooplankton steeply increased in their ^{210}Po concentrations. Although the ^{210}Po concentrations in zooplankton in these two samples do fall within the range of values previously measured in the French Polynesian zooplankton within the biomass range of 1–0.14 mg m^{-3}, we regard these as inadequate numbers from the southwestern Pacific to properly test the hypothesis of increasing ^{210}Po concentration in zooplankton with declining biomass.

Based on our albeit limited data sets, the oligotrophic waters of the southwestern Pacific and French Polynesia are associated with enhanced ^{210}Po levels in zooplankton. We conjecture that as oligotrophy is predicted to increase in the southern hemisphere with warming ocean temperatures (Sarmiento et al., 2004; Henson et al., 2016; Bindoff et al., 2019) ^{210}Po levels in zooplankton may also be enhanced over a greater oceanic region. Potentially enhanced ^{210}Po concentrations in zooplankton could be expected to be transferred to higher trophic-level organisms, exposing coastal communities who consume them to higher levels of this radiologically most important radionuclide in seafood (Aarkrog et al., 1997).

6.5 CONCLUSION

The results of this study have supported two hypotheses, namely, the enhancing effect of oligotrophy on ^{210}Po concentrations in zooplankton, and zooplankton's ability to substantially influence dissolved ^{210}Po concentrations in these southwestern Pacific oligotrophic waters by way of its vertical transport through the production of ^{210}Po-enriched fecal pellets. The results for ^{210}Pb are also consistent with these two tested hypotheses. The heightened ^{210}Po levels in zooplankton fecal pellets suggest that their dietary phytoplankton and particulate material are also elevated under oligotrophy. This hypothesis now warrants further investigation in support of a better understanding of the factors responsible for heightened ^{210}Po, and now ^{210}Pb, levels in zooplankton under oligotrophy.

ACKNOWLEDGMENTS

This study was conducted under a bilateral collaborative agreement between the Australian Nuclear Science & Technology Organization (ANSTO) and the French Institute of Research for Development (IRD), Noumea, New Caledonia. The authors acknowledge the support provided by IRD scientific colleagues and the crew of the RV *Alis*, to allow us to undertake this study. Professor Pere Masque is thanked for comments on an earlier draft.

REFERENCES

Aarkrog, A., Baxter, M.S., Bettencourt, A.O., Bojanowski, R., Bologa, A., Charmasson, S., Cunha, I., Delfanti, R., Duran, E., Holm, E., Jeffree, R., Livingston, H.D., Mahapanyawong, S., Nies, H., Osvath, I., Pingyu, Li, Povinec, P.P., Sanchez, A., Smith, J.N., and Swift, D. 1997. A comparison of doses from ^{137}Cs and ^{210}Po in marine food: A major international study. *Journal of Environmental Radioactivity* 34: 69-90.

Beasley, T.M., Heyraud, M., Higgo, J.J.W., Cherry, R.D., and Fowler, S.W. 1978. ^{210}Po and ^{210}Pb in zooplankton fecal pellets. *Marine Biology* 44: 325–328.

Behrenfeld, M.J., O'Malley, R.T., Siegel, D.A., McClain, C.R., Sarmiento, J.L., Feldman, G.C., Milligan, A.J., Falkowski, P.G., Letelier, R.M., Boss, E.S. 2006. Climate-driven trends in contemporary ocean productivity. *Nature* 444: 752–755.

Bindoff, N.L., Cheung, W.W.L., Kairo, J.G., Arístegui, J., Guinder, V.A., Hallberg, R., Hilmi, N., Jiao, N., Karim, M.S., Levin, L., O'Donoghue, S., Purca Cuicapusa, S.R., Rinkevich, B., Suga, T., Tagliabue, A., and Williamson, P. 2019. Changing ocean, marine ecosystems, and dependent communities. In: *IPCC Special Report on the Ocean and Cryosphere in a Changing Climate* [H.-O. Pörtner, D.C. Roberts, V. Masson-Delmotte, P. Zhai, M. Tignor, E. Poloczanska, K. Mintenbeck, A. Alegría, M. Nicolai, A. Okem, J. Petzold, B. Rama, and N.M. Weyer (eds.)]. In press.

Bopp, L., Monfray, P., Aumont, O., Dufresne, J-L., Le Treut, H., Madec, G., Terray, L., and Orr, J.C. 2001. Potential impact of climate change on marine export production. *Global Biogeochemical Cycles* 15: 81–99.

Caffin, M., Moutin, T., Foster, R.A., Bouruet-Aubertot, P., Doglioli, A.M., Berthelot, H., Grosso, O., Helias-Nunige, S., Leblond, N., Gimenez, A., Petrenko, A.A., de Verneil, A., and Bonnet, S. 2018. Nitrogen budgets following a Lagrangian strategy in the Western Tropical South Pacific Ocean: The prominent role of N2 fixation (OUTPACE cruise), *Biogeosciences Discussions*. doi:10.5194/bg-2017-468.

Cheng, L., Abraham, J., Hausfather, Z., and Trenberth, K.E. 2019. How fast are the oceans warming? *Science* 363: 128–129.

Cherry, R.D., Fowler, S.W., Beasley, T.M., and Heyraud, M. 1975. Polonium-210: Its vertical oceanic transport by zooplankton metabolic activity. *Marine Chemistry* 3: 105–110.

Chung, Y., and Wu, T. 2005. Large ^{210}Po deficiency in the northern South China Sea. *Continental Shelf Research* 25: 1209–1224.

Farber Lorda, J., Fowler, S.W., Miquel, J-C., Rodriguez y Baena, A., and Jeffree, R.A. 2013. 210Po/210Pb dynamics in relation to zooplankton biomass and trophic conditions during an annual cycle in northwestern Mediterranean coastal waters. *Journal of Environmental Radioactivity* 115: 43–52.

Fowler, S.W. 1977. Trace elements in zooplankton particulate products. *Nature* 269: 51–53.

Fowler, S.W. 2011. ^{210}Po in the marine environment with emphasis on its behaviour within the biosphere. *Journal of Environmental Radioactivity* 102: 448–461.

Fowler, S.W., and Fisher, N.S. 2004. Radionuclides in the biosphere. In: *Radioactivity in the Environment*, Vol. 6, Marine Radioactivity, Livingston, H.D. (Ed.), 167–203. Elsevier Ltd.: Oxford.

Henson, S., Beaulieu, C., and Lampitt, R. 2016. Observing climate change trends in ocean biogeochemistry: When and where. *Global Change Biology* 22(4). doi:10.1111/gcb.13152.

Heyraud, M., Fowler, S.W., Beasley, T.M., and Cherry, R.D. 1976. Polonium-210 in euphausiids: A detailed study. *Marine Biology* 34: 127–136.

IAEA. 2004. *Sediment Distribution Coefficients and Concentration Factors for Biota in the Marine Environment*, Technical Report Series No. 422, International Atomic Energy Agency: Vienna, Austria.

Jeffree, R.A., Carvalho, R., Fowler, S.W., and Farber-Lorda, J. 1997. Mechanism for enhanced uptake of radionuclides by zooplankton in French Polynesian oligotrophic waters. *Environmental Science & Technology* 31: 2584–2588.

Jeffree, R.A., and Szymczak, R. 2000. Enhancing effect of marine oligotrophy on environmental concentrations of particle-reactive trace elements. *Environmental Science & Technology* 34: 1966–1969.

Kim, G. 2001. Large deficiency of polonium in the oligotrophic ocean's interior. *Earth and Planetary Science Letters* 192: 15–21.

Krishnaswami, S., Baskaran, M., Fowler, S.W., and Heyraud, M. 1985. Comparative role of salps and other zooplankton in the cycling and transport of selected elements and natural radionuclides in Mediterranean waters. *Biogeochemistry* 1: 353–360.

La Rosa, J. 1976. A simple system for recovering zooplanktonic fecal pellets in quantity. *Deep-Sea Research* 23: 995–997.

Nozaki, Y., Ikuta, N., and Yashima, M. 1990. Unusually large ^{210}Po deficiencies relative to ^{210}Pb in the Kuroshio Current of the East China and Philippine Seas. *Journal of Geophysical Research* C4: 5321–5329.

Nozaki, Y., Thomson, J., and Turekian, K.K. 1976. The distribution of ^{210}Pb and ^{210}Po in the surface waters of the Pacific Ocean. *Earth and Planetary Science Letters* 32: 304–312.

Outola, I., Filliben, J., Inn, K.G.W., La Rosa, J., McMahon, C.A., Peck, G.A., Twining, J., Tims, S.G., Fifield, L.K., Smedley, P., Antón, M.P., Gascó, C., Povinec, P., Pham, M.K., Raaum, A., Wei, H.-J., Krijger, G.C., Bouisset, P., Litherland, A.E., Kieser, W.E., Betti, M., Aldave de las Heras, L., Hong, G.H., Holm, E., Skipperud, L., Harms, A.V., Arinc, A., Youngman, M., Arnold, D., Wershofen, H., Sill, D.S., Bohrer, S., Dahlgaard, H., Croudace, I.W., Warwick, P.E., Ikäheimonen, T.K., Klemola, S., Vakulovsky, S.M., and Sanchez-Cabeza, J.A. 2006. Characterization of the NIST seaweed Standard Reference Material. *Applied Radiation and Isotopes* 64(10–11): 1242–1247.

Peck, G.A. 2000. *The Determination and Behaviour of Radionuclides in Aquatic Systems*, PhD Thesis, University of Melbourne.

Peck, G.A., and Smith, J.D. 2000. Distribution of dissolved and particulate ^{226}Ra, ^{210}Pb and ^{210}Po in the Bismarck Sea and western equatorial Pacific Ocean. *Marine and Freshwater Research* 51: 647–658.

Poletico, C., Twining, J.R., and Jeffree, R.A. 1994. Comparison of concentrations of natural and artificial radionuclides in plankton from French Polynesian and Australian coastal waters. *Transactions of the American Nuclear Society* 70(Suppl. 1): 989–993.

Rodriguez y Baena, A.M., Fowler, S.W., and Miquel, J.C. 2007. Particulate organic carbon: Natural radionuclide ratios in zooplankton and their freshly produced fecal pellets from the NW Mediterranean (MedFlux 2005). *Limnology and Oceanography* 52: 966–974.

Rutgers van der Loeff, M.M., and Geibert, W.U. 2008. U- and Th-series nuclides as tracers of particle dynamics, scavenging and biogeochemical cycles in the oceans. In: *Radioactivity in the Environment,* Vol. 13, U—Th Series Nuclides in Aquatic Systems, Krishnaswami, S., and Cochran, J.K. (Eds.), 227–268. Elsevier: Oxford.

Sarmiento, J.L., Slater, R., Barber, R., Bopp, L., Doney, S.C., Hirst, A.C., Kleypas, J., Matear, R., Mikolajewicz, U., Monfray, P., Soldatov, V., Spall, S.A., and Stouffer, R. 2004. Response of ocean systems to climate warming. *Global Biogeochemical Cycles* 18: GB3003. doi:10.1029/2003GB002134.

Small, L.F., Fowler, S.W., Moore, S.A., and La Rosa, J. 1983. Dissolved and fecal pellet carbon and nitrogen release by zooplankton in tropical waters. *Deep-Sea Research* 30(12A): 1199–1220.

Small, L.F., Fowler, S.W., and Unlu, M.Y. 1979. Sinking rates of natural copepod fecal pellets. *Marine Biology* 51(3): 233–241.

Smith, J.D., and Hamilton, T.F. 1984. Improved technique for recovery and measurement of polonium-210 from environmental materials. *Analytica Chimica Acta* 160: 69–77.

Stewart, G.M., Cochran, J.K., Miquel, J.C., Masque, P., Szlosek, J., Rodriguez y Baena, A.M., Fowler, S.W., Gasser, B., and Hirschberg, D.J. 2007. Comparing POC export from ^{234}Th/^{238}U and ^{210}Po/^{210}Pb disequilibria with estimates from sediment traps in the northwest Mediterranean. *Deep-Sea Research* I(54): 1549–1570.

Stewart, G.M., and Fisher, N.S. 2003. Bioaccumulation of polonium-210 in marine copepods. *Limnology and Oceanography* 48: 2011–2019.

Stewart, G.M., Fowler, S.W., and Fisher, N.S. 2008. The bioaccumulation of U- and Th-series radionuclides in marine organisms. In: *Radioactivity in the Environment*, Vol. 13, U—Th Series Nuclides in Aquatic Systems, Krishnaswami, S., and Cochran, J.K. (Eds.), 269–305. Elsevier Ltd.: Oxford.

Stewart, G.M., Fowler, S.W., Teyssie, J.-L., Cotret, O., Cochran, J.K., and Fisher, N.S. 2005. Contrasting the transfer of polonium-210 and lead-210 across three trophic levels in marine plankton. *Marine Ecology Progress Series* 290: 27–33.

Stewart, G.M., Moran, S.B., and Lomas, M.W. 2010. Seasonal POC fluxes at BATS estimated from ^{210}Po deficits. *Deep-Sea Research* I 57: 113–124.

Tang, Y., Castrillejo, M., Roca-Marti, M., Masque, P., Lemaitre, N., and Stewart, G. 2018a. Distributions of ^{210}Po and ^{210}Pb activities along the North Atlantic GEOTRACES GA01 (GEOVIDE) cruise: Partitioning between the particulate and dissolved phase. *Biogeosciences*. doi:10.5194/bg-2018-210.

Tang, Y., Lemaitre, N., Castrillejo, M., Roca-Marti, M., Masque, P., and Stewart, G. 2018b. The export flux of particulate organic carbon derived from $^{210}Po/^{210}Pb$ disequilibria along the North Atlantic GEOTRACES GA01 (GEOVIDE) transect. *Biogeosciences*. doi:10.5194/bg-2018-309.

Towler, P.H., Smith, J.D., and Dixon, D.R. 1996. Magnetic recovery of radium, lead and polonium from seawater samples after preconcentration on a magnetic adsorbent of manganese dioxide coated magnetite. *Analytica Chimica Acta* 328: 53–59.

Turekian, K.K., Nozaki, Y., and Benninger, L.K. 1977. Geochemistry of atmospheric radon and radon products. *The Annual Review of Earth and Planetary Sciences* 5: 227–255.

Uddin, S., Behbehani, M., Alghadban, A., Sajid, S., Al-Zekri, W., Ali, M., Al-Jutaili, S., Al-Musallam, L., Vinod, V., and Al-Murad, M. 2018. ^{210}Po concentration in selected calanoid copepods in the northern Arabian Gulf. *Marine Pollution Bulletin* 133: 861–864.

Van Wambeke, F., Gimenez, A., Duhamel, S., Dupouy, C., Lefevre, D., Pujo-Pay, M., and Moutin, T. 2018. Dynamics and controls of heterotrophic prokaryotic production in the western tropical South Pacific Ocean: Links with diazotrophic and photosynthetic activity. *Biogeosciences* 15: 2669–2689.

Wang, X., Murtugudde, R., Busalacchi, A.J., and Le Borgne, R. 2005. De-coupling of net community production and new production in the euphotic zone of the equatorial Pacific: A model study. *Geophysical Research Letters* 32: L21601. doi:10.1029/2005GL024100.

Whicker, F.W., and Schultz, V. 1982. *Radioecology: Nuclear Energy and the Environment*. Vol. 2. CRC Press: Cleveland, USA.

Zar, J.H. 1996. *Biostatistical Analysis*. 3rd Ed. Prentice-Hall: NJ.

7 Marine Natural Products
Bioactive Isoprenoids and Provitamin A Carotenoids

J. Paniagua-Michel

CONTENTS

7.1 INTRODUCTION

The marine environment provides a cornucopia of natural products from marine organisms. Moreover, the oceans' immensity represents probably the most valuable natural resource and the greatest planetary coverage of nearly 70% of the earth's surface and more than 90% of the volume of its crust. The marine environment also holds the highest and unique biodiversity of habitats and richness, offering untapped sources and promising opportunities for drug discovery with superior chemical novelty (van der Westhuyzen 2018), as well as new bioactivities and functionalities (Reverter et al. 2020). Likewise, the biological and chemical biodiversity in the oceans is much greater than the biodiversity on land, of which the species variety has no terrestrial counterparts. This condition places oceans as unique for the discovery of biotechnologically important and novel bioactive metabolites. The particularity and ability of marine organisms to thrive and succeed in hostile, complex, and extreme habitats enable marine microorganisms to develop unique physiol-metabolic properties, namely, halophily, barophily, and most of the existing microbial nutrition modes. Therefore, these organisms in their *ad hoc* environments are responsible for producing a wide variety of specific and powerful active

DOI: 10.1201/9781003303909-7

substances that cannot be found anywhere else and may not be produced by their terrestrial counterparts (Hamed et al. 2015). The recent development of high-throughput screening in the bioprospecting of marine natural products and the respective new genomics and metabolomics tools maintain great expectations in the speed of the discovery of natural products and metabolites and their respective biosynthetic pathways (Reverter et al. 2020; Mohanty et al. 2020). Since 1970, marine natural product chemistry has played an important role in the discovery of new biologically active compounds and has continued to inspire the innovation and marine biotechnology of the future. Specifically, natural products of marine origin have proved effective as potent and affordable sources of novel and unique compounds with antioxidant, antitumor, immodulatory, and anti-inflammatory activities, as well as against pain effects and are promissory anticancer and anti-infective agents (Reverter et al. 2020; Mayer et al. 2020). Recently, the discovery of Aplidin, obtained from sea squirts called *Aplidium albicans* originally studied in samples from the Caribbean Sea, is under intense investigation by scientists around the world to treat the COVID-19 pandemic. The main goal is to seek treatments that protect specific human proteins from the disease that the coronavirus is targeting. Furthermore, the possible broad-spectrum antiviral activity of the didemnin compound plitidepsin extracted from *A. albicans* is in advanced clinical trials and is a promising therapeutic candidate against COVID-19 (White et al. 2021). On the other hand, the ubiquitous presence of marine natural products, such as isoprenoids isolated from marine organisms, has encouraged interest in many fields of biotechnologically important and industrial compounds. Generally, isoprenoids are also called terpenoids and are recognized as a structurally diverse and biofunctional group of metabolites produced in the three kingdoms of life, namely, bacteria, archaea, and eukaryotic microorganisms (Pérez-Gil et al. 2017). Isoprenoids are the largest class of metabolites found in nature, where the most widespread metabolites are in plants, fungus, algae, and marine organisms. These ancient molecules hold a peculiar characteristic of marine sources, beside the structural presence of halogen atoms, is the occurrence of nitrogen atoms, which are unusual or unknown functionally in terrestrial isoprenoids (Le Bideau et al. 2017). In recent decades, isoprenoids have been utilized in different industries as food supplements and in the pharmaceutical and chemical industries as advanced biofuels, as well as in foods and feeds. Many recent initiatives have been undertaken to assess the therapeutic potential of isoprenoids against cancer, and SARS-CoV-2 and as an anti-inflammatory with promissory results (Parveen 2021). During the last decade, orientated studies to test the efficacy of *cannabis* isoprenoids for the treatment of these new viral infections causing pandemics have been underway. In this regard, marine-derived isoprenoids from microalgae have led to the discovery of very interesting therapeutic candidates. Two such examples include trabectedin and aplidin. Likewise, marine natural products, mainly represented by provitamin A-carotenoid isoprenoids, are a reality and in continuous progress in the current development of pharmaceuticals, nutraceuticals, and environmental applications, as well as in various biotechnological developments. In this chapter, the main properties of isoprenoids, carotenoids, and provitamin A carotenoids from marine resources are presented and updated in basic and applied developments.

7.2 MARINE NATURAL PRODUCTS: A CORNUCOPIA OF ISOPRENOIDS (TERPENES)

The marine environment provides a cornucopia of natural products and metabolites from marine organisms, mainly represented by isoprenoids. These secondary metabolites had a relevant role in the early evolution of life on earth by protecting photosynthetic membranes of excess energy (Penuelas and Munne-Bosch 2005). Isoprenoids or terpenoids are extraordinarily diverse in chemistry, structure and function. These abundant molecules (estimates >70 000 distinct compounds) represent about 65% of all known natural products, and hundreds of new structures are registered every year. Isoprenoids otherwise known as terpenoids are defined as the hydrocarbons of plant origin of the general formula $(C_5H_8)_n$ as well as their oxygenated, hydrogenated and dehydrogenated derivatives. This prolific class of naturally occurring compounds derived from five carbon isoprene units whose evolutionary success is based in part on the simplicity of building molecules of different sizes. Isoprenoids are found in almost all classes of living organisms and are differentiated from one another by their basic skeletal structure associated with functional groups (Reyes and Leung 2018). According to Wallach and Rutzicka-Kubeczka (2010), all isoprenoids are derived from the universal five-carbon building blocks, isopentenyl diphosphate and its allylic isomer dimethylallyl diphosphate (Tholl 2015). They are synthesized in two compartmentalized pathways, namely, the mevalonic acid (MEV) that mostly occurs in cytosol and the methylerythritol (MEP) that is developed in chloroplasts. Isoprenoids are made up of different numbers of units (C_5H_8), which typically govern their classification into hemiterpenoids (C5), monoterpenoids (C10), sesquiterpenoids (C15), diterpenoids (C20), triterpenoids (C30), and tetraterpenoids (C40; Wang et al. 2021). Hence, isoprenoids are classified into six major classes, namely, steroids, taxanes, tocopherols, artemisinins, cannabinoids, and ingenanes (Reyes and Leung 2018). The relevant environmental presence of isoprenoids contributes to the colors of plants and algae, flavors, and fruits and are considered as essential for the development of plants and algae. These secondary metabolites are particularly interesting in terms of diversity and originality and have wide applications in several bioeconomies at local, regional, and global levels. These fascinating molecules in most cases display particular features from marine organisms, with prominent biological activity apart of their ecological role in the marine environment. Their main role in human and animal health is mainly as additives in food, feed, and pharmaceutical products, including beneficial properties as anticancer, antioxidant, and anti-inflammatory functions. Among the commercial and industrial isoprenoid products intended for pharmaceutical, chemical, food, environmental, and biomedical applications, the ones such as flavorings, fragrances, and mainly those used as food coloring agents are currently benefited. The ubiquitous presence of isoprenoids in almost all classes of living organisms can be identified as contributing to the flavor, scents, colors or volatile compounds of plants leaves, flowers, and fruits, as well as in important functions in extremophile marine organisms as in algae and bacteria. These molecules also play protective functions against plants diseases and infestations and in animals are important precursors of steroids and sterols (Reyes and Leung 2018). For example, the sesquiterpenoid artemisinin is used to cure malaria, as well as the diterpenoid taxol and the meroterpenoid vincristine used for the treatment of cancer. Of the marine isoprenoids, cortistatin has potent antiangiogenic activity and may be useful in the treatment of cancer and macular degeneration (Reyes and Leung 2018). Among the carotenoids, α-carotene, β-carotene, β-cryptoxanthin, lutein, lycopene, and zeaxanthin are the most common dietary carotenoids. Carotenoids can be also further classified into provitamin A, which may exist in the forms: beta-carotene, alfa-carotene, gamma-carotene, and beta-cryptoxanthin that can each be converted by the body to retinol (vitamin A). Lutein, zeaxanthin, and lycopene have no vitamin A activity. This fat-soluble vitamin has a relevant role in reproduction, vision, and immune function in different forms as retinol, retinal, and retinoic acid or retinyl ester. In this context, marine natural products, mainly represented by isoprenoids provitamin A carotenoids are considered relevant in the current development of pharmaceuticals, nutraceuticals, and environmental applications.

7.3 MARINE ISOPRENOIDS AND PROVITAMIN A CAROTENOIDS

Among isoprenoids, carotenoids are a subfamily of a very diverse class of pigments that are primarily synthesized by all photosynthetic organisms, including plants, microalgae, some non-photosynthetic bacteria, and fungi (Saini and Keum 2019). Carotenoids are lipophilic isoprenoid compounds comprising a polyene backbone containing a conjugated chain of alternating double and single carbon–carbon bonds that may or may not be cyclized at one or both ends (Canniffe and Hitchcock 2021). Because of their structural composition of 9 to 11 conjugated double bonds, carotenoids functions as an antioxidant by exerting quenching singlet oxygen and scavenging free radicals (Eggersdorfer and Wyss 2018). The all-trans form more commonly occurs in nature and is characteristic of photosynthetic carotenoids (Paniagua-Michel et al. 2012), but other mono- and poly-*cis* geometric isomers can be products of enzymatic activity or abiotically (e.g., by light, oxygen, or heat). Most carotenoids are tetraterpenes (they are produced from four C_{10} terpenes, each of which is composed of two C_5 isoprene building blocks) containing 40 carbon atoms and eight isoprene units (Canniffe and Hitchcock 2021). The most characteristic feature of carotenoids is the long series of conjugated double bonds forming the central part of the molecule. This gives them their shape, chemical reactivity, and light-absorbing properties (Paniagua-Michel 2020). Based on their structure and functions, these carotenoids can be divided into two major groups: (1) Hydrocarbon carotenoids are referred to as carotenes (such as phytoene, lycopene, α-carotene, and β-carotene) or primary carotenoids, including violaxanthin, antheraxanthin, and zeaxanthin, lutein, and neoxanthin, which are associated with structural and functional components in the photosynthetic apparatus and primarily function as light-harvesting pigments (Mussagy et al. 2019). (2) The secondary carotenoids are oxygenated derivatives of hydrocarbon carotenoids are known as xanthophylls (e.g., lutein, canthaxanthin, and astaxanthin), which also contain oxygen and are thus more polar. More than 1100 carotenoids with different structures and biological functions have been discovered from various sources (Liu et al. 2021). The main properties of carotenoids are largely due to its structure and lay the bases for their complex and unique biological functions, β-carotene has a provitamin A function, while lutein/zeaxanthin constitute macular pigment in the eye (Eggersdorfer and Wyss 2018). Marine natural carotenoids have main advantages when compared to the controversial synthetic pigments that are associated to serious toxic effects in the human body. These fat-soluble pigments possess several bioactivities, including anticancer, antioxidant, anti-inflammatory, and antiviral activities. Recently, the use of natural pigments has advanced in several applications as nutraceuticals and healthy functional foods, natural food colorants, and cosmetic products for safety and human health. For instance, the lutein benefit of reducing the progression of age-related macular eye disease and cataracts is strengthening in its reality (Eggersdorfer and Wyss 2018). In addition, there is evidence that carotenoids have beneficial effects in addition to eye health, also producing improvements in cognitive function and cardiovascular health, and may help prevent some types of cancer (Eggersdorfer and Wyss 2018). Among all carotenoids, humans utilize more than 30 of them to protect various organs against oxidative stress and inflammation that cause accelerated aging, cancers, cardiovascular diseases, diabetes, metabolic syndrome, and various age-related pathologies, such as Alzheimer's disease and age-related macular degeneration (Eggersdorfer and Wyss 2018). Furthermore, carotenoids have several physiological effects, such as those on the immune system, reproduction, lipid metabolism, and photoprotection in skin; diseases such as adiposity, obesity, diabetes, cardiovascular disease, hypertension, atherosclerosis, and cancer; and inflammation (Nakano and Wiegertjes 2020)

7.4 THE PROVITAMIN A CAROTENOIDS: THE KEY ROLE OF THEIR RING STRUCTURE

Carotenoids have very important functions in human health. Because of their potent antioxidant properties and provitamin A, carotenoids are considered important molecules of vital importance in protecting against chronic degenerative diseases, such as aging, cancer, cataract, cardiovascular, and neurodegenerative diseases. Because humans and animals cannot synthesize carotenoids, their diet is vital to ingesting them and then using them for important functions, such as visual pigments, antioxidants, or colorants. Beta-carotene, alpha-carotene, and beta-cryptoxanthin have provitamin A activity.

Zeaxanthin and lutein are the major carotenoids in the macular region (yellow spot) of the retina in humans.

Humans' consumption of foodborne carotenoids in their diet is mainly represented by β-carotene, besides lycopene, lutein, zeaxanthin, β-cryptoxanthin, and α-carotene, which roughly account for 90% of circulating carotenoids. During circulation, carotenoids are absorbed and bound to lipoproteins and then targeted to different tissues (liver, lung, macula, prostate, skin, adipose, brain, skin) playing functional and bioactive performance. Carotenoids can be divided into vitamin A precursor compounds and nonvitamin A precursor compounds according to whether carotenoids can be broken to form vitamin A (Liu et al. 2021). For instance, less than 10% of the carotenoids have been reported as metabolized to retinol and act as vitamin A precursors. In blood plasma, the predominant carotenoids represent about 90%, namely, β-carotene, lycopene, lutein, β-cryptoxanthin, and α-carotene. Carotenoids can be further classified into provitamin A carotenoids and exist in four forms: beta-carotene, alpha-carotene, gamma-carotene, and beta-cryptoxanthin that can each be converted to retinol (vitamin A). The other group or non-provitamin A carotenoids (e.g., lycopene, lutein, zeaxanthin, astaxanthin) cannot be converted to retinal or retinol because they lack the nonsubstituted β-ionone ring structure. The provitamin A activity of carotenoids confers them the ability to prevent and resist serious human health disorders, as well as other nutritional value (Liu et al. 2021). Provitamin A carotenoids mean that they can be converted in the body to vitamin A. Among provitamin A carotenoids, β-carotene may be used to provide all or part of the vitamin A in multivitamin supplements, which vitamin A activity from supplements may be much higher than that of β-carotene from foods. This vitamin is recognized as essential for normal growth and development, immune system function, and vision. Currently, the only essential function of carotenoids recognized in humans is that of provitamin A carotenoids (α-carotene, β-carotene, β-cryptoxanthin) to serve as a source of vitamin A.

7.5 MARINE CAROTENOIDS: USES AND APPLICATIONS

Until now, carotenoids are ecologically friendly functional feed additives in the aquaculture and health food industries. Marine pigmented biomass has recently emerged as a natural source of valuable carotenoids (Torregrosa et al. 2018), mainly in aquaculture considered a major user of pigments to improve organoleptic properties and health of livestock. In salmon industry, carotenoids have been used successfully as a feed ingredient, which mostly is dependent on natural and synthetic carotenoid sources. In crustaceans' aquaculture, such as in *Penaeidae* (*Penaeus vannamei* and *Penaeus japonicus monodon*), astaxanthin is the main and predominant carotenoid commercially exploited. Fish and crustaceans uptake dietary carotenoids, which follow a series of metabolic transformations following oxidative or reductive pathways. In aquaculture activities, astaxanthin-enriched feeds are used to improve the antioxidative state and immune system, which represents disease resistance, growth performance, survival, and improved egg quality in cultured fish and crustaceans without exhibiting any cytotoxicity, deleterious, or side effects (Nakano and Wiegertjes 2020). Moreover, this red pigment is reported as an effective modulator of gene expression accompanying alterations in signal transduction by regulating reactive oxygen species production (Nakano and Wiegertjes 2020). The ubiquitous presence of carotenoids in the marine environment is represented by algae. Particularly, marine microalgae are considered the supreme source of carotenoids, including their bioactive members. The multiple metabolic modes of microalgae, including autotrophic, heterotrophic, and mixotrophic, confer their ability to produce a wealth of biological and value-added products according to their origin and culture conditions. Among the strategies to assure a reliable and constant production of carotenoids, the cultivation and respective downstream processes of marine microorganisms, especially microalgae, offer several advantages over the other approaches that have been explored previously (Torregrosa et al. 2018). Marine natural microalgae are preferred because of its sustainability and rapid turnover rates, which makes them a suitable alternative to synthetic analogs and to terrestrial sources. Compared with plants microalgae grow faster, particularly those strains belonging to chlorophyceae, namely, *Dunaliella salina* and the freshwater species *Haematococcus pluvialis*. Both strains represent the higher content of natural β-carotene and astaxanthin, respectively. The microalgae *D. salina* is the supreme source of natural β-carotene widely utilized as provitamin A and in multiple applications as a coloring agent, cosmoceutical, nutraceutical and cancer prevention programs. Astaxanthin, a red hydrosoluble C40 carotenoid, is one of the most abundant marine carotenoids that is currently used as a food and feed additive and is an important component for pharmaceutical and cosmetic applications with antioxidant activities. The commercial production of vital microbial carotenoids at an industrial scale can be consulted in recent updated reviews (Saini and Keum 2019).

7.6 ECONOMIC AND POTENTIAL MARKETING VALUE OF BIOACTIVE CAROTENOIDS

The biotechnological demand for these pigments has received considerable attention because of their potential beneficial uses in human health care, food processing, pharmaceuticals, and cosmetics (Torregrosa et al. 2018). Likewise, the industrial production of carotenoids has been reported currently in 80–90%, which are synthetically produced by chemical synthesis. However, the high demand for these pigments, mainly for naturally produced carotenoids, is increasing due to the health concern of synthetic counterparts (Saini and Keum 2019). The total amount of carotenoids available from nature has been estimated to be more than 100 million tons (Nakano and Wiegertjes 2020). The world's high demand for these bioactive compounds is mainly covered by Europe and the United States, which are the most important and highest markets for carotenoids (Torregrosa et al. 2018). The global market for carotenoids has been recently calculated to be higher than US$1.5 billion (Nakano and

Wiegertjes 2020). According to Liu et al. (2021), β-carotene, canthaxanthin, and astaxanthin are the most widely used pigments in the market and have the highest added value (Liu et al. 2021). In the last decade, significant progress in the production of carotenoids due to their powerful antioxidant, vitamin A, anti-inflammatory and cell-signaling activities have been developed to meet the exceptionally high demands of dietary supplements, food coloring, pharmaceuticals, cosmetics, and animal feed. The global market value for carotenoids in 2017 was calculated at nearly US$1.5 billion, with an estimated growth projection in 2022 to US$2.0 billion and with a compound annual growth rate (CAGR) of 5.7% for the 2017–2022 period (Saini and Keum 2019). This market value according to Saini and Keum (2019) is presented by capsanthin, astaxanthin, β-carotene, lutein, lycopene, and canthaxanthin, which together share nearly 90% of the total market value. On the other hand, based on their projections, animal feed based on poultry and aquatic animals is the most prominent and demanding sector, about 41% of the total revenue share, which comes after food and dietary supplements. In the case of capsanthin, it alone is projected to reach US$385 million by 2022 at a compound annual growth rate of 5.1%. In the case of the astaxanthin market, which was estimated to be US$288.7 million in 2017, it is projected to reach US$426.9 million by 2022 at a CAGR of 8.1% (Saini and Keum 2019; BBC Research 2018).

7.7 CONCLUSION

The biotechnology and industrial applications of marine natural carotenoids are under intense research and development because of their potent biological activities. Likewise, biomedical research on these fat-soluble compounds is very scarce, mainly in those already reported as bioactive as well as in new and nonbioactive carotenoids. The development of bioprospection and bioscreening studies leading to the exploration of new extremophiles environments for carotenoids with new properties must be encouraged as well as the metabolic pathways associated with their production. The recent developments in exploring new nutrition modes in the producer's species of bioactive carotenoids must contribute to increasing production yields, as well as the progressive studies in genomics and metabolomics approaches.

REFERENCES

Ameri, A. 2014. Marine Microbial Natural Products. *Jundishapur Journal of Natural Pharmaceutical Products* 9(4): e24716.

BBC Research. 2018. The Global Market for Carotenoids, FOD025F. *Dialog*, https://www.bcre search.com/title of subordinate document.

Canniffe, D.P., and Hitchcock, A. 2021. Photosynthesis | Carotenoids in Photosynthesis – Structure and Biosynthesis. In J. Jess (ed.), *Encyclopedia of Biological Chemistry* (3rd ed.). Elsevier, pp. 163–185.

Eggersdorfer, M., and Wyss, A. 2018. Carotenoids in Human Nutrition and Health. *Archives of Biochemistry and Biophysics* 15: 18–26.

Hamed, I., Ozoqul, F., Ozoqul, Y., and Regenstein, J.M. 2015. Marine Bioactive Compounds and Their Health Benefits: A Review. *Comprehensive Reviews in Food Science and Food Safety* 14: 446–465.

Le Bideau, F., Kousara, M., Chen, L., Wei, L., and Dumas, F. 2017. Tricyclic Sesquiterpenes from Marine Origin. *Chemical Reviews* 117: 6110–6159.

Liu, C.H., Hu, B., Cheng, Y., Guo, Y., Yao, W., and Qian, H. 2021. Carotenoids From Fungi and Microalgae: A Review on Their Recent Production, Extraction, and Developments. *Bioresource Technology* 337: 125–398.

Mayer, M.S., Guerrero, A.J., Rodriguez, A.D., Taglialatela-Scafati, O., Nakamura, F., and Fusetani, N. 2020. Marine Pharmacology in 2014–2015: Marine Compounds with Antibacterial, Antidiabetic, Antifungal, Anti-inflammatory, Antiprotozoal, Antituberculosis, Antiviral and Anthelmintic Activities; Affecting the Immune and Nervous Systems, and other Miscellaneous Mechanisms of Action. *Marine Drugs* 18: 5 doi: 10.3390/md1801000.

Mohanty, I., Podell, S., Biggs, J.S., Garg, N., Allen, E.E., and Agarwal, V. 2020. Multi-omic Rofiling of *Melophlus* Ssponges Reveals Diverse Metabolomic and Microbiome Architectures that are Non-overlapping with Ecological Neighbors. *Marine Drugs* 18: 124.f

Mussagy, C.U., Winterburn, J., Carvalho, V.S., and Pereira, J.F.B. 2019. Production and Extraction of Carotenoids Produced by Microorganisms. *Applied Biochemistry and Biotechnology* 103: 1095–1114.

Nakano, T., and Wiegertjes, G. 2020. Marine Pigment Diversity. *Marine Drugs* 18: 1–17.

Paniagua-Michel, J. 2020. Dunaliella salina: Sustainable Source of B-Carotene. In G.A. Ravinshankar and A. Ranga Rao (ed.), *Handbook of Algal Technologies and Phytochemicals*. Boca Raton, FL: CRC Press, pp. 1–9.

Paniagua-Michel, J., Olmos, J., and Ruiz, M. 2012. Pathways of Carotenoids Biosynthesis in Bacteria and Microalgae. In J.L. Barredo (ed.), *Microbial Carotenoids From Bacteria and Microalgae*. New York: Springer Humana Press, pp. 1–12.

Parveen, S. 2021. *Introductory Chapter: Terpenes and Terpenoids*. doi: 10.5772/ *Intechopen*.98261.

Penuelas, J., and Munne-Bosch, S. 2005. Isoprenoids: An Evolutionary Pool for Photoprotection. *Trends in Plant Science* 10: 166–169. doi: 10.1016/j.tplants.2005.02.005.

Pérez-Gil, J., Rodríguez-Concepción, M., and Vickers, C.E. 2017. Biogenesis of Fatty Acids, Lipids and Membranes. In O. Geiger (ed.), *Handbook of Hydrocarbon and Lipid Microbiology*. doi: 10.1007/978-3-319-43676-0_6-1.

Reverter, M., Rohde, S., Parchemin, C.H., Tapissier-Bontemps, N., and Schupp, J.P. 2020. Metabolomics and Marine Biotechnology: Coupling Metabolite Profiling and Organism Biology for the Discovery of New compounds. *Frontiers in Marine Science* 7: 1–8.

Reyes, B.A.S., and Leung, A.B. 2018. Selected Phyto and Marine Bioactive Compounds: Alternatives for the Treatment of Type 2 Diabetes. *Studies in Natural Products Chemistry* 55: 111–143.

Saini, R.K., and Keum, Y.S. 2019. Microbial Platforms to Produce Commercially Vital Carotenoids at Industrial Scale: An Updated Review of Critical Issues. *Journal of Industrial Microbiology & Biotechnology* 46: 657–674.

Tholl, D. 2015. Biosynthesis and Biological Functions of Terpenoids in Plants. *Advances in Biochemical Engineering/Biotechnology* 148: 63–106.

Torregrosa, J., Montero, Z., Fuentes, J.L., Reig, M., Garbayo, I., Vílchez, C., and Martínez-Espinosa, R.M. 2018. Exploring the Valuable Carotenoids for the Large-Scale Production by Marine Microorganisms. *Marine Drugs* 16: 203. https://doi.org/10.3390/md16060203.

van der Westhuyzen, A.E., Frolova, L.V., Kornienko, A., and van Otterlo, W.A.L. 2018. The Rigidins: Isolation, Bioactivity, and Total Synthesis-Novel Pyrrolo[2,3-d] Pyrimidine Analogues Using Multicomponent Reactions. *Alkaloids: Chemistry & Biology* 79: 191–220.

Wang, Z., Zhang, R., Yang, Q., Zhang, J., Zhao, Y., Zheng, Y., and Yang, J. 2021. Recent Advances in the Biosynthesis of Isoprenoids in Engineered Saccharomyces cerevisiae. *Advances in Applied Microbiology* 114: 1–35.

White, K.M., Rosales, M., Yildiz, S., Kehrer, T., Miorin, L., Moreno, E., . . . García-Sastre, A. 2021. Plitidepsin has Potent Preclinical Efficacy against SARS-CoV-2 by Targeting the Host Protein eEF1A. *Science* 371: 926–931.

8 Utilization of Fisheries' By-Products for Functional Foods

Muhamad Darmawan, Nurrahmi Dewi Fajarningsih, Sihono, and Hari Eko Irianto

CONTENTS

8.1 INTRODUCTION

Fisheries sectors keep demonstrating a growing and significant role in providing food and nutrition for the world. According to the Food and Agriculture Organization (FAO), in 2018, the total world fisheries and aquaculture production reached 178.5 million tons, comprising 96.4 million and 82.1 million tons from capture fisheries and aquaculture, respectively. Asian countries produced 69% of the global fisheries, of which 35% was produced by China as the biggest fish producer in the world, followed by the Americas (14%), Europe (10%), Africa (7%) and Oceania (1%). About 87% (156 million tons) of production was used for human consumption while the rest was ordained for nonfood uses (FAO, 2020).

The growth of fisheries production in the world has led to the increasing amounts of fisheries by-products, which is about 20 million tons, that if not being utilized optimally will lead to the environmental problem of waste (Khawli et al., 2020). The fish by-products amounts vary depending on the species, fish size, fishing ground and season (Rustad et al., 2011). In order to provide a wide variety of fisheries' products for human consumption, the fisheries'

DOI: 10.1201/9781003303909-8

product-processing industries have significantly grown globally (Nawaz et al., 2020). Besides by-catch fish, a great number of by-products are derived from the fisheries' processing, that is, heads, skin, bones, scales and viscera, which is between 30–85% of the processed fish (Khawli et al., 2019).

Historically, fisheries' by-products were used directly as feed for aquaculture or livestock, fertilizers or used for silage production (Hardy, 1992). However, about two decades ago, the utilization of fish by-products as a significant source for various value-added products has been gaining much attention (FAO, 2020; Khawli et al., 2019). Fisheries' by-products that are still very rich in several valuable materials, such as high-quality proteins, lipids, vitamins, minerals, enzymes and bioactive compounds, may be employed in various products that are not only nutritious but also associated with many health benefits such as antioxidant, antihypertension, antihyperglycemic and anti-inflammatory (Khawli et al., 2020; Caruso et al., 2020; Kim and Mendis, 2006; Rustad, 2003). Currently, research and development activities regarding utilizing fisheries' by-products in various functional foods are emerging. Furthermore, utilizing the by-products will not only convert what was once underutilized or even waste into high-value-added products but also minimize their negative impact on the environment.

8.2 SOURCE OF FISHERIES' BY-PRODUCTS

According to Kaanane and Hind (2020), there is no one definition of *marine by-products*. In general, the term *by-products* is designated for all unused parts during the production that can be recovered into something useful. Thus, due to their economic potential, fisheries' by-products should be distinguished from fisheries' waste (Irianto et al., 2014). According to Ramírez (2013), by-products of the fisheries processing industry are material that still can be utilized into secondary products for human consumption (co-products), while the term *waste* is normally used when the secondary products were intended for nonhuman consumption.

The valorization of fisheries by-products is an encouraging step from all aspects. Converting by-products into value-added goods not only will create a friendlier industry to the environment but also will give more economic benefit. It provides a better solution for the blue sector and an initiative toward zero waste (Irianto et al., 2014; Nawaz et al., 2020). The enormous amount of fishery by-catch or processing by-products that are continuously being produced needs to be managed to avoid environmental problems and keep resource sustainability (Caruso et al., 2020).

Irianto et al. (2014) classified fisheries' by-product sources into five different categories. The first source is by-products from utilizing fisheries resources. The capture fisheries donate a massive amount of by-product material in the form of whole fish from many kinds of species, which all depend on the type of fishing gear and mesh size used. On the other hand, the donation from the aquaculture fisheries is considered to be none. According to an FAO publication concerning the third assessment of global marine fisheries discards, an estimate of annual discards by marine commercial fisheries between 2010–2014 reached about 9.1 million tons (10.1% of yearly catches; FAO, 2019). The second fisheries' by-products source is from the processing industries. The remaining processing industries of shrimp, frog, and fish fillet for frozen products may reach up to 35 % of the initial weight. According to Nawaz et al. (2020), the growth of the processing sector resulted in massive by-product or waste production, which is about 20–80% of the landed weight depending on the fish species and processing types. According to Caruso et al. (2020), the fish processing industry produces a range from more than 25–70% of by-products made of viscera, heads, scales, bones, and the like. That considerable amount represents a significant waste worthy of being valorized (Caruso et al., 2020). The third source is by-products produced from secondary products processes. The secondary products are goods co-produced by the industry to valorize all the materials involved in the processing. The fourth source of by-products is by-products from surplus during the high fishing season. Usually, this condition happens when the industry could not absorb the high volume of fish landed. The last source of fisheries' by-products comes from the remainder of distribution or marketing where the quality of the fish product was dropped due to unproperly handling during the distribution or marketing. This kind of product could be used as raw materials to produce fish meals or other products.

Nawaz et al. (2020) categorized the source of by-products into two categories: organic waste and inorganic waste. Organic-source by-products include skin, viscera, fish meat (red and white), and a part of the scales, bones and fins, while inorganic by-products include the fishbone or shrimp and crab shell. Fishbones contain 60–70% minerals, including calcium, phosphorous and hydroxyapatite (Kim and Mendis, 2006). Both sources of by-products, either organic or inorganic waste, could significantly benefit the industry if they were utilized properly. In addition to that, the environmental issues coming from fisheries' waste could be overcome or minimized to achieve sustainability in fisheries' activities.

8.3 UTILIZATION OF FISH MARINE BY-PRODUCTS AS FUNCTIONAL FOODS

8.3.1 Gelatin and Collagen

Gelatin is a fibrous protein derived from hydrolytic degradation of collagen, which is obtained from cartilages, skin, bones and connective tissues of various animals (Akbar et al., 2017). As a functional biopolymer, gelatin has very extensive applications in industries ranging from pharmaceuticals, cosmetics, materials, foods and photography depend on their rheological properties, that is, transparency, solubility, viscosity, gel strength and thermal stability. Moreover, the application of gelatin as functional foods is also expanding (Karim and Bhat, 2009; Gomez-Guillen et al., 2011). The global demand of gelatin is constantly increasing. World gelatin production reached 450,000 tons in 2018 with an estimated value of US$4.52 billion (Tkaczewska et al., 2018).

In 2010, the world's major gelatin sources are pig skin (46%), bovine hide (29.4%), and pork and cattle bones (23.1%), while gelatin from fish only contributed less than 1.5% of the total gelatin production. However, due to religious and ethical issues, the interest of fish gelatin has risen. Therefore, researchers have been exploring different fish species, utilizing fisheries' by-product and optimizing the extraction of fish gelatin (Karim and Bhat, 2009). Fisheries' by-products are potential sources to be utilized in gelatin, not only because it is a halal source but also because it is a safer and healthier material with bioactive properties (Cheng et al., 2009).

Collagen and gelatin-derived peptides show various bioactivities, such as a capacity for binding minerals, being antibacterial, having lipid-lowering effect, immunomodulatory activities and beneficial effects on bone and skin (Gomez-Guillen et al., 2011). Antioxidant and antihypertensive activity were the most studied area of the gelatin-derived peptides. Gelatin peptides have shown better antioxidant and antihypertensive activity compared to fish muscle peptides (Kim and Mendis, 2006). The gelatin antioxidant activities have been associated with their repeated unique Gly-Pro-Hyp peptide sequences. The antioxidant activity of fish gelatin hydrolysates was reported to exhibit the protection of mice skin against ultraviolet radiation damage (Hou et al., 2009). Peptide that was hydrolized using alcalase from gelatin of Skipjack *Katsuwonus pelamis* exhibits higher antioxidant activities compare to peptides hydrolized using other enzime (Aberoumand, 2010). The strong antioxidant activities, mainly in the 2,2-azinobis (3-ethylbenzothiazoline-6-sulfonic acid) (ABTS) assay, have been demonstrated by derived peptides from both collagen and gelatin obtained from yellowfin tuna skin. The gelatin peptide fractions of 10–30 kDa and collagen peptide fractions of 3–10 kDa exhibited the highest antioxidant activity (Nurilmala et al., 2020).

The high concentration of hydrophobic amino acids may relate to the antihypertensive activity of collagen and gelatin and their hydrolysate peptides. Angiotensin-converting-enzyme (ACE) peptide inhibitors are naturally contained in various fish materials and marine collagen such as tuna (Hwang, 2010), Alaska pollack (Nakajima et al., 2009), salmon (Ohta et al., 1997), bonito (Fujita et al., 2000), shark (Wu et al., 2008), sardine (Otani et al., 2009), fish skin gelatin (Park et al., 2009), fish cartilage (Nagai et al., 2006) and squid tunic gelatin (Aleman et al., 2011). Peptides with high ACE inhibitory activity obtained from gelatin extract of Alaska pollack skin by serial hydrolysis using protease in the order alcalase, pronase E and collagenase (Byun and Kim, 2002). The high-ACE inhibitory activity was also exhibited by pepsin hydrolysate, which was obtained from Pacific cod skin gelatin using several enzymes (Ngo et al., 2016). Purified peptides from Thornback ray skin gelatin also showed ACE inhibitor activity (Lassoued et al., 2015).

A diet of fish collagen hydrolysates may be useful in controlling the temporal increase of plasma triglycerides. The in vivo studies showed that metabolism and lipid absorption in rats were affected by the diet with fish skin collagen hydrolysates. Plasma triglycerides of rats were reported to be decreased significantly by fish collagen peptide 2 h after the intake (Saito et al., 2009). Clinical investigations have also reported that the pain of patients suffering from osteoarthritis was reduced by the intake of collagen/gelatin hydrolysate. The synthesis of cartilage matrix, which is used as a gene delivery promoting bone agent, also involves hydrolyzed collagen (Moskowitz, 2000; Nakagawa and Tagawa, 2000)

The antimicrobial activities of gelatin depend on their sequence of peptides, amino acid composition, hydrophobicity, attributes of molecular weight, and charge state (Kouhdasht et al., 2021). Gelatin from salmon showed antimicrobial activities due to its oligopeptide content that is obtained during the extraction of gelatin from salmon skin collagen. Oligopeptides, short peptides built of 2–10 residues of amino acid with less than 1 kDa molecular weight, have been reported to exhibit antimicrobial activity (Gomez-Guillen et al., 2010; Matiacevich et al., 2013; Wang et al., 2018). Additionally, the ability of collagen to penetrate a lipid-free interface and easily absorbable make it has fungicidal and bactericidal properties, low antigenicity and superior biocompatibility. According to those attributes, collagen could be used as a good surface-active agent (Jus et al., 2009). The biodegradable films and active packaging materials can be formulated by enriching the gelatin films with natural antioxidant or antimicrobial substances (Kavoosi et al., 2013; Kim and Mendis, 2006).

8.3.2 Fish Protein Hydrolysate

Fish processing by-products can be a great source of high-quality protein, amino acids, and enzymes. Its protein content, which varies from 8–35% (Sila and Bougatef, 2016), has a good potential to be utilized in a variety of high-value products for human consumption. One of the effective ways to add the value of the fish by-products is by hydrolyzing its protein content into different sized small peptides, about 2–12 amino acids, which then will increase the bioavailability of the functional peptide sequence (Jensen et al., 2019). As a result, due to the presence and accessibility of the peptides functional active side chain, fish protein hydrolysates (FPHs) were reported to exert superior bioactivities compare to the whole protein (Kim and Wijesekara, 2010; Nirmal et al., 2021).

There are several ways to hydrolyze the fish protein such as chemically hydrolysis, thermal hydrolysis and enzymatic hydrolysis (Zamora-Sillero et al., 2018). Among the methods, chemically hydrolysis using alkali and acid is relatively low cost and simple to operate; thus, the method is widely used at an industrial scale. However, compared to biochemical methods, the chemical and thermal hydrolysis approach is harsher, which caused amino acid profile damage, a loss of nutrients, and produced very heterogeneous peptides due to the nonspecific peptide bond cleaving. Biochemical hydrolysis can be done using endogenous proteolytic enzymes of the fish proteins or using exogenous enzymes. According to Zamora-Sillero et al. (2018), the hydrolyzing process of the fish protein is much more controllable when an exogenous enzyme is used compared to the endogenous enzyme. Moreover, an exogenous enzyme will produce a better homogeneous protein hydrolysate. Therefore, an enzymatic approach is currently considered as the most effective protein hydrolysis method available to produce FPH. Enzyme selection has a key role in hydrolyzing fish protein.

Every enzyme acts on site-specific cleavage to obtain specific peptides; however, it also requires an optimum pH and temperature to perform optimally. Several enzymes usually used to produce FPHs are alcalase, pepsin, papain, trypsin, flavourzyme, neutrase, bromelain, protamex and α-chymotrypsin (Ennaas et al., 2015; Ngo et al., 2014; Sae-leaw et al., 2016; Sampath Kumar et al., 2012). The nutritional and functional levels of the FPHs depend on the hydrolysis degree and the amino acid compositions, including the presence of essential amino acids, protein structures and their stability during digestion (Halim et al., 2016; Nirmal et al., 2021).

FPHs have been reported as a superior source for various biological activities. One of the highlighted bioactivities of the fish protein hydrolysates is antihypertensive activity. ACE is able to catalyze the conversion of angiotensin I to the active angiotensin II (a vasoconstrictor) and inactivate bradykinin (a vasodilator), which results in the blood pressure increasing (Lee and Hur, 2017). The efficacy of ACE inhibitors in treating hypertension is well reported. It is also reported that natural ACE inhibitors are safer for long-term use than synthetic medicines (Kim and Wijesekara, 2010; Korczek et al., 2018). Thus, food-derived ACE inhibitors such as FPHs become important alternatives to these medicines (Zamora-Sillero et al., 2018).

A lower molecular weight of FPHs peptides exhibits a higher antihypertensive property. It was reported that the amino acid chain length, molecular weight, and the molecular interaction of the FPH's peptides very much affect its antihypertensive activity (Yathisha et al., 2019). A study of ACE inhibitor activity of FPHs from salmon by-products conducted by Ahn et al. (2012) showed that FPHs produced using alcalase showed the highest ACE inhibitor activity compare to FPHs made using neutrase, pepsin, flavourzyme, trypsin and protamex enzymes. Their study also suggests that the presence of Phe, Leu and Tyr amino acid residues at the C-terminal plays an important role on antihypertensive activity. There are many studies reporting the antihypertensive activity of FPHs from various fisheries' by-products, such as from visceral parts of houndfish; the heads of salmon, silver warehou, flathead and barramundi; and the backbones of ribbon fish, as well as cuttlefish waste (Abdelhedi et al., 2016; Amado et al., 2014; Bougatef et al., 2008; Nurdiani et al., 2016; Ping et al., 2014), suggesting the potential of fish by-products to be sources of value-added antihypertensive products.

Strong evidence showed that many human diseases such as cancer, cardiovascular, stroke, diabetes, rheumatoid arthritis and dementia are associated with the accumulation of free radicals (Florence, 1995). Antioxidants are molecules that are able to donate an electron to a free radical and neutralize it, hence reducing the free radical damage capacity (Lobo et al., 2010). The potential of fish by-products' protein hydrolysate as a natural antioxidant has also been highlighted and reported by many studies, such as Lassoued et al. (2015), who reported the high antioxidant activity of the pentapeptides Ala-Val-Gly-Ala-Thr purified from the *Raja clavate* (thornback ray) skin protease-hydrolyzed

FPH. The antioxidant potential of fisheries' by-product FPH was also reported by other studies, such as by Chi et al. (2015), who purified antioxidant peptides of Phe-Ile-Gly-Pro, Gly-Pro-Gly-Gly-Phe-Ile and Gly-Ser-Gly-Gly-Leu from protein hydrolysates of bluefin leatherjacket (*Navodon septentrionalis*) heads.

Several published studies also reported the antimicrobial potential of the fisheries' by-products' protein hydrolysates. One of the studies is conducted by Ennaas et al. (2015), who isolated and characterized antibacterial peptides from protamex hydrolysates of Atlantic mackerel by-products. From the four peptides—AKPGDGAGSGPR, GLPGPLGPAGPK, RKSGDPLGR, and SIFIQRFTT—isolated, the smallest peptides showed the highest antimicrobial activity against gram-positive (*Listeria innocua*) and gram-negative (*Escherichia coli*). Other potential bioactivities of fisheries' by-product protein hydrolysates have been explored and reported in many studies, such as antidiabetic (Li-Chan et al., 2012; Roblet et al., 2016; Chi *et al.*, 2015), immunomodulatory activities (Karnjanapratum et al., 2016; Sae-leaw et al., 2016) and antitumor (Hsu et al., 2011; Hung et al., 2014).

8.3.3 Chitin and Chitosan

The seafood processing industry discards large amounts of crustacean shellfish wastes. The exoskeletons are converted into solid residue, which then accumulates in landfills and becomes an environmental pollutant. Beneficially, those crustacean shellfish waste can be converted into value-added goods that will reduce its environmental impact (Gallo et al., 2016). Crustacean shells are the primary source of production for chitin and chitosan; shrimp cuticle reveals 30–40% of chitin, followed by 15–30% crab; the rest is composed of proteins and minerals (Dima et al., 2017; Elieh-Ali-Komi and Hamblin, 2016).

Chitin and chitosan are natural polymers that are well known as mucopolysaccharide, which is abundant in the shell wall of marine invertebrates (Ganesan et al., 2020). Chitin is the most abundant and widely distributed amino polysaccharide polymer found in nature. Chitin performs as the building material that gives strength to the exoskeletons of crustaceans, insects, and the cell walls of fungi (Azuma and Ifuku, 2016; Elieh-Ali-Komi and Hamblin, 2016; Philibert et al., 2017). Chitin is a poly-β- [1, 4]-N-acetyl-d-glucosamine or poly (N-acetyl D-glucosamine). Due to the solid hydrogen bonding between the amide groups and the carbonyl groups of the nearby chains, chitin occurs in three polymorphic forms: α-, β-, and γ- (Philibert et al., 2017). Chitosan (poly β-(1–4)-D-glucosamine) is a cationic linear polysaccharide obtained by partial deacetylation of chitin. It is composed of randomly distributed β-(1–4)-linked D-glucosamine (deacetylated unit) and N-acetyl-D-glucosamine (acetylated unit; Dima et al., 2017).

Due to their physicochemical and biological properties, chitin, chitosan and its derivatives have the potential to be used as functional substrates for needs in the biomedical, pharmaceutical, food and environmental industries (Philibert et al., 2017). Chitin and chitosan show excellent biological properties, such as being nontoxic, biocompatible and biodegradable in the human body, and display various bioactivities, including immuno-stimulant, anticancer, antibacterial, wound healing and hemostatic (Dash et al., 2011). Thus, chitin is used for various applications, such as tissue engineering and drug delivery and as an excipient and drug carrier in film, gel or powder form for applications involving mucoadhesive (Philibert et al., 2017). One of the most well-known applications of chitosan is in the dietetic field, where it acts as a dietary fiber. Due to its chemical composition, chitosan is able to bind fatty and oily substances, favoring their elimination (Gallo et al., 2016). Moreover, due to the important characteristics of chitosan, such as its gel-forming capability, high adsorption capacity physicochemical characteristics, chemical stability, high reactivity and excellent chelation behavior, it has become an attractive alternative to other biomaterials available in the market (Dash et al., 2011; Thirunavukkarasu and Shanmugam, 2009).

8.3.4 Fish Oil

Several studies have reported that fatty acids are not only contained in edible parts of fish, such as muscle, belly flap, and subcutaneous tissue, but also are present in fish by-products, such as the skin, liver, guts, head, gills and fishbone. In this regard, a high content of fatty acids was found in liver, head, and mesenteric tissue (Ciriminna et al., 2017). Fat content in fish varied from 2–30% depending on the type of species, reproductive, dietary, environmental, seasonal variations, and geography. The large amount of by-product generated from fish processing, especially from fatty fish, would be a potential source of good-quality fish oil. The main compound of fish oil is unsaturated fatty acids docosahexaenoic acid (DHA) and eicosapentaenoic acid (EPA), which are classified as omega-3 fatty acids, making fish oil superior and different from other types of oils. The proportion of DHA in omega-3 fatty acids is up to 2–3 times higher than EPA (Chaijan *et al.*, 2006).

Fish oils are readily digested to produce the energy and have been reported to exhibit several bioactivities (Kim and Mendis, 2006). The benefit of omega-3 fatty acids in fish oil is very essential for cardiovascular diseases (Durmus, 2018). Thus, consuming DHA and EPA can prevent the formation of cardiovascular disorders by reducing risk factors, such as blood pressure, platelet aggregation, triglyceride concentrations and heart arrhythmias (Fung et al., 2009; Raatz and Bibus, 2016). In addition, a diet enriched with fish oil has been reported to prevent cognitive disorders for older people yet induce brain development for children (Pinel et al., 2014; Graciano et al., 2016; He et al., 2017). It is reported that a diet with high doses of fish oil (>6 g/day) contained omega-3 fatty acids (4 g/day) in patients with hypertriglyceridemia significantly lowered the triglyceride content in blood (Asztalos et al., 2016). Furthermore, a study on fortified foods with fish oil has correlated it with health benefits, including antihyperlipidemic, anti-inflammatory, antithrombotic and antihypertensive (Gonzalez-Sarrias et al., 2013).

Numerous studies have been conducted to extract oil from fish-processing by-products, especially from viscera (guts and liver) of fish. Most fish oils are extracted from the livers of fatty fish species (Bechtel et al., 2010). Omega-3 fatty acids contained in different fisheries by-products, that is, fishbone, gills, guts, head, liver and skin of pelagic species such as Alaska pink salmon and Alaska walleye pollock (Oliveira and Bechtel, 2005), black rockfish (Oliveira et al., 2011), Black Sea anchovy (Gencbay and Turhan, 2016), Pacific Ocean perch (Bechtel et al., 2010), sardine (Khoddami et al., 2009), salmon (Sun et al., 2006), sea bream (Pateiro et al., 2020), sea bass (Munekata et al., 2020) and tuna (Khoddami et al., 2012). The hydrolyzed fish oil extracted from the heads and bones of various fish species contained 32.78% polyunsaturated fatty acids (PUFAs), with 10.36% EPA and 9.12% DHA (Nascimento et al., 2015). Refined oil from *Sardinella lemuru* in the by-products of the canning industry contained an EPA concentration of 650.65 μg/mL (Nurbayasari et al., 2015). Oil extraction from bigeye tuna skin, scales and bones resulted in fish oil with 27.7–31.5% PUFAs, including 24.7–28.3% EPA and DHA (Ahmed et al., 2017).

8.3.5 Minerals

The mineral content of fish products obtained from the whole fish is approximately 10%, which is mainly composed of calcium and phosphorus (Toppe et al., 2007). Fish-processing activities generate fishbone by-product approximately 10–15% of the weight, which is often regarded as waste (Malde et al., 2010). Fishbone is a marine by-product that is potentially a source of calcium, an essential mineral for human health (Kim and Mendis, 2006). The average mineral content in fishbone range from 50–74%, depending on the species (Fratzl et al., 2004). Likewise, the composition of minerals in marine fish was affected by the season, biological characteristics (species, dark/white muscle, size, sexual maturity), food source, processing method, catching area and environmental factors such as salinity, temperature, water chemistry and contaminant (Turhan et al., 2004). Fishbone also contains hydroxyapatite, calcium phosphates that take part in bonding the bone, which does not break under physiological conditions (Ozawa and Suguru, 2002).

Calcium has an important role in numerous health functions such as strengthening the bone and teeth, nerve function and cofactor of many enzymatic reactions (Kim and Jung, 2007). The phosphopeptide content in fishbone could increase the bioavailability of calcium by inhibiting the formation of insoluble calcium phosphate (Hoang et al., 2003). The intake of fishbone phosphopeptides increased Ca bioavailability and prevent Ca deficiency (Jung et al., 2006). It was also reported that tuna bone powder supplementation was able to improve the microstructure of maternal bone and increase the density of bone minerals in lactating rats and their offspring (Suntornsaratoon et al., 2018).

Calcium phosphate and hydroxyapatite [$Ca_{10}(OH)_2(PO_4)_6$] are crystalline inorganic substances that are mainly composed of fishbone (60–70%) and are attached to fibrillar collagen (Kim and Mendis, 2006; Malde et al., 2010). Generally, calcium in fishbone flour could not optimally absorb by the metabolism system due to its macro-calcium form (Tongchan et al., 2009).

Macro-calcium should be converted into an edible form using a hot acetic acid solution and a hot-water treatment to soften its structure. In addition, an extraction method using superheated steam resulted the better bone recovery within short period, because the loss of soluble component from tissue was reduced (Ishikawa et al., 1990). Isolating hydroxyapatite from fishbone using a very high-temperature treatment (1300°C) produced an excellent biocompatibility and higher strength hydroxyapatite structure (Choi et al., 1999; Kim and Park, 2000). The alkaline extraction of tuna frame and tuna bone powder (yellowfin tuna waste) resulted in a calcium content of 24.56% and 38.16%, respectively (Nemati et al., 2016). The content and profile of minerals in sea bream by-products were reported by Pateiro et al. (2020). The calcium content in head, gills, and fishbone with values of 2389.24, 1873.24, and 1618.83 mg/100 g, respectively. The lower values were obtained in cuts such as guts and liver (19.44 and 12.81 mg/100 g, respectively). A similar result was shown in phosphorous content, where the values for fishbone, gills and head were five times higher than the other by-products.

8.4 CHALLENGES

It can be predicted that the production of the world's fisheries, either from capture fisheries or aquaculture sectors, will increase significantly in the future. The increase in fisheries' production will also impact the augmentation of the fisheries' by-products, which, if not appropriately managed, will cause a significant impact on the environment. Utilizing fisheries' by-products or even waste as various value-added products can be a solution in the future. Valorizing fisheries' by-products in multiple products in many fields, such as food and nonfood, becomes one alternative for handling the risk of environmental problems. Utilizing fisheries' by-products for pharmaceutical or functional food is expected to significantly increase these by-products' added value. However, one of the main challenges relates to the technological readiness needed to utilize fisheries' by-products in pharmaceutical or functional food products. The right technology supported by the readiness of all supporting facilities and infrastructure is a proper way to manage fisheries' by-products surplus in the years to come.

REFERENCES

Abdelhedi, O., Jridi, M., Jemil, I., Mora, L., Toldrá, F., Aristoy, M. C., and Nasri., R. 2016. Combined biocatalytic conversion of smooth hound viscera: Protein hydrolysates elaboration and assessment of their antioxidant, anti-ACE and antibacterial activities. *Food Resesearch International,* 86: 9–23.

Aberoumand, A. 2010. Isolation and characteristics of collagen from fish waste material. *World Journal. Fish Marine Science*, 2: 471–474.

Ahmed, R., Haq, M., Cho, Y. J., and Chun, B. S. 2017. Quality evaluation of oil recovered from by-products of bigeye tuna using supercritical carbon dioxide extraction. *Turkish Journal of Fisheries and Aquatic Sciences,* 17: 663–672. https://doi.org/10.4194/1303-2712-v17_4_02.

Ahn, C. B., Jeon, Y. J., Kim, Y. T., and Je, J. Y. 2012. Angiotensin I converting enzyme (ACE) inhibitory peptides from salmon byproduct protein hydrolysate by Alcalase hydrolysis. *Process Biochemistry*, 47: 2240–2245.

Akbar, L., Jaswir, I., Jamal, P., and Oktavianti, F. 2017. Fish gelatin nanoparticles and their food applications: A review. *International Food Research Journal*, 24: 255–264.

Aleman, A., Gimenez, B., Perez-Santín, E., Gomez-Guillen, M. C., and Montero, P. 2011. Contribution of Leu and Hyp residues to antioxidant and ACE-inhibitory activities of peptides sequences isolated from squid gelatin hydrolysate. *Food Chemistry*, 125: 334–341.

Amado, I. R., Vázquez, J. A., González, P., Esteban-Fernández, D., Carrera, M., and Piñeiro., C. 2014. Identification of the major ACE-inhibitory peptides produced by enzymatic hydrolysis of a protein concentrate from cuttlefish wastewater. *Marine Drugs*, 12: 1390–1405.

Asztalos, O. B., Gleason, J. A., Sever, S., Gedik, R., Asztalos, B. F., Horvath, K. V., Dansinger, M. L., Lamon-Fava, S., and Schaefer, E. J. 2016. Effects of eicosapentaenoic acid and docosahexaenoic acid on cardiovascular disease risk factors: A randomized clinical trial. *Metabolism: Clinical and Experimental*, 65(11): 1636–1645.

Azuma, K., and Ifuku, S. 2016. Nanofibers based on chitin: A new functional food. *Pure and Applied Chemistry*, *88*(6): 605–619. https://doi.org/10.1515/pac-2016-0504.

Bechtel, P. J., Morey, A., Oliveira, A. C. M., Wu, T. H., Plante, S., and Bower, C. K. 2010. Chemical and nutritional properties of Pacific Ocean perch (*Sebastes alutus*) whole fish and by-products. *Journal of Food Processing and Preservation,* 34: 55–72.

Bougatef, A., Nedjar-Arroume, N., Ravallec-Plé, R., Leroy, Y., Guillochon, D., Barkia, A., and Nasri, M. 2008. Angiotensin I-converting enzyme (ACE) inhibitory activities of sardinelle (*Sardinella aurita*) by-products protein hydrolysates obtained by treatment with microbial and visceral fish serine proteases. *Food Chemistry,* 111: 350–356.

Byun, H. G., and Kim, S. K. 2002. Structure and activity of angiotensin-I converting enzyme inhibitory peptides derived from Alaskan pollack skin. *Journal of Biochemistry and Molecular Biology,* 35: 239–243.

Caruso, G., Floris, R., Serangeli, C., and Di Paola, L. 2020. Fishery wastes as a yet undiscovered treasure from the sea: Biomolecules sources, extraction methods, and valorization. *Marine Drugs*, 18(12). https://doi.org/10.3390/md18120622.

Chaijan, M., Benjakul, S., Visessanguan, W., and Faustman, C. 2006. Changes of lipids in sardine (*Sardinella gibbosa*) muscle during iced storage. *Food Chemistry*, 99: 83–91.

Cheng, F. Y., Wan, T. C., Liu, Y. T., Chen, C. M., Lin, L. C., and Sakata, R. 2009. Determination of angiotensin-I converting enzyme inhibitory peptides in chicken leg bone protein hydrolysate with alcalase. *Animal Science Journal,* 80: 91–97.

Chi, C. F., Wang, B., Wang, Y. M., Zhang, B., and Deng, S. G. 2015. Isolation and characterization of three antioxidant peptides from protein hydroly- sate of bluefin leatherjacket (*Navodon septentrionalis*) heads. *Journal of Functional Foods*, 12: 1–10.

Choi, J. S., Lee, C. K., Jeon, Y. J., Byun, H. G., and Kim, S. K. 1999. Properties of the ceramic composites and glass ceramics prepared by using the natural hydroxyapatite derived from tuna bone. *Journal of Korean Industrial and Engineering Chemistry,* 10: 394–399.

Ciriminna, R., Meneguzzo, F., Delisi, R., and Pagliaro, M. 2017. Enhancing and improving the extraction of omega-3 from fish oil. *Sustainable Chemistry and Pharmacy*, 5: 54–59.

Dash, M., Chiellini, F., Ottenbrite, R. M., and Chiellini, E. 2011. Chitosan—A versatile semi-synthetic polymer in biomedical applications. *Progress in Polymer Science (Oxford)*, 36(8): 981–1014. https://doi.org/10.1016/j.progpolymsci.2011.02.001.

Dima, J. B., Sequeiros, C., and Zaritzky, N. 2017. Chitosan from marine crustaceans: Production, characterization, and applications. In *Biological Activities and Application of Marine Polysaccharides*. InTech. https://doi.org/10.5772/65258.

Durmus, M. 2018. Fish oil for human health: Omega-3 fatty acid profiles of marine seafood species. *Food Science and Technology,* 39(2): 454–461.doi https://doi.org/10.1590/fst.21318.

Elieh-Ali-Komi, D., and Hamblin, M. R. 2016. Chitin and Chitosan: Production and application of versatile biomedical nanomaterials. *International Journal of Advanced Research (Indore)*, 4(3): 411–427.

Ennaas, N., Hammami, R., Beaulieu, L., and Fliss, I. 2015. Purification and characterization hydrolysate of four antibacterial peptides from protamex Biochem, of Atlantic mackerel (*Scomber scombrus*) by-products. *Biochemistry and Biophysical Research Communications*, 462: 195–200.

FAO. 2019. A third assessment of global marine fisheries discards. In *FAO Fisheries and Aquaculture Technical Paper*. No. 633 (Vol. 633). http://www.fao.org/3/CA2905EN/ca2905en.pdf.

FAO. 2020. The State of World fisheries and aquaculture 2020. *Sustainability in Action.* Rome. https://doi.org/https://doi.org/10.4060/ca9229en.

Florence, T. M. 1995. The role of free radicals in disease. *Australian and New Zealand Journal of Ophthalmology*, 23: 3–7. https://doi.org/10.1111/j.1442-9071.1995.tb01638.x

Fratzl, P., Gupta, H. S., Paschalis, E. P., and Roschger, P. 2004. Structure and mechanical quality of the collagen—mineral nano-composite in bone. *Journal of Materials Chemistry,* 14(14): 2115–2123.

Fujita, H., Yokoyama, K., and Yoshikawa, M. 2000. Classification and antihypertensive activity of angiotensin I-converting enzyme inhibitory peptides derived from food proteins. *Journal of Food Science,* 65: 564–569.

Fung, T., Rexrode, K. M., Mantzoros, C. S., Manson, J. E., Willett, W. C., and Hu, F. B. 2009. Mediterranean diet and incidence of and mortality from coronary heart disease and stroke in women. *Circulation*, 119(8): 1093–1100.

Gallo, M., Naviglio, D., Caruso, A. A., and Ferrara, L. 2016. Applications of chitosan. In *Novel Approaches of Nanotechnology in Food,* ed. A. M. Grumezescu, 425–464. Academic Press. http://dx.doi.org/10.1016/B978-0-12-804308-0/00013-3.

Ganesan, A. R., Saravana Guru, M., Balasubramanian, B., Mohan, K., Chao Liu, W., Valan Arasu, M., Abdullah Al-Dhabi, N., Duraipandiyan, V., Ignacimuthu, S., Sudhakar, M. P., and Seedevi, P. 2020. Biopolymer from edible marine invertebrates: A potential functional food. *Journal of King Saud University-Science*, 32(2): 1772–1777. https://doi.org/10.1016/j.jksus.2020.01.015.

Gencbay, G., and Turhan, S. 2016. Proximate composition and nutritional profile of the black sea anchovy (*Engraulis encrasicholus*) whole fish, fillets, and by-products. *Journal of Aquatic Food Product Technology*, 25: 864–874.

Gomez-Guillen, M. C., Gimenez, B., Lopez-Caballero, M. E., and Montero, M. P. 2011. Functional and bioactive properties of collagen and gelatin from alternative sources: A review. *Food Hydrocolloids*, 25: 1813–1827. https://doi.org/10.1016/j.foodhyd.2011.02.007.

Gomez-Guillen, M. C., Lopez-Caballero, M. E., Lopez de Lacey, A., Aleman, A., Gimenez, B., and Montero, P. 2010. Antioxidant and antimicrobial peptide fractions from squid and tuna skin gelatin. In *Sea By-products as a Real Material: New Ways of Application,* ed. E. Le Bihan, and N. Koueta, 89–115. Kerala, India: Transworld Research Network Signpost, Chapter 7.

Gonzalez-Sarrias, A., Larrosa, M., García-Conesa, M. T., Tomas-Barberan, F. A., and Espin, J. C. 2013. Nutraceuticals for older people: Facts, fictions and gaps in knowledge. *Maturitas,* 75(4): 313–334. https:// dx.doi.org/10.1016/j.maturitas.2013.05.006. PMid:23791247.

Graciano, M. F., Leonelli, M., Curi, R., and Carpinelli, A. R. 2016. Omega-3 fatty acids control productions of superoxide and nitrogen oxide and insulin content in ONS-1E cells. *Journal of Physiology and Biochemistry*, 72(4): 699–710. https://dx.doi.org/10.1007/s13105-016-0509-1.PMid:27474043.

Halim, N., Yusof, H., and Sarbon, N. 2016. Functional and bioactive properties of fish protein hydolysates and peptides: A comprehensive review. *Trends in Food Science & Technology*, 51: 24–33.

Hardy, R. W. 1992. Fish processing by-products and their reclamation. In *Edible Meat By-products*, ed. A. M. Pearson, and T. R. Dutson, 199–216. https://doi.org/10.1007/978-94-011-7933-1_9.

He, Y., Li, J., Kodali, S., Chen, B., and Guo, Z. 2017. Rationale behind the near-ideal catalysis of *Candida antarctica* lipase A (CAL-A) for highly concentrating ω-3 polyunsaturated fatty acids into monoacylglycerols. *Food Chemistry*, 219: 230–239.

Hoang, Q. Q., Sicheri, F., Howard, A. J., and Yang, D. S. C. 2003. Bone recognition mechanism of porcine osteocalcin from crystal structure. *Nature,* 425: 977–980.

Hou, H., Li, B., Zhao, X., Zhuang, Y., Ren, G., and Yan, M. 2009. The effect of pacific cod (*Gadus macrocephalus*) skin gelatin polypeptides on UV radiation induced skin photoaging in ICR mice. *Food Chemistry*, 115(3): 945–950.

Hsu, K. C., Li-Chan, E. C. Y., and Chia, L. J. 2011. Antiproliferative activity of peptides prepared from enzymatic hydrolysates of tuna dark muscle on human breast cancer cell line MCF-7. *Food Chemistry*, 126: 617–622.

Hung, C. C., Yang, Y. H., Kuo, P. F., and Hsu, K. C. 2014. Protein hydrolysates from tuna cooking juice inhibit cell growth and induce apoptosis of hu- man breast cancer cell line MCF-7. *Journal of Functional Foods*, 11: 563–570.

Hwang, J. S. 2010. Impact of processing on stability of angiotensin I-converting enzyme (ACE) inhibitory peptides obtained from tuna cooking juice. *Food Research International,* 43: 902–906.

Irianto, H. E., Dewi, A. S., and Giyatmi. 2014. Prospective utilization of fishery by-products in Indonesia. In *Seafood Processing By-products: Trends and Applications*, ed. S. K. Kim. New York: Springer Science+Business Media. https://doi.org/10.1007/978-1-4614-9590-1.

Ishikawa, M., Kato, M., Mihori, T., Watanabe, H., and Sakai, Y. 1990. Effect of vapor pressure on the rate of softening of fish bone by super-heated steam cooking. *Nippon Suisan Gakkaishi*, 56: 1687–1691.

Jensen, C., Dale, H., Hausken, T., Lied, E., Hatlebakk, J., Brønstad, I., Lied, G., and Hoff, D. 2019. Supplementation with cod protein hydrolysate in older adults: A dose range cross-over study. *Journal of Nutritional Science*, 8.

Jung, W. K., Lee, B. J., and Kim, S. K. 2006. Fish-bone peptide increases calcium solubility and bioavailability in ovariectomised rats. *British Journal of Nutrition,* 95: 124–128.

Jus, S., Kokol, V., and Guebitz, G. M. 2009. Tyrosinase-catalysed coating of wool fibres with different protein-based biomaterials. *Journal of Biomaterials Science*, 20: 253–269. https://doi.org/101163/156856209X404523.

Kaanane, A., and Hind, M. 2020. Valorization Technologies of Marine By-Products. In *Innovation in the Food Sector through the Valorization of Food and Agro-food By-products*, ed. A. N. de Barros. IntechOpen. https://doi.org/10.5772/intechopen.91078.

Karim, A. A., and Bhat, R. 2009. Fish gelatin: Properties, challenges, and prospects as an alternative to mammalian gelatins. *Food Hydrocolloids,* 23(3): 563–576.

Karnjanapratum, S O'Callaghan, Y. C., Benjakul, S., and O'Brien, N. 2016. Antioxidant, immunomodulatory and antiproliferative effects of gelatin hydrolysate from unicorn leatherjacket skin. *Journal of the Science of Food and Agriculture*, 96: 3220–3226.

Kavoosi, G., Dadfar, S. M. M., Purfard, A. M., and Mehrabi, R. 2013. Antioxidant and antibacterial properties of gelatin films incorporated with carvacrol. *Journal of Food Safety*, 33: 423–432. https://doi.org/10.1111/jfs.12071.

Khawli, F. Al, Martí-Quijal, F. J., Ferrer, E., Ruiz, M. J., Berrada, H., Gavahian, M., Barba, F. J., and de la Fuente, B. 2020. Aquaculture and its by-products as a source of nutrients and bioactive compounds. *Advances in Food and Nutrition Research*, 92: 1–33. https://doi.org/10.1016/bs.afnr.2020.01.001.

Khawli, F. Al, Pateiro, M., Domínguez, R., Lorenzo, J. M., Gullón, P., Kousoulaki, K., Ferrer, E., Berrada, H., and Barba, F. J. 2019. Innovative green technologies of intensification for valorization of seafood and their by-products. *Marine Drugs*, 17(12): 1–20. https://doi.org/10.3390/md17120689.

Khoddami, A., Ariffin, A. A., Bakar, J., and Ghazali, H. M. 2009. Fatty acid profile of the oil extracted from fish waste (head, intestine and liver) (*Sardinella lemuru*). *World Applied. Science Journal,* 7: 127–131.

Khoddami, A., Khoddami, A., Ariffin, A. A., Bakar, J., and Ghazali, H. M. 2012. Quality and fatty acid profile of the oil extracted from fish waste (head, intestine and liver) (*Euthynnus affinis*). *African Journal of Biotechnology*, 11: 1683–1689.

Kim, S. K., and Jung, W. K. 2007. Fish and bone as a calcium source. Maximizing the value of marine by-products. In *Maximising the Value of Marine By-products*, ed. F. Shahidi, 328–336. Boca Raton, FL: CRC Press.

Kim, S. K., and Mendis, E. 2006. Bioactive compounds from marine processing byproducts: A review. *Food Research International,* 39: 383–393. https://doi.org/10.1016/j.foodres.2005.10.010.

Kim, S. K., and Park, P. J. 2000. Evaluation of mucous membrane irritation by hydroxyapatite sinter produced from tuna bone in Syrian hamsters. *Korean Journal of Life Science*, 10: 605–609.

Kim, S. K., and Wijesekara, I. 2010. Development and biological activities of marine-derived bioactive peptides: A review. *Journal of Functional Foods,* 2: 1–9.

Korczek, K., Tkaczewska, J., and Migdał, W. 2018. Antioxidant and antihypertensive protein hydrolysates in fish products—A review. *Czech Journal of Food Sciences,* 36: 195–207. https://doi.org/10.17221/283/2017-CJFS

Kouhdasht, A. M., Nasab, M. M., Lee, C. W., Yun, H., and Eun, J. B. 2021. Structure—function engineering of novel fish gelatin-derived multifunctional peptides using high-resolution peptidomics and bioinformatics. *Nature*, 11(7401): 1–15. https://doi.org/10.1038/s41598-021-86808-9.

Lassoued, I., Mora, L., Nasri, R., Jridi, M., Toldra, F., and Aristoy, M. C. 2015. Characterization and comparative assessment of antioxidant and ACE inhibitory activities of thornback ray gelatin hydrolysates. *Journal of Functional Foods*, 13: 225–238.

Lee, S. Y., and Hur, S. J. 2017. Antihypertensive peptides from animal products, marine organisms, and plants. *Food Chemistry*, 228: 506–517.

Li-Chan, E. C. Y., Hunag, S. L., Jao, C. L., Ho, K. P., and Hsu, K. C. 2012. Peptides derived from atlantic salmon skin gelatin as dipeptidyl-peptidase IV inhibitors. *Journal of Agricultural and Food Chemistry,* 60: 973–978.

Lobo, V., Patil, A., Phatak, A., and Chandra, N. 2010. Free radicals, antioxidants and functional foods : Impact on human health. *Pharmacognosy Review*, 4: 118–126. https://doi.org/10.4103/0973-7847.70902

Malde, M. K., Graff, I. E., Siljander-Rasi, H., Venalainen, E., Julshamn, K., Pedersen, J. I., and Valaja, J. 2010. Fish bones—a highly available calcium source for growing pigs. *Journal of Animal Physiology and Animal Nutrition*, 94: 66–76.

Matiacevich, S., Cofre, D. C., Schebor, C., and Enrione, J. 2013. Physicochemical and antimicrobial properties of bovine and salmon gelatin-chitosan films. *CyTA-Journal of Food*, 11(4): 366–378. http://dx.doi.org/10.1080/19476337.2013.773564.

Moskowitz, R. W. 2000. Role of collagen hydrolysate in bone and joint disease. *Seminars in Arthritis and Rheumatism*, 30: 87–99.

Munekata, P. E. S., Pateiro, M., Domínguez, R., Zhou, J., Barba, F. J., and Lorenzo, J. M. 2020. Nutritional characterization of sea bass processing by-products. *Biomolecules*, 10: 232.

Nagai, T., Nagashima, T., Abe, A., and Suzuki, N. 2006. Antioxidative activities and angiotensin I-converting enzyme inhibition of extracts prepared from chum salmon (*Oncorhynchus keta*) cartilage and skin. *International Journal of Food Properties*, 9(4): 813–822.

Nakagawa, T., and Tagawa, T. 2000. Ultrastructural study of direct bone formation induced by BMPs-collagen complex implanted into an ectopic site. *Oral Diseases*, 6: 172–179.

Nakajima, K., Yoshie-Stark, Y., and Ogushi, M. 2009. Comparison of ACE inhibitory and DPPH radical scavenging activities of fish muscle hydrolysates. *Food Chemistry*, 114: 844–851.

Nascimento, V. L. V., Bermudez, V. M. S., Oliveira, A. L. L., Kleinberg, M. N., Ribeiro, R. T M., Abreu, R. F. A., and Carioca, J. O. B. 2015. Characterization of a hydrolyzed oil obtained from fish waste for nutraceutical application. *Food Science and Technology*, 35(2): 321–325. http://dx.doi.org/10.1590/1678-457x.6583.

Nawaz, A., Li, E., Irshad, S., Xiong, Z., Xiong, H., Shahbaz, H. M., and Siddique, F. 2020. Valorization of fisheries by-products: Challenges and technical concerns to food industry. *Trends in Food Science & Technology*, 99: 34–33. https://doi.org/https://doi.org/10.1016/j.tifs.2020.02.022.

Nemati, M., Huda, N., and Ariffin, F. 2016. Development of calcium supplement from fish bone wastes of yellowfin tuna (*Thunnus albacares*) and characterization of nutritional quality. *International Food Research Journal*, 24(6): 2419–2426.

Ngo, D. H., Ryu, B., and Kim, S. K. 2014. Active peptides from skate (*Okamejei kenojei*) skin gelatin diminish angiotensin-I converting enzyme activity and intracellular free radical-mediated oxidation. *Food Chemistry*, 15: 246–255.

Ngo, D. H., Vo, T. S., Ryu, B. M., and Kim, S. K. 2016. Angiotensin-I-converting enzyme (ACE) inhibitory peptides from Pacific cod skin gelatin using ultrafiltration membranes. *Process Biochemistry*, 51: 1622–1628.

Nirmal, N. P., Santivarangkna, C., Benjakul, S., and Maqsood, S. 2021. Fish protein hydrolysates as a health promoting ingredient- recent update. *Nutrition Reviews*, 00: 1–14. https://doi.org/10.1093/nutrit/nuab065

Nurbayasari, R., Utomo, B. S. B., Basmal, J., and Kusumawati, R. 2015. Optimization of fish oil *Sardinella lemuru* from canning industry by products. *JPHPI*, 18(3): 276–286. https://doi.org/10.17844/jphpi.2015.18.3.276.

Nurdiani, R., Dissanayake, M., Street, W. E., Donkor, O. N., Singh, T. K., and Vasiljevic, T. 2016. In vitro study of selected physiological and physicochemical properties of fish protein hydrolysates from 4 Australian fish species. *Food Research Institute*, 23: 2029–2040.

Nurilmala, M., Hizbullah, H. H., Karnia, E., Kusumaningtyas, E., and Ochiai, Y. 2020. Characterization and antioxidant activity of collagen, gelatin, and the derived peptides from Yellowfin tuna (*Thunnus albacares*) skin. *Marine Drugs*, 18(98):1–12. https://doi.org/10.3390/md18020098.

Ohta, T., Iwashita, A., Sasaki, S., and Kawamura, Y. 1997. Antihypertensive action of the orally administered protease hydrolysates of chum salmon head and their angiotensin I-converting enzyme inhibitory peptides. *Food Science and Technology International*, 4: 339–343.

Oliveira, A. C. M., and Bechtel, P. J. 2005. Lipid composition of Alaska pink salmon (*Oncorhynchus gorbuscha*) and Alaska walleye pollock (*Theragra chalcogramma*) byproducts. *Journal of Aquatic Food Product Technology,* 14: 73–91.
Oliveira, A. C. M., Bechtel, P. J., Lapis, T. J., Brenner, K. A., and Ellingson, R. 2011. Chemical composition of black rockfish (*Sebastes melanops*) fillets and byproducts. *Journal of Food Processing and Preservation*, 35: 466–473.
Otani, L., Ninomiya, T., Murakami, M., Osajima, K., Kato, H., and Murakami, T. 2009. Sardine peptide with angiotensin I-converting enzyme inhibitory activity improves glucose tolerance in stroke-prone spontaneously hypertensive rats. *Bioscience, Biotechnology and Biochemistry,* 73: 2203–2209.
Ozawa, M., and Suguru, S. 2002. Microstructural development of natural hydroxyapatite originated from fish-bone waste through heat treatment. *Journal of the American Ceramic Society*, 85: 1315–1317.
Park, C. H., Kim, H. J., Kang, K. T., Park, J. W., and Kim, J. S. 2009. Fractionation and angiotensin I-converting enzyme (ACE) inhibitory activity of gelatin hydrolysates from by-products of Alaska pollock surimi. *Fisheries and Aquatic Science*, 12(2): 79–85.
Pateiro, M., Munekata, P. E. S., Domínguez, R., Wang, M., Barba, F. J., Bermúdez, R., and Lorenzo, J. M. 2020. Nutritional profiling and the value of processing by-products from gilt head sea bream (*Sparus aurata*). *Marine Drugs*, 18: 101.
Philibert, T., Lee, B. H., and Fabien, N. 2017. Current Status and New Perspectives on Chitin and Chitosan as Functional Biopolymers. *Applied Biochemistry and Biotechnology*, 181(4): 1314–1337. https://doi.org/10.1007/s12010-016-2286-2.
Pinel, A., Morio-Liondore, B., and Capel, F. 2014. N-3 Polyunsaturated fatty acids modulate metabolism of insulin sensitive tissues: Implication for the prevention of type 2 diabetes. *Journal of Physiology and Biochemistry*, 70(2): 647–658. http://dx.doi.org/10.1007/s13105-013-0303-2.PMid:24371037.
Ping, Z., Jin-Ling, W., Guo-Qing, H., and Jianping, W. 2014. Purification, identification, and vivo activity of Angiotensin I-converting enzyme inhibitory peptide, from ribbonfish (Trichiurushaumela) backbone. *Journal of Food Science*, 79.
Raatz, S., and Bibus, D. 2016. *Fish and Fish Oil in Health and Disease Prevention.* USA: Academic Press. https://doi.org/10.1016/C2014-0-02727-X.
Ramírez, A. 2013. Innovative uses of fisheries by-products. *GLOBEFISH Research Programme FAO*, 110: 1–53.
Roblet, C., Akhtar, M., Mikhaylin, S., Pilon, G., Gill, T., and Marette, A. L. B. 2016. Enhancement of glucose uptake in muscular cell by peptide fractions separated by electrodialysis with filtration membrane from salmon frame protein hydrolysate. *Journal of Functional Foods,* 22: 337–346.
Rustad, T. 2003. Utilisation of marine by-products. *Electric Journal of Enviromental, Agricultural and Food Chemistry*, 4(2): 458–463.
Rustad, T., Storrø, I., and Slizyte, R. 2011. Possibilities for the utilisation of marine by-products. *International Journal of Food Science & Technolog*, 46: 2001–2014. https://doi.org/10.1111/j.1365-2621.2011.02736.x
Sae-leaw, T., O'Callaghan, Y. C., Benjakul, S., and O'Brien, N. M. 2016. Antioxidant, immunomodulatory and antiproliferative effects of gelatin hydrolysates from seabass (*Lates calcarifer*) skins. *International Journal of Food Science & Technology,* 51: 1545–1551.
Saito, M., Kiyose, C., Higuchi, T., Uchida, N., and Suzuki, H. 2009. Effect of collagen hydrolysates from salmon and trout skins on the lipid profile in rats. *Journal of Agricultural and Food Chemistry*, 57(21): 10477–10482.

Sampath Kumar, N. S., Nazeer, R. A., and Jaiganesh, R. 2012. Purification and identification of antioxidant peptides from the skin protein hydrolysate of two marine fishes, horse mackerel (*Magalaspis cordyla*) and croaker (*Otolithes ruber*). *Amino Acids,* 42: 1641–1649.

Sila, A., and Bougatef, A. 2016. Antioxidant peptides from marine by-products: Isolation, identification and application in food systems: A review. *Journal of Functional Foods,* 21: 10–26.

Sun, T., Xu, Z., and Prinyawiwatkul, W. F. A. 2006. Composition of the oil extracted from farmed Atlantic salmon (*Salmo salar* L.) viscera. *Journal of American Oil Chemistry Society,* 83: 615–619.

Suntornsaratoon, P., Charoenphandhu, N., and Krishnamra, N. 2018. Fortified tuna bone powder supplementation increases bone mineral density of lactating rats and their offspring. *Journal of the Science of Food and Agriculture,* 98(5): 2027–2034.

Thirunavukkarasu, N., and Shanmugam, A. 2009. Extraction of chitin and chitosan from mud crab *Scylla tranquebarica* (*Fabricius*, 1798). *International Journal on Applied Bio-Engineering*, 3(2): 31–33. https://doi.org/10.18000/ijabeg.10048.

Tkaczewska, J., Morawska, M., Kulawik, P., and Zajac, M. 2018. Characterization of carp (*Cyprinus carpio*) skin gelatin extracted using different pretreatments method. *Food Hydrocolloids*, 81: 169–179. https://doi.org/10.1016/j.foodhyd.2018.02.048.

Tongchan, P., Prutipanlai, S., Niyomwas, S., and Thongraung, S. 2009. Effect of calcium compound obtained from fish byproduct on calcium metabolism in rats. *Asian Journal of Food and Agro-Industry,* 2(04): 669–676.

Toppe, J., Lbrektsen, S., Hope, B., and Aksnes, A. 2007. Chemical composition, mineral content and amino acid and lipid profiles in bones from various fish species. *Comparative Biochemistry and Physiology,* Part B 146: 395–401.

Turhan, S., Ustun, S. N., and Altunkaynak, B. 2004. Effect of cooking methods on total and heme iron contents of anchovy (*Engraulis encrasicholus*). *Food Chemistry*, 88(2): 169–172.

Wang, L., Sun, J., Ding, S., and Qi, B. 2018. Isolation and identification of novel antioxidant and antimicrobial oligopeptides from enzymatically hydrolyzed anchovy fish meal. *Process Biochemistry,* 1359–5113. https://doi.org/10.1016/j.procbio.2018.08.021.

Wu, H., He, H. L., Chen, X. L., Sun, C. Y., Zhang, Y. Z., and Zhou, B. C. 2008. Purification and identification of novel angiotensin-I-converting enzyme inhibitory peptides from shark meat hydrolysate. *Process Biochemistry,* 43: 457–461.

Yathisha, U. G., Bhat, I., Karunasagar, I., and Mamatha, B. S. 2019. Antihypertensive activity of fish protein hydrolysates and its peptides. *Critical Reviews in Food Science and Nutrition*, 59: 2363–2374. https://doi.org/10.1080/10408398.2018.1452182

Zamora-Sillero, J., Gharsallaoui, A., and Prentice, C. 2018. Peptides from fish by-product protein hydrolysates and its functional properties: An overview. *Marine Biotechnology,* 20: 118–130. https://doi.org/10.1007/s10126-018-9799-3

9 A Review on L-Asparaginase

Important Sources and Its Applications

B. Deivasigamani and R. Sharmila

CONTENTS

DOI: 10.1201/9781003303909-9

9.1 MARINE MICROORGANISMS

Marine microbes have extremely supreme power forr human health, welfare and environment. In the regulation of Earth's climate, microbes play a vital role by their abundance and diversity. They help in the production and release of carbon products particularly CO_2 and CH_4. The multifaceted quality of marine microorganisms providesthe essential requirements of our society. They ensure oxygen, food, and the fitness of the marine environment. They also provide for a maximum unique source of genomic information and biomolecules that can be used in medical and industrial applications. Thus, the employment of marine microorganisms in the production of antibiotics, antitumor compounds and enzymes and the role of bioremediation has unlocked a new episode for the profit of humankind (Roychowdhury et al., 2018).

Marine microorganisms are the genesis of new enzymes when compared to plant- and animal-derived enzymes. Marine and terrestrial enzymes are highly differentiated in the factors, including temperature, salinity, pressure and lighting. Microbial enzymes from marine sources have got its diverse potential in industrial applications (Nguyen and Nguyen, 2017).

Marine microbes encompass a comparative prime reservoir for commercially important compounds. They have peculiar characteristics of acclimatizing themselves to the extreme marine environmental conditions, such as variation in temperature, high pressure, limited substrate and acidic or alkaline water conditions. These eccentric properties have provoked many researchers to initiate an in-depth search of the marine environment (Baharum et al., 2010).

According to UN Environment Programme (UNEP), Global Marine Assessments (G.M.A.) saline water covers nearly 71% of the earth's surface. The depth of the water ranges from 3.8 km, covering a volume of 1370×10^6 km. This is the marine ecosystem that serves as the reservoir for food, natural gas, oil, minerals and many bioactive compounds that have abundant importance in the pharmaceutical field. Millions of people rely on the marine ecosystem to sustain their livelihood. The ocean remains the main stabilizer of the world's climate. There are around 178,000 marine species in 34 phyla in the oceanic water, according to the Global Biodiversity Assessment (GBA) produced by the UNEP.

9.2 MARINE SOURCES ACT AS THERAPEUTIC AGENTS

9.2.1 Bacteria

Exploitation of the marine microorganisms leads to the development of novel drugs and therapeutic methods. Isolating all marine bacteria is not possible under laboratory conditions, and hence, it is highly necessary to develop new culturing techniques for the slow-growing bacteria that are exclusive to unique natural products, for example, anticancer agents, such as bryostatins, discodermolide, sarcodictyin and eleutherobin; antibiotics, such as marinone; and antiparasitic compounds, such as valinomycinis from *Streptomyces sp.* strains from the Mediterranean.

9.2.2 Cyanobacteria

About 150 genera and 2000 species of the diversified population can be seen in cyanobacteria. The capability of anticancerous property of marine cyanobacteria has been discovered for deriving maximum marine-based chemicals. *Nostoc*, *Calothrix*, *Lynfbya* and *Symploca* are the well-studied species of marine cyanobacteria. Thus, several compounds are used as templates for the production of new drugs with anticancerous properties.

9.2.3 Fungi

Plenty of unique structures and active secondary metabolites have been derived from marine fungi. Endophytic fungi such as *Halorosellinia sp.* and *Guignardia sp.* habituating in the mangrove were isolated, and they act as potent anticancer agents that are derivatives of anthracenedione. Antileukemia drugs Cytarabine and trabectedin, which help in treating soft tissue sarcoma, are obtained from marine fungal sources. Acremonin A, which is isolated from *Acremonium sp.*, and a xanthone derivative isolated from *Wardomycesanomalus* are known to be a natural antioxidative product of marine fungi origin (Faivre et al., 2005).

9.2.4 Sponge

Nearly 10,000 sponge varieties have been observed worldwide (Broggini et al., 2003), and they are mostly seen in the marine environment (Armand et al., 2001). About 11 sponge genera have been found in a range of bioactive compounds, of which three genera (*Petrosia*, *Haliclona*, *Discodemia*) promisingly produce anticancer and anti-inflammatory compounds (Bai et al., 1992).

9.2.5 Bryozoan

The natural lactone that is isolated from the marine Bryozoan is bryostatin-1, which has both antitumor activity and immunomodulatory effects (Wall et al.,

1999). It has in vitro cytotoxicity effect over various leukemia and tumor cell lines. Bryostatin-1 proved antitumor activity against leukemia, melanoma, ovarian cancer and lymphoma. It also has an stimulatory effect on several anti-cancer agents such as cisplatin, cytosine arabinoside, vincristine, paclitaxel, melphalan and others (Wall et al., 1999).

9.2.6 Aplidine

Aplidine induces the rapid arrest of the cell cycle at G1-G2 and causes inhibition of protein synthesis, thus resulting in the apoptosis of cancer cells (Faivre et al., 2005). It has shown to have more activity than Didemnin and has not shown any neuromuscular toxicity of cells (Faivre et al., 2005).

9.2.7 Dolostatins

In vitro antileukemic activity has been recorded in Dolastatin against various cell lines of lymphoma, human leukemia and on solid tumors. There is documentation of its antitumor activity against LOX#IMVI melanoma, OVCAR-3 and NCI-H522 NSCLC cell lines (Pettit *et al.,* 1987).

9.2.8 Ecteinascidins

Ecteinascidia turbinate is a Caribbean tunicate from which ecteinascidins (ETS) is derived. The promising activity in human and marine tumor models led to ecteinascidin 743 (ET 743) being in early clinical development. It induces a broad inhibition of activated transcription with no effect on constitutive transcription (Minuzzo et al., 2000).

9.3 THERAPEUTIC ENZYMES AND ITS APPLICATIONS

In the 1960s, a therapeutic enzyme was regarded as a part of replacement therapy for the deficiencies caused by genetic disorders which have been in use for about 40 years. Enzymes are much needed in our bodies for metabolic purposes. Consuming enzyme-rich food will accelerate the immune response and will prevent a lot of diseases and will act as antiaging factor. Some of the enzymes medicinally important in both digestive and metabolic processes may be used alone or in association with other therapies for treating several diseases. The two crucial characteristics of the enzymes are (1) their affinity and specificity toward binding their target molecules and (2) the catalytic property by which they convert the target to required products. These properties in enzymes allow them to be the most potent drugs to treat several disorders (Cooney and Rosenbluth, 1975).

Asparaginase is recently engaged in the treatment of acute lymphoblastic leukemia. Due to a lack of capability of synthesizing L-asparagine, an essential amino acid, the tumor cells starve for exogenous asparagine and ultimately will be forced to die (Gurung et al., 2013).

Chitinase has antimicrobial activity. The chitin, which is present in the cell wall of most pathogenic organisms like fungi, protozoa and helminths, is a good target for antimicrobials (Fusetti et al., 2002). The lytic enzymes derived from bacteriophages can be utilized for the treatment of several infections. They also exhibit activity against new resistant bacterial species. Proteolytic enzymes isolated from bacterial sources have anti-inflammatory properties. These proteolytic enzymes are also employed to eliminate burned skin (Gurung et al., 2013).

Collagenase aids in the treatment of burns and skin ulcers. It helps to lyse and remove dead skin and dead tissue, thereby helping the repair mechanism. This ultimately improves the action of antibiotics to work better in improving the individual's healing process (Ostlie et al., 2012). Lipase helps in the treatment of digestive disorders. By activating the tumor necrosis factor, they can be used in the treatment of malignant tumors. Disorders like dyspepsia, gastrointestinal disturbances and digestive allergies are treated with lipase enzymes. Lipase obtained from *Candida rugosa* is used to produce lovastatin, a drug that has the ability to decrease serum-level cholesterol. The hydrolysis of 3-phenylglycidic acid ester is a key intermediate in the production of diltiazem hydrochloride. It is more commonly used as a coronary vasodilator and is synthesized from *Serratia marcescens* lipase.

Numerous applications have been derived directly from enzymes because of their pharmaceutical properties. There are enzymes that have therapeutic uses, including L-arginase (antitumor); L-asparaginase (antitumor), L-tyrosinase (antitumor), L-glutaminase (antitumor), uricase (gout), urokinase (blood clots), glucosidase (antitumor), lysozyme (antibiotic), trypsin (inflammation), urokinase (anticoagulant), ribonuclease (antiviral), rhodanase (cyanide poisoning), hyaluronidase (heart attack), serratiopeptidase (anti-inflammatory), β-lactamase (penicillin allergy), dornase α (cystic fibrosis), collagenase (skin ulcers), streptokinase (anticoagulant), laccase (detoxifier), lipase (digest lipids), rasburicase (hyperuricemia), sacrosidase (congenital sucraseisomaltase deficiency) and peptidase (celiac disease). Currently, enzyme replacement therapy has been permitted for six lysosomal storage diseases, and rDNA human enzymes are under clinical trials (Kaur et al., 2012).

The therapeutic and industrial values of marine microbes to produce bioactive compounds would have arisen due to the fact of competition for space and nutrition that might have put pressure on them for this production. The recent development of importance related to the isolation of bioactive compounds from the marine environment is due to their potential applications in the pharmacological industry. Because of the enzymes' unique advantages, several manufacturers showed profound interest in following enzymatic methods according to their requirements for their clinical products.

9.4 MICROFLORA AND THE FISH GUT

About 28,000 fish species comprise nearly half of all vertebrate diversity and thus represent a wide range of physiologies, ecologies and natural history

(Nelson, 2006). For acquiring knowledge about the evolution and ecology of host–microbiota interactions, fish represent an important vertebral group (Nayak, 2010). Currently culture-based methods are used in obtaining information regarding intestinal microflora, which often show only a limited range of microbial diversity, and the culture-independent DNA sequence-based methods are deployed to define the bacterial diversity (Roeselers et al., 2011).

The health status of the fish is mostly dependent on the environment. They are in proximity with a wide range of microorganisms, which include various pathogenic and opportunistic bacteria that may inhabit the external and internal surfaces of the body (Ellis, 2001). Immediately after hatching, the colonization of the gastrointestinal tract of larvae starts and completes within a few hours. The expression of genes in the fish digestive tract can be modulated by the colonizing bacteria, thus building a promising habitat for themselves and restricting invasion by other bacteria into the ecosystem (José Luis *et al.,* 2006).

9.5 ACUTE LYMPHOBLASTIC LEUKEMIA

Acute lymphoblastic leukemia (ALL) is a type of blood cancer and bone marrow. The disease advances rapidly and produces immature cells of the blood rather mature ones so the term *acute* is used. The term *lymphocytic* addresses the white blood cells called lymphocytes. The other name for ALL is lymphoblastic leukemia. Children are the maximal target for this kind of cancer, and there are good chances for recovery, although this is not the case with adults. Symptoms may include pale skin color, feeling tired, fever, inflamed lymph nodes and easy bleeding or bruising (National Cancer Institute, 2017). Lymphoblasts were overproduced in the people with ALL in the bone marrow; they constantly multiply, thereby restricting the formation of normal cells in the bone marrow (Seiter et al., 2014).

Per Rytting (2012), acute leukemia occurs when there is malevolent transformation happens in the hematopoietic cells with abnormal existence. The two different cells such as ALL or acute myelocytic leukemia (AML) proliferate unusually, substituting normal bone marrow tissue and hematopoietic cells, which induce anemia, granulocytopenia and thrombocytopenia.

Many of the investigations focused on the variability in the genetic level in xenobiotic metabolism. The development of ALL may be affected by the DNA repair pathways and functions of cell-cycle checkpoints that might depend on dietary, environmental and other external factors. Although there are a limited number of investigations, reports exist to support a possible role for polymorphisms in the genes that are coding for cytochrome P450, glutathione S-transferases, nicotinamide adenine dinucleotide phosphate (NAD(P)H) quinone oxidoreductase, serine hydroxymethyltransferase, thymidylate synthase and cell-cycle inhibitors.

Although the sample sizes and number of investigations are limited, there are data to validate a possible causal role for polymorphisms in genes encoding glutathione S-transferases, cytochrome P450, methylenetetrahydrofolate reductase, NAD(P)H quinone oxidoreductase, thymidylate synthase, cell-cycle inhibitors and serine hydroxymethyltransferase (Krajinovic et al., 2002, Skibola et al., 2002; Lanciotti et al., 2005; Healy et al., 2007; Gast et al., 2007).

At least five multistep mutational pathways have been witnessed in T-cell ALL, which leads to frank leukemia, and in certain circumstances, these pathways involve five or more documented genetic lesions (Ferrando et al., 2002; Armstrong and Look, 2005; Grabher et al., 2006). Current research on antileukemic experiments depends mostly on animal models that accurately mimic the molecular pathogenesis of B-cell precursor or T-cell lymphoblastic leukemia (O'Neil et al., 2004). New models from zebrafish for T-cell ALL provide a promising alternative vertebrate system for researching leukemia (Amatruda et al., 2002). Ongoing research on ALL in the zebrafish model includes a *myc* transgene-driven system, in which lymphoblasts hopefully reproduce the multistep tumorogenic pathway observed in up to 60% of human T-cell ALL (Langenau et al., 2003; Langenau et al., 2005), and a transgenic zebrafish model, of which TEL-AML1 oncoprotein stimulates B-cell precursor leukemia (Sabaawy et al., 2006).

9.6 L-ASPARAGINASE: IMPORTANT SOURCES AND MEDICAL APPLICATIONS

In 1978, the United States approved L-asparaginase for medical use (Distasio, 1976). It is listed as an essential medicine by the World Health Organization (WHO), and it is also regarded as one of the most prominent and safest medicines that are in high demand for healthy living. In 1953, the anticancer effect of L-asparaginase was found by Kidd. The lymphomas in rats and mice were witnessed to regress after treating them with the guinea pig serum (Dolowy et al., 1966). Later, it was confirmed that the enzyme L-asparaginase was responsible for the tumor regression (Drainas and Pateman, 1977). Dolowy was the one who first used this enzyme in a human patient (Dunlop et al., 1978).

Recently great prominence has been laid on usage of L-asparaginase as a chemotherapeutic and antileukemic agent for the treatment of ALL (Warangkar et al., 2008; Siddalingeshwara et al., 2010). In 1961, Broom successfully demonstrated that the antileukemic activity of the guinea pig serum was due to L-asparaginase. L-asparaginase was also found in numerous animal sources namely, rat liver, fish tissue, liver, brain, pancreas, ovary, spleen, testes, lung and kidney of many birds and mammals.

There are two types of L-asparaginase. Humankind expresses Type I asparaginase, which is expressed in the cytoplasm and is involved in the hydrolysis activity of both L-Asn and L-Gln. Type II L-asparaginase is expressed in the periplasmic space of the membranes of the bacteria under anaerobic conditions, and they show greater specificity for hydrolysis of L-asparagine (Abdel and Olama, 2002).

The synthesis of L-asparagine a nonessential amino acid is stopped in the tumor cells owing to the fact that it lacks aspartate-ammonia ligase activity. The normal cells are never affected by the L-asparaginase enzyme because they have the potentiality to synthesize L-asparagine for their own metabolism. But the presence of L-asparaginase may minimize the free exogenous concentration of L-asparagine that causes the tumor cells to starve and ultimately results in death. The intravenous administration of the enzyme is effective only when the levels of L-asparagine in the bloodstream are exceptionally low (Gurung et al., 2013).

The irreversible conversion of blood glutamine into glutamic acid and ammonia is because of the presence of glutaminase activity of asparaginase. The resulting glutamate reacts with sodium in blood and gives rise to the production of monosodium glutamate (Kurtzberg et al., 2003). There are reports suggesting that due to the presence of glutaminase activity of asparaginase, leukemia patients suffer from many life-threatening side effects, such as leucopenia, acute pancreatitis, immunosuppression, hyperglycemia, thromboembolysis and neurological seizures (Devi et al.,2012; Ramya et al.,2012). Therefore, it is necessary to make a search for glutaminase-free asparaginase from the native microorganisms. In the control of leukemia, the pharmacodynamics of asparaginase differs by formulation (Pinheiro and Boos, 2004). The treatment mainly depends on the intensity of the dose and duration of the treatment of asparaginase rather than the type of asparaginase used (Silverman et al., 2001; Pui and Evans, 2006). Currently, L-asparaginase obtained from *Erwinia carotovora* and *Escherichia coli* is of commercial importance. Other microbes, such as *Bacillus sp., Corneybacterium glutamicum, Enterobacter sp., Pseudomonas stutzeri* and others, also produce a feasible amount of enzyme. As far as fungi are concerned, *Aspergillus oryzae* was found to synthesis large amount of enzyme.

Comparatively, L-asparaginase from *Erwinia* was found with lesser antileukemic activity than *E. coli*, which has got fewer side effects (Duvalet al., 2002; Moghrabi et al., 2007). L-asparaginase is greatly dispersed among plants, animals and microorganisms. Numerous microorganisms are known to synthesize L-asparaginase, which includes *E. carotovora*, *E. coli* (Warangkar et al., 2008), actinomycetes (Dharmaraj, 2011), *Bacillus sp.* (Ebrahiminezhad et al., 2011), *Bacillus licheniformis* (Jain et al., 2012), *Cladosporium sp.* (Kumar et al., 2013), *Thermococcus kodakaraensis* (Chohan et al., 2013), *Bacillus tequilensis* and *Bacillus subtilis* (Pradhan et al., 2013), among others. Commercially available L-asparaginase is mostly produced from bacteria, and there are reports of hypersensitivity, which leads to the existence of allergic reactions and anaphylaxis owing to the long-term usage of bacterial L-asparaginase. Eukaryotic L-asparaginase obtained from yeast and filamentous fungi have proved to show fewer side effects than the asparaginase from a bacterial source. The persistence of a mechanism known as nitrogen catabolic repression makes fungal asparaginase more advantageous than bacterial asparaginase, which represses the production of unwanted catabolic enzymes during enzyme production. L-asparaginase is presently utilized from three sources: *E.coli* L-asparaginase, *E. crysanthemi* L-asparaginase and polyethylene glycol (PEG)–conjugated L-asparaginase (Rytting, 2012). Recently, the initial treatment for leukemia started with PEG-conjugated asparaginase than the native product because of the long-acting and less allergic capability (Avramis et al., 2002).

9.7 DIFFERENT TYPES OF L-ASPARAGINASE

9.7.1 *E. coli* L-Asparaginase

The most studied form of asparaginase is from the bacteria *E. coli* and is currently administered as a drug against acute leukemia. The intravenous or intramuscular administration of the drug is presently practiced for the adult patient with leukemia at the doses of 25000IU/m^2, and children were given 2500–5000IU/m^2 on a schedule of 10 for eight doses. The drug successfully depletes the L-asparagine in the spinal fluid by not crossing the blood-brain barrier (Riccardi et al.,1981; Ahlke et al.,1997). The half-life of enzymatic activity is reported to be approximately 1.24 days in children.

9.7.2 *Erwinia* L-Asparaginase

Worldwide treatment of ALL currently involves the asparaginase produced from *E. chrysanthemi*. The United Nations has approved *Erwinia* for the treatment of leukemia. When compared to bacterial L-asparaginase, *Erwinia* L-asparaginase has a decreased half-life; hence, the drug must be given in greater doses and more often so as to remove asparagine completely (Asselin et al.,1999). According to Boos et al. (1996), the patients receiving *Erwinia* recover their asparagine level more quickly. Allergic reactions caused by the long-term use of bacterial L-asparaginase can be effectively overcome by the L-asparaginae from *Erwinia*, which makes it a prominent drug (Panosyan et al., 2004).

9.7.3 PEG-Conjugated L-Asparaginase

Avramis and Panosyan (2005) reported that PEG-conjugated by asparaginase produced from *E. coli* showed increased circulation time for the enzyme together with decreased immunogenicity of asparaginase. Patients inheriting antibodies to PEG removed the drug more swiftly than did those with no antibodies to PEG.

9.8 STRUCTURE OF L-ASPARAGINASE

Two distinct L-asparaginases with distinct properties have been studied in *E. coli*, especially in their affinities toward the substrate asparagine. Documentation of crystallographic structure of L-asparaginase II was done by Swain et al. (1993). The enzyme was observed to be a homo-tetramer which belongs to α/β class of proteins consisting of 222-symmetry. There are two subunits present on each domain that has specific topological features. Ramya et al. (2012) proposed that the majority of the bacterial L-asparaginase share the same tertiary and quaternary structures and some common biochemical characteristics. The tetramer of identical subunits designated as A–D (Kozak et al., 2002) of the functional form of *E. coli* L-asparaginase II was bound by the noncovalent forces with the molecular weight ranging between 140–160 kDa

(Aung et al., 2000; Kozak et al., 2002). Every monomer consists of about 330 amino acids residues, which form 14 β strands and 8 α-helices that are arranged into two easily identifiable domains; the bigger N-terminal domain and the smaller C terminal domain are connected by a linker molecule consisting of approximately 20 residues (Pourhossein and Korbekandi, 2014).

Asparagine synthetase is a large enzyme composed of two similar subunits. The given structure (Figure 9.1) has been referred to be the enzyme isolated from bacteria. This enzyme is solely responsible for the production of asparagine by combining with ammonia molecule directly to aspartic acid. The enzyme uses glutamine in the case of humans to give the amine instead of ammonia. L-asparaginase (Figure 9.2) is obtained in the purified form from the bacterial cells and is used in chemotherapy. It is composed of four identical subunits.

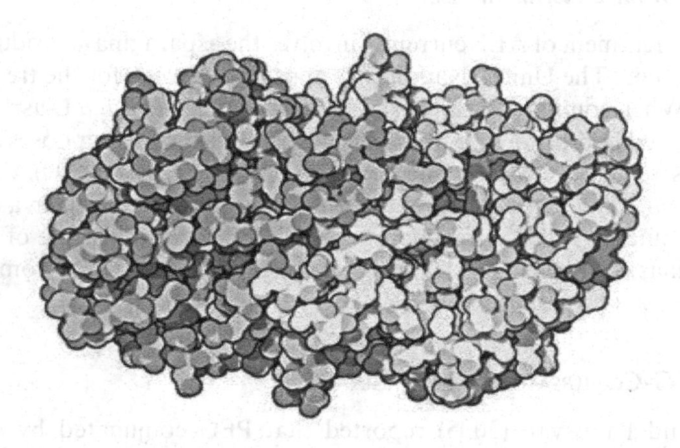

FIGURE 9.1 Structure of (dimer) asparagine synthetase (Kumar Jha *et al.*).

FIGURE 9.2 Structure of (tetramer) asparaginase (Kumar Jha *et al.*).

Grip asparagine, which is the active site of the enzyme (red), makes use of a well-placed amino acid threonine (green) to perform the reaction of cleavage. L-asparaginase tetramer can be termed as a dimer of dimers because of the presence of each of the four active sites located in between the N- and C-terminal domains of two adjacent monomers. In spite of the existence of structural elements and functional groups to form a complete active site environment, the active L-asparaginase enzyme is regarded to be a tetramer with each domain containing one active center and catalysis of the hydrolysis of L-asparagine to L-aspartic acid and ammonia (Khushoo et al., 2004). Conserved residues strictly inhabit the formation of the active site. A flexible loop that is a part of the active site (between 10–40 residues) contains the residues threonine-12 and tyrosine-25 that are important. Threnine-12 is the nucleophile actively involved in the acylation reaction (Aung et al., 2000). It is this flexible loop that controls the access to the active site cavity that opens and closes in a ligand-dependent fashion (Aung et al., 2000; Kozak et al., 2002).

9.9 EXTRACELLULAR AND INTRACELLULAR L-ASPARAGINASE

The bacterial enzymes are broadly classified into two groups: (1) intracellular enzymes or endozymes and (2)extracellular enzymes. Intracellular enzymes function inside the bacterial cells. They are duly responsible for synthesizing new protoplasmic substances and in the production of the cellular energy into the cells through cell membranes. The extracellular enzymes hydrolyze the complex compounds into their respective simple forms. Due to these processes, metabolic products are formed and excreted by the cells in the environment. Determining these end products helps identify bacteria (Duly and Nannipieri, 1998).

The majority of the microbial L-asparaginase is exceptionally intracellular in nature except for a few microbial enzymes that are extracellular, that is, secreted outside the cell (Narayana et al., 2008). Comparatively extracellular enzymes are more preferred to intracellular because of greater collection of enzyme in the culture broth under normal environment, simple extraction and downstream processing. Moreover, the extracellular bacterial L-asparaginase is deficient in protease, and the released protein carried to the medium is generally soluble, biologically active and has a promising N-terminus, free from endotoxins, which ultimately results in a reduction of adverse effects (Amena et al., 2010). *Citrobacter sp.* has been reported to produce L-asparaginase intracellularly (Bascomb et al., 1975). In the study performed by Hari Krishnan et al. (2016), the isolated marine bacteria obtained from the coastal regions of Kerala showed both intracellular and extracellular L-asparaginase activity. Thus, the screened *Bacillus sp.* showed extracellular activity while *Shewanella sp.* showed greater intracellular asparaginase activity.

9.10 PRODUCTION OF L-ASPARAGINASE

Commercially producing microbial enzymes involves various fermentation technologies (Sabu et al., 2000). They are solid-state fermentation, submerged-type fermentation and immobilization, among others. These techniques are mainly preferred for bulk production for commercial purpose (Lozano et al., 2012). When compared to submerged fermentation, solid-state fermentation is more advantageous. It needs minimum control and ease of product recovery; the possibility of contamination is also less and it involves simpler methods to treat the fermented residues. Both the processes involve extraction, centrifugation, precipitation, evaporation, filtration and concentration in order to obtain the pure enzyme (Saleem Basha et al., 2009). A significant variation in enzyme production was observed between fermentation of solid and submerged state in the *Lactobacillus sp.* from a marine water sample. They suggest that the difference may have developed due to the accumulated intermediate metabolites between the substrate and product formed in submerged fermentation. The result may be probably due to the change in the physiological condition of the microorganism in solid-state and submerged fermentation (Bhargavi and Jaya Madhuri, 2017).

The enormous amount of therapeutic enzyme production is made simpler due to progress of recombinant DNA technology. This provokes the enzyme activity and its stability for a longer time. Enzyme manufacturers are able to produce adequate amounts of enzymes from any microbial source through the technology involved in genetic engineering. The properties of enzymes can be adjusted prior to its production by the means of protein engineering (Kaul, 2008).

9.11 ACTION MECHANISM OF L-ASPARAGINASE

Asparagine is necessary for DNA synthesis and survival cells; however, most cells have the ability to synthesize asparagine from glutamine. Asparaginase is highly particular for the cell cycle for G1 phase (Rose, 2006; Rytting, 2012). The chemistry and mechanism of action of L-asparaginase (Figure 9.3) are reported by many workers (Avramis and Panosyan, 2005 and Zeidan *et al.*, 2009). Normal cells synthesize asparagine by the activity of asparagine synthetase for their survival, which does not take place in malignant cells. Tumor cells need asparagine, which is completely dependent on the exogenous environment in the circulation of blood (Swainet al., 1993).

9.12 ROLE OF L-ASPARAGINASE IN THE METABOLISM OF AMINO ACIDS

In the biosynthesis of the aspartic family of aminoacids, L-asparaginase plays a very important role (Figure 9.4). Great industrial interest was developed in the *Cornybacteria*-producing aminoacids since they excrete large amounts of different aminoacids. The industrially important aminoacids produced by *Cornybacterium glutamicum* are methionine, threonine and lysine, which, under normal physiologic conditions, might limit for the biosynthesis of lysine and/or threonine biosynthesis. The Kreb's cycle involves converting asparagine to aspartate (using glutamic acid as an aminoacid donor) by the action of

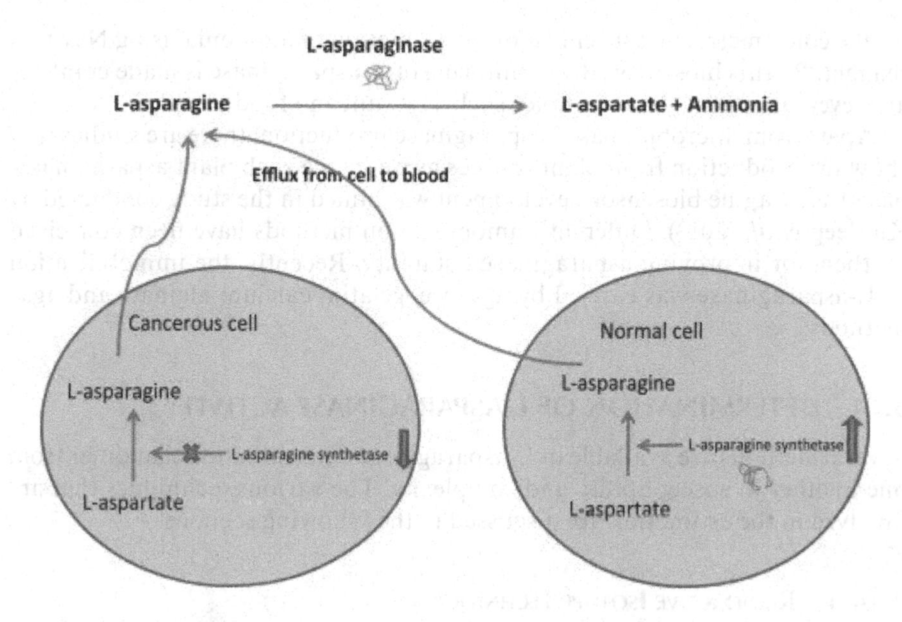

FIGURE 9.3 Mode of action of L-asparaginase (Noura et al., 2014).

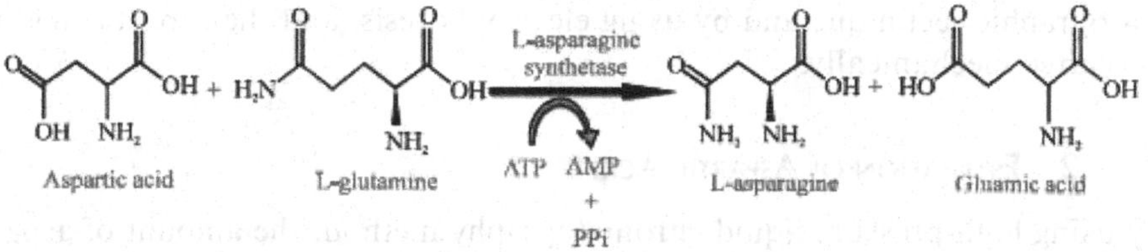

Aspartic acid + L-glutamine → (L-asparagine synthetase; ATP → AMP + PPi) → L-asparagine + Glutamic acid

FIGURE 9.4 L-asparaginase biosynthesis: aspartic acid is trans-aminated to asparagine and glutamine (Noura et al., 2014).

asparaginase. L-asparaginase production takes place as a result, controlling the surplus enzyme that converts extra asparagine to aspartic acid, which is the direct precursor of lysine and threonine (Mesas et al., 1990).

9.13 IN MAKING BIOSENSORS

The extent of ammonia produced by the enzymatic action of L-asparaginase is used for making a diagnostic biosensor, which directly interprets the level of L-asparagine in the patient's blood (Verma et al., 2007). Nikolaos and Mohamed Muharram (2016), in their study, used the enzyme to assemble a microplate of 96-well format-based biosensor for the estimation of L-asparagine in the biological samples. In this technique, the immobilization of the enzyme was done by crosslinking with glutaraldehyde, and the sensory scheme was based on the colorimetric measurement of the formation of ammonia using Nessler's reagent. By this biosensor, the monitoring of L-asparaginase is made comfortable even in small-volume samples such as serum and food samples.

Apart from microbial-based asparaginase production, there are studies that show the production from plant sources as well. One such plant asparaginase-based asparagine biosensor development was aimed in the study conducted by Kuldeep *et al.* (2013). Different immobilization methods have been conceived by them for improving asparaginase's stability. Recently, the immobilization of L-asparaginase was carried by agarose, gelatin, calcium-alginate and agar methods.

9.14 DETERMINATION OF L-ASPARAGINASE ACTIVITY

Several methods are available in L-asparaginase determination that differ from one another in susceptibility and complexity. The various techniques that are involved in the estimation are discussed in the following sections.

9.14.1 Radioactive Isotope Technique

In this technique, radio-labeled asparaginase ($_{14}C$-asparaginase) was used as a substrate by Ehrman *et al.* (1970). The aspartic acid that forms is estimated at the end of the reaction. The formed aspartic acid is separated by a paper chromatographic technique and by using electrophoresis, and the aspartic acid is measured mechanically.

9.14.2 Estimation of Aspartic Acid

By using high-pressure liquid chromatography method, the amount of aspartic acid produced by the hydrolysis of asparagine by asparaginase is estimated when the reaction terminates by methanol after the free amino acids are labeled. They are available in the mixture with O-phthaldialdehyde (Barry Jones *et al.,* 1981).

9.14.3 Indophenol Assay

This assay involves calorimetric quantification using a hypochloride reagent. The liberated ammonia from the asparagine due to hydrolysis of asparaginase reacts with hypochloride in the presence of phenol to form indophenols, which can be estimated at A_{625nm} (Tetsuya Tosa *et al.*, 1972).

9.14.4 Fluorometric Assay

The activity of the asparaginase is estimated by using fluorometric compound L-aspartic acid-β-7-amino-4-methylcoumarin (Asp AMC), which releases 7-amino-4-methlcoumarin by the hydrolysis of L-asparaginase and measured by a spectrofluorometer at the wavelengths of 465nm and 360nm for the process of excitation and emission, respectively (Ylikangas et al., 2000)

9.14.5 Hydroxylamine Method

By this method, the amount of asparagine present in the medium is estimated using hydroxylamine hydrochloride, which converts the asparagine into aspartyl β-hydroxamate by the absorbance at A_{515nm} (Mesas et al., 1990).

9.14.6 Nesslerization Assay

This is a widely used method for the estimation of asparaginase because of its rapid and easy procedure. This method involves estimating ammonia liberated by asparaginase by hydrolysis of asparagine. The estimated ammonia is directly proportionate to the amount of asparaginase (Imada *et al.*, 1973).

9.14.7 Plate Method

Semi-quantitative plate method developed by Gulati *et al.* (1997) is the one that involves estimating based on the production of ammonia due to hydrolysis of L-asparagine. It is degraded by glutamate dehydrogenase subsequently with the oxidation of β-NADH. The reduction in β-NADH is estimated spectrometrically at 340nm.

9.15 ROLE OF L-ASPARAGINASE IN FOOD-PROCESSING INDUSTRY

L-asparaginase is being used in the food industry to reduce acrylamide formation, which is suspected to be a carcinogen produced in starchy food products (Amrein et al., 2004; Kornbrust et al., 2010). Cases are reported in potato chips, in which the enzyme has reduced the formation of acrylamide in fried and overcooked food materials (Rosen et al., 2002; Tareke *et al.,* 2002; Rosen and Hellenas, 2002). Maillard reaction is the process involved in the acrylamide formation in heated food due to the reaction of asparagine and reducing sugars (1912), which is the causal agent for the appearance of brown color in the fried and baked foods. The acrylamide formation can be minimized by the deamination mechanism of asparagine, which is available in the food substance (Ciesarová et al., 2006), whereas the complete elimination of the acrylamide is not at all possible due to other asparagine pathways (Kornbrust et al.,2010). Hendriksen et al. (2009) observed the effective reduction of acrylamide to a significant amount, up to 90% in starch-containing food items, without changing the taste and appearance. In this regard, asparaginase obtained from *Aspergillus oryzae* and *Aspergillus niger* is used in the baking industry (Morales et al., 2008). The optimum working temperature and pH of this enzyme is 40–60°C and 6.0–7.0, respectively. Generally, baking temperatures reach up to

120°C; hence, it is necessary to have an enzyme that is active at a wide range of temperatures and pH. Therefore, L-asparaginase from various sources has been studied for both therapeutic and industrial applications.

9.16 FACTORS AFFECTING THE PRODUCTION OF L-ASPARAGINASE

9.16.1 Effect of Carbon and Nitrogen Sources

The physio-chemical conditions of the production of L-asparaginase differ with the microorganisms. The production of the enzyme is directly proportionate to the components used in the medium. Various carbon sources (sucrose, fructose, maltose, mannitol, lactose) and nitrogen sources (soya bean, beef extract, yeast extract, peptone, ammonium sulfate) are used for producing L-asparaginase (Rati Kumari *et al.*, 2014).

9.16.2 Effect of pH and Temperature on L-Asparaginase Production

Krishna and Nibh (2012), in a study on the production of L-asparaginase from *Penicillium sp.*, recorded that the optimum pH for L-asparaginase production using a 0.05M Tris-HCl buffer ranges from 3–10 showed a maximum enzyme activity at pH 7, and the purified L-asparaginase was recorded maximum activity at 37°C. Normally L-asparaginase from the serum of guinea pig has pH 7.5–8.5. At reduced pH and a temperature of 20°C, it maintained its stability for a period of 6 months during repeated freezing and thawing, and it was subjected to surface denaturation when heating to 55°C (Mishra, 2006).

9.17 CYTOTOXIC EFFECT

Treating cells with the cytotoxic compound may lead to various cell fates. Due to the necrosis that the cells undergo, they lose membrane integrity and die quickly as a result of cell lysis. The cells will lose their vigorously growing and dividing capacity (a reduction in cell viability), or the cells can begin a program of controlled cell death (apoptosis).

The cells undergoing necrosis typically reveal quick swelling, lose the integrity of the membrane, hinder metabolism and discharge their contents into the environment. Cells that undergo rapid necrosis *in vitro* do not have enough time or energy to activate the machinery of apoptosis and so will not express apoptotic markers. Apoptosis is characterized by unique cytological and molecular actions that include a modification in the refractive index of the cell, nuclear condensation, cytoplasmic shrinkage and cleavage of DNA into regularly sized fragments. Cells in a culture that endure apoptosis eventually undergo secondary necrosis. They will stop metabolism, lose membrane integrity and lyse (Riss et al., 2004).

Assessing cell membrane integrity is one of the most common ways to determine the viability of the cell and its cytotoxic effects. Compounds that possess cytotoxic effects often compromise cell membrane integrity. Vital dyes, such as propidium iodide or trypan blue, are normally excluded from the inside of healthier cells; however, if the cell membrane is negotiated, they easily cross the membrane and stain the intracellular components (Riss et al., 2004). The extracellular enzyme possessed a cytotoxic effect on HL60 cell line (Hari Krishnan et al., 2016). The traditional methods of identifying the bacterial strain rely on culturing technique. Such techniques have certain limitations. In order to overcome those limitations, molecular characterization came into force. 16s rRNA sequencing can give much more information than culturing methods. Metagenomics is another field that involves the study and analysis of microbial communities from the surrounding environment without culturing. It also affords the potential to determine novel enzymes through function-based screening. Hence, metagenomics has given the scientific community with a range of novel enzymes (Zhang and Kim, 2010).

REFERENCES

Abdel, F.Y.R. and Olama, Z.A. 2002. L-asparaginase production by *Pseudomonas aeroginosa* by solid state culture: Evaluation and optimization culture conditions using factorial designs. *Process Biochemistry.*, 38: 115–122.

Ahlke, E., Nowak-Gottl, U., Schulze-Westhoff, P., Werber, G., Borste, H., Wurthwein, G., Jurgens, H. and Boos, J. 1997. Dose reduction of asparaginase under pharmacokinetic and pharmacodynamic control during induction therapy in children with acute lymphoblastic leukaemia. *British Journal of Haemtology* 96: 675–681.

Amatruda, J.F., Shepard, J.L., Stern, H.M. and Zon, L.I. 2002. Zebrafish as a cancer model system. *Cancer Cell*, 1: 229–231.

Amena, S., Vishalakshi, N., Prabhakar, M., Dayanand, A. and Lingappa, K. 2010. Production, purification and characterization of L-asparaginase from *Streptomyces gulbargensis*. *Brazilian Journal of Microbiology*, 41: 173–178.

Amrein, T.M., Schönbächler, B., Escher, F. and Amado, R. 2004. Acrylamide in gingerbread: Critical factors for formation and possible ways for reduction. *Journal of Agriculture and. Food Chemistry*, 52(13): 4282–4288.

Armand, J.-V., Ady-Vago, N. and Faivre, S. 2001. Phase I and pharmacokinetic study of aplidine (apl) given as a 24-hour continuous infusion every other week (q2w) in patients (pts) with solid tumor (st) and lymphoma (NHL). In *Proceeding of 2001 ASCO Annual Meeting*. American Society of Clinical Oncology, San Francisco, CA, USA.

Armstrong, S.A. and Look, A.T. 2005. Molecular genetics of acute lymphoblastic leukemia. *Journal of Clinical Oncology*, 23: 6306–6315.

Asselin, B.L., Kreissman, S., Coppola, D.J., Bernal, S.D., Leavitt, P.R., Gelber, R.D., Sallan, S.E. and Cohen, H.J. 1999. Prognostic significance of early response to a single dose of asparaginase in childhood acute lymphoblastic leukaemia. *Journal of Pediatric Hematology/Oncology*, 21: 6–12.

Aung, H.P., Bocola, M., Schleper, S. and Rohm, K.H. 2000. Dynamics of a mobile loop at the active site of *Escherichia coli* asparaginase. *Biochimica Biophysica Acta (BBA)-Protein Structure Molecular Enzymology*, 1481: 349–359.

Avramis, V.I. and Panosyan, E.H. 2005. Pharmacokinetic/pharmacodynamic relationships of asparaginase formulations. *Clinical Pharmacokinetics*, 44(4): 367–393.

Avramis, V.I., Sencer, S., Periclou, A.P., *et al.* 2002. A randomized comparison of native *Escherichia coli* asparaginase and polyethylene glycol conjugated asparaginase for treatment of children with newly diagnosed standard-risk acute lymphoblastic leukemia: A Children' s Cancer Group study. *Blood*, 99: 1986–1994.

Badr, E.I. and Foda, M.S. 1976. Kinetics and properties of L-asparaginase and L-glutaminase activities of *Pseudomonas ovalis*. *Zentralbl Bakteriol Parasetenkd Infektionskr Hyg*., 131: 489–496.

Baharum, S.N., Beng, E.K. and Mokhtar, M.A.A. 2010. Marine microorganisms: Potential application and challenges. *Journal of Biological Sciences*, 10: 555–564.

Bai, R., Friedman, S.J., Pettit, G.R. and Hamel, E. 1992. Dolastatin-15, a potent antimitotic depsipeptide derived from *Dolabella auricularia*: Interaction with tubulin and effects on cellular microtubules. *Biochemica Pharmacology*, 43: 2637–2645.

Barry, Jones, Svante, Pääbo, and Stanley, Stein. 1981. Amino acid analysis and enzymatic sequence determination of peptides by an improved o-phthaldialdehyde precolumn labeling procedure. *Journal of Liquid Chromatography*, 4(4): 565–586.

Bascomb, S., Banks, G.T. and Skarstedt, M.T. 1975. The properties and large-scale production of L-asparaginase from *Citrobacter*. *Journal of General Microbiology*, 91: 1–16.

Bell, T.L. and Adams, M.A. 2004. Ecophysiology of ectomycorrhizal fungi associated with *Pinus spp*. *Plant Ecology*, 171(1–2); 35–52.

Bhargavi, M. and Jaya Madhuri, R. 2017. Evaluation of cytotoxic activity of l-asparaginase from marine BKJM2 against JURKAT J6 and PA1 Celllines. *International Journal of Pharma and Bio Sciences*, 8(4): 269–274.

Boos, J., Werber, G., Ahlke, E., Schulze-Westhoff, P., Nowak-Gottl, U., Wurthwein, G., Verspohl, E.J., Ritter, J. and Jurgens, H. 1996. Monitoring of asparaginase activity and asparagine levels in children on different asparaginase preparations. *European Journal of Cancer*, 32A: 1544–1550.

Borek, D., Michalska, K., Brzezinski, K., Kisiel, A., Podkowinski, J., Bonthron, D.T., Krowarsch, D., Otlewski, J. and Jaskolski, M. 2004. Expression, purification and catalytic activity of *Lupinus luteus* asparagines ß-amidohydrolase and its *Escherichia coli* homolog. *European Journal of Biochemistry*, 271(15): 3215–3226.

Broggini, M., Marchini, S.V., Galliera, E., Borsotti, P., Taraboletti, G., Erba, E., Sironi, M., Jimeno, J., Faircloth, G.T., Giavazzi, R., *et al*. 2003. Aplidine, a new anticancer agent of marine origin, inhibits vascular endothelial growth factor (VEGF) secretion and blocks VEGF-VEGFR-1 (flt-1) autocrine loop in human leukemia cells MOLT-4. *Leukemia*, 17: 52–59.

Bruneau, L., Chapman, R. and Marsolais, F. 2006. Co-occurrence of both L-asparaginase subtypes in *Arabidopsis*: At3g16150 encodes a K^+-dependent L-asparaginase. *Planta*, 224: 668–679.

Chohan, S.M. and Rashid, N. 2013. TK1656, a thermostable l-asparaginase from *Thermococcus kodakaraensis*, exhibiting highest ever reported enzyme activity. *Journal of Bioscience and. Bioengineering*, 116: 438–443.

Ciesarová, Z., Kiss, E. and Boegl, P. 2006. Impact of L-asparaginase on acrylamide content in potato products. *Journal of Food and Nutrition Research*, 45: 141–146.

Cooney, D.A. and Rosenbluth, R.J. 1975. Enzymes as therapeutic agents. *Advances in Pharmacology*, 12: 185–289.

Devi, S. and Azmi, W. 2012. One step purification of glutaminase free L-asparaginase from *Erwinia carotovora* with anti-cancerous activity. *International Journal of Life Science and Pharma Research*, 2: 56–61.

Dharmaraj, S. 2011. Study of L-asparaginase production by *Streptomyces noursei* MTCC 10469, isolated from marine sponge *Callyspongia diffusa*. *Iranian Journal of Biotechnology*, 9: 102–108.

Distasio, J.A., Niederman, R.A., Kafkewitz, D. and Goodman, D. 1976. Purification and characterization of L-asparaginase with anti-lymphoma activity from *Vibrio succinogenes*. *Journal of Biological Chemistry*., 251: 6929–6933.

Dolowy, W.C., Henson, D., Cornet, J. and Sellin, H. 1966. Toxic and antineoplastic effects of L-Asparaginase: Study of mice with lymphoma and normal monkeys and report on a child with leukemia. *Cancer*, 19: 1813–1819.

Drainas, C. and Pateman, J.A. 1977. L-Asparaginase activity in the fungus *Aspergillus nidulans*. *Biochemical Society Transactions*, 5: 259–261.

Duly, O. and Nannipieri, P. 1998. Intracellular and extracellular enzyme activity in soil with reference to elemental cycling. *Journal of Plant Nutrition and Soil Science*, 243–248.

Dunlop, P.C., Meyer, G.M., Ban, D. and Roon, R.J. 1978. Characterization of two forms of asparaginase in *Saccharomyces crevisiae*. *Journal of Biological Chemistry*, 253(4): 1297–1304.

Duval, M., Suciu, S., Ferster, A., *et al.* 2002. Comparison of *Escherichia coli*-asparaginase with *Erwinia*-asparaginase in the treatment of childhood lymphoid malignancies: Results of a randomized European Organisation for Research and Treatment of Cancer-Children's Leukemia Group phase 3 trial. *Blood*, 99:2734–2739.

Ebrahiminezhad, A., Rasoul-Amini, S. and Ghasemi, Y. 2011. L-Asparaginase Production by Moderate Halophilic Bacteria Isolated from Maharloo Salt Lake. *Indian Journal of Microbiology*, 51: 307–311.

Ehrman, Mark, Howard, Cedar and James, H. Schwartz. 1970, January 10. L-Asparaginase II of *Escherichia coli*: Studies of the enzymatic mechanism of action. *The Journal of Biological Chemistry*, 246(1): 88–94.

Ellis, A.E. 2001. Innate host defense mechanisms of fish against viruses and bacteria. *Developmental & Comparative Immunology*, 25(8–9): 827–839.

Faivre, S., Chieze, S., Delbaldo, C., Ady-Vago, N., Guzman, C., Lopez-Lazaro, L., Lozahic, S., Jimeno, J., Pico, F., Armand, J., *et al.* 2005. Phase I and pharmacokinetic study of aplidine, a new marine cyclodepsipeptide in patients with advanced malignancies. *Journal of Clinical Oncology*, 23: 7871–7880.

Ferrando, A.A., Neuberg, D.S., Staunton, J., *et al.* 2002. Gene expression define novel oncogenic pathways in T-cell acute lymphoblastic leukemia. *Cancer Cell*, 1: 75–87.

Fusetti, F., Moeller, H.V. and Houston, D. 2002. Structure of human chitotriosidase: Implications for specific inhibitor design and function of mammalian chitinase-like lectins. *Journal of Biological Chemistry*, 277: 25537–25544.

Gast, A., Bermejo, J.L., Flohr, T., *et al.* 2007. Folate metabolic gene polymorphisms and childhood acute lymphoblastic leukemia: A case-control study. *Leukemia*, 21: 320–325.

Pettit, George R., Kamano, K. et al. 1987. The isolation and structure of a remarkable marine animal antineoplastic constituent: Dolastatin 10. *Journal of American Chemical Society*, 109(22): 6883–6885.

Grabher, C., Von Boehmer, H. and Look, A.T. 2006. Notch 1 activation in the molecular pathogenesis of T-cell acute lymphoblastic leukaemia. *Nature Reviews Cancer*, 6: 347–359.

Gulati, R., Saxena, R.K., Gupta, R. 1997. A rapid plate assay for screening L-asparaginase producing micro-organisms. *Letters in Applied Microbiology*, 24(1): 23–26.

Gurung, N., Ray, S., Bose, S. and Rai, V. 2013. A broader view: Microbial enzymes and their relevance in industries, medicine, and beyond. *BioMed Research International*, 18. Hindawi Publishing Corporation.

HariKrishnan, K., Arjun, J.K., Aneesha, B. and Kavitha, T. 2016. Therapeutic L-asparaginase activity of bacteria isolated from marine sediments. *International Journal of Pharmaceutical Sciences and Drug Research*, 8(4): 229–234.

Healy, J., Bélanger, H., Beaulieu, P., *et al.* 2007. Promoter SNPs in G1/S checkpoint regulators and their impact on the susceptibility to childhood leukemia. *Blood*, 109: 683–692.

Hendriksen, H.V., Kornbrust, B.A., Ostergaard, P.R. and Stringer, M.A. 2009. Evaluating the potential for enzymatic acrylamide mitigation in a range of food products using an asparaginase from *Aspergillus oryzae. Journal of Agricultural and Food Chemistry*, 57: 4168–4176.

Hymavathi, M., Sathish, T., Rao, C.S. and Praksham, S. 2009. Enhancement of L-asparaginase production by isolated *Bacillus circulans* (MTCC 8574) using response surface methodology. *Applied Biochemistry and Biotechnology*, 159(1): 191–198.

Imada, A., Igarasi, S., Nakahama, K., Isono, M. 1973. Asparaginase and glutaminase activities of micro-organisms. *Journal of General Microbiology*, 76(1): 85–99.

Jain, R., Zaidi, K.U., Verma, Y. and Saxena, P. 2012. L-Asparaginase: A promising enzyme for treatment of acute lymphoblastic leukiemia. *People's Journal of Scientific Research,* 5: 29–35.

José Luis, Balcázar, Oliver, Decamp, Daniel, Vendrell, Ignacio, De Blas and Imanol, Ruiz-Zarzuela. 2006. Health and nutritional properties of probiotics in fish and shellfish. *Microbial Ecology in Health and Disease*, 18(2): 65–70.

Joner, P.E., Kristiansen, T. and Einasson, M. 1976. Purification and properties of L-asparaginase B from *Acinetobacter calcoaceticus. Biochimica et Biophysica Acta (BBA)—Enzymology,* 438(1): 287–295.

Kaul, R.H. 2008. Enzyme production. Encyclopedia of life support systems(EOLSS). *Biotechnology*, 5. www.eolss.net/Sample-Chapters/C17/E6-58-05-01.pdf.

Khushoo, A., Pal, Y., Singh, B.N. and Mukherjee, K.J. 2004. Extracellular expression and single step purification of recombinant *Escherichia coli* L-asparaginase II. *Protein Expression and Purification*, 38: 29–36.

Kornbrust, B., Stringer, M., Lange Na and Hendriksen, H. 2010. Asparaginase—an enzyme for acrylamide reduction in food products. In *Enzymes in Food Technology*, 2nd ed. Wiley-Blackwell, Hoboken, NJ.

Kozak, M., Borek, D., Janowski, R. and Jaskolski, M. 2002. Crystallization and preliminary crystallographic studies of five crystal forms of *Escherichia coli* L-asparaginase II (Asp90Glu mutant). *Acta Crystallographica Section D Biological Crystallography*, 58: 130–132.

Krajinovic, M., Labuda, D. and Sinnett, D. 2002. Glutathione S-transferase P1 genetic polymorphisms and susceptibility to childhood acute lymphoblastic leukaemia. *Pharmacogenetics*, 12: 655–658.

Krishna Raju, Patro and Nibha, Gupta. 2012. Extraction, purification and characterization of L-asparaginase from *Penicillium* sp. by submerged fermentation. *International Journal for Biotechnology and Molecular Biology Research*, 3(3): 30–34.

Kuldeep, Kumar, Mandeep Kataria and Neelam Kumar. 2013. Plant asparaginase-based asparagine biosensor for leukemia. *Artificial Cells, Nanomedicine, and Biotechnology*, 41(3): 184–188.

Kumar, M., Ramasamy, R. and Manonmani, H.K. 2013. Production and optimization of L-asparaginase from *Cladosporium sp*. using agricultural residues in solid state fermentation. *Industrial Crops and Products*, 43: 150–158.

Kurtzberg, J., Yousem, D. and Beauchamp, N. Jr. 2003. Asparaginase. In D.W. Kufe, R.E. Pollock, R.R. Weichselbaum, et al. (ed.), *Holland-Frei Cancer Medicine*, 6th ed., Chap. 55. BC Decker, Hamilton, ON.

Lanciotti, M., Dufourm, C., Corral, L., *et al.* 2005. Genetic polymorphism of NAD(P)H: Quinone oxidoreductase is associated with an increased risk of infant acute lymphoblastic leukemia without MLL gene rearrangements. *Leukemia*, 19: 214–216.

Langenau, D.M., Feng, H., Berghmans, S., *et al.* 2005. Cre/lox-regulated transgenic zebrafish model with conditional myc-induced T cell acute lymphoblastic leukemia. *Proceedings of National Academic Sciences USA*, 102:6068–6073.

Langenau, D.M., Traver, D., Ferrando, A.A., *et al.* 2003. Myc-induced T cell leukemia in transgenic zebrafish. *Science,* 299: 887–890.

Lozano, S.V., Sepulveda, T.V. and Torres, E.F. 2012. Lipases production by solid fermentation: The case of *Rhizopushomothallicus* in perlite. Lipases and phospholipases: Methods and protocols. *Series: Methods of Molecular Biology,* 861: 227–237.

Manna, S., Sinha, A., Sadhukhan, R. and Chakrabarty, S.L. 1995. Purification, characterization and antitumor activity of L-asparaginase isolated from Pseudomonas stuzeri MB-405. *Current Mocrobiology*, 30: 198–291.

Mesas, J.M., Gill, J.A. and Martin, J.F. 1990. Characterization and partial purification of L-asparaginase from *Corynebacterium glutamicum. Journal of General Microbiology*, 136: 515–519.

Michalska, K., Bujacz, G. and Jaskolski, M. 2006. Crystal structure of plant asparaginase. *Journal of Molecular. Biology*, 360: 105–116.

Minuzzo, M., Marchini, S., Broggini, M., Faircloth, G., D'incalci, M. and Mantovani, R. 2000. Interference of transcriptional activation by the anti-neoplastic drug ET-743. *Proceedings of National Academic Sciences USA,* 97: 6780–6784.

Mishra, A. 2006. Production of L-asparaginase, an anticancer agent from *Aspergillus niger*using agricultural waste in solid state fermentation. *Applied Biochemistry and Biotechnology*, 135: 33–42.

Moghrabi, A., Levy, D.E., Asselin, B., *et al.* 2007. Results of the Dana-Farber Cancer Institute ALL Consortium Protocol 95–01 for children with acute lymphoblastic leukemia. *Blood,*109:896–904.

Mohapatra, B.R., Sani, R.K. and Banerjee, U.C. 1995. Characterization of L-asparaginase from *Bacillus* sp. isolated from an intertidal marine alga (*Sargassum* sp.). *Letters in Applied Microbiology*, 21: 380.

Morales, F., Capuano, E. and Fogliano, V. 2008. Mitigation strategies to reduce acrylamide formation in fried potato products. *Annals of New York Academy of Sciences*, 1126: 89–100.

Mukherjee, J., Majumdar, S. and Scheper, T. 2000. Studies on nutritional and oxygen requirements for production of L-Asparaginase by *Enterobacter aerogenes. Applied Microbiology and Biotechnology,* 53: 180–184.

Narayana, K.J.P., Kumar, K.G. and Vijayalakshmi, M. 2008. L-asparaginase production by *Streptomyces albidoflavus. Indian Journal of Microbiology*, 48: 331–336.

Narta, U., Roy, S., Kanwar, S.S. and Azmi, W. 2011. Improved production of L-asparaginase by *Bacillus brevis* cultivated in the presence of oxygen-vectors. *Bioresource Technology*, 102(2): 2083–2085.

National Cancer Institute. 2017, December 8. Childhood acute lymphoblastic leukemia treatment. *Developmental & Comparative Immunology*, 25(8–9): 827–839.

Nayak, S.K. 2010. Role of gastrointestinal microbiota in fish. *Aquaculture Research*, 41: 1553–1573.

Nelson, J.S. 2006. *Fishes of the World.* 4th ed. John Wiley & Sons, Inc., Hoboken, NJ.

Nguyen, T.H. and Nguyen, V.D. 2017. Characterization and applications of marine microbial enzymes in biotechnology and probiotics for animal health. *Advances in Food and Nutrition Research,* 80: 37–74.

Nikolaos, E. Labrou and Magdy Mohamed Muharram. 2016. Biochemical characterization and immobilization of Erwinia carotovora L-asparaginase in a microplate for high-throughput biosensing of L-asparagine. *Enzyme and Microbial Technology*, 92: 86–93.

Noura, El-Ahmady El-Naggar, El-Ewasy, S.M. and El-Shweihy, N.M. 2014. Microbial L-asparaginase as a potential therapeutic agent for the treatment of acute lymphoblastic leukemia: The pros and cons. *International Journal of Pharmacology,* 10(4): 182–199.

O'Neil, J., Shank, J., Cusson, N., Murre, C. and Kelliher, M. 2004. TAL1/SCL induces leukemia by inhibiting the transcriptional activity of E47/HEB. *Cancer Cell*, 5:587–596.

Ostlie, D.J., Juang, D., Aguayo, P., Pettiford-Cunningham, J.P., Erkmann, E.A. and Rash, D.E. 2012. Topical silversulfadiazine vs. collagenase ointment for the treatment of partial thickness burns in children: A prospective randomized trial. *Journal of Pediatric Surgery,* 47: 1204–1207.

Oza, V.P., Trivedi, S.D., Parmar, P.P. and Subramanian, R.B. 2009. *Withania somnifera* L. (Ashwagandha): A novel source of l-asparaginase. *Journal of Integrative Plant Biology*, 51: 201–206.

Panosyan, E.H., Seibel, N.L., Martin-Aragon, S., Gaynon, P.S., Avramis, I.A., Sather, H., Franklin, J., Nachman, J., Ettinger, L.J., La, M., Steinherz, P., Cohen, L.J., Siegel, S.E. and Avramis, V.I. 2004. Asparaginase antibody and asparaginase activity in children with higher-risk acute lymphoblastic leukemia: Children's Cancer Group Study CCG-1961. *Journal of Pediatric Hematology and Oncology*, 26: 217–226.

Pinheiro, J.P. and Boos, J. 2004. The best way to use asparaginase in childhood acute lymphoblastic leukemia: Still to be defined? *British Journal of Haematology*, 125: 117–127.

Pourhossein, M. and Korbekandi, H. 2014. Cloning, expression, purification and characterisation of *Erwinia carotovora* L-asparaginase in *Escherichia coli*. *Advanced Biomedical Research*, 3: 2277–9175.

Pradhan, B., Dash, S.K. and Sahoo, S. 2013. Screening and characterization of extracelluar L-asparaginase producing *Bacillus subtilis* strain hswx88, isolated from Taptapani hotspring of Odisha, India. *Asian Pacific Journal of Tropical Biomedicine,* 3: 936–941.

Pui, C.H. and Evans, W.E. 2006. Treatment of acute lymphoblastic leukemia. *New England Journal of Medicine*, 354: 166–178.

Rati Kumari Sinha, Hare Ram Singh, Santosh Kumar Jha. 2014. Production, purification and kinetic characterization of L-asparaginase from *Pseudomonas fluorescens. International Journal of Pharmacy and Pharmaceutical Sciences*, 7(1): 135–138.

Ramandeep, K. and Bhupinder, S. 2012. Enzymes as drugs: An overview. *Journal of Pharmaceutical Education and Research,* 3: 29–41.

Ramya, L.N., Doble, M., Rekha, V.P.B. and Pulicherla, K.K. 2012. L-Asparaginase as potent anti-leukemic agent and its significance of having reduced glutaminase side activity for better treatment of acute lymphoblastic leukaemia *Applied Biochemistry and Biotechnology*, 167: 2144–2159.

Riccardi, R., Holcenberg, J.S., Glaubiger, D.L., Wood, J.H. and Poplack, D.G. 1981. Asparaginase pharmacokinetics and asparagine levels in cerebrospinal fluid of rhesus monkeys and humans. *Cancer Research,* 41: 4554–4558.

Riss, T.L. and Moravec, R.A. 2004. Use of multiple assay endpoints to investigate the effects of incubation time, dose of toxin, and plating density in cell-based cytotoxicity assays. *Assay Drug Development Technology*, 2(1): 51–62.

Roeselers, G., Mittge, E.K., Stephens, W.Z., Parichy, D.M., Cavanaugh, C.M., Guillemin, K. and Rawls, J.F. 2011. Evidence for a core gut microbiota in the zebrafish. *ISME Journal*, 5(10): 1595–1608.

Rose, B.D. Ed. 2006. *Asparaginase: Drug information*, 14.2 ed. Elsevier, Waltham, MA.

Rosen, J. and Hellenas, K.E. 2002. Analysis of acrylamide in cooked foods by liquid chromatography tandem mass spectrometry. *Analyst*, 127: 880–882.

Roychowdhury, R., Roy, M., Zaman, S. and Mitra, A. 2018, August. Marine microbes: A unique group for the benefit of mankind. *Journal of Emerging Technologies and Innovative Research,* 5(8).

Rytting, M. 2012. Role of L-asparaginase in acute lymphoblastic leukemia: Focus on adult patients. *Blood and Lymphatic Cancer: Targets and Therapy*, 2: 117–124.

Sabaawy, H.E., Azuma, M., Embree, L.J., *et al.* 2006. TEL-AML1 transgenic zebrafish model of precursor B cell acute lymphoblastic leukemia. *Proceedings of National Academic Sciences USA,* 103(15): 166–171.

Sabu, A., Chandrasekaran, M. and Pandey, A. 2000. Biopotential of microbial glutaminases. *Chemistry. Today,* 18: 21–25.

Sahu, M.K., Poorani, E., Sivakumar, K., Thangaradjou, T. and Kannan, L. 2007. Partial purification and anti-leukemic activity of L-asparaginase enzyme of the actinomycete strain LA-29 isolated from the estuarine fish. *Mugil cephalus* (*Linn.*), 28(3): 645–650.

Saleem Basha, N., Rekha, R., Komala, M. and Ruby, S. 2009. Production of extracellular anti-leukaemic enzyme lasparaginase from marine actinomycetes by solid state and submerged fermentation: Purification and characterisation. *Tropical Journal of Pharmaceutical Research*, 8(4): 353–360.

Seiter, K., Sarkodee-Adoo, C., Talavera, F., Sacher, R.A. and Besa, E.C., Eds. 2014. *Acute Lymphoblastic Leukemia*. Medscape Reference, WebMD.

Siddalingeshwara, K.G. and Lingappa, K. 2010. Key fermentation factors for the synthesis of L-asparaginase: An anti-tumor agent through SSF methodology. *Pharma Science Monitor*, 1(1): 60–64.

Silverman, L.B., Gelber, R.D., Dalton, V.K., *et al.* 2001. Improved outcome for children with acute lymphoblastic leukemia: Results of Dana-Farber Consortium Protocol 91–01. *Blood,* 97: 1211–1218.

Skibola, C.F., Smith, M.T., Hubbard, A., *et al.* 2002. Polymorphisms in the thymidylate synthase and serine hydroxymethyltransferase genes and risk of adult acute lymphocytic leukemia. *Blood*, 99: 3786–3791.

Stams den Boer, M.L., Holleman, A., Appel, T.M., Beverloo, H.B., van Wering, E.R., Janka-Schaub, G.E., Evans, W.E. and Pieters, R. 2005. Asparagine synthetase expression is linked with L-asparaginase resistance in TEL-AML1-negative but not TEL-AML1-positive pediatric acute lymphoblastic leukemia. *Blood*, 105: 4223–4225.

Swain, A.L., Jaskolski, M., Housset, D., Rao, J. and Wlodawer, A. 1993. Crystal structure of *Escherichia coli* L-asparaginase, an enzyme used in cancer therapy. *Proceedings of the National Academy of Sciences*, 90(4): 1474–1478.

Tareke, E., Rydberg, P., Karlsson, P., Eriksson, S. and Tornqvist, M. 2002. Analysis of acrylamide, a carcinogen formed in heated foodstuffs. *Journal of Agricultural and Food Chemistry*, 50: 4998–5006.

Tetsuya, Tosa, Ryujiro, Sano, Kozo, Yamamoto, et al., 1972. L-Asparaginase form Proteus vulgaris. Purification, crystallization, and enzymic properties. *Biochemistry*, 11(2): 217–222.

Verma, N., Kumar, K., Kaur, G. and Anand, S. 2007. L-asparaginase: A promising chemotherapeutic agent. *Critical Review of Biotechnology*, 27: 45–62.

Wall, N.R., Mohammed, R.M., Nabha, S.M., *et al.* 1999. Modulation of Ciap-1 by novel antitubulin agents when combined with bryostatin 1 results in increased apoptosis in human early pre-B acute lymphoblastic leukemia cell line REH. *Biochemical and Biophysical Research Communications,* 266: 76–80.

Warangkar, S.C. and Andkhobragade, C.N. 2008. An optimized medium for screening of L-asparaginase production by *Escherichia coli. American Journal of Biochemistry & Biotechnology,* 4: 422–424.

Wriston, J.C. 1985. Asparaginase. *Methods Enzymology,* 113: 608–618.

Yao, M., Yasutake, Y., Morita, H. and Tanaka, I. 2005. Structure of the type I L-asparaginase from the hyperthermophilic archaeon Pyrococcushorikoshii at 2.16

angstroms resolution. *Acta Crystallographica Section D Biological Crystallography*, 61(3): 294–301.

Ylikangas, P. and Mononen, I. 2000. A fluorometric assay for L-asparaginase activity and monitoring of L-asparaginase therapy. *Analytical Biochemistry*, 280: 42–45.

Zeidan, A., Wang, E.S. and Wetzler, M. 2009. Pegasparaginase: Where do we stand? *Expert Opinion on Biological Therapy*, 9: 111–119.

Zhang, C. and Kim, S.-K. 2010. Research and application of marine microbial enzymes: Status and prospects. *Marine Drugs*, 8(6): 1920–1934.

10 A Comparative Study of Organic Pollutants in Seawater, Sediments, and Oyster Tissues at Hab River Delta, Balochistan Coast, Pakistan

Sadar Aslam, Malik Wajid Hussain Chan, Grzegorz Boczkaj, and Ghazala Siddiqui

CONTENTS

10.1 INTRODUCTION

Bivalve mollusks have been used extensively to monitor chemical contamination. As the relationship between the concentration of chemical contaminants and the biological responses in bivalve mollusks continued to explore (Wang and Lu, 2017). As a result, the toxic action of specific compounds and groups of compounds have been identified (Abdel-Shafy and Mansour, 2016). Lipophilic organic contaminants such as polycyclic aromatic hydrocarbons (PAHs),

DOI: 10.1201/9781003303909-10

polychlorinated biphenyls (PCBs) and other synthetic compounds for which PCBs and PAHs are environmental pollutants generated during the incomplete combustion of organic materials (e.g., oil, coal, wood, and petrol). They are organic compounds that are colorless, pale yellow, or white solids (Abdel-Shafy and Mansour, 2016).

PAHs' nature in the environment has been well documented by Menzie et al. (1992). They serve in part as model compounds and are highly resistant to degradation in the marine environment. Thus, such compounds and their metabolites may accumulate in high concentrations in animal tissues and interfere with normal metabolic processes, causing adverse effects on, for example, immunity, tumors, growth, reproduction, and development (Capuzzo et al., 1988; Abdel-Shafy and Mansour, 2016). PAHs strongly bind to marine sediments, which serves as a pollution reservoir, and then are released under specific conditions (Zhang and Zheng, 2003). As a result, sediment-dwelling and filter-feeder organisms are most susceptible to such contamination (Abdel-Shafy and Mansour, 2016).

Most marine organisms have a high bio-transformation potential for PAH, resulting in insignificant biomagnification in the aquatic food chain (Bleeker and Verbruggen, 2010). Nevertheless, filter-feeding bivalves (e.g., oysters and mussels) filter large volumes of water and have a low metabolic capacity for PAH. Consequently, PAHs accumulate in their tissues (Wang and Lu, 2017). The water-soluble low-molecular-mass PAHs are rapidly degraded in water, but their continuous release of PAHs through wastewater discharged in the marine environment can result in elevated concentrations in many bivalves growing in the vicinity of industrialized areas (Abdel-Shafy and Mansour, 2016). The bioavailability, bioconcentration and toxic effects of lipophilic contaminants through the food chain, causing neurological impairments, neurodegenerative diseases and neurodevelopmental disorders in humans (Widdows et al., 1987; Donkin et al., 1990; Zeliger, 2013).

The concentration of PAHs found in shellfish and fish is much higher than in the environment (Abdel-Shafy and Mansour, 2016). The limited capacity of bivalve mollusks to degrade organic contaminants results in their uptake and accumulation in high concentrations (Wang and Lu, 2017). In bivalve mollusks, exposure to organic contaminants results in impairment of physiological mechanisms (Widdows, 1985), histopathological disorders (Lowe 1988; Moore, 1988) and loss of reproductive potential (Neff and Haensly, 1982; Berthou et al., 1987).

Uptake and bioaccumulation of organic contaminants by marine bivalves depends on the bioavailability of specific compounds, the duration of exposure and the physiological condition of populations (Wang and Lu, 2017). Variation in filtration rates may affect the volume of water exchanged by an organism and contaminants in both aqueous and particulate phases (Capuzzo, 1996). The lipid content of bivalve mollusks may vary seasonally, influencing both total body burden of contaminants and the distribution to specific tissues (Wang and Lu, 2017). Differences in contaminant concentrations among species of bivalves from different habitats may be the result of physico-chemical differences in the availability of sediment-bound contaminants (Capuzzo, 1996; Wang and Lu, 2017).

The uptake of petroleum hydrocarbons by oysters has been documented in numerous studies and is consistent with observations made on other species of bivalve mollusks. Bieri and Stamoudis (1977) observed the accumulation and release of hydrocarbons by eastern oysters exposed to fuel oil in a semi-natural exposure system. Alkanes, branched alkanes, and olefins were accumulated and released first, followed by alkylnaphthalenes (with up to five alkyl carbons) and biphenyles (with up to two alkyl carbons), as well as the fluorenes (with up to one methyl group). Highly substituted naphthalenes, biphenyls and fluorenes, in addition to dibenzothiophenes and phenanthrenes, were accumulated and released later.

The identification of volatile compounds is generally carried out by gas chromatography (GC) due to influential detection performance and strong separation ability. Nonpolar (volatiles) to moderately polar (less volatile) compounds can be detected because these compounds maintain vapor form in the closed, heated system of GC. Mass spectra are obtained exploiting GC-MS, which are interpreted and identified in individuals from the complex samplesthat are possible. GC-MS is very reliable and accurate method to detect contaminants (Pandey et al., 2011). These spectra are matched with electronic libraries and their retention indices are calculated (Siddiquee et al., 2015).

This chapter describes a few important aspects relating to (1) the analysis and profiling by GC-MS technique of volatile organic compounds (VOCs) in two oyster species (*Magallana bilineata* and *Magallana cuttackensis*), (2) the investigation of hydrocarbons/organic pollutants in seawater and sediments by GC-MS technique and (3) comparison of the possible bioaccumulation of organic pollutants/hydrocarbons.

10.2 METHODOLOGY

10.2.1 Samples Collection and Sample Preparation

10.2.1.1 Oyster Samples

The oyster samples were collected from Hab River Delta (Figure 10.1) water and sediments were collected from the vicinity of an oyster reef during March 2018. The samples were stored in icebox at 4°C while placing them in zip-locked bags and transported to the laboratory immediately after sampling. Three oysters were shucked, and their tissues were immediately frozen in polyethylene bags; the frozen tissues were homogenized by a blender and soaked in a suitable quantity of *n*-Hexane for 24 h. The mixture was placed in a separating funnel, and the extract was collected for GCMS analysis. Prior to GC-MS analysis, anhydrous sodium sulfate was used to remove the water from the extract.

10.2.1.2 Water and Sediment Samples

The seawater and sediments samples were collected from the same place in sterile glass bottles and jars, respectively. After collection, the samples were stored

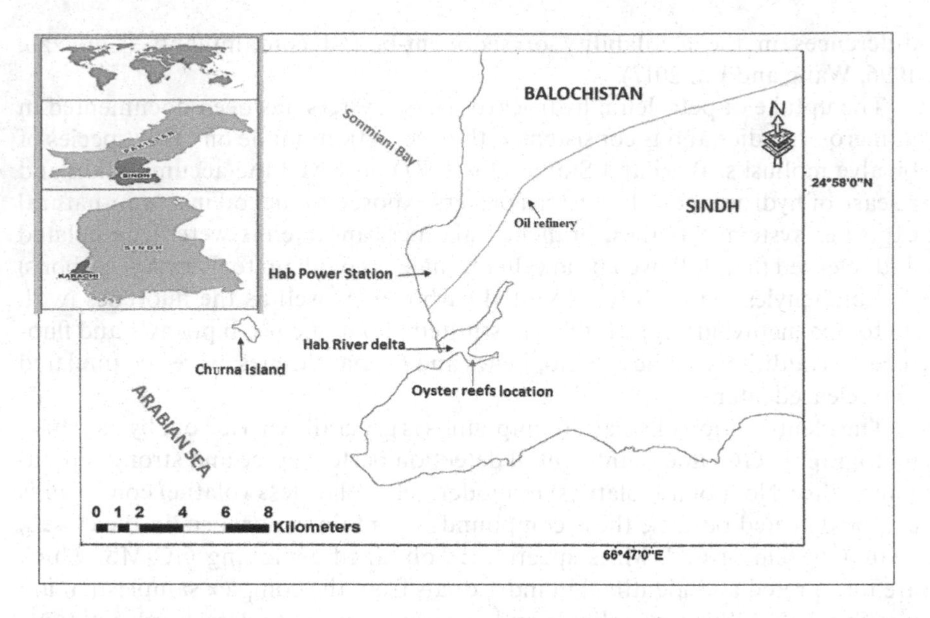

FIGURE 10.1 Location of the study area, the Hab River Delta, Balochistan, Pakistan.

in an icebox before transportation to the lab. After transportation the samples were stored in refrigerator (−20 *n*-HexaneC). For the extraction of hydrocarbons from water and sediments, the liquid–liquid extraction technique was used. An *n*-Hexane (a solvent of choice based on previous papers in this field, Weisman, 1998; Johnson, 2016) was used for extraction. Anhydrous sodium sulfate was used for removing water from the extract. The same approach was used by Adeniji et al., 2017.

10.3 GS-MS ANALYSIS

The experiments were performed by means of a 7890A gas chromatograph (Agilent technology) coupled to a 5975C inter XLEI/CI MSD mass spectrometer with triple-axis detector (Agilent technology). A HP-5MS 5% phenyl-methyl-siloxane capillary column was used in these studies (30m × 0.25 mm i.d., 0.25 μm film thickness). The injection (0.4 μL) was made in a split-less mode at a temperature of 300°C. The following oven temperature program was used: the initial temperature was set at 60°C (isothermally for 1 min) followed by a ramp of 5°C/min to 180°C (isothermally for 15 min), followed by a ramp of 7°C/min to 300°C for 15 min. Helium was used as carrier gas with a flow of 1 mL/min (constant flow mode). The mass spectra were obtained at 50 Hz frequency, the m/z range for full SCAN mode was 50–700, the temperature of the transfer line was 280°C, and the temperature of the ion source was 250°C. Prior to analysis, the *n*-Hexane extracts were filtered through a 0.45μm polytetrafluoroethylene fluoropore syringe–driven filterunit (Milipore, Bedford, USA) and then injected in a splitless mode. The compounds were identified on the basis of their mass spectra and retention indices using a NIST11database library.

10.4 LINEAR RETENTION INDEX CALCULATION

Molecules in GC analyses can be confirmed by comparing with co-injecting standards. Molecules with resembling chemical structures produce similar mass spectra and hence cannot be differentiated by mass spectrum alone. The equation propounded by van den Dool and Kratz (1963) for calculation for the linear retention index (LRI; used in temperature programming conditions) by following formula:

$$LRI = 100n + 100\ (t_x - t_n / t_{n+1} - t_n),$$

where

t_{n}, t_{n+1}, and t_x are net retention times, and n = carbon number.

10.5 RESULTS AND DISCUSSION

This study is the first to report the presence of organic pollutants at the Hab River mouth. No previous literature has been published on the environment pollution of this place. However, in previous findings of heavy-metal contamination by Aslam et al. (2020a) in the same area (conducted as a second part of the same research project) indicated the oysters in Hab River mouth were facing the metal pollution threat due to fast-growing industries and urbanization. The pollution status of oyster reefs was monitored in seawater, sediments, tissues and shells of present oysters. High levels of metals indicate the alarming condition of the Hab River Delta and the need for increased attention to coastal management.

Churna Island is situated about 9km outside Hab River outfall. The Churna and associated land along with Hab River Delta is known for its high biodiversity (Aslam et al., 2018, 2019a, 2019b, 2019c, 2019d, 2020b, 2020c; Dayrat et al., 2020) and a variety of wetland habitats. It is now famous for tourism, diving and recreational activities, contributing to the man-made pollution input. In the lower reaches of Hab River the HUBCO (Hub Power Company), one of the largest private-sector power plants (Figure 10.2a) is located in its southern part area adjacent to the oil refinery (Figure 10.2b) discharging effluents near the mouth of Hab River. The combustion of fossil fuels in HUBCO is also contributing in terms of thermal and other effluents causing severe pollution in the area. It is also interesting to note (Tables 10.1 and 10.2) that many substituted aromatic hydrocarbons are present, and most of them are also very likely deriving from petroleum manufacturing.

Close to oyster reefs at Hab, there is a jetty, resulting in boat traffic (Figure 10.2c) which is also a major source of water pollution with oil fractions, forming surface oil slicks (Figure 10.2d). The human population near

FIGURE 10.2 (a) HUBCO; (b) oil refinery; (c) boat traffic near oyster reefs; (d) oil slick over the reefs.

the studied area is increasing as the Sindh government established *Shaheed Mohtarma Benazir Bhutto Towns*. Home heating, automotive sources and associated highway runoff are contributing to the anthropogenic pollution. Numerous substituted aromatic hydrocarbons in analyzed samples indicate that these anthropogenic sources affectthe environment, which, in turn, results in uptake of petroleum hydrocarbons by oysters.

TABLE 10.1
Hydrocarbons/Organic Pollutants in Seawater and Sediments at Hab River Delta through GC-MS Analytical Technique

S. No.	RT.	RI Cal.	RI Lit.	Cas No.	Compounds	Water	Sediments
1	8.02	1000	1000	124–18–5	Decane[1]	✓	✓
2	8.32	1031	1019	526–73–8	1,2,3-Trimethylbenzene[1]	✓	X
3	8.67	1040	1032	496–11–7	Indane[1]	X	✓
4	9.70	1068	1056	493–02–7	trans-Decalin[1]	X	✓
5	10.36	1142	1070	105–05–5	1,4-Diethylbenzene[2]	X	✓
6	11.53	1100	1100	1120–21–4	Undecane[1]	✓	✓
7	23.68	1400	1400	629–59–4	Tetradecane[1]	✓	X
8	25.89	1445	1448	1795–15–9	Octylcyclohexane[1]	X	✓
9	26.67	1465	1467	2189–60–8	Octylbenzene[1]	X	✓
10	27.47	1485	1484	643–93–6	3-Methylbiphenyl[1]	✓	X
11	28.25	1500	1500	629–62–9	Pentadecane[1]	X	✓
12	29.07	1525	1524	135–19–3	2-Naphthalenol[1]	X	✓
13	33.82	1635	1635	119–61–9	Benzophenone[3]	✓	X
14	34.83	1655	1657	103–50–4	Dibenzyl ether[1]	X	✓
15	35.59	1670	1675	104–72–3	Decylbenzene[1]	X	✓
16	37.21	1700	1700	629–78–7	Heptadecane[1]	X	✓
17	43.44	1779	1786	120–12–7	Anthracene[1]	✓	✓
18	43.24	1791	1789	85–01–8	Phenanthrene[1]	✓	X
19	43.49	1792	1794	112–88–9	1-Octadecene[3]	✓	X
20	43.51	1800	1800	593–45–3	Octadecane[1]	✓	X
21	45.23	1833	1835	14237–73–1	2E)-3,7,11,15-Tetramethyl-2-hexadecene[3]	✓	X
22	46.13	1855	1858	605–02–7	1-Phenylnaphthalene[1]	✓	✓
23	47.90	1900	1900	629–92–5	Nonadecane[1]	X	✓
24	49.52	1947	1959	610–48–0	1-Methylanthracene[1]	✓	X
25	51.26	2036	2024	1576–67–6	3,6-Dimethylphenanthrene[3]	✓	X
26	52.30	2054	2060	206–44–0	Fluoranthene[1]	X	✓
27	53.32	2100	2100	629–94–7	Heneicosane[1]	✓	X
28	53.62	2109	2113	129–00–0	Pyrene[1]	X	✓
29	53.99	2127	2136	781–43–1	9,10-Dimethylanthracene[1]	X	✓
30	54.84	2146	2158	92–06–8	m-Terphenyl[3]	✓	X
31	55.43	2195	2191	92–94–4	p-Terphenyl[1]	X	✓
32	57.53	2300	2300	638–67–5	Tricosane[1]	✓	✓

Note: Literature references: (1) Lai and Song, 1995 (2) Xu et al., 2003 (3) Leffingwell et al., 2013

TABLE 10.2
Profiling of Volatile Compounds, Organic Pollutants/Hydrocarbons Tissues of *M. cuttackensis* and *M. bilineata*

S. No.	RT.	RI		Cas No.	Compounds	*M. cuttakensis*	*M. bilineata*
		Cal.	Lit.				
1	8.17	1023	1023	5989–27–5	D-Limonene[1]	X	✓
2	16.13	1256	1231	103–82–2	Benzeneacetic acid[1]	✓	X
3	16.19	1256	1232	122–99–6	2-Phenoxyethanol[1]	✓	X
4	16.33	1258	1236	13466–78–9	3-carene[1]	✓	X
5	16.75	1263	1250	89–81–6	3-Carvomenthenone[2]	✓	X
6	19.76	1299	1294	120–72–9	Indole[3]	✓	X
7	22.11	1353	1355	334–48–5	Decanoic acid[1]	✓	X
8	23.68	1400	1400	629–59–4	Tetradecane[4]	✓	X
9	24.34	1407	1405	121–33–5	Vanillin[3]	✓	X
10	24.56	1412	1409	17398–16–2	2,3,5-Trimethyl-6-ethylpyrazine[5]	✓	X
11	24.94	1422	1445	91–64–5	Coumarin[1]	X	✓
12	24.94	1422	1434	10522–26–6	2-Methyl-1-undecanol[2]	✓	X
13	25.76	1442	1436	87–44–5	β-Caryophyllen[3]	✓	X
14	27.37	1483	1476	112–53–8	1-Dodecanol[1]	✓	✓
15	27.48	1485	1494	14901–07–6	β-Ionone[1]	✓	X
16	28.71	1516	1524	1700–10–3	1,3-Cyclooctadiene[5]	✓	X
17	29.90	1546	1543	14905–56–7	2,6,10-trimethyl-tetradecane[2]	✓	X
18	30.72	1567	1565	143–07–7	Dodecanoic acid[1]	✓	✓
19	33.34	1625	1622	3892–00–0	2,6,10-trimethyl-pentadecane[2]	✓	X
20	33.76	1634	1635	28882–58–8	Isopropylmethylnaphthalene[6]	X	✓
21	35.48	1668	1676	18679–18–0	γ-6-(Z)-Dodecenolactone[1]	X	✓
22	40.573	1751	1750	544–63–8	Tetradecanoic acid[7]	✓	✓
23	42.456	1780	1776	85–01–8	Phenanthrene[8]	✓	✓
24	43.22	1791	1786	120–12–7	Anthracene[8]	X	✓
25	45.81	1848	1846	36653–82–4	1-Hexadecanol[9]	✓	✓
26	46.24	1858	1860	1002–84–2	Pentadecanoic acid[7]	X	✓
27	47.64	1892	1888	5129–60–2	Pentadecanoic acid, 14-methyl-, methyl ester[7]	X	✓
28	48.279	1910	1910	373–49–9	9-Hexadecenoic acid[7]	✓	✓
29	48.782	1926	1930	112–39–0	Methyl hexadecanoate[10]	X	✓
30	50.241	1975	1960	57–10–3	Hexadecanoic acid[7]	X	✓
31	50.78	2000	2000	112–95–8	Eicosane[4]	X	✓
32	51.59	2025	2026	91–20–3	Naphthalene[5]	✓	X
33	52.95	2080	2077	206–44–0	Fluoranthene[8]	X	✓
34	54.15	2134	2132	129–00–0	Pyrene[8]	✓	X
35	54.83	2167	2165	112–80–1	9-Octadecenoic acid (Z)-[7]	✓	X
36	55.21	2184	2180	6114–18–7	9-Octadecenoic acid, ethyl ester[7]	✓	X
37	55.89	2217	2210	57–11–4	Octadecanoic acid[7]	✓	X
38	64.81	2491	2472	218–01–9	Chrysene[8]	X	✓

Note: Literature references: (1)Mahmoud & Buettner, 2017 (2) Huang et al. 2019 (3) Niimi et al. 2010 (4) Goodner, 2008 (5) Fan et al. 2017 (6) Leffingwell et al. 2013 (7) Bouzidi et al. 2016 (8) Lai & Song, 1995 (9) Muchtaridi et al. 2014 (10) Payo et al. 2011.

Despite the discharge of effluents from the power plant and oil refinery in the area, a community of oysters and associated reef fauna and flora exists with significant ecological importance. The main industrial activity along the coast of Hab is the Hab Industrial and Trading Estate (HITE). HITE is located in western Karachi near the border between Sindh and Balochistan Provinces. It signifies the most current and fast-emerging industrial area. The major assemblage of industries at HITE include chemicals, textile, industrial gases, food and beverages, automobile poultry farms, engineering and metallurgical and cement and glass industries. The wastes discharged by these sources are untreated and eventually find their way into the Arabian Sea through the Hab River (Rizvi, 1997), a main source of pollution.

The preliminary chemical profiling by GC-MS of seawater and sediments samples from the Hab River Delta revealed the presence of the following seven major groups based on their chemical structure to commonly known classes of compounds:

1. Alkanes (decane, undecane, tetradecane, pentadecane, heptadecane, octadecane, nonadecane, heneicosane, tricosane)
2. Alkenes (decylbenzene, 1-octadecene and 2E)-3,7,11,15-Tetramethyl-2-hexadecene)
3. Ketones (benzophenone)
4. Other organic pollutants (dibenzyl ether)
5. Aromatic hydrocarbons (1,2,3-trimethylbenzene, indane, 1,4-diethylbenzene, octylbenzene, m-terphenyl; p-terphenyl;3-methylbiphenyl)
6. Cycloalkanes (trans-decalin, octylcyclohexane)
7. Polycyclic aromatic hydrocarbons (PAHs; 2-naphthalenol; 1-phenylnaphthalene; 1-methylanthracene; 3,6-dimethylphenanthrene; anthracene; fluoranthene; pyrene; phenanthrene;9,10-dimethylanthracene; Table 10.1).

Many compounds that have not been detected in the water and sediments samples might have been present in lower concentrations. The presence or absence of organic pollutants in sediments should not be judged without some knowledge about the mineralogical sediment characteristics. Since most pollutants of low solubility are likely to be associated with fine particular matter (organic detritus, silt, clay), a sample consisting of almost pure sand under equivalent rates of pollutant input can be expected to contain lower concentrations of pollutants than a sample that has more silt and clay, although in a previous study by Aslam et al. (2020d) at the same site it was concluded that the fine sand has the highest percentage (71%) followed by silt/clay fraction (18%). It is not surprising, therefore, that most of these samples contain very low pollutant levels. However, the sources of these pollutants came from river mouth and are characterized by silt/clay contents; for this reason, they have the potential to contain pollutants at higher levels.

The main problem encountered in the analysis of oyster tissues was the presence of biogenic compounds and their derivatives, often commanding a major presence on the chromatograms. All organic acids are of known biogenic origin. Unidentifiable compounds of probable biogenic origin may also be present. Most of the tissues compounds that are not present in sediment extracts because they were masked by high hydrocarbon levels. The pollutants in sediments near large population places, industrial complexes and dense transportation networks or at the mouths of rivers that connect to such places must be taken as evidence of man-made input.

Profiling of organic pollutants in tissues of oysters was performed also by means of GC-MS technique. Due to the high number of identified compounds, they were grouped according to their chemical structure; several major groups of pollutants were reported in following nine classes of compounds:

1. Alkanes (2,6,10-trimethyl-tetradecane; tetradecane; 2,6,10-trimethyl-pentadecane; eicosane)
2. Alkenes (D-limonene; 3-carene; β-caryophyllene; 1,3-cyclooctadiene)
3. Alcohols (2-phenoxyethanol; 2-methyl-1-undecanol; 1-dodecanol, 1-hexadecanol)
4. Aldehydes (vanillin)
5. Ester (methyl hexadecanoate; 14-methyl-pentadecanoic acid; methyl ester; 9-octadecenoic acid; ethyl ester)
6. Ketones (3-carvomenthenone; coumarin; β-ionone; γ-6-(Z)-dodecenolactone)
7. Organic acids (benzeneacetic acid; decanoic acid; dodecanoic acid; 9-hexadecenoic acid; hexadecanoic acid; tetradecanoic acid; pentadecanoic acid; 9-octadecenoic acid (Z); octadecanoic acid)
8. PAHs (naphthalene; fluoranthene; pyrene; phenanthrene; anthracene; Chrysene)
9. Other organic pollutants (indole; 2,3,5-trimethyl-6-ethylpyrazine; isopropylmethylnaphthalene; Table 10.2).

A major part of these compounds is most likely related to endogenous processes in the oyster. Several compounds that were identified in oysters are characteristic secondary plant metabolites that are not synthesized by oyster.

The only possible explanation for their presence in oyster (limonene, benzene acetic acid, 3-carene) is that they were accumulated from exogenous sources. Currently, d-limonene is also widely used as a flavor and a fragrance and is listed as "generally recognized as safe" in food by the U.S. Food and Drug Administration (21 CFR 182.60). 2-phenoxyethanol, a drug that dissolves well in water and oils, provides effective anesthesia with short induction time (13 min), and over 95% survival rate is used for oysters (Mamangkey et al., 2009). Seafood contains rich polyunsaturated fatty acids (Ackman, 1990). Many oyster flavor volatiles arise from the oxidation of these fatty acids (Cruz-Romero et al. 2008; Piveteau et al., 2000). The oyster alkenes most identified were C8 alkenes, and some of them possessed conjugated structures, such as (Z,Z)-3,5-octadiene, 1,3-trans-5-cisoctatriene and 1,3-cyclooctadiene. The alcohols were identified in the volatile compositions of oyster derived from the autoxidation of unsaturated fatty acids (Cruz-Romero et al., 2008).

In a study conducted by Bieri et al. (1981), naphthalene, fluorantheneand phenanthrene were identified in oyster tissues. These are due to surface oil slicks (Figure 10.2d); after prolonged exposure, these oysters indeed could have been exposed periodically to surface oil slicks.

In both oyster samples, several PAHs were identified. The PAHS are present in petroleum fractions, and they are formed during thermal cracking taking place in some processes of crude oil processing (Gilgenast et al., 2011; Boczkaj et al., 2014; Makoś et al., 2018a, 2018b). This indicates a health risk for humans in this region because PAHs are carcinogenic. However, Gardner et al. (1991) has well documented the concentrations of organic contaminates in eastern oyster, *Crassostrea virginica*, and in sediments with reference to evidence scale as carcinogens. He gave a scale of carcinogens to different contaminants. According to his scale, Chrysene has limited evidence as carcinogens. Phenanthrene has inadequate evidence while anthracene and fluoranthene have no evidence of being carcinogenic. The only cancer promoter we detected in our results is pyrene. Bender et al. (1988) examined the distribution of PAHs in eastern oysters from the Elizbeth River, Virginia, and conducted laboratory studies compared the uptake and depuration of PAHs by eastern oysters and hard clam, *Mercenaria mercenaria*, with exposure to contaminated sediments from the Elizabeth River. Animals were exposed to contaminated sediments for 28 days, followed by a 28-day depuration phase. Oysters accumulated three to four times more total PAH than clams with similar uptake rates. Bioconcentration factors for oysters ranged from 1600 for phenanthrene to 36,000 for methyl-pyrene (Capuzzo, 1996).

Wade et al. (1988) examined the distribution of PAHs in sediments and eastern oysters at 153 stations in the Gulf of Mexico. They observed comparable concentrations of total PAHs in sediments and oysters, but the molecular distribution of PAHs differed. Oyster samples showed a predominance of 2-, 3- and 4-ring PAH compounds, whereas sediment samples had low concentrations of low-molecular-weight PAHs (2 and 3ring) and higher concentrations of 5-ring compounds. Based on ratios of phenanthrene to anthracene in both oyster and sediment samples, the authors suggested that the dominant source of PAHs at most sampling sites was pollution by fossil fuels. The ingestion of detrital particles is a possible mechanism of uptake of hydrocarbons (Bieri and Stamoudis, 1977). Berthou et al. (1987) reviewed hydrocarbon accumulation data for the European flat oyster, *Ostrea edulis*, and the Pacific oyster, *Crassostrea gigas*, after the wreck of the oil tanker *Amoco Cadiz* off the coast of Brittany, France. Berthou et al. (1987) confirmed the rapid accumulation of hydrocarbons by both species. Uptake and depuration of specific hydrocarbons reflected differences in hydrocarbon availability in the sediments resulting from weathering process in addition to physico-chemical properties of specific compounds. Rapid accumulation of a wide range of hydrocarbons was followed by an initial rapid loss of alkanes and low-molecular-weight hydrocarbons, but the persistence of polycyclic aromatic hydrocarbons and dibenzothiophenes was also noted (Friocourt et al., 1982).

The long-term effects of petroleum hydrocarbons and PAHs are lesions of the digestive tract epithelium, interstitial tissues, gonads and gills. Moore et al. (1988) have well reviewed the cellular and physiological responses of bivalve mollusk to specific PAHs. The most important physiological changes associated with exposure to petroleum hydrocarbons or other lipophilic organic contaminants are those responses that may affect an organism's growth and survival and thus its potential to contribute the population gene pool. Alterations in growth potential may take place as a result of changes in feeding behavior, respiratory metabolism or digestive efficiencies (Capuzzo, 1996). Donkin et al. (1990) suggested that reductions in the scope of growth in *M. edulis* were related to the accumulation of 2- and 3-ring aromatic hydrocarbons, because these compounds induce narcotizing effects on the ciliary feeding mechanism.

10.6 CONCLUSION

The compounds which were showing the presence in both sediments and water were decane, undecane, 1-phenylnaphthalene, anthracene and tricosane. While the compounds that were present in both oyster tissues were 1-dodecanol, dodecanoic acid, tetradecanoic acid, 1-hexadecanol, 9-hexadecenoic acid and phenanthrene. The presence of compounds identified in oysters results mainly from the uptake from the water and sediments. The present study revealed that oysters contain unsubstituted PAHs, which are carcinogenic. These data put a new spotlight on the health-friendly nature of oysters, which source should be provided along with certification of the state of the environment. Furthermore, oysters should be routinely subjected to quality control protocols, including tests of PAH content. On the other hand, it would be possible to use these tissues as a bio-indicator in future research. The oyster reefs in Pakistan are on the verge of extinction as a consequence of environmental and anthropogenic factors. We recommend this place be converted into a Marine Protected Area (MPA) under the 2010 Aichi Biodiversity Targets for the conservation plan leading to the restoration goal of present oyster reefs at Hab River Delta (Balochistan coast) in Pakistan with a sizable ecologic sanctuary (Aslam et al., 2022).

ACKNOWLEDGMENTS

This study is a part of PhD research work by the first author on oysters' reef at Hab River Delta, Balochistan, Pakistan. The authors are thankful to Higher Education Commission, Pakistan for providing funds during research.

FUNDING

SA was supported by "Access to Scientific Instrumentation Program (ASIP)" grants (No. 20–2(14)/ASIP/R&D/HEC/18/001054(HEJ)/24 funded by national funds through the Higher Education Commission, Pakistan.

COMPLIANCE WITH ETHICAL STANDARDS

Conflict of interest: The authors declare no conflict of interest.

Ethical approval: This article does not contain any studies with live animals performed by any of the authors.

Sampling and field studies: All necessary permits for sampling and observational field studies have been obtained by the authors from the competent authorities and are mentioned in the acknowledgments, if applicable. The study is compliant with CBD and Nagoya protocols.

REFERENCES

Abdel-Shafy, H. I., & Mansour, M. S. (2016). A review on polycyclic aromatic hydrocarbons: Source, environmental impact, effect on human health and remediation. *Egyptian Journal of Petroleum*, 25(1), 107–123. https://doi.org/10.1016/j.ejpe.2015.03.011.

Ackman, R. G. (1990). Seafood lipids and fatty acids. *Food Review International*, 6, 617–646. https://doi.org/10.1080/87559129009540896.

Adeniji, A. O., Okoh, O. O., & Okoh, A. I. (2017). Analytical methods for the determination of the distribution of total petroleum hydrocarbons in the water and sediment of aquatic systems: A review. *Journal of Chemistry*, 2017. https://doi.org/10.1155/2017/5178937.

Aslam, S., Siddiqui, G., Chan, M. W. H., Kazmi, J. H., & Afsar, N. (2022). Field and GIS validation of oyster stocks and the reef ecosystem of Hab River Delta, Balochistan, Pakistan: a conservation approach. *The Newsletter of the IUCN/SSC Mollusc Specialist Group Species Survival Commission, International Union for Conservation of Nature. TENTACLE*, 30, 46–48.

Aslam, S., Chan, M. W. H., Siddiqui, G., Boczkaj, G., Kazmi, S. J. H., &Kazmi, M. R. (2020a). A comprehensive assessment of environmental pollution by means of heavy metal analysis for oysters' reefs at Hab River Delta, Balochistan, Pakistan. *Marine Pollution Bulletin,* 153, 110970. https://doi.org/10.1016/j.marpolbul.2020.110970.

Aslam, S., Siddiqui, G., Dekker, H., Mustaquim, J., & Kazmi, S. J. H. (2020b). Biodiversity on intertidal oyster reefs in Hab River mouth: 35 new records from Pakistan. *Regional Studies in Marine Science*, 39, 101415. https://doi.org/10.1016/j.rsma.2020.101415.

Aslam, S., Mustaquim, J., & Siddiqui, G. (2020c). First record of the polychaete worm Ceratonereis (Composetia) burmensis (Phyllodocida: Nereididae) from Pakistan. *Pakistan Journal of Scientific and Industrial Research (Series B: Biological Sciences)*, 63B(2), 132–134. https://v2.pjsir.org/index.php/biological-sciences/article/view/1279.

Aslam, S., Siddiqui, G., & Kazmi, S. J. H. (2020d). A preliminary study on spatial assessment using restoration metrics for intertidal oyster reefs at the Hab River mouth in Pakistan. *Regional Studies in Marine Science*, 33, 100956. https://doi.org/10.1016/j.rsma.2019.100956.

Aslam, S., Chan, M. W. H., Siddiqui, G, Kazmi, S. J. H., Shabbir, N., & Ozawa, T. (2019a). A near-round natural pearl discovered in the edible oyster Magallana bilineata. *Gems & Gemology*, 55(3), 439–440.

Aslam, S., Oskars, T. R., Siddiqui, G., & Malaquias, Manuel A. E. (2019b). Beyond shells: First detailed morphological description of the mangrove-associated gastropod Haminoea fusca (A. Adams, 1850) (Cephalaspidea: Haminoeidae), with a COI phylogenetic analysis. *Zoosystema*, 41, 313–326. https://doi.org/10.5252/zoosystema2019v41a16.

Aslam, S., Siddiqui, G., Kazmi, S. J. H., & Moura, C. J. (2019c). First occurrence of the non-indigenous bryozoan Amathia verticillata (della Chiaje, 1882) at the Hab river mouth in Pakistan. *Regional Studies in Marine Science*, 30, 100706. https://doi.org/10.1016/j.rsma.2019.100706.

Aslam, S., Pfingstl, T., Siddiqui, G., Arbea, J. I., & Kazmi, S. J. H. (2019d). First confirmed record of the littoral genus Fortuynia Hammen (Acari: Oribatida: Fortuyniidae) from Pakistan (Northern Arabian Sea). *Entomological News*, 128, 535–539. https://bioone.org/journals/entomological-news/issues.

Aslam, S., Arbea, J. I., & Siddiqui, G. (2018). First record of Psedanurida Schott (Collembola, Pseudachorutinae) from the Hab River delta, Balochistan, Pakistan, an area with a high potential for the conservation of oysters biodiversity. *Boletín de la Sociedad Entomológica Aragonesa (S.E.A.)*, 62, 167–170. https://www.researchgate.net/publication/326271478.

Bender, M. E., Hargis Jr, W. J., Huggett, R. J., & Roberts Jr, M. H. (1988). Effects of polynuclear aromatic hydrocarbons on fishes and shellfish: An overview of research in Virginia. *Marine Environmental Research*, 24(1–4), 237–241. https://doi.org/10.1016/0141-1136(88)90307-8.

Berthou, F., Balouet, G., Bodennec, G., & Marchand, M. (1987). The occurrence of hydrocarbons and histopathological abnormalities in oysters for seven years following the wreck of the Amoco Cadiz in Brittany (France). *Marine Environmental Research*, 23(2), 103–133.https://doi.org/10.1016/0141-1136(87)90041-9.

Bieri, R. H., DuFur, P. O., Huggett, R. J., MacIntyre, W., Shou, P., Smith, C. L., & Su, C. W. (1981). Organic compounds in surface sediments and oyster tissues from the Chesapeake Bay. https://doi.org/10.25773/4da6-x649.

Bieri, R. H., & Stamoudis, V. C. (1977). The fate of petroleum hydrocarbons from a No. 2 fuel oil spill in a seminatural estuarine environment. In *Fate and effects of petroleum hydrocarbons in marine ecosystems and organisms* (pp. 332–344). Pergamon. https://doi.org/10.1016/B978-0-08-021613-3.50039-5.

Bleeker, E. A. J., & Verbruggen, E. M. J. (2010). Bioaccumulation of polycyclic aromatic hydrocarbons in aquatic organisms. RIVM Report 601779002/2009 (pp. 1–54).

Boczkaj, G., Przyjazny, A., & Kamiński, M. (2014). Characteristics of volatile organic compounds emission profiles from hot road bitumens. *Chemosphere*, 107, 23–30.https://doi.org/10.1016/j.chemosphere.2014.02.070.

Bouzidi, N., Seridi, H., Daghbouche, Y., Piovetti, L., & El Hattab, M. (2016). Comparison of the Chemical Composition of "Cystoseira sedoides (Desfontaines) C. Agardh" Volatile Compounds Obtained by Different Extraction Techniques. *Records of Natural Products*, 10(1), 58.

Capuzzo, J. M. (1996). The bioaccumulation and biological effects of lipophilic organic contaminants. In V. S. Kennedy, R. I. E. Newell, & A. F. Eble (eds.), *The Eastern Oyster Crassostrea virginica* (Chapter 15). Maryland Sea Grant College Publication, College Park, MD (pp. 539–557).

Capuzzo, J. M., & Leavitt, D. (1988). Lipid composition of the digestive glands of *Mytilus edulis* and *Carcinus maenas* in response to pollutant gradients. *Marine Ecology Progress Series,* 46(1/3), 139–145.

Cruz-Romero, M. C., Kerry, J. P., & Kelly, A. L. (2008). Fatty acids, volatile compounds and colour changes in high-pressure-treated oysters (*Crassostrea gigas*). *Innovative Food Science and Emerging Technologies*, 9, 54–61. https://doi.org/10.1016/j.ifset.2007.05.003.

Dayrat, B., Goulding, T. C., Apte, D., Aslam, S., Bourke, A., Comendador, J., . . . & Tan, S. H. (2020). Systematic revision of the genus *peronia* fleming, 1822 (Gastropoda, euthyneura, pulmonata, onchidiidae). *ZooKeys*, 972, 1. https://dx.doi.org/10.3897%2Fzookeys.972.52853.

Donkin, P., Widdows, P., Evans, S. V., Worrall, C. M., & Carr, M. (1990). Quantitative structure -activity relationships for the effect of hydrophobic chemicals on rate of feeding by mussels (*Mytilus edulis*). *Aquatic Toxicology*, 14, 277–294.

Fan, Y., Li, Z., Xue, Y., Hou, H., & Xue, C. (2017). Identification of volatile compounds in Antarctic krill (Euphausia superba) using headspace solid-phase microextraction and GC-MS. *International Journal of Food Properties,* 20(sup1), S820–S829. https://doi.org/10.1080/10942912.2017.1315589.

Friocourt, M. P., Berthou, F., & Picart, D. (1982). Dibenzothiophene derivatives as organic markers of oil pollution. *Toxicological & Environmental Chemistry*, 5(3–4), 205–215. https://doi.org/10.1080/02772248209356979.

Gardner, G. R., Yevich, P. P., Harshbarger, J. C., & Malcolm, A. R. (1991). Carcinogenicity of Black Rock Harbor sediment to the eastern oyster and trophic transfer of Black Rock Harbor carcinogens from the blue mussel to the winter flounder. *Environmental Health Perspectives*, 90, 53–66. https://doi.org/ 10.1289/ehp.90-1519497.

Gilgenast, E., Boczkaj, G., Przyjazny, A., & Kamiński, M. (2011). Sample preparation procedure for the determination of polycyclic aromatic hydrocarbons in petroleum vacuum residue and bitumen. *Analytical and Bioanalytical Chemistry*, 401(3), 1059–1069.https://doi.org/10.1007/s00216-011-5134-9.

Goodner, K. L. (2008). Practical retention index models of OV-101, DB-1, DB-5, and DB-Wax for flavor and fragrance compounds. *LWT-Food Science and Technology,* 41(6), 951–958. https://doi.org/10.1016/j.lwt.2007.07.007.

Huang, X. H., Zheng, X., Chen, Z. H., Zhang, Y. Y., Du, M., Dong, X. P., Lei, Qin., & Zhu, B. W. (2019). Fresh and grilled eel volatile fingerprinting by e-Nose, GC-O, GC—MS and GC× GC-QTOF combined with purge and trap and solvent-assisted flavor evaporation. *Food Research International,* 115, 32–43. https://doi.org/10.1016/j.foodres.2018.07.056.

Johnson, B. (2016). *Determining trace amounts of contaminants in water*. Water Conditioning and Purification. http://www.horizontechinc.com/PDF/WCP_article_2011.pdf.

Lai, W. C., & Song, C. (1995). Temperature-programmed retention indices for gc and gc-ms analysis of coal-and petroleum-derived liquid fuels. *Fuel*, 74(10), 1436–1451.

Leffingwell, J. C., Alford, E. D., Leffingwell, D., Penn, R., & Mane, S. A. (2013). Identification of the volatile constituents of cyprian latakia tobacco by dynamic and static headspace analyses. *Leffingwell Rep*, 5(2), 1–29.

Lowe, D. (1988). Alterations in cellular structure of Mytilus edulis resulting from exposure to environmental contaminants under field and experimental conditions. *Mar. Ecol. Prog. Series, 46*(1/3), 91–100.

Mahmoud, M. A. A., & Buettner, A. (2017). Characterisation of aroma-active and off-odour compounds in German rainbow trout (*Oncorhynchus mykiss*). Part II: Case of fish meat and skin from earthen-ponds farming. *Food Chemistry*, 232, 841–849. https://doi.org/10.1016/j.foodchem.2016.09.172.

Makoś, P., Fernandes, A., & Boczkaj, G. (2018a). Method for the simultaneous determination of monoaromatic and polycyclic aromatic hydrocarbons in industrial effluents using dispersive liquid—liquid microextraction with gas chromatography—mass spectrometry. *Journal of Separation Science*, 41(11), 2360–2367.https://doi.org/10.1002/jssc.201701464.

Makoś, P., Przyjazny, A., & Boczkaj, G. (2018b). Hydrophobic deep eutectic solvents as "green" extraction media for polycyclic aromatic hydrocarbons in aqueous samples. *Journal of Chromatography A,* 1570, 28–37.https://doi.org/10.1016/j.chroma.2018.07.070.

Mamangkey, N. G. F., Acosta-Salmon, H., & Southgate, P. C. (2009). Use of anaesthetics with the silver-lip pearl oyster, Pinctada maxima (Jameson). *Aquaculture*, 288(3–4), 280–284. https://doi.org/10.1016/j.aquaculture.2008.12.008.

Menzie, C. A., Potocki, B. B., & Santodonato, J. (1992). Exposure to carcinogenic PAHs in the environment. *Environmental Science &Technology*, 26(7), 1278–1284.https://doi.org/10.1021/es00031a002.

Moore, M. N. (1988). Cytochemical responses of the lysosomal system and NADPH-ferrihemoprotein reductase in molluscan digestive cells to environmental and experimental exposure to xenobiotics. *Marine Ecology Progress Series,* 46(1), 81–89.

Muchtaridi, M., Musfiroh, I., Subarnas, A., Rambia, I., Suganda, H., & Nasrudin, M. E. (2014). Chemical composition and locomotors activity of essential oils from the rhizome, Stem, and leaf of Alpinia malaccencis (Burm F.) of Indonesian Spices. *Journal of Applied Pharmaceutical Science*, 4(1), 52. https://doi.org/10.7324/JAPS.2014.40108.

Neff, J. M., & Haensly, W. E. (1982). Long-term impact of the Amoco Cadiz oil spill on oysters *Crassostrea gigas* and plaice *Pleuronectes platessa* from Aber-Benoit and Aber-Wrach, Brittany, France (pp. 269–328), Ecological study of the Amoco Cadiz Oil Spill. NOAA-CNEXO Report.

Niimi, J., Leus, M., Silcock, P., Hamid, N., & Bremer, P. (2010). Characterisation of odour active volatile compounds of New Zealand sea urchin (Evechinus chloroticus) roe using gas chromatography—olfactometry—finger span cross modality (GC—O—FSCM) method. *Food Chemistry*, 121(2), 601–607. http://dx.doi.org/10.1016/j.foodchem.2009.12.071.

Pandey, S. K., Kim, K. H., & Brown, R. J. (2011). A review of techniques for the determination of polycyclic aromatic hydrocarbons in air. *TrAC Trends in Analytical Chemistry*, 30(11), 1716–1739.https://doi.org/10.1016/j.trac.2011.06.017.

Payo, D. A., Colo, J., Calumpong, H., & Clerck, O. D. (2011). Variability of non-polar secondary metabolites in the red alga Portieria. *Marine Drugs*, 9(11), 2438–2468.

Piveteau, F., Le Guen, S., Gandemer, G., Baud, J. P., Prost, C., & Demaimay, M. (2000). Aroma of fresh oysters *Crassostrea gigas*: Composition and aroma notes. *Journal of Agricultural and Food Chemistry,* 48, 4851–4857.https://doi.org/10.1021/jf991394k.

Rizvi, S. N. (1997). Status of marine pollution in the context of coastal zone management in Pakistan. In *Coastal Zone Management Imperative for Maritime Developing Nations*, Springer, Dordrecht (pp. 347–370). https://doi.org/10.1007/978-94-017-1066-4_19.

Siddiquee, S., Al Azad, S., Bakar, F. A., Naher, L., & Kumar, S. V. (2015). Separation and identification of hydrocarbons and other volatile compounds from cultures of *Aspergillus niger* by GC—MS using two different capillary columns and solvents. *Journal of Saudi Chemical Society*, 19(3), 243–256.https://doi.org/10.1016/j.jscs.2012.02.007.

Van den Dool, H., & Kratz, P. D. (1963). A generalization of the retention index system including linear temperature programmed gas-liquid partition chromatography. *Journal of Chromatography A,* 11, 463–471. https://doi.org/10.1016/S0021-9673(01)80947-X.

Wade, T. L., Atlas, E. L., Brooks, J. M., Kennicutt, M. C., Fox, R. G., Sericano, J., Garcia-Romero, B., & DeFreitas, D. (1988). NOAA Gulf of Mexico status and trends program: trace organic contaminant distribution in sediments and oysters. *Estuaries*, 11(3), 171–179. https://doi.org/10.2307/1351969.

Wang, W. X., & Lu, G. (2017). Heavy metals in bivalve mollusks. In *Chemical contaminants and residues in food*, 2nd ed., pp. 553–594. https:// doi.org/10.1016/B978-0-08-100674-0.00021-7.

Weisman, W. (Ed.). (1998). *Analysis of petroleum hydrocarbons in environmental media (Total Petroleum Hydrocarbon Criteria Working Group Series)* (Vol. 1). Amherst Scientific Publishers, Amherst, MA.

Widdows, J., Donkin, P., & Evans, S. V. (1985). Recovery of *Mytilus edulis* L. from chronic oil exposure. *Marine Environmental Research*, 17(2–4), 250–253.https://doi.org/10.1016/0141-1136(85)90098-4.

Widdows, J., Donkin, P., & Evans, S. V. (1987). Physiological responses of Mytilus edulis during chronic oil exposure and recovery. *Marine Environmental Research*, 23(1), 15–32.https://doi.org/10.1016/0141-1136(87)90014-6.

Xu, X., Stee, L., Williams, J., Beens, J., Adahchour, M., Vreuls, R. J. J., Brinkman, U. A., & Lelieveld, J. (2003). Comprehensive two-dimensional gas chromatography (GC× GC) measurements of volatile organic compounds in the atmosphere. *Atmospheric Chemistry and Physics*, 3(3), 665–682.

Zeliger, H. I. (2013). Exposure to lipophilic chemicals as a cause of neurological impairments, neurodevelopmental disorders and neurodegenerative diseases. *Interdisciplinary Toxicology*, 6(3), 103–110. https://dx.doi.org/10.2478%2Fintox-2013-0018.

Zhang, C., & Zheng, G. (2003). Characterization and desorption kinetics of PAHs from contaminated sediment in Houston Ship Channel. *2003 Annual Report*, 39–41.

11 Marine Adhesive Proteins for Novel Applications

Smit P. Bhavsar, Maushmi S. Kumar, and Vandana B. Patravale

CONTENTS

11.1 INTRODUCTION

Bioadhesives are natural polymeric materials that act as adhesives and studies have proved the possibility of their usage for biomedical and other applications (Zhao et al., 2017; Zhu et al., 2017). The requirements for medical adhesives are crucial; for instance, they should be biocompatible, strongly adhere to numerous surfaces being elastic, and show results in wet conditions as well. Most adhesives comprise one or more of these criteria (Vinters et al., 1985; Li et al., 2017). Synthetic adhesives might be toxic, carcinogenic, allergenic, and fail to exist in severe environmental conditions and/or comply with legislative restrictions. To the contrary, adhesives that are present in nature are biocompatible,

DOI: 10.1201/9781003303909-11

nontoxic and capable of sticking to numerous surfaces that are dry, wet, or underwater. This can be temporary or permanent and might not instill an exothermic reaction. Hence, there has been an increase in research focusing on the characterization and biomimetic utilization of biological adhesive systems. Biological attachment is an extremely common feature among several species (Peled-Bianco and Davidovich Pinhas, 2015; Smith et al., 2016). Since conditions for adhesion in aquatic and terrestrial environments are severely different (Ditsche and Summers, 2014), we discuss underwater adhesion. Many attachment devices have evolved uniquely and consist of many biological functions. For example, interstitial meiofauna (organisms living between sand granules of marine or freshwater beaches) have to secure themselves to a substrate to avoid displacement from their environment. Meanwhile, many species show a highly mobile lifestyle. In recent times, research on adhesive secretions has primarily focused on marine invertebrates and permanently attached animals, such as mussels and barnacles (Maier and Butler, 2017; Waite et al., 2017). We have summarized the protein structures and arrangements of various marine adhesives along with their relevant features, in comparison to a unique model system, known as the competitive model of adhesion. Byssus formed by mussels is a prominent adhesive and is widely studied to elucidate its mechanism of adhesion. An outlook for biotechnological production of mussel-inspired materials is being discussed with some unique applications involving the role of the catecholic amino acid l–3,4-dihydroxyphenylalanine (Dopa) along with specific industries of relevance.

11.2 MARINE ADHESIVE PROTEINS AND THEIR STRUCTURE

Adhesion has been a crucial property for survival and the implementation of basic functions of a large number of marine and freshwater animals; for example, the organic fraction of echinoderm's footprints consists of mainly proteins and carbohydrates, which play a major role in adhesion (Flammang et al., 1998; Santos et al., 2009). Proteins are important for adhesion and cohesion, which has also been demonstrated by the removal of footprints after an experimental treatment with the enzyme trypsin (Thomas and Hermans, 1985; Flammang et al., 1996). One prevalent feature of adhesive protein is the presence of domains that mediate protein–protein and protein–carbohydrate interactions (Hennebert et al., 2014, 2015; Rodrigues et al., 2016). The presence of lectin-binding domain is the most crucial among all (Toubarro et al., 2016). Presently, many temporary adhesive proteins and protein domains in various species have been identified based on their characteristics.

In *Asterias rubens* (*A. rubens*) commonly known as starfish, the entire length sequence of proteins involved in temporary adhesion was extracted (Hennebert et al., 2014). The sequence of a large protein consisted of 3853 predicted amino acids named sea star footprint (Sfp-1). Immunohistochemistry also localized Sfp-1 in Type 1 adhesive (Type 1 is a premium-grade, traditional, nonflammable acrylic adhesive) vehicles form a fibrillary meshwork of footprints. Apart from Sfp-1, 34 footprint-specific proteins were observed in *A. rubens*. Forty-one more proteins were found in footprints, as well as in the mucus secretion of this organism, which could be demonstrated for adhesive application (Hennebert et al., 2014). In addition to the sea star *A. rubens*, the sea urchin *Paracentrotus lividus* is a highly studied echinoderm species in terms of temporary adhesion. The protein fraction of its footprints is highly adhered to its amino acid composition, with excessive domination of glycine, alanine, valine, serine, threonine, and asparagine. The levels of proline and half-cysteine residues are greater compared to an average eukaryotic protein helping it with the role of adhesion (Santos et al., 2009). Barnacles are one of the most prominent biofouling organisms, which cause a lot of economic damage during transport (Schultz et al., 2011). Therefore, there is high demand to understand their massive settlement technique to overcome this problem. The barnacle cyprid larvae find surfaces and adhere to a suitable site and settle down. During this discovery phase, they secrete a temporary adhesive before producing permanent cement and go through metamorphosis to the sessile form (Nott and Foster, 1969). The temporary adhesive is proteinaceous and works as a settlement pheromone (Clare and Matsumura, 2009). In the barnacle *Amphibalanus amphitrite*, the glycoprotein "settlement-inducing protein complex" (SIPC) was described as the key to gregarious settlement (Dreanno et al., 2006). Furthermore, SIPC absorbs to various surfaces, highlighting its role as an adhesive protein. In the freshwater polyp *Hydra magnipapillata*, adhesive proteins are characterized by the combination of next-generation sequencing and mass spectrometry (Rodrigues et al., 2016). By using area-specific RNA sequencing, a list of transcripts certainly expressed in the foot of the animals can be curated. For 40 of these transcripts, the observation within the basal disc was further evaluated with whole-mount *in situ* RNA hybridization. From these, 21 proteins were confirmed with mass spectrometry of the adhesive footprints (Rodrigues et al., 2016). In the flatworm *Macrostomum lignano*, *in situ* hybridization screening conducted of tail-specific transcripts resulted in 20 transcripts expression in adhesive organs of intact animals and while they undergo tail regeneration (Lengerer et al., 2018). Research on the nature and specific details of these transcripts is currently in progress.

11.3 COMPETITIVE MODEL OF ADHESION

The adhesive area of temporary adhering animals is usually covered with a distinctive glycocalyx, also known as a "fuzzy coat" (Lengerer et al., 2016; Schröder and Bosch 2016). In theory, the attachment to the substrate happens through a thin layer of footprints that is homogeneous in nature, in which the meshwork at the top surface results in cohesive strength and networks the adhesive material to the glycocalyx of the animals. Figure 11.1 indicates the adhesion phenomenon with the parts involved.

In the process, the adhesive footprint remains attached to the substrate and the detachment should always occur either between the adhesive material and the glycocalyx or within the glycocalyx layer (Flammang et al., 1998). Hermans et al. (1983) were among the first to propose the fact that the de-adhesive

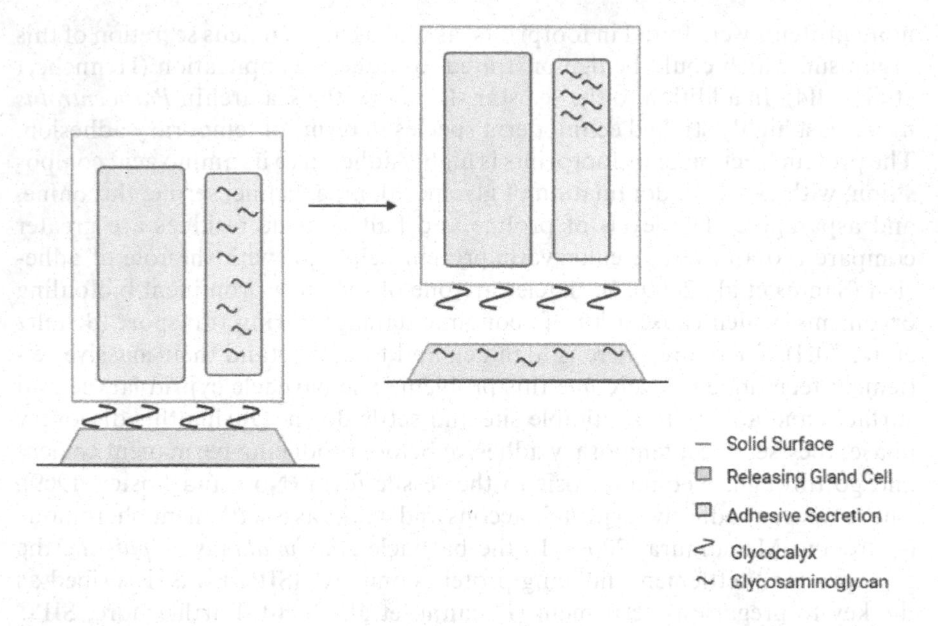

FIGURE 11.1 Adhesion and de-adhesion in the competitive model of marine bioadhesive.

secretion competes with the glycocalyx for binding sites on the adhesive surface. He had concluded that the de-adhesive material consists of glycosaminoglycans and reacts highly with the adhesive, thereby releasing the animal from the substrate (Figure 11.1) (Hermans et al., 1983). In the sea star *Leptasterias hexactis*, a well-known glycosaminoglycan, which is a supplement of heparin, reduces the attachment of the animals. The authors came to consensus that glycosaminoglycans are much similar to heparin, which is released during detachment (Thomas and Hermans, 1985). Although later studies hinted at an enzymatic release (Flammang et al., 1998), the competition theory has never been completely disregarded. In addition to it, a synergy of different detachment models is also possible.

Adhesion is the most important event for the survival and implementation of basic functions of large number of marine and freshwater animals. For many organisms, the morphology of adhesive organs has been studied, but knowledge about the composition of the adhesive and releasing substances is still limited. Hence, to study high performance adhesives and their prospects, we refer to mussel byssus adhesive and their individual structure's adhesive applications.

11.4 MUSSEL BYSSUS ADHESIVE

Man-made polymers such as Kevlar, Teflon and nylon show similar characteristics as mussel byssus, a biopolymer synthesized from proteins present in organisms like mussels, spiders and worms (Baer et al., 2017). Current age biotechnological procedures demand ecologically sustainable materials. Biopolymers such as mussel byssus consist of biodegradable protein-based fibers, which best suits this requirement. This high-performance fiber has applications in several industries as wet adhesives, self-healing materials, and coatings. The structure of mussel byssus complements the mechanical strength property of the fiber. The properties of the fiber complement the mechanical and structural integrity and for the purpose of studying the protein molecule configuration and amino acid sequence. The structure–function relationships will enable us to understand the ideal design of the biopolymer for our requirement and to understand their application.

11.4.1 Morphology and Structure–Function Relationships

A typical byssus from marine mussels (blue mussel) owns 50–100 byssal threads. Each thread is approximately 2–5 cm in length and is further divided into three structurally and functionally separate parts known as plaque, core, and cuticle. The plaque is a versatile underwater adhesive glue that serves as a boundary between the byssal thread and the surface (Waite et al., 2017). The core of the thread is a fibrous biopolymer fastened into the plaque that delivers the energy-damping function (Reinecke et al., 2016, Carrington et al., 2004). The cuticle is a thin sheath adjacent to the core and plaque projected to function as a protective coating (Holten-Andersen et al., 2007). The core, cuticle, and plaque are briefed distinctly, providing an overview of multiscale structure–function relationships and their applicative uses. Byssus core is the fibrous core of byssal threads and primarily functions to figure out its tensile and mechanical properties (Schmitt et al., 2015; Krauss et al., 2013). Threads have two separate regions known as the distal and proximal thread, which are adjacent to the surface and mussel, individually. These two regions have dissimilar morphology and mechanics with the distinct region exhibiting a fibrous morphology and later one exhibiting a corrugated structure (Carrington et al., 2004). The summation of high strength and extensibility in the distal region results in toughness that usually exceeds most synthetic polymers (Gosline et al., 2002), as well as a large hysteresis. Most noteworthy, threads are able to recover initial properties (Harrington et al., 2009), which is highly desirable from the view of eligible biomimicry use. Around an extensible core is a thin (5–10-μm) rough bump layer known as the cuticle. Mechanically, the cuticle is tighter and harder compared to the core. Surprisingly, even after having high hardness and stiffness, the cuticle is extremely extensible, which is highly appreciable since hard materials are usually nonelastic. Thus, the cuticle has the potential to inspire novel engineering polymers for extensible coatings in applications such as flexible electronics and biomedical devices. Another part known as the byssus plaque has received maximum attention from researchers wishing to duplicate its outstanding ability to adhere to almost every surface in salty seawater environments, which has relatable biomedical and industrial relevance (Lee et al., 2011). Bound water, ionic double layers, and biofilms present various challenges for wet adhesion, even though mussels have evolved a protein-based mechanism to counter this (Lee et al., 2011; Waite et al., 2017). Lead by the Waite group, about 40 years of research has gone into unraveling the super-complex adhesive mechanism in mussels, resulting in a very detailed understanding of the biochemistry behind their stickiness and applications (Waite et al., 2017).

11.4.2 Mussel-Inspired Materials

On the basis of the structure-function relationships described earlier, various groups have attempted to study extracted design principles from the byssus into man-made materials. However, due to the underlying complex circumstances of the natural system, it is difficult to work on it. While traditional polymers are a result of long chains with repetition of monomeric subunits, protein function is identified due to specific sequences of amino acid lengths presenting side chains along with different chemical properties. The arrangement and length of the amino acids in the proteins identified by natural selection are linked to the higher order hierarchical structure of the materials that determines the material properties for various uses, and recent mussel-inspired materials have considered minimalist approach, resulting in a reduced complexity of byssal proteins to single amino acid moieties (Holten-Andersen et al., 2011; Fullenkamp et al., 2013). The ability of a biotechnological approach for mussel-inspired materials was implemented more than ten years ago with purified proteins from the byssus and mussel soft tissue, which included mfp-1 and the preCols (family of modular collagenous proteins). For example, preCols extracted and purified from mussel foot tissue of *Mytilus californianus* were conveniently drawn into fibers with ultrastructure and mechanical properties similar to native byssal threads (Harrington et al., 2009). Similarly, various groups utilized both basic and acidic partial hydrolysis of crude byssus, which led to a larger amount of relatively impure fibrous byssus material, which forms freestanding water-insoluble films with a fibrillar structure and tenable properties (Byette et al., 2016, Montroni et al., 2017).

Recombinant technology offers promising approach toward producing higher yields of byssus proteins. Efforts to produce preCols in bacteria have met numerous challenges due to the high degree of post-translational modification. preCol extraction from mussel foot tissue results in nanofibrils under native conditions; however, this approach presently suffers from the equal low yields of native extraction and will have to be improved. It was proposed that both proteins influence the mechanical properties of byssal threads via their direct interactions with preCols (Sagert et al., 2009, Suhre et al., 2014). But it does improve mouse osteoblast cell adhesion and its reaction with type I collagen, emphasizing its role in thread assembly, function, and significant use as a biomedical adhesive (Yoo et al., 2016).

A great deal of progress has been made in the development of mussel-inspired materials and adhesives in the previous 15 years, both in synthetic chemistry and biotechnological means as compiled in Table 11.1. New efforts toward improving mussel-inspired materials are aimed to achieve an improved

TABLE 11.1
Mussel-Inspired Materials Achieved through Biotechnology (Applicative Properties)

Mussel Thread Protein	Recombinant Peptide	Resulting Material	Properties	Reference
preCols	NA	Hand-drawn fiber	Similar structure and properties to native threads	Harrington and Waite, 2008
Byssal-thread hydrolysate	NA	Water insoluble films	pH-tunability, dye removal	Byette, 2014; Montroni et al., 2017
mfp-3	NA	Self-coacervate	Dependent on pH and ionic strength	Wei et al., 2014
preCol-D	prCol-D	Nanofibrils	Defined secondary structure, presence of triple helices	Golser and Scheibel, 2017
ptmp-1	ptmp-1	Adhesive	Enhancing mouse osteoblast cell adhesion, collagen interaction	Yoo et al., 2016
mfp-1	rfp-1	Surface coating, adhesive hydrogel, polymeric nanoparticles, coacervate	Flexible viscoelastic and self-healing behavior, pH-responsive and anti-breast-cancer drug release	Zeng, 2010; Kim et al., 2014
mfg-5	rfp-5	Adhesive	Suitable for cell biological application, immobilization of antibodies	Hwang, 2007
mfp-1, mfp-5	fp-151	Adhesive	Greater production yield and improved purification	Choi et al., 2014; Hwang et al., 2007
mfp-1, mfp-5	fp-151-peptide	Adhesive	Enhanced adhesion, proliferation, and spreading of mammalian cells	Choi et al., 2010
mfp-1, mfp-3, mfp-5	fp-131, fp-151	Complex coacervate	Enhanced adhesive strength	Lim et al., 2010
preCol-D	HRD-DN	Freestanding films	Metal-tunable architecture	Jehle et al., 2018
mfp-3	mfp-3S-pep	Coacervate adhesive	Improved adsorption kinetics and quantity	Wei et al., 2016

biological, physical, and chemical understanding of byssus formation (Priemel et al., 2017).

Proteinaceous biopolymeric materials similar to silk and mussel byssus play a crucial role in inspiring sustainable production of next-generation advanced polymers. To be precise, byssus has already laid many mussel-inspired materials and glues, which tap into the metal coordination chemistry of the byssal threads to achieve wet adhesion, self-healing capacity and high toughness (Krogsgaard et al., 2016). It has been clearly established that the biochemical setting of the natural protein sequences of the byssal proteins is important for material performance. A biotechnological approach for creating mussel-inspired polymers must be united with a deep understanding of the natural byssus fabrication process. Fresh investigations in this direction have tinted the importance of stimuli-responsive protein sequences (Waite et al., 2017), which can be willingly investigated via biotechnological approaches. However, this will likely require appropriate technology development to imitate this complex fabrication process, which is indeed most closely approximated with microfluidic devices and three-dimensional (3D) printing. While much remains to be done, the outlook for biotechnological production of mussel-inspired materials is optimistic. Since the unique and quick adhesion of marine mussels to diverse surfaces of solid in wet environments happens due to the secretion of Dopa containing mussel adhesive proteins (MAPs), several efforts have been made in the development of synthetic mussel-inspired adhesives with water-resistant adhesion and cohesion properties by modifying polymer systems with Dopa and its derivatives. The newer developments of novel Dopa-containing adhesives with significant mechanical properties and other functionalities as viscous coacervated adhesives, soft adhesive hydrogels, smart adhesives, and stiff adhesive polyesters are summarized in the following section. Furthermore, their upcoming applications in bioengineering, biological and biomedical fields are also highlighted.

11.5 NOVEL APPLICATIONS

11.5.1 Dopa-Inspired Design of Synthetic Underwater Adhesives

Marine adhesive proteins are in great demand in engineering and related biomedical fields. Numerous advancements in engineering bio-inspired or biomimetic adhesive materials that have the capacity to perform in high-moisture surroundings are being made. The current adhesive design system utilizes mussel protein–inspired Dopa-based adhesives based on the peculiar molecular features of Dopa residues in the diverse adhesive components of marine organisms. These Dopa groups are frequently used to govern the synthesis of functional adhesives with high underwater bonding strength and tenacity. The bonding strength is greatly linked to the catechol content, degree of crosslinking, and molecular weight of the copolymer. The strength of adhesion is seen to be stronger in saltwater as compared to deionized water, which indicates its universality in harsh environmental water conditions. Inspired by the "one-two punch" synergistic action between catechol Dopa and lysine residues occurring in mussel foot proteins, a synthetic equivalent of cyclic trichrysobactin, Tren-Lys-Cam (TLC), adhesive molecule has been designed, which can be an adequate plan to explore further molecular synergies in bioadhesion. Synthetic TLC adhesive molecules have also been utilized to suggest the reasons for the robustness of natural mussel in harsh environmental conditions like high salt concentration, pH levels, and conditions of hydration. 2,3-dihydroxy catechol in the TLC molecules have actually shown enhanced resistance to oxidation as compared to bare Dopa (Shuanhong et al., 2020).

11.5.2 Coacervate Dopa-Functionalized Adhesives

During the liquid protein adsorption process, mussel foot proteins appear in the form of coacervates (Wei et al., 2016). Coacervation is the liquid-liquid phase separation of a colloidal system consisting of ionic polymers or proteins (De Kruif et al., 2004) and the dense fluid of concentrated polymers referred to as a coacervate plays an important role in the outstanding wet adhesion of marine organisms such as mussels and sandcastle worms (B.P. Lee et al., 2011; Shao et al., 2010). The formation of the coacervate is generally driven by weak and nonspecific interactions including electrostatic attraction, hydrophobic interaction and hydrogen bonding, followed by the entropic gains by releasing ions and water molecules, which is susceptible to the solution and environmental conditions (e.g., concentration, pH, ionic strength) (Kizilay et al., 2011, Van der Gucht et al., 2011). Taking advantage of their relatively high viscosity, reduced interfacial energy and high diffusion coefficients of the solute and solvent molecules, the Dopa-functionalized coacervates with great wetting properties and adhesiveness have become promising candidates as effective underwater adhesives (Waite et al., 2005). Although Dopa shows a slight influence on coacervation, it is indispensable for promoting peptide adsorption by removing interfacial water from the target surfaces. Waite and coworkers developed a simple wet adhesive primer from low-molecular-weight catecholic zwitterionic surfactants compromising amine, phosphate, hydrophobic, and catechol functionalities (Ahn et al., 2015). The coacervated zwitterionic platform with reduced complexity offered smooth and thin coatings (<4 nm) on various substrates with strong wet adhesion up to approximately 50 mJ/m^2, implicating great promise for nanofabrication. Besides zwitterionic systems, self-coacervation occurred in cationic and polyphenolic rmfp-1, which was attributed to strong cation-π interaction at short range (S. Kim et al., 2017). The coacervate formation was triggered by salt concentration similar to nature seawater condition (>0.7 M), since salt screened the electrostatic repulsion while still maintaining the attractive cation−π interaction. The friction coefficient (<0.03) of the formed coacervate was remarkably lower than that of conventional coacervates, expanding their applications as lubricants.

11.5.3 Adhesive Hydrogel

As a highly water-swollen 3D network, hydrogels possess structural resemblance to biological tissues and high permeability to oxygen and nutrients, which have been designed as platforms for diverse biomedical applications, such as drug delivery, tissue engineering, and implantable devices (Hoffman, 2012; Caló and Khutoryanskiy, 2015). Particularly, intense efforts have been spent on the development of adhesive hydrogels with biocompatibility and sufficient wet adhesion, holding considerable promise as tissue adhesives to replace surgical sutures and tapes for wound management (Annabi et al., 2014). Owing to the robust adhesion and cohesion in wet conditions, hydrogels functionalized with Dopa and its analogues have aroused great interest as a new family of tissue adhesives over the past two decades (Li et al., 2017). Poly (ethylene glycol) (PEG) is one of the most widely used synthetic polymers for the development of hydrogels with outstanding hydrophilicity and biocompatibility, and the Messersmith group has incorporated Dopa or catechol functionality as terminal or side groups to a family of PEG polymers with different structures (e.g., linear, hyperbranched, block; Lee et al., 2006). The in vivo performance of the catechol-modified PEG adhesive hydrogel was further investigated, in which the in situ gelation was induced within 1 min by oxidation and the transplanted islet was immobilized on the extrahepatic tissues up to 1 year with a minimal inflammatory response (Brubaker et al., 2010) When an Ala-Ala dipeptide substrate was incorporated between the catechol group and the branched PEG, the adhesive hydrogel displayed enzymatic degradation on the addition of neutrophil elastase over several months, showing improved cellular infiltration after dorsal subcutaneous implantation in mice. Owing to their outstanding biocompatibility and biodegradability (Liu et al., 2008), naturally occurring polymers, especially polysaccharides, have been extensively explored for the development of injectable adhesive hydrogels in combination with Dopa derivatives.

11.5.4 Smart Adhesives

Smart adhesives that can undergo on-demand bonding/debonding transition to substrates in response to external stimuli (e.g., temperature, pH, light) have gained growing interest in developing advanced functional adhesives, because the precise control of the adhesiveness is intriguing for numerous practical applications, such as removable wound dressings, recyclable packaging, and reusable structural components (Donnelly et al., 2005). Mussel-inspired smart adhesives have been generated taking advantage of the strong wet adhesion of Dopa and the responsiveness of catechol chemistry. Furthermore, newly engineered biofilms have been proposed for functioning as cellular glues, with enhanced environmental tolerance and self-regenerating ability (Zhang et al., 2019). Since Dopa-incorporated recombinant protein adhesives have attracted lots of attention, understanding the description of the expression of

Dopa-incorporated engineered recombinant proteins is recommended (Hauf et al., 2017; Yang et al., 2014).

11.5.5 Adhesive Polyesters

Synthetic polyesters have been extensively applied in diverse biomedical applications, especially in tissue engineering due to their good mechanical property, easy producibility and controllable biodegradation (Park et al., 2011). Combining adhesive Dopa derivatives with biocompatible polyesters has led to strong tissue adhesives successfully served as bone glue. There has been synthesis of hyperbranched poly (β-amino ester) with dopamine and triacrylate monomers through a Michael addition reaction, which exhibited 37-kPa adhesion strength to wet tissue surface after crosslinking by fibrinogen within 15 min (Zhang et al., 2014). The mechanical properties of the prepared adhesive were further improved by reinforcement with nano-sized hydroxyapatite particles, and it was demonstrated that the nanocomposite was able to act as a efficient bone adhesive for sternal closure with tunable curing speed and sufficient load-bearing capacity (Zhang et al., 2014). A poly (ester urea)–based adhesive showing adhesive strength comparable with fibrin glue was developed by introducing pendant catechol groups (Zhou et al., 2015), and after the incorporation of poly (propylene glycol) into the backbone, the adhesive was rendered ethanol solubility that is favorable for clinical applications (Zhou et al., 2016). When a plant-based poly (lactic acid) (PLA) was modified with catechol functionality, the biomimetic adhesive possessed strong adhesion strength of 2.6 MPa in air and 1.0 MPa under wet conditions to aluminum substrates (Jenkins et al., 2012).

11.5.6 Biomedical Applications

Practitioners continue to look for applicable biomaterial adhesives that are biodegradable and nontoxic. Wound healing, tissue engineering, recovery from surgery, dental composites, and orthopedic cements are a few areas where bioadhesives are required. Fibrin sealants, albumin-based compounds, glutaraldehyde glues, cyanoacrylates, hydrogels, and collagen-based compounds are some general adhesives used in the medical industry (Silverman et al., 2007). Cell-Tak™ and MAP™ are commercially accessible medical adhesives developed from extracted MAPs. However, low extraction yield and efforts in purification make both adhesives inefficient to be used in tissue engineering and cell culture (Gim et al., 2008). Extracted MAPs have the distinguishing feature to adhere to porcine skin and collagen membranes (Ninan et al., 2007). Recombinant MAPs are investigated for biomedical applications because of their high production volumes. The biomimetics of MAPs were explored in many studies to improve the biomedical applications of adhesives on various surfaces. Dopa-K motifs, formed in mussels by oxidative polymerization of Dopa, have the ability to build adlayers on any surface and to improve cell adhesion to any surface, including antiadhesive surfaces like poly(tetrafluoroethylene). Polydopamine-assisted hydroxyapatite formation, inspired by mussel adhesion, has been developed for use in biomineralization of any surface, which is important in tissue engineering (Ryu et al., 2010). There is a cumulative interest in developing various biomimetic polymers and natural MAPs in biomedical applications with better-quality functionalities at lower cost.

11.5.7 Industrial and Other Novel Applications

Petroleum-based adhesives that are cheap and easy to handle, with good functional properties, are widely used for bioengineering applications. Concerns over the use of these adhesives are formaldehyde emission, nondegradability, nonrenewability, and fluctuation in cost with petroleum prices (Cha et al., 2008; Liu et al., 2002). Bio-based adhesives and biomimetic adhesive analogs are alternatives to petroleum-based adhesives and are now well recognized as a solution to these concerns. There are great interests in emerging adhesives from renewable resources, such as proteins and fibers; however, adhesives developed from renewable resources are associated with poor water resistance and weak adhesion (Liu et al., 2002). Many efforts have been taken to develop water-resistant wood adhesives using the knowledge of mussel adhesion mechanisms (Liu et al., 2004). For example, a chitosan-based adhesive has been developed using a tyrosinase-catalyzed dopamine approach, which led to improved adhesion and water resistance (Yamada et al., 2000). Mussel-inspired Dopa-grafted soy protein isolate (SPI) and cysteamine-grafted SPI (Liu et al., 2004), prepared via amide linkages similar to mussel protein, exhibited higher water resistance and higher bond strength than alkali-treated SPI. Conjugation of linear monomethoxy-terminated PEG to Dopa residues has shown antifouling properties on surfaces.

11.6 CONCLUSION

The rapid and robust wet adhesion of marine mussels has been gaining considerable interest in the development of advanced adhesives. The improved understanding of the adhesion mechanisms of mussel foot proteins reveals that Dopa plays a critical role in the universal adhesion under wet conditions via various interactions. Diverse synthetic and natural polymers, a series of mussel-inspired adhesives with on-demand mechanical properties (e.g., viscous coacervates, soft hydrogels and stiff polyesters), and other functionalities have been successfully designed for a broad range of biomedical and engineering applications, such as tissue/bone adhesives, drug carriers, and surgical implants, as well as pollutant adsorbents and plastic adhesives. Multifunctional materials have attracted growing attention nowadays, and adhesive catechol moieties have been combined with polymer networks exhibiting other intriguing properties such as biodegradability, stimuli-responsivity, antibacterial activity, and self-healing ability. Since most adhesive systems are complicated and require tedious synthesis processes, it still remains a challenge to develop a facile and universal approach to manufacturing multifunctional adhesives on demand.

Besides the catechol group, other phenolic compounds (e.g., gallic acid, epigallocatechin gallate, tannic acid) also hold promise as building blocks for adhesive materials due to their structural similarity. Compared to catechol-based polymers, these phenol-containing materials are much less reported and are worth further studying due to their easy accessibility and unique properties. Until now, the development of mussel-mimetic adhesives has generally focused on catechol chemistry, but the biological system is far more complex than Dopa. More efforts are still needed to incorporate important finding into the synthetic adhesive systems, which can further promote the design of next-generation adhesives for a broad range of applications.

ACKNOWLEDGMENT

The authors would like to acknowledge Institute of Chemical Technology and SVKM'S NMIMS for all the support and required facility to carry out this work. We would also like to acknowledge Dr. Sreeranjini Pulakkat (ICT, Mumbai) for her assistance in proofreading the book chapter.

REFERENCES

Ahn, B. K., Das, S., Linstadt, R., Kaufman, Y., Martinez-Rodriguez, N. R., Mirshafian, R., Kesselman, E., Talmon, Y., Lipshutz, B. H., Israelachvili, J. N. and Waite, J. H. 2015. High-performance mussel-inspired adhesives of reduced complexity. *Nature Communications*, 6: 8663

Annabi, N., Tamayol, A., Shin, S. R., Ghaemmaghami, A. M., Peppas, N. A., Khademhosseini. 2014. A surgical materials: Current challenges and nano-enabled solutions. *Nano Today*, 9: 574–589

Baer, A., Schmidt, S., Haensch, S., Eder, M., Mayer, G. and Harrington, M. J. 2017. Mechanoresponsive lipid-protein nanoglobules facilitate reversible fibre formation in velvet worm slime. *Nature Communications*, 8: 974

Brubaker, C. E., Kissler, H., Wang, L., Kaufman, D. B. and Messersmith, P. B. 2010. Biological performance of mussel-inspired adhesive in extrahepatic islet transplantation. *Biomaterials*, 31: 420–427.

Byette, F., Laventure, A., Marcotte, I. and Pellerin, C. 2016. Metal-ligand interactions and salt bridges as sacrificial bonds in mussel byssus-derived materials. *Biomacromolecules*, 17: 3277–3286

Byette, F., Pellerin, C. and Marcotte, I. 2014. Self-assembled pH-responsive films prepared from mussel anchoring threads. *Journal of Materials Chemistry*, B, 2: 6378–6386

Caló, E. and Khutoryanskiy, V. V. 2015. Biomedical applications of hydrogels: a review of patents and commercial products. *European Polymer Journal*, 65: 252–267

Carrington, E. and Gosline, J. 2004. Mechanical design of mussel byssus: Load cycle and strain rate dependence. *American Malacological Bulletin*, 18: 135–142

Cha, H. J., Hwang, D. S. and Lim, S. 2008. Development of bio adhesives from marine mussels. *Biotechnology Journal*, 3

Choi, B.-H., Cheong, H., Jo, Y. K., Bahn, S. Y., Seo, J. H. and Cha, H. J. 2014. Highly purified mussel adhesive protein to secure biosafety for in vivo applications. *Microbial Cell Factories*, 13

Choi, B.-H., Choi, Y. S., Kane, D. G., Kim, B. J., Song, Y. H. and Cha, H. J. 2010. Cell behavior on extracellular matrix mimic materials based on mussel adhesive protein fused with functional peptides. *Biomaterials*, 31: 8980–8988

Clare, A. S. and Matsumura, K. 2009. Nature and perception of barnacle settlement pheromones. *Biofouling*, 15: 57–71
De Kruif, C. G., Weinbreck, F. and de Vries, R. 2004. Complex coacervation of proteins and anionic polysaccharides. *Current Opinion in Colloid and Interface Science*, 9: 340–349
Ditsche, P. and Summers, A. P. 2014. Aquatic versus terrestrial attachment: Water makes a difference. *Beilstein Journal of Nanotechnology*, 5: 2424–2439
Dreanno, C., Matsumura, K., Dohmae, N., Takio, K., Hirota, H., Kirby, R. R. and Clare, A. S. 2006. An alpha2-macroglobulin-like protein is the cue to gregarious settlement of the barnacle Balanus amphitrite. *Proceedings of the National Academy of Sciences of the United States of America*, 103: 14396–14401
Flammang, P. 1996. Adhesion in echinoderms. *Echinoderm Studies*, 5: 1–60
Flammang, P., Michel, A., Cauwenberge, A. V., Alexandre, H. and Jangoux, M. 1998. A study of the temporary adhesion of the podia in the sea star asterias rubens (Echinodermata, asteroidea) through their footprints. *The Journal of Experimental Biology*, 201: 2383–2395
Fullenkamp, D. E., He, L., Barrett, D. G., Burghardt, W. R. and Messersmith, P. B. 2013. Mussel-inspired histidine-based transient network metal coordination hydrogels. *Macromolecules*. doi: 10.1021/ma301791n
Gim, Y., Hwang, D. S., Lim, S., Song, Y. H. and Cha, H. J. 2008. Production of fusion mussel adhesive fp-353 in Escherichia coli. *Biotechnology Progress*, 24: 1272–1277
Golser, A. V. and Scheibel, T. 2017. Biotechnological production of the mussel byssus derived collagen presold. *RSC Advances*, 7: 38273–38278
Gosline, J., Lillie, M., Carrington, E., Guerette, P., Ortlepp, C. and Savage, K. 2002. Elastic proteins: Biological roles and mechanical properties. *Philosophical Transactions of the Royal Society, London, Ser. B*, 357: 121–132
Harrington, M. J., Gupta, H. S., Fratzl, P. and Waite, J. H. 2009. Collagen insulated from tensile damage by domains that unfold reversibly: In situ X-ray investigation of mechanical yield and damage repair in the mussel byssus. *J. Struct. Biol*, 167: 47–54
Harrington, M. J. and Waite, J. H. 2008. pH-dependent locking of giant mesogens in fibers drawn from mussel byssal collagens. *Biomacromolecules*, 9: 1480–1486
Hauf, M., Richter, F., Schneider, T., Faidt, T., Martins, B. M., Baumann, T., Durkin, P., Dobbek, H., Jacobs, K., Möglich, A. and Budisa, N. 2017. Photoactivatable mussel-based underwater adhesive proteins by an expanded genetic code. ChemBioChem, 18: 1819–1823
Hennebert, E., Leroy, B., Wattiez, R. and Ladurner, P. 2015. An integrated transcriptomic and proteomic analysis of sea star epidermal secretions identifies proteins involved in defence and adhesion. *Journal of Proteomics*, 128: 83–91
Hennebert, E., Wattiez, R., Demeuldre, M., Ladurner, P., Hwang, D. S., Waite, J. H. and Flammang, P. 2014. Sea star tenacity mediated by a protein that fragments, then aggregates. *Proceedings of the National Academy of Sciences of the United States of America*, 111: 6317–6322
Hermans, C. O. 1983. The duo-gland adhesive system. *Oceanography and Marine Biology: An Annual Review*, 21: 283–339
Hoffman, A. S. 2012. Hydrogels for biomedical applications. *Adv Drug Del Rev*, 64: 18–23
Holten-Andersen, N., Fantner, G. E., Hohlbauch, S., Waite, J. H. and Zok, F. W. 2007. Protective coatings on extensible biofibres. *Nature Materials*, 6: 669–672
Holten-Andersen, N., Harrington, M. J., Birkedal, H., Lee, B. P., Messersmith, P. B., Lee, K. Y. C. and Waite, J. H. 2011. pH-induced mussel metal-ligand cross-links yield self-healing polymer networks with near-covalent elastic moduli. *Proceedings of the National Academy of Sciences of the United States of America*, 108: 2651–2655

Hwang, D. S., Gim, Y., Kang, D. G., Kim, Y. K. and Cha, H. J. 2007. Recombinant mussel adhesive protein Mgfp 5 as cell adhesion biomaterial. *Journal of Biotechnology*, 127: 727–735

Jehle, F., Fratzl, P. and Harrington, M. J. 2018. Metal-tunable self-assembly of hierarchical structure in mussel inspired peptide films. *ACS Nano*, 12: 2160–2168

Kim, B. J., Oh, D. X., Kim, S., Seo, J. H., Hwang, D. S., Masic, A., Han, D. K. and Cha, H. J. 2014. Mussel-mimetic protein-based adhesive hydrogel. *Biomacromolecules,* 15: 1579–1585

Kim, S., Yoo, H. Y., Huang, J., Lee, Y., Park, S., Park, Y., Jin, S., Jung, Y. M., Zeng, H., Hwang, D. S. and Jho, Y. 2017. Salt triggers the simple coacervation of an underwater adhesive when cations meet aromatic π electrons in seawater. *ACS Nano*, 113: E847–853

Kizilay, E., Kayitmazer, A. B. and Dubin, P. L. 2011. Complexation and coacervation of polyelectrolytes with oppositely charged colloids. *Advances in Colloid and Interface Science*, 167: 24–37

Krauss, S., Metzger, T. H., Fratzl, P. and Harrington, M. J. 2013. Self-repair of a biological fiber guided by an ordered elastic framework. *Biomacromolecules*, 14: 1520–1528

Krogsgaard, M., Nue, V. and Birkedal, H. 2016. Mussel-inspired materials: Self-healing through coordination chemistry. *Chemistry—A European Journal*, 22: 844–857

Lee, B. P., Messersmith, P. B., Israelachvili, J. N. and Waite, J. H. 2011. Mussel-inspired adhesives and coatings. *Annual Review of Materials Research*, 41: 99–132

Lee, H., Scherer, N. F. and Messersmith, P. B. 2006. Single-molecule mechanics of mussel adhesion. *Proc Natl Acad Sci U S A,* 103: 12999–13003

Lengerer, B., Hennebert, E., Flammang, P., Salvenmoser, W. and Ladurner, P. 2016. Adhesive organ regeneration in *Macrostomum lignano*. *BMC Developmental Biology*, 16

Lengerer, B., Wunderer, J., Pjeta, R., Carta, G., Kao, D., Aboobaker, A., Beisel, C., Berezikov, E., Salvenmoser, W. and Ladurner, P. 2018. Organ specific gene expression in the regenerating tail of *Macrostomum lignano*. *Developmental Biology*, 433: 448–460

Li, J., Celiz, A. D., Yang, Q., Wamala, I., Whyte, W., Seo, B. R., Vasilyev, N. V., Vlassak, J. J. and Suo, Z. 2017. Tough adhesives for diverse wet surfaces. *Science*, 357: 378–381

Lim, S., Choi, Y. S., Kang, D. G., Song, Y. H. and Cha, H. J. 2010. The adhesive properties of coacervated recombinant hybrid mussel adhesive proteins. *Biomaterials*, 31: 3715–3722

Liu, Y. and Li, K. 2002. Chemical modification of soy protein for wood adhesives." *Macromolecular Rapid Communications*, 23: 739–742

Liu, Y. and Li, K. 2004. Modification of soy protein for wood adhesives using mussel protein as a model: The influence of a mercapto group. *Macromolecular Rapid Communications*, 25: 1835–1838

Liu, Z., Jiao, Y., Wang, Y., Zhou, C. and Zhang, Z. 2008. Polysaccharides-based nanoparticles as drug delivery systems. *Advanced Drug Delivery Reviews*, 60: 1650–1662

Maier, G. P. and Butler, A. 2017. Siderophores and mussel foot proteins: The role of catechol, cations, and metal coordination in surface adhesion. *Journal of Biological Inorganic Chemistry*, 22: 739–749

Montroni, D., Piccinetti, C., Fermani, S., Calvaresi, M., Harrington, M. J. and Falini, G. 2017. Exploitation of mussel byssus mariculture waste as a water remediation material. *RSC Advances,* 7: 36605–36611

Ninan, L., Stroshine, R., Wilker, J. and Shi, R. 2007. Adhesive strength and curing rate of marine mussel protein extracts on porcine small intestinal submucosa. *Acta Biomaterialia,* 3: 687–694

Nott, J. A. and Foster, B. A. 1969. On the structure of the antennular attachment organ of the cypris larva of *Balanus balanoides*. *Philosophical Transactions of the Royal Society Lond. B*, 115–134

Park, S. A., Lee, S. H. and Kim, W. D. 2011. Fabrication of porous polycaprolactone/hydroxyapatite (PCL/HA) blend scaffolds using a 3D plotting system for bone tissue engineering. *Bioprocess and Biosystems Engineering*, 34: 505–513

Peled-Bianco, H. and Davidovich-Pinhas, M. 2015. *Bioadhesion and Biomimetics*. Singapore: Pan Stanford Publishing Pte. Ltd.

Priemel, T., Degtyar, E., Dean, M. N. and Harrington, M. J. 2017. Rapid self-assembly of complex biomolecular architectures during mussel byssus bio fabrication. *Nature Communications*, 8

Reinecke, A., Bertinetti, L., Fratzl, P. and Harrington, M. J. 2016. Cooperative behaviour of a sacrificial bond network and elastic framework in providing self-healing capacity in mussel byssal threads. *Journal of Structural Biology*, In Press. doi: 10.1016/j.jsb.2016.1007.1020.

Rodrigues, M., Ostermann, T., Kremeser, L., Lindner, H., Beisel, C., Berezikov, E., Hobmayer, B. and Ladurner, P. 2016. Profiling of adhesive-related genes in the freshwater cnidarian Hydra magnipapillata by transcriptomics and proteomics. *Biofouling*, 32: 1115–1129

Ryu, J., Ku, S. H., Lee, H. and Park, C. B. 2010. Mussel-inspired polydopamine coating as a universal route to hydroxyapatite crystallization. *Advanced Functional Materials*, 20: 2132–2139

Sagert, J. and Waite, J. H. 2009. Hyperunstable matrix proteins in the byssus of Mytilus galloprovincialis. *The Journal of Experimental Biology*, 212: 2224–2236

Santos, R., da Costa, G., Franco, C., Gomes-Alves, P., Flammang, P. and Coelho, A. V. 2009. First insights into the biochemistry of tube foot adhesive from the sea urchin Paracentrotus lividus (Echinoidea, Echinodermata). *Marine Biotechnology*, 11: 686–698

Schmitt, C. N. Z., Politi, Y., Reinecke, A. and Harrington, M. J. 2015. Role of sacrificial protein-metal bond exchange in mussel byssal thread self-healing. *Biomacromolecules*, 16: 2852–2861

Schröder, K. Bosch, T. C. G. 2016. The origin of mucosal immunity: Lessons from the holobiont hydra. *MBio*, 7: e01184–16

Schultz, M. P., Bendick, J. A., Holm, E. R. and Hertel, W. M. 2011. Economic impact of biofouling on a naval surface ship. *Biofouling*, 27: 87–98

Shao, H. and Stewart, R. J. 2010. Biomimetic underwater adhesives with environmentally triggered setting mechanism. *Advanced Materials*, 22: 729–733

Shuanhong, M., Yang, W. and Feng, Z. 2020. Bioinspired synthetic wet adhesives: From permanent bonding to reversible regulation. *Current Opinion in Colloid & Interface Science*, 47: 84–98

Silverman, H. G. and Roberto, F. F. 2007. Understanding marine mussel adhesion. *Marine Biotechnology*, 9: 661–681

Smith, A. M. 2016. *Biological Adhesives*. Springer Cham, Springer International Publishing Switzerland 2016. ISBN 978-3-319-46081-9

Suhre, M. H., Gertz, M., Steegborn, C. and Scheibel, T. 2014. Structural and functional features of a collagen binding matrix protein from the mussel byssus. *Nature Communications*, 5

Thomas, L. A. and Hermans, C. O. 1985. Adhesive interactions between the tube feet of a starfish, *Leptasterias hexactis*, and substrata. *The Biological Bulletin*, 169

Toubarro, D., Gouveia, A., Ribeiro, R. M., Simões, N., da Costa, G., Cordeiro, C. and Santos, R. 2016. Cloning, characterization, and expression levels of the nectin gene from the tube feet of the sea urchin Paracentrotus lividus. *Marine Biotechnology*, 18: 372–383

Van der Gucht, J., Spruijt, E., Lemmers, M. and Cohen Stuart, M. A. 2011. Polyelectrolyte complexes: Bulk phases and colloidal systems. *Journal of Colloid and Interface Science*, 361: 407–422

Vinters, H. V., Galil, K. A., Lundie, M. J. and Kaufmann, J. C. 1985. The histotoxicity of cyanoacrylates. A selective review. *Neuroradiology*, 27: 279–291

Waite, J. H. 2017. Mussel adhesion—essential footwork. *The Journal of Experimental Biology*, 220: 517–530

Waite, J. H., Andersen, N. H., Jewhurst, S. and Sun, C. 2005. Mussel adhesion: finding the tricks worth mimicking. *The Journal of Adhesion*, 81

Wei, W., Petrone, L., Tan, Y., Cai, H., Israelachvili, J. N., Miserez, A. and Waite, J. H. 2016. An underwater surface-drying peptide inspired by a mussel adhesive protein. *Advanced Functional Materials*, 26: 3496–3507

Yamada, K., Chen, T., Kumar, G., Vesnovsky, O., Topoleski, L. D. T. and Payne, G. F. 2000. Chitosan based water resistant adhesive. Analogy to mussel glue. *Biomacromolecules*, 1: 252–258

Yang, B., Ayyadurai, N., Yun, H., Choi, Y. S., Hwang, B. H., Huang, J., Lu, Q., Zeng, H. and Cha, H. J. 2014. In vivo residue-specific dopa-incorporated engineered mussel bioglue with enhanced adhesion and water resistance. *Angewandte Chemie International Edition*, 53: 13360–13364

Yoo, H. Y., Song, Y. H., Foo, M., Seo, E., Hwang, D. S. and Seo, J. H. 2016. Recombinant mussel proximal thread matrix protein promotes osteoblast cell adhesion and proliferation. *BMC Biotechnology*, 16

Zeng, H., Hwang, D. S., Israelachvili, J. N. and Waite, J. H. 2010. Strong reversible Fe3+-mediated bridging between dopa-containing protein films in water. *Proc. Natl. Acad. Sci. U.S.A.*, 107: 12850–12853

Zhang, H., Bré, L., Zhao, T., Newland, B., Da Costa, M. and Wang, W. 2014. A biomimetic hyperbranched poly (amino ester)-based nanocomposite as a tunable bone adhesive for sternal closure. *Journal of Materials Chemistry B*, 2: 4067–4071

Zhang, C., Huang, J., Zhang, J., Liu, S., Cui, M., An, B., Wang, X., Pu, J., Zhao, T., Fan, C., Lu, T. K., Zhong, C. 2019. Engineered Bacillus subtilis biofilms as living glues. *Materials Today*, 28: 40–48.

Zhao, Y., Wu, Y., Wang, L., Zhang, M., Chen, X., Liu, M., Fan, J., Zhou, F. and Wang, Z. 2017. Bio-inspired reversible underwater adhesive. *Nat. Commun*, 8: 2218.

Zhou, J., Bhagat, V. and Becker, M. L. 2016. Poly (ester urea)-based adhesives: Improved deployment and adhesion by incorporation of poly (propylene glycol) segments. *ACS Applied Materials & Interfaces*, 8: 33423–33429

Zhou, J., Defante, A. P., Lin, F., Xu, Y., Yu, J., Gao, Y., Childers, E., Dhinojwala, A. and Becker, M. L. 2015. Adhesion properties of catechol-based biodegradable amino acid-based poly (ester urea) copolymers inspired from mussel proteins. *Bio macromolecules*, 16: 266–274

Zhu, W., Peck, Y., Iqbal, J. and Wang, D.-A. 2017. A novel DOPA-albumin based tissue adhesive for internal medical applications. *Biomaterials*, 147: 99–115

12 Hepatoprotective Marine Phytochemicals
Recent Perspectives

Annapoorna BR, Vasudevan S, Sindhu K, Vani V, Nivya V, Venkateish VP, and Madan Kumar P

CONTENTS

DOI: 10.1201/9781003303909-12

12.1 INTRODUCTION

The liver is the largest solid organ and the only organ capable of natural regeneration of lost tissue as little as 25% of liver can regenerate into the whole liver (Häussinger 2011). The liver is composed of hepatocytes (parenchymal cells) that occupy almost 80% of the total liver volume and non-parenchymal cells, with such cells as sinusoidal endothelial cells, Kupffer cells, hepatic stellate cells (HSCs), and pit cells contributing only 6.5% to the liver volume. Any injury to the hepatocytes lead to acute or chronic liver disease. Hepatocytes chiefly function in glucose release via glycogenolysis, the production of glucose via gluconeogenesis and the detoxification process (Kmieć 2001). Sinusoidal endothelial cells constitute the closed lining or wall of the capillary. They contain small fenestrations to allow the free diffusion of substances, but not of particles like chylomicrons, between the blood and the hepatocyte (Trefts et al. 2017). They also clear potent bacterial endotoxins. HSCs are of mesenchymal origin situated in the space of Disse and represent the chief source of hepatic extracellular matrix (ECM) components. They contain numerous lipid droplets containing retinoids, triglycerides, cholesterol, and free fatty acids (Lee et al. 2016). Damage to the liver leads to activation of HSCs leading to increased proliferation and progressive loss of vitamin A. HSCs are also responsible for deposition and organization of collagen in the injured liver (Trefts et al. 2017). Pit cells represent liver-specific natural-killer cells and have the capacity to kill tumor cells also play in the antiviral defense (Bouwens et al. 1992).

12.1.1 Liver Cancer

Liver cancer is an inflammation-driven liver disease preceded chiefly by hepatitis viral infection and nonalcoholic steatohepatitis (NASH). Based on the types of cells that become cancerous, liver cancer is categorized into primary liver tumors, which include hepatocellular carcinoma (HCC), fibrolamellar carcinoma, cholangiocarcinoma, hepatoblastoma, and mesenchymal cancers of the liver. Liver cancer is the sixth-most common cancer and the third-largest cause of cancer mortality worldwide in 2020, with approximately 906,000 new cases and 830,000 deaths. It is two to three times higher among men than in women and liver cancer ranks fifth in terms of global incidence and second in terms of mortality for men. Liver cancer is the leading cause of cancer death in Mongolia, Thailand, Cambodia, Egypt, and Guatemala among both men and women and in an additional 18 countries among men. Primary liver cancer includes HCC and intrahepatic cholangiocarcinoma. Globally, HCC is the dominant type of liver cancer, accounting for approximately 75% of all liver cancers (Sung et al. 2021). Most HCC cases (>80%) occur in either sub-Saharan Africa or in Eastern Asia (El-Serag and Rudolph 2007). The highest incidence rates in the world are found in Asia and Africa (Petrick et al. 2020). Most Asian countries are in the intermediate- to high-incidence zones of HCC. However, India falls in the low incidence zone. About 42,230 U.S. adults are expected to be diagnosed with liver cancer in 2021, according to the American Society of Clinical Oncology.

12.1.1.1 Epidemiology

The epidemiology of liver cancer is influenced by some of the key risk factors involving chronic infection with hepatitis B virus (HBV) or hepatitis C virus (HCV), aflatoxin-contaminated foods, heavy alcohol intake, excess body weight, type 2 diabetes, and smoking (Mohamed-Alaa-Eldeen H. Mohamed 2018). The major risk factors vary from region to region. In most high-risk HCC countries (China, the Republic of Korea, sub-Saharan Africa), the key determinants are chronic HBV infection, aflatoxin exposure, or both, whereas, in other countries (Japan, Italy, Egypt), HCV infection is likely the predominant cause. In Mongolia, HBV and HCV and coinfections of HBV carriers with HCV or hepatitis delta viruses, as well as alcohol consumption, all contribute to the high burden (Chimed et al. 2017). Although risk factors tend to vary substantially by geographic region, major risk factors for cholangiocarcinoma include liver flukes (e.g., in the northeastern region of Thailand, where *Opisthorchis viverrini* is endemic; Prueksapanich et al. 2018) metabolic conditions (including obesity, diabetes, and nonalcoholic fatty liver disease), excess alcohol consumption, and HBV or HCV infection (Welzel et al. 2007; Petrick et al. 2020). HBV infection and HCV infection account for 56% and 20% of liver cancer deaths worldwide, respectively (Donato et al. 2001). The risk of liver cancer due to aflatoxin B1 consumption is higher in India. However, in comparison to the high-aflatoxin-incidence countries such as China and Taiwan, the aflatoxin level among Indians are relatively low. In conclusion, HBV is the major risk factor for HCC in India followed by HCV (Asim et al. 2013).

HCC develops in a stepwise manner from chronic liver inflammation caused by viral infections with HBV or HCV to liver damage, which disrupts the hepatic vasculature and perturbs proper blood flow and O_2 supply, creating a hypoxic microenvironment. Hypoxic hepatocytes activate HSCs in the liver, which robustly deposit collagen, leading to fibrosis and cirrhosis and finally to HCC (Yuen and Wong 2020). Other nonviral risk factors include heavy alcohol consumption; metabolic diseases such as being overweight, obesity, and diabetes; and aflatoxin exposure. NASH patients who do not have cirrhosis also develop HCC (Tokushige et al. 2011; Yasui et al. 2011). Several cohort and case-control studies have shown that diabetes mellitus is significantly associated with HCC. It has been demonstrated that pro-inflammatory cytokines like tumor necrosis factor-α (TNF-α) and interleukin-6 (IL-6) mediate the progression from NASH to HCC through their action on the IKK and JNK signaling pathways (Shoelson et al. 2007; Park et al. 2010). Besides the indirect effects of HCV on HCC formation, viral proteins and the RNA genome directly interact with and block tumor suppressors like retinoblastoma tumor suppressor protein (Rb), the epidermal growth factor receptor (EGFR), the DEAD-box RNA helicase (DDX3), p53, and miR-122. It has been demonstrated that the HCV NS5B protein, which is the RNA-dependent RNA polymerase, binds to Rb and targets it for degradation via the ubiquitin–proteasome pathway. The resultant lower abundance progresses the cell-cycle transition from Gl to S phase, thus stimulating hepatocellular proliferation (Munakata et al. 2007). Meta-analysis study from six prospective cohort studies suggested that coffee consumption significantly reduces liver cancer risk as compared to caffeine-containing beverages other than coffee, such as green tea, in the Japanese population (Tamura et al. 2019).

12.1.1.2 Diagnosis

Screening for liver cancer at a minimum of every 6 months in high-risk populations enables early diagnosis and treatment to avoid progression to cirrhosis and liver cancer. Liver ultrasonography and serum alpha-fetoprotein (AFP) levels are often used for the early screening of HCC. A serum AFP level of ≥400 μg/L is highly suggestive of HCC. Liver ultrasonography is a liver imaging method that helps detect different stages of liver cancer. Combining ultrasonography with magnetic resonance imaging (MRI) through advanced volume navigation software provides an effective means for accurate localization of liver tumors (Dong et al. 2016). Integration of multiple ultrasonography techniques helps in preoperative diagnosis, intraoperative localization and postoperative evaluation of HCC (Berzigotti et al. 2018). Dynamic contrast-enhanced computed tomography and multimodal MRI scans are the first-choice imaging methods for diagnosing patients with abnormal liver ultrasonography and serum AFP screening results. Multimodal MRI enables the detection of smaller liver tumors (diameter ≤2.0 cm) than dynamic contrast-enhanced computed tomography (Lee et al. 2015; Liu et al. 2017). The characteristic "wash in and wash out" enhancement pattern is used to diagnose HCC from imaging data. On dynamic contrast-enhanced computed tomography and MRI images, liver tumors exhibit a distinct homogeneous or inhomogeneous "wash in" enhancement in the arterial phase (mainly in the late arterial phase) and a "wash out" in the portal venous phase and/or equilibrium phase (Omata et al. 2017; Marrero et al. 2018). A liver biopsy is usually not necessary in patients with space-occupying lesions with typical imaging characteristics and is eligible for a clinical diagnosis of HCC (Nagtegaal et al. 2020). Digital subtraction angiography provides information on vascular anatomic variation and the anatomic relationship between liver tumor(s) and important blood vessels.

12.1.1.3 Treatment

Surgical intervention and hepatectomy, radiation therapy, liver transplantation, and chemotherapeutic drugs are the main treatment strategies for liver cancer (Lin et al. 2012). Laparoscopic hepatectomy involves reduced invasiveness and more rapid recovery (Jiang and Cao 2015). HCC usually has a long latent period, meaning that most patients are in an intermediate, advanced, or terminal stage when they are diagnosed. Chemotherapy is a treatment option for late or very late HCC (Veereman et al. 2015). Among nonsurgical treatments, transarterial chemoembolization (TACE) is one of the most commonly used for HCC. However, the nonspecific distribution and nonspecific mode of action of anticancer drugs leads to a high intrinsic toxicity and low survival profile of patients followed by more serious systemic effects when it kills cancer cells (Li et al. 2015; Veereman et al. 2015; Abou-Alfa et al. 2018) Currently, the tyrosine kinase inhibitors (TKIs) such as sorafenib, lenvatinib, and cabozantinib are the Food and Drug Administration (FDA)–approved first-line targeted drugs for advanced HCC, with a modest survival benefit (Ikeda et al. 2016, Abou-Alfa et al. 2018). Nivolumab, an immune checkpoint inhibitor targeting PD-1, was recently approved by the FDA as a second-line treatment for sorafenib-resistant HCC. A clinical trial demonstrated that nivolumab has a response rate of only 20% in HCC patients (El-Khoueiry et al. 2017). Antiviral treatment with nucleoside analogs is associated with reduced recurrence and prolonged survival in HCC patients with HBV infection (Huang et al. 2015).

Side effects are common during treatment in patients with HCC. Therefore, the timely and appropriate use of hepatoprotective agents is essential. Some functions of these hepatoprotective agents include antioxidant, anti-inflammatory, enzyme-reducing, detoxifying, hepatocyte membrane repair. Most commonly used hepatoprotective drugs include magnesium isoglycyrrhizinate injection, diammonium glycyrrhizinate, compound glycyrrhizin, bicyclol, silymarin, reduced glutathione, ademetionine, ursodeoxycholic acid, polyene phosphatidylcholine, and ulinastatin (Zhou et al. 2020).

12.2 MARINE PHYTOCHEMICALS

The marine environment consists of mainly three varieties of flora with distinctive biological characters, making them a huge source of biological resources. They are phytoplankton, seaweeds/marine algae, and seagrass. The very common marine phytochemicals found are proteins, peptides, polysaccharides, lipids and fatty acids, polyphenols, phytosterols, and pigments, among others.

12.2.1 Proteins

Seaweeds are gaining much interest in recent years due to the presence of high amounts of proteins that can be used as a potent nutrient source. However, protein levels are maximum in red seaweeds (47% on a dry-weight [DW] basis), moderate in green algae (9–26%), and lowest in brown seaweeds (3–15%) (Biris-Dorhoi et al. 2020). Lectins and phycobiliproteins constitute the major classes of algal proteins. Lectins are a diverse group of carbohydrate proteins, which bind reversibly to sugar residues. Lectins obtained from algae perform major pharmacological activities such as mitogenic, cytotoxic, antibacterial, antinociceptive, anti-inflammatory, antiviral (HIV-1), platelet aggregation inhibition, and anti-adhesion (Bleakley and Hayes 2017; Fontenelle et al. 2018). Agglutinin and isoagglutinin glycoproteins are mucin-binding proteins obtained from the species of red algae which anti-inflammatory and mitogenic activity and cytotoxic effects on mice and human cancer cell lines (Holdt and Kraan 2011). Phycobiliproteins are hydrophilic proteins having a role in photosynthesis in algal species. Algal phycobiliproteins perform antioxidant, anti-inflammatory, neuroprotective, hypocholesterolemic, hepatoprotective, antiviral, antitumor, liver-protecting, atherosclerosis treatment, and serum lipid-reducing and lipase inhibition activities. Classification is done based on the color and absorption characteristics of phycoerythrin, phycocyanin, allophycocyanin, and phycoerythrocyanin (Holdt and Kraan 2011; Bleakley and Hayes 2017). Phytotransferrins and iron starvation–induced protein 2a (ISIP2a) are proteins present in cell membranes which possess iron-chelating properties (Turnšek et al. 2019).

Light-harvesting proteins have been proved to show photoprotection activity in the diatom *Phaeodactylum tricornutum* (Hao et al. 2018). LCIP63 are abundantly found in most diatoms, at weights of 63KDa, CO_2-inducible protein plays an important part in photosynthesis (Jensen et al. 2019).

12.2.2 Peptides

The marine ecosystem represents 70% of the earth and therefore considered a reservoir of bioactives, such as peptides, polysaccharides, phenolic compounds, minerals, carotenoids, polyunsaturated fatty acids, and so on, with various pharmacological properties (Kim et al. 2008). Bioactive peptides are short-chain protein fragments that act as a source of nitrogen and amino acids and possess pharmacological properties. The functionality of bioactive peptides depends on the composition and sequence of amino acid residues, which usually contain 3–20 amino acids. Low-molecular-weight peptides can easily pass through the intestinal wall and can exert their pharmacological effects. Major pharmacological properties that marine bioactive peptides possess are antioxidant, anticancer, hepatoprotective, immunomodulatory, antimicrobial and anticoagulant, among others. Major sources of bioactive peptides include sponges (phylum Porifera), mollusks, ascidians, and seaweeds (Cho et al. 2008). Bioactive peptides can be used as a natural alternative for synthetic drugs, but purifying these peptides is difficult as they are found in complexes with polysaccharides. Bioactive peptides occur in various forms, such as linear peptides, cyclic peptides, depsipeptides, peptide derivatives, and amino acids, among others (Venugopal 2008).

12.2.3 Polysaccharides

In recent years, marine phytochemicals are explored for various applications in the biomedical and biotechnological fields. Among them, marine polysaccharides have been gradually gaining attention for their potential use for pharmacological, nutraceuticals, and cosmetics purposes. Polysaccharides derived from the marine flora and fauna possess various biological efficacy, such as antioxidant, antibacterial, antiviral, anti-inflammatory, and anticancer (Ruocco et al. 2016). Also, marine polysaccharides are used in drug delivery systems due to their biocompatibility and biodegradability. Marine polysaccharides have a major advantage as compared to others; they are cost-effective and have strong interactions with other bioactive compounds. Due to responses against external stimuli like pH, temperature and electric field marine polysaccharides tend to be used in tissue engineering applications (Liu et al. 2008; Silva et al. 2012). Fucoidan, carrageenan, alginate, chitosan, and laminarian are potential marine polysaccharides extracted from marine algae and seaweeds. Fucoidan, a marine polysaccharide, has been demonstrated as an anticancer agent that is extracted from brown seaweed. Fucoidan exhibits anticancer efficacy against hepatocellular carcinoma by cell-cycle arrest and induced cell death. (Lin et al. 2020; Jin et al. 2021) Alginate is a marine polysaccharide that is the most abundant biopolymer obtained from various of marine brown algae. Alginate is an ideal biopolymer for the drug delivery system. Reports showed that alginate-based drug carriers in different forms increased the cellular uptake of drugs in cancer treatment (He et al. 2020).

12.2.4 Lipids and Fatty Acids

Lipids are microbiomolecules and are the building blocks for the unicellular and multicellular organisms. Based on their biochemical subunit origins, they are classified into ketoacyl (fatty acids, glycerolipids, glycerophospholipids, sphingolipids, saccrolipids, polyketides, and isoprenes (sterol lipids, prenol lipids). Fatty acids are diverse group of molecules and part of a lipid. Fatty acids are essential nutrients that influence early growth and development in humans. Fatty acid containing more than one carbon double bond are known as polyunsaturated fatty acids. Omega-3 fatty acids are unsaturated fatty acids that play an important role in several biological activities and provides certain health benefits. Omega-3 fatty acids such as docosahexaenoic acid (DHA), eicosapentaenoic acid (EPA), arachidonic acids are biologically important fatty acids. Several marine microalgae are rich in DHA and EPA next to fatty fishes (Sun et al. 2018; Harwood 2019). Like plant phytosterol, marine microalgae and diatoms also contain various types of sterols such as fucosterols from brown algae, sitosterol from red algae, and stigmasterols from diatoms possess certain biological functions (Tang et al. 2002; Hannan et al. 2020). Glycerophospholipids are the major components of the biological membranes. Polyketides are composed of secondary metabolites and secondary products from animals, plants, and microorganisms. All eight categories of lipids play different roles in biological systems, such as storing energy, acting as structural components, and signaling, among others; although various lipids and fatty acids are synthesized on their own, some of them must be gained from diet. The dietary lipids and fatty acids mostly obtained from marine organisms not only fulfill a role as dietary supplements, but they are also used as therapeutics, nutraceuticals, and more. The past 20 years of marine and biotechnological research exhibited several products including products under clinical trials for various pathological conditions.

12.2.5 Polyphenols

The marine ecosystem is a reservoir of novel bioactives containing a large concentration of antioxidant compounds, such as polyphenols and carotenoids

(Cardoso et al. 2015; Wells et al. 2017). Marine phytochemicals, like other polyphenolic compounds, possess potential health benefits in numerous human diseases due to their antioxidant (García-Casal et al. 2009), antimicrobial, antiviral, antidiabetic (Lopes et al. 2016), anticancer (Cao et al. 2016), anti-allergic, and anti-inflammatory activities. Marine macroalgae (seaweeds) are a rich source of polyphenolic compounds such as catechins, gallic, chlorogenic, caffeic, acanthophorin A, acanthophorin B, tiliroside, protocatechuic, p-hydroxybenzoic, 2,3-dihydroxybenzoic, chlorogenic, caffeic, p-coumaric, salicylic, ferulic acid, cinnamic acid, myricetin, quercetin, and phlorotannins (Kim et al. 2009; Lee et al. 2016).

Phlorotannins (1,3,5-trihydroxy benzene) are polymers of phloroglucinol exclusively found in brown algae and biosynthesized through the acetate–malonate pathway. Different types of phlorotannins have been identified from different marine species, a few of which include Phlorofucofuroeckol A, dieckol, dioxinodehydroeckol, eckstolonol, triphlorethol-A, fucosterol, phloroglucinol, eckol, phlorofucofuroeckol-A, 2-phloroeckol, 7-phloroeckol (Kim et al. 2009). Among marine brown algae, *Ecklonia cava*, *Ecklonia stolonifera*, *Ecklonia kurome*, *Eisenia bicyclis, Ishige okamurae, Sargassum thunbergii, Hizikia fusiformis, Undaria pinnatifida*, and *Laminaria japonica* have been reported for phlorotannins with beneficial health biological activities (Li et al. 2011). A few of the carotenoids isolated from marine sources such as algae, fungi, and bacteria include astaxanthin (*Hematococcus pluvialis*), fucoxanthin (*Sargassum siliquastrum, Hijikia fusiformis, Undaria pinnatifida, Laminaria japonica*), tedaniaxanthin, lutein (*Dunaliella salina*), siphonaxanthin, lycopene (*haloarchaea*), antheraxanthin, zeaxanthin (*Halophila stipulacea*), violaxanthin, neoxanthin, peridinin (*Heterocapsa triquetra*), β-cryptoxanthin β-carotene (*Dunaliella salina*), ketocarotenoids, canthaxanthin (Thraustochytrium strains ONC-T18 and CHN-1), echinenone, diadinoxanthin, dinoxanthin, and alloxanthin (Galasso et al. 2017).

12.2.6 Pigments

Recent research highlighted the importance of marine pigments over synthetic pigments owing to their serious health hazards caused in humans. Marine organisms like algae, microorganisms and animals are the major sources of marine pigments which include chlorophylls, carotenoids and phycobiliproteins (Pangestuti and Kim 2011). Marine pigments are known to exert antioxidant, anti-inflammatory, anti-obesity, anticancer, and anti-angiogenic effects, as well as drug delivery and wound-healing properties (Pereira et al. 2014).

12.2.6.1 Chlorophyll and Its Derivatives

Chlorophyll is a blue/green color pigment belonging to the class tetrapyrroles, isolated from algae, plants, and cyanobacteria (Pereira et al. 2014). Studies have shown that chlorophyll exerted antioxidant, antiviral, anticancer and antimutagenic properties. Chlorophyll a, chlorophyll b, chlorophyll c, and chlorophyll d are the major types of chlorophyll molecules spread across marine biosystems

(Hosikian et al. 2010). Porphyrins, chlorins, bacteriochlorins, pheophorbides, texaphyrins, porphycenes, and phthalocyanines are the major pigments derived from chlorophyll molecules from marine algae and cyanobacteria (Ormond and Freeman 2013). These derivatives also possess various biological activities, such as antibacterial (Alenezi et al. 2017), antioxidant (Lanfer-Marquez et al. 2005) anti-inflammatory (Jelić et al. 2012), antimutagenic (Ferruzzi and Blakeslee 2007), and more.

12.2.6.2 Carotenoids

Carotenoids are secondary metabolites that are yellow to orange-red color pigments found in nature and play an important role in photoprotection and light harvesting during photosynthesis (Bandaranayake 2006). They possess various biological activities such as antioxidant, antiviral, anti-inflammatory, and anticancer activities (Rao and Rao 2007). Major carotenoids present in marine organisms include astaxanthin, fucoxanthin, lutein, canthaxanthin, zeaxanthin, neoxanthin, and violaxanthin (Lichtenthaler 1987).

12.2.6.3 Phycobiliproteins

Phycobiliproteins are fluorescent proteins found in red algae and cyanobacteria. They are hydrophilic and stable in physiological conditions (Viskari and Colyer 2003). Phycobiliproteins are divided into three classes such as (bright pink), phycocyanin's (dark blue), allophycocyanin's (aqua-blue) pigments (Stadnichuk 1995).

12.3 HEPATOPROTECTIVE EFFECTS OF MARINE PHYTOCHEMICALS

12.3.1 Proteins

Liver injury is caused by various reasons like excess of xenobiotics (alcohol, CCl4, bromobenzene). Oxidative stress causes liver diseases such as fatty liver, viral hepatitis, and hepatic fibrosis (Li et al. 2019). The two peptides rich in Glu and Asp, from protein hydrolysates (SPH) of the marine fungus *Schizochytrium sp.* obtained by Alcalase and Flavourzyme sequentially, are SPH-I and SPH-II. These peptides exhibited antioxidant activities, inhibition of lipid peroxidation, mitigates alcohol-induced hepatotoxicity in mice (Cai et al. 2017). The pigment-protein complex (PPC) obtained from the marine algae *Chlorella vulgaris* through thylakoid protein solubilization and chromatography showed antioxidant activity, reduced lipid peroxidation, lower hepatic malondialdehyde (MDA) and hepatotoxic activity against CCl4-induced liver damage in Wistar rats. PPC thus has the potential to be used against oxidative stress and free radical oxidation in the human body (Cai et al. 2015). Phycobiliproteins obtained from blue-green algae (cyanobacteria) *Spirulina platensis* when administered in adequate amounts, increased the levels of antioxidant enzymes, lowered the levels of hepatic injury biomarkers, and maintained structural integrity in alloxani-mediated hepatotoxicity (Aissaoui et al. 2017). Phycocyanin, a phycobiliprotein, exerted hepatoprotective effect by lowering levels of serum levels of aspartate aminotransferase (AST), alanine aminotransferase (ALT), triglycerides, total cholesterol, low-density lipoprotein and increased the levels of superoxide dismutase (SOD) enzyme and MDA in the liver in CCl4-induced liver damage. Phycocyanin also reduced oxidative stress, exerted anti-inflammatory effect, and inhibited non-alcoholic fatty liver disease progression in mice (Li et al. 2019). Paracetamol toxicity causes significant adverse effects on hematological, serum biochemical parameters, and oxidant–antioxidant status. Phycobiliproteins obtained from algae *Chlorella vulgaris*, along with a pretreatment of thiamine, offered the most protection from the lipid peroxidation acting as reactive oxygen species (ROS) scavenger, increasing the activities of antioxidant enzymes and oxidative stress in hepatic tissues of rats (Latif et al. 2021).

12.3.2 Bioactive Peptides

Phycobiliproteins present in spirulina and porphyridium are reported to exert anticancer and hepatoprotective properties. Phycocyanin caused cell cycle arrest at the G1 phase and downregulated cyclin E and CDK-2 and upregulated p21 levels. Phycocyanin suppresses COX-2 expression and prostaglandin E (2) production (Ravi et al. 2015; Jiang et al. 2017). Hepatoprotective effect of C- phycocyanin extracted from *Spirulina platensis* was studied in CCl4 and R-(/)-Pulegone induced hepatotoxicity in rats. It was found that in treatment groups, ALT activity was reduced and liver enzymes such as cytochrome P450, glucose-6-phosphatase and aminopyrine-N-demethylase were restored. In another study conducted on aluminum chloride-induced liver damage in Swiss albino mice, spirulina showed a reduction in lipid peroxidation and an increase in the activity of SOD and catalase in the liver. The administration of spirulina also showed a reduction in the activities of ALT and AST (Vadiraja et al. 1998; Adel and Aita 2014). Kahalalides are cyclic peptides synthesized from *Elysia rufescens*, a Sacoglossan mollusk species. Kahalalide F is a peptide that contains dehydroaminobutyric acid and has anticancer properties. Kahalalide F has been shown to have efficacy against prostate cell lines and to decrease tumor proliferation in both *in vitro* and *in vivo* experiments. Although its exact mode of action is unknown, it is known to target lysosomes and cause intracellular acidification. By inhibiting the activity of particular genes that promote cell proliferation and DNA replication, this protein also aims to mitigate tumor spreading and proliferation. It was found to be effective in the treatment of prostate cancer, melanoma, and hepatocellular carcinoma (Rademaker-Lakhai et al. 2005; Martín-Algarra et al. 2009). Didemnins are obtained from tridem solidum (*Caribbean tunicate*), which belong to acyclic or depsipeptides family. Didemnins have been proved to have many pharmacological properties, such as antitumor, antiviral, and immunosuppressive activities. Didemnin B has been studied in prostate cancer cells and found that it showed antiproliferative activity towards cancer cells. Tamandaris A, depsipeptide from marine ascidian of Didemnidae family has been reported to exhibit potent towards cancer cell lines (Vervoort and Fenical 1969). Antioxidant peptides PIIVYWK (1004.57 Da, P1), TTANIEDRR (1074.54 Da, P2) and FSVVPSPK (860.09 Da, P3) from *Mytilus edulis*, blue mussel, showed hepatoprotective effects against H2O2 induced hepatic damage in cultured hepatocytes. Peptides P1 and P3 increased the expression of heme oxygenase-1 and increased cell viability (Young-SangKimSoo et al. 2016). Novel peptides NIPP-1 and NIPP-2 have been extracted from microalgae; *Navicula incerta* has shown protectivity against TGF-β1 stimulated fibrogenesis. *In vitro* study conducted in human HSCs showed that NIPP-1 inhibited fibril formation of collagen-I and increased MMP levels and prevented tissue inhibitors of matrix metalloproteinase (TIMP) production in a dose-dependent manner. Both NIPP-1 and NIPP-2 alleviated liver fibrosis by reducing α-SMA, TIMPs, collagen and PDGF in treated groups (Kang et al. 2013). Protein hydrolysate from *Schizochytrium sp.* showed hepatoprotective effects in acute alcohol-induced liver injury in mice. Treatment with protein hydrolysate significantly reduced serum ALT, AST activities and restored antioxidant enzyme activities (Cai et al. 2017).

12.3.3 Polysaccharides

The most biologically active components of marine polysaccharides are sulfated polysaccharides. Fucoidans are sulfated polysaccharides isolated from various species of brown algae like *Sargassum filipendula*, *Saccharina (Laminaria) cichorioides*, *Fucus vesiculosus*, *Fucus evanescens*, *Undaria pinnatifida*, *Ascophyllum nodosum*, and *Laminaria japonica* exhibited the antifibrotic effect through the regulation of the TGF-β1 pathway in the CCl4-induced rats. Fucoidan showed anticancer efficacy against HCC through downregulation of CXCL2 (Nagamine et al. 2009). Also, reports evidenced that fucoidan led to overexpression of lncRNA LINC00261 and inhibited the progression of HCC (Ma et al. 2021). Fucoidans exhibited hepatoprotective effects against alcohol-intoxicated rats through its antioxidant efficacy (Lin et al. 2020; Jin et al. 2021). Galacto fucoidan extracted from *Undaria pinnatifida* exhibited anticancer effect in different cancer cell lines including HepG2 cells. Fucoidan isolated from *Ascophyllum nodosum* induced activation of caspases-9, -3 leading to altered mitochondrial membrane permeability. The sulfated polysaccharide from the *Ecklonia cava* affected caspases-7 and -8 and upregulated Bax and Bcl-xL levels. Thus, fucoidan exhibited anticancer effects through different modes of action like inducing caspase-independent apoptosis; also, it is shown to regulate different signal transduction pathways, such as the PI3k/AKT, VEGF, MAPK pathways, to inhibit cell proliferation of liver, breast, melanoma, colon, lung, and prostate cancer cell lines (Kim et al. 2010). Reports evidence that fucoidan acts as a chemopreventive agent by decreasing clonogenic growth and inhibiting cell transformation after the treatment (Kim and Nam 2018). Fucoidan also enhanced the anti-proliferative action of resveratrol at a nontoxic dose. Some fucoidan has been demonstrated as cytoprotective along with commercial chemotherapeutic agents like 5-fluorouracil.

Laminarans are low-molecular-weight brown algal polysaccharides that have been shown to have cancer preventive properties. The fermented form of the lamarians from marine exhibits hepatoprotective in lipopolysaccharide (LPS)-induced hepatotoxicity through modulation of the immune system in hepatic tissues. Carrageenan is sulfated galactan extracted from marine red algae such as Chondrus, Gigartina, Hypnea, and Eucheuma. Based on the sulfated positions and molecular weight, carrageenans are classified as; ι-carrageenan, κ-carrageenan, and λ-carrageenan all the three groups of carrageenan are biologically and commercially important. k-carrageenan exhibits anticancer effects, along with epirubicin against hepatocellular carcinoma in both *in vitro* and *in vivo* experiments; the results of that study showed the potential therapeutic efficacy in H22-implanted mice models and HepG2 cells treated with kappa selano carrageenan in combination with epirubicin (Silva et al. 2012).

12.3.4 Lipids and Fatty Acids

Marine algae and diatoms are major sources of lipids and fatty acids, which have been explored for food supplements and various therapeutics. Marine microalgae such as *Isochrysis*, *Tetraselmis*, *Chaetoceros*, *Thalassiosira*, and *Nannochloropsis* are rich in polyunsaturated fatty acids such as DHA and EPA (Adarme-Vega et al. 2012). Several reports revealed that DHA- and EPA-induced apoptosis in cancer cells both *in vitro* and *in vivo*. In HCC cells (Bel-7402), DHA induced caspase-dependent cell death through damage to the mitochondria, leading to decreased mitochondrial membrane potential and cell migration (Lee and Bae 2007; Sun et al. 2013). Omega fatty acids showed hepatoprotective effects against acute paracetamol-induced liver injury in Wistar rats (El-Gendy et al. 2021). Several studies have reported that phytosterols possess anticancer efficacy through induction of apoptosis in different cancer cells. Stigmasterol isolated from the marine diatom *Navicula incerta* exhibited anticancer effects in HepG2 cells through regulating apoptosis (Kim et al. 2014). Brown algae *Sargassum carpophyllum*, *Turbinaria conoides*, and derived sterol fucosterol also exhibited anticancer effects in different cancer cells (Sheu et al. 1999; Tang et al. 2002). Collectively, marine-derived lipids and fatty acids showed hepatoprotective and anticancer effects against HCC.

12.3.5 Polyphenols

Dieckol, a phlorotannin of *Ecklonia cava*, is known to have antioxidant, anticancer, antidiabetic, and anti-inflammatory effects. Dieckol-rich phlorotannins from *Ecklonia cava* alleviated ethanol-induced liver injury by reducing MDA formation and total cholesterol and modulating the apoptosis pathway by upregulating the expression of Bcl-xL and downregulating Bax (Kang et al. 2012). Dieckol-rich phlorotannins also upregulated cleaved caspases-3, 7, 8, and 9 expression levels and truncated Bid and Bim in Hep3B cells (Yoon et al. 2013). Dieckol-enriched fraction ameliorated liver steatosis by stimulating hepatic fatty acid β-oxidation through the AMPK-CPT-1/PPAR-α pathway

(Liu et al. 2019). Sprygina and group (2017) reported that polyphenol enriched *S. pallidum* extract exhibited antioxidant and hepatoprotective effect by reducing liver enzyme markers and lipid peroxidation levels. Dieckol attenuated pro-caspase-3 and pro-PARP expression downregulated α-SMA, and TGF-β1 expression decreased phosphorylation of ERK, p38, AKT, NF-kB, and IkB and enhanced miR-134 level and JNK phosphorylation in LX-2 and HSC-T6 cells (Lee et al. 2016). 7-Phloro-Eckol from *Ecklonia Cava* alleviated alcohol-induced oxidative stress in HepG2 cells by significantly reducing ROS and NO levels (Lin et al. 2021). Phloroglucinol and dieckol are potent inhibitors of glucose-induced angiogenesis both *in vitro* and *in vivo* (Hwang et al. 2021). Astaxanthin administration to diethylnitrosamine-induced animals significantly inhibited the development of HCC by downregulating cyclin D1 mRNA expression and increasing serum adiponectin levels (Ohno et al. 2016). Dietary fucoxanthin inhibited oxidative stress and inflammation in liver and prevented early phase fibrosis in diet-induced NASH model (Takatani et al. 2020).

12.3.6 Pigments

Recent research is focused on using pigments extracted from marine sources as anticancer agents due to their significant pharmaceutical applications (Manivasagan et al. 2018). Studies have shown the anticancer effects of chlorophyll and its derivatives via assessment of its antimutagenic activities. Fucoxanthin was reported to be effective against various cancers, such as liver, prostate, leukemia, breast, and others. In HepG2 cells, the anticancer activity of fucoxanthin was studied in which fucoxanthin reduced the cell viability by arresting the cell cycle at the G0/G1 phase (Das et al. 2008).

Kelman et al. (2009) studied the effect of fucoxanthin using obese/diabetic models (also a model for liver disease) and demonstrated the antioxidant activity of fucoxathin. The hepatoprotective effect of echinochrome, a marine pigment isolated from sea urchins, was tested in a sepsis-induced liver damage model. The study showed echinochrome treatment significantly improved liver function and liver architecture (Mohamed et al. 2019). A PPC isolated from marine algae *Chlorella vulgaris* showed hepatoprotective effect in CCl4-induced rats. The results demonstrated PPC treatment normalized liver enzyme levels that were upregulated by CCl4 induction (Cai et al. 2017). Scytonemin, another marine pigment isolated from cyanobacteria, inhibited cell cycle regulatory kinases with a potential for curing hyper-proliferative disorders (Stevenson et al. 2002).

12.4 ANTICANCER MARINE PHYTOCHEMICALS UNDER CLINICAL TRIALS

Cancer therapeutics leads to an increased rate of cancer patients' survival, but for many cancer patients, even the advanced and expensive therapies extend life for a few months. Therefore, there is a need for promising new therapeutic agents to treat cancer. Current available therapeutic agents are mostly from natural sources and naturally derived agents. Among these few of the agents such as Cytarabine, Eribulin mesylate, Brentuximab, Vedotin and Trabectidine are marine-based. At present numerous marine-derived phytochemicals are studied in various developmental stages of clinical investigations. Among 28 marine drugs in clinical trials, six drugs are in phase 3, 14 are in phase 2, and eight in phase 1. Marine-derived bioactive compounds such as pliditepsin, kahalalida F, spisulosine, pseudopterosin A, marizomib (aalinosporamide A, tetrodotoxin, neovastat) are under the phase I clinical trial for cancer treatment revealing the anticancer potential of marine bioactive compounds (Rawat et al. 2006; Alves et al. 2018).

12.5 CONCLUSION

Liver cancer is a deadly malignancy with very limited treatment options. The current cancer therapeutics for liver cancer treatment showed side effects in patients. Therefore, timely and appropriate use of hepatoprotective agents is

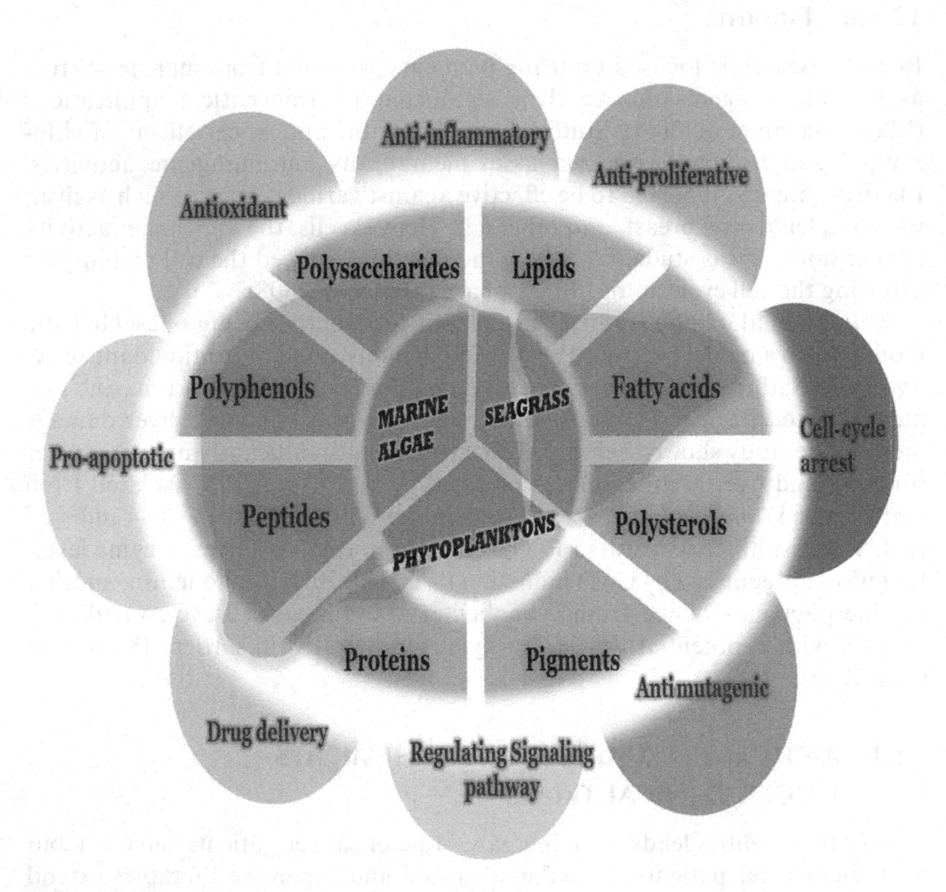

FIGURE 12.1 Overall schematic diagram of various marine phytochemicals exhibiting hepatoprotective and anticancer effects in experimental models.

essential. With various new communicable and noncommunicable diseases surfacing, there is a need for new drugs with novel modes of action, which demands potential alternative natural sources of polyphenols and carotenoids from different sources besides terrestrial plants; thus, considerable research has been conducted in the marine ecosystem. The marine environment has a rich-natural treasure for humankind, such as food and a lot of pharmacological agents for various pathological conditions. Marine environment fulfills major food supplements for humans such as proteins, carbohydrates, lipids, fatty acids, and micronutrients.

REFERENCES

Abou-Alfa, Ghassan K., Tim Meyer, Ann-Lii Cheng, Anthony B. El-Khoueiry, Lorenza Rimassa, Baek-Yeol Ryoo, Irfan Cicin, et al. 2018. “Cabozantinib in Patients with Advanced and Progressing Hepatocellular Carcinoma.” *New England Journal of Medicine* 379 (1): 54–63. https://doi.org/10.1056/NEJMoa1717002.

Adarme-Vega, T. Catalina, David K. Y. Lim, Matthew Timmins, Felicitas Vernen, Yan Li, and Peer M. Schenk. 2012. “Microalgal Biofactories: A Promising Approach towards Sustainable Omega-3 Fatty Acid Production.” *Microbial Cell Factories* 11 (July): 96. https://doi.org/10.1186/1475-2859-11-96.

Adel, Nashwa, and Nashwa A. Abu Aita. 2014. “Hepatoprotective Effect of Spirulina Platensis Against Aluminum Chloride Induced Liver Damage in Rats.” *Global Veterinaria* 13 (4): 552–559. https://doi.org/10.5829/idosi.gv.2014.13.04.85235.

Aissaoui, Ourida, Malek Amiali, Nora Bouzid, Khaled Belkacemi, and Arezki Bitam. 2017. “Effect of Spirulina Platensis Ingestion on the Abnormal Biochemical and Oxidative Stress Parameters in the Pancreas and Liver of Alloxan-Induced Diabetic Rats.” *Pharmaceutical Biology* 55 (1): 1304–1312. https://doi.org/10.1080/13880209.2017.1300820.

Alenezi, Khazna, Artak Tovmasyan, Ines Batinic-Haberle, and Ludmil T. Benov. 2017. “Optimizing Zn Porphyrin-Based Photosensitizers for Efficient Antibacterial Photodynamic Therapy.” *Photodiagnosis and Photodynamic Therapy* 17 (March): 154–159. https://doi.org/10.1016/j.pdpdt.2016.11.009.

Alves, Celso, Joana Silva, Susete Pinteus, Helena Gaspar, Maria C. Alpoim, Luis M. Botana, and Rui Pedrosa. 2018. “From Marine Origin to Therapeutics: The Antitumor Potential of Marine Algae-Derived Compounds.” *Frontiers in Pharmacology* 9 (August): 777. https://doi.org/10.3389/fphar.2018.00777.

Asim, Mohammad, Manash Pratim Sarma, and Premashis Kar. 2013. “Etiological and Molecular Profile of Hepatocellular Cancer from India.” *International Journal of Cancer* 133 (2): 437–445. https://doi.org/10.1002/ijc.27993.

Bandaranayake, Wickramasinghe M. 2006. “The Nature and Role of Pigments of Marine Invertebrates.” *Natural Product Reports* 23 (2): 223–255. https://doi.org/10.1039/b307612c.

Berzigotti, Annalisa, Giovanna Ferraioli, Simona Bota, Odd Helge Gilja, and Christoph F. Dietrich. 2018. “Novel Ultrasound-Based Methods to Assess Liver Disease: The Game Has Just Begun.” *Digestive and Liver Disease: Official Journal of the Italian Society of Gastroenterology and the Italian Association for the Study of the Liver* 50 (2): 107–112. https://doi.org/10.1016/j.dld.2017.11.019.

Biris-Dorhoi, Elena-Suzana, Delia Michiu, Carmen R. Pop, Ancuta M. Rotar, Maria Tofana, Oana L. Pop, Sonia A. Socaci, and Anca C. Farcas. 2020. “Macroalgae—A Sustainable Source of Chemical Compounds with Biological Activities.” *Nutrients* 12 (10): 3085. https://doi.org/10.3390/nu12103085.

Bleakley, Stephen, and Maria Hayes. 2017. "Algal Proteins: Extraction, Application, and Challenges Concerning Production." *Foods* 6 (5): 33. https://doi.org/10.3390/foods6050033.

Bouwens, L., P. De Bleser, K. Vanderkerken, B. Geerts, and E. Wisse. 1992. "Liver Cell Heterogeneity: Functions of Non-Parenchymal Cells." *Enzyme* 46 (1–3). https://doi.org/10.1159/000468782.

Cai, Xixi, Ana Yan, Nanyan Fu, and Shaoyun Wang. 2017. "In Vitro Antioxidant Activities of Enzymatic Hydrolysate from Schizochytrium Sp. and Its Hepatoprotective Effects on Acute Alcohol-Induced Liver Injury In Vivo." https://doi.org/10.3390/md15040115.

Cai, Xixi, Qian Yang, and Shaoyun Wang. 2015. "Antioxidant and Hepatoprotective Effects of a Pigment—Protein Complex from Chlorella Vulgaris on Carbon Tetrachloride-Induced Liver Damage in Vivo." *RSC Advances* 5 (116): 96097–96104. https://doi.org/10.1039/C5RA17544E.

Cardoso, Susana M., Olívia R. Pereira, Ana M. L. Seca, Diana C. G. A. Pinto, and Artur M. S. Silva. 2015. "Seaweeds as Preventive Agents for Cardiovascular Diseases: From Nutrients to Functional Foods." *Marine Drugs* 13 (11): 6838–6865. https://doi.org/10.3390/md13116838.

Chimed, Tuvshinjargal, Tuvshingerel Sandagdorj, Ariana Znaor, Mathieu Laversanne, Badamsuren Tseveen, Purevsuren Genden, and Freddie Bray. 2017. "Cancer Incidence and Cancer Control in Mongolia: Results from the National Cancer Registry 2008–12." *International Journal of Cancer* 140 (2): 302–309. https://doi.org/10.1002/ijc.30463.

Cho, San Soon, Hyo Ku Lee, Chang Yeon Yu, Myong Jo Kim, Eun Soo Seong, Bimal Kumar Ghimire, Eun Hwa Son, Myoung Gun Choung, and Jung Dae Lim. 2008. "Isolation and Charaterization of Bioactive Peptides from Hwangtae (Yellowish Dried Alaska Pollack) Protein Hydrolysate." *Journal of Food Science and Nutrition* 13 (3): 196–203. https://doi.org/10.3746/JFN.2008.13.3.196.

Das, Swadesh K., Takashi Hashimoto, and Kazuki Kanazawa. 2008. "Growth Inhibition of Human Hepatic Carcinoma HepG2 Cells by Fucoxanthin Is Associated with Down-Regulation of Cyclin D." *Biochimica Et Biophysica Acta* 1780 (4): 743–749. https://doi.org/10.1016/j.bbagen.2008.01.003.

Donato, F., U. Gelatti, A. Tagger, M. Favret, M. L. Ribero, F. Callea, C. Martelli, A. Savio, P. Trevisi, and G. Nardi. 2001. "Intrahepatic Cholangiocarcinoma and Hepatitis C and B Virus Infection, Alcohol Intake, and Hepatolithiasis: A Case-Control Study in Italy." *Cancer Causes & Control: CCC* 12 (10): 959–964. https://doi.org/10.1023/a:1013747228572.

Dong, Yi, Wen-Ping Wang, Feng Mao, Zheng-Biao Ji, and Bei-Jian Huang. 2016. "Application of Imaging Fusion Combining Contrast-Enhanced Ultrasound and Magnetic Resonance Imaging in Detection of Hepatic Cellular Carcinomas Undetectable by Conventional Ultrasound." *Journal of Gastroenterology and Hepatology* 31 (4): 822–828. https://doi.org/10.1111/jgh.13202.

El-Gendy, Zeinab A., Seham A. El-Batran, Sah Youssef, A. Ramadan, Walid El Hotaby, Rofanda M. Bakeer, and Rania F. Ahmed. 2021. "Hepatoprotective Effect of Omega-3 PUFAs against Acute Paracetamol-Induced Hepatic Injury Confirmed by FTIR." *Human & Experimental Toxicology* 40 (3): 526–537. https://doi.org/10.1177/0960327120954522.

El-Khoueiry, Anthony B., Bruno Sangro, Thomas Yau, Todd S. Crocenzi, Masatoshi Kudo, Chiun Hsu, Tae-You Kim, et al. 2017. "Nivolumab in Patients with Advanced Hepatocellular Carcinoma (CheckMate 040): An Open-Label, Non-Comparative, Phase 1/2 Dose Escalation and Expansion Trial." *Lancet (London, England)* 389 (10088): 2492–2502. https://doi.org/10.1016/S0140-6736(17)31046-2.

El-Serag, Hashem B., and K. Lenhard Rudolph. 2007. "Hepatocellular Carcinoma: Epidemiology and Molecular Carcinogenesis." *Gastroenterology* 132 (7): 2557–2576. https://doi.org/10.1053/j.gastro.2007.04.061.

Ferruzzi, M. G., and J. Blakeslee. 2007. "Digestion, Absorption, and Cancer Preventative Activity of Dietary Chlorophyll Derivatives." *Nutrition Research*. https://doi.org/10.1016/j.nutres.2006.12.003.

Fontenelle, Thais Pontes Carvalho, Glauber Cruz Lima, Jacilane Ximenes Mesquita, José Luiz de Souza Lopes, Tarcísio Vieira de Brito, Francisco das Chagas Vieira Júnior, Adriano Bezerra Sales, et al. 2018. "Lectin Obtained from the Red Seaweed Bryothamnion Triquetrum: Secondary Structure and Anti-Inflammatory Activity in Mice." *International Journal of Biological Macromolecules* 112 (June): 1122–1130. https://doi.org/10.1016/j.ijbiomac.2018.02.058.

Galasso, Christian, Cinzia Corinaldesi, and Clementina Sansone. 2017. "Carotenoids from Marine Organisms: Biological Functions and Industrial Applications." *Antioxidants (Basel, Switzerland)* 6 (4): E96. https://doi.org/10.3390/antiox6040096.

García-Casal, Maria N., José Ramírez, Irene Leets, Ana C. Pereira, and Maria F. Quiroga. 2009. "Antioxidant Capacity, Polyphenol Content and Iron Bioavailability from Algae (Ulva Sp., Sargassum Sp. and Porphyra Sp.) in Human Subjects." *The British Journal of Nutrition* 101 (1): 79–85. https://doi.org/10.1017/S0007114508994757.

Ghareeb, M. A., M. A. Tammam, A. El-Demerdash, and A. G. Atanasov. 2020. "Insights about Clinically Approved and Preclinically Investigated Marine Natural Products." *Current Research in Biotechnology* 2: 88–102. https://doi.org/10.1016/j.crbiot.2020.09.001.

Hannan, Md Abdul, Abdullah Al Mamun Sohag, Raju Dash, Md Nazmul Haque, Md Mohibbullah, Diyah Fatimah Oktaviani, Md Tahmeed Hossain, Ho Jin Choi, and Il Soo Moon. 2020. "Phytosterols of Marine Algae: Insights into the Potential Health Benefits and Molecular Pharmacology." *Phytomedicine: International Journal of Phytotherapy and Phytopharmacology* 69 (April): 153201. https://doi.org/10.1016/j.phymed.2020.153201.

Hao, Ting-Bin, Tao Jiang, Hong-Po Dong, Lin-jian Ou, Xiang He, and Yu-Feng Yang. 2018. "Light-Harvesting Protein Lhcx3 Is Essential for High Light Acclimation of Phaeodactylum Tricornutum." *AMB Express* 8 (1): 174. https://doi.org/10.1186/s13568-018-0703-3.

Harwood, John L. 2019. "Algae: Critical Sources of Very Long-Chain Polyunsaturated Fatty Acids." *Biomolecules* 9 (11): E708. https://doi.org/10.3390/biom9110708.

Häussinger, D. 2011. *Liver Regeneration*. Berlin: De Gruyter. https://public.ebookcentral.proquest.com/choice/publicfullrecord.aspx?p=765889.

He, Lili, Zhenghui Shang, Hongmei Liu, and Zhi-Xiang Yuan. 2020. "Alginate-Based Platforms for Cancer-Targeted Drug Delivery." *BioMed Research International* 2020: 1487259. https://doi.org/10.1155/2020/1487259.

Holdt, Susan Løvstad, and Stefan Kraan. 2011. "Bioactive Compounds in Seaweed; Functional Food Applications and Legislation." *Journal of Applied Phycology* 23 (3): 543–597. https://doi.org/10.1007/s10811-010-9632-5.

Hosikian, Aris, Su Lim, Ronald Halim, and Michael K. Danquah. 2010. "Chlorophyll Extraction from Microalgae: A Review on the Process Engineering Aspects." *International Journal of Chemical Engineering* 2010 (June): e391632. https://doi.org/10.1155/2010/391632.

Huang, Gang, Wan Yee Lau, Zhen-Guang Wang, Ze-Ya Pan, Sheng-Xian Yuan, Feng Shen, Wei-Ping Zhou, and Meng-Chao Wu. 2015. "Antiviral Therapy Improves Postoperative Survival in Patients with Hepatocellular Carcinoma: A Randomized Controlled Trial." *Annals of Surgery* 261 (1): 56–66. https://doi.org/10.1097/SLA.0000000000000858.

Hwang, Jin, Hye-Won Yang, Yu An Lu, Jun-Geon Je, Hyo-Geun Lee, K. H. N. Fernando, You-Jin Jeon, and BoMi Ryu. 2021. "Phloroglucinol and Dieckol Isolated from Ecklonia Cava Suppress Impaired Diabetic Angiogenesis; A Study of in-Vitro and in-Vivo." *Biomedicine & Pharmacotherapy = Biomedecine & Pharmacotherapie* 138 (June): 111431. https://doi.org/10.1016/j.biopha.2021.111431.

Ikeda, Masafumi, Takuji Okusaka, Shuichi Mitsunaga, Hideki Ueno, Toshiyuki Tamai, Takuya Suzuki, Seiichi Hayato, Tadashi Kadowaki, Kiwamu Okita, and Hiromitsu Kumada. 2016. "Safety and Pharmacokinetics of Lenvatinib in Patients with Advanced Hepatocellular Carcinoma." *Clinical Cancer Research: An Official Journal of the American Association for Cancer Research* 22 (6): 1385–1394. https://doi.org/10.1158/1078-0432.CCR-15-1354.

Jelić, Dubravko, Iva Tatić, Marija Trzun, Boška Hrvačić, Karmen Brajša, Donatella Verbanac, Marija Tomašković, et al. 2012. "Porphyrins as New Endogenous Anti-Inflammatory Agents." *European Journal of Pharmacology* 691 (1–3): 251–260. https://doi.org/10.1016/j.ejphar.2012.05.049.

Jensen, Erik L., Romain Clement, Artemis Kosta, Stephen C. Maberly, and Brigitte Gontero. 2019. "A New Widespread Subclass of Carbonic Anhydrase in Marine Phytoplankton." *The ISME Journal* 13 (8): 2094–2106. https://doi.org/10.1038/s41396-019-0426-8.

Jiang, Hai-tao, and Jing-yu Cao. 2015. "Impact of Laparoscopic Versus Open Hepatectomy on Perioperative Clinical Outcomes of Patients with Primary Hepatic Carcinoma." *Chinese Medical Sciences Journal = Chung-Kuo I Hsueh K'o Hsueh Tsa Chih* 30 (2): 80–83. https://doi.org/10.1016/s1001-9294(15)30016-x.

Jiang, Liangqian, Yujuan Wang, Qifeng Yin, Guoxiang Liu, Huihui Liu, Yajing Huang, and Bing Li. 2017. "Phycocyanin: A Potential Drug for Cancer Treatment." *Journal of Cancer* 8 (17): 3416. https://doi.org/10.7150/JCA.21058.

Jin, Jun-O., Pallavi Singh Chauhan, Ananta Prasad Arukha, Vishal Chavda, Anuj Dubey, and Dhananjay Yadav. 2021. "The Therapeutic Potential of the Anticancer Activity of Fucoidan: Current Advances and Hurdles." *Marine Drugs* 19 (5): 265. https://doi.org/10.3390/md19050265.

Kang, Kyong-Hwa, Zhong-Ji Qian, BoMi Ryu, Fatih Karadeniz, Daekyung Kim, and Se-Kwon Kim. 2013. "Hepatic Fibrosis Inhibitory Effect of Peptides Isolated from Navicula Incerta on TGF-β1 Induced Activation of LX-2 Human Hepatic Stellate Cells." *Preventive Nutrition and Food Science* 18 (2): 124. https://doi.org/10.3746/PNF.2013.18.2.124.

Kang, Min-Cheol, Ginnae Ahn, Xiudong Yang, Kil-Nam Kim, Sung-Myung Kang, Seung-Hong Lee, Seok-Chun Ko, et al. 2012. "Hepatoprotective Effects of Dieckol-Rich Phlorotannins from Ecklonia Cava, a Brown Seaweed, against Ethanol Induced Liver Damage in BALB/c Mice." *Food and Chemical Toxicology: An International Journal Published for the British Industrial Biological Research Association* 50 (6): 1986–1991. https://doi.org/10.1016/j.fct.2012.03.078.

Kelman, Dovi, Ami Ben-Amotz, and Ilana Berman-Frank. 2009. "Carotenoids Provide the Major Antioxidant Defence in the Globally Significant N2-Fixing Marine Cyanobacterium Trichodesmium." *Environmental Microbiology* 11 (7): 1897–1908. https://doi.org/10.1111/j.1462-2920.2009.01913.x.

Kim, A.-Reum, Tai-Sun Shin, Min-Sup Lee, Ji-Young Park, Kyoung-Eun Park, Na-Young Yoon, Jong-Soon Kim, et al. 2009. "Isolation and Identification of Phlorotannins from Ecklonia Stolonifera with Antioxidant and Anti-Inflammatory Properties." *Journal of Agricultural and Food Chemistry* 57 (9): 3483–3489. https://doi.org/10.1021/jf900820x.

Kim, Eun Ji, So Young Park, Jae-Yong Lee, and Jung Han Yoon Park. 2010. "Fucoidan Present in Brown Algae Induces Apoptosis of Human Colon Cancer Cells." *BMC Gastroenterology* 10 (August): 96. https://doi.org/10.1186/1471-230X-10-96.

Kim, In-Hye, and Taek-Jeong Nam. 2018. "Fucoidan Downregulates Insulin-like Growth Factor-I Receptor Levels in HT-29 Human Colon Cancer Cells." *Oncology Reports* 39 (3): 1516–1522. https://doi.org/10.3892/or.2018.6193.

Kim, Se Kwon, Y. Dominic Ravichandran, Sher Bahadar Khan, and Young Tae Kim. 2008. "Prospective of the Cosmeceuticals Derived from Marine Organisms." *Biotechnology and Bioprocess Engineering* 13 (5): 511–523. https://doi.org/10.1007/S12257-008-0113-5.

Kim, Young-Sang, Xi-Feng Li, Kyong-Hwa Kang, BoMi Ryu, and Se Kwon Kim. 2014. "Stigmasterol Isolated from Marine Microalgae Navicula Incerta Induces Apoptosis in Human Hepatoma HepG2 Cells." *BMB Reports* 47 (8): 433–438. https://doi.org/10.5483/bmbrep.2014.47.8.153.

Kmieć, Zbigniew. 2001. *Cooperation of Liver Cells in Health and Disease*. Vol. 161. Advances in Anatomy Embryology and Cell Biology. Berlin, Heidelberg: Springer. https://doi.org/10.1007/978-3-642-56553-3.

Lanfer-Marquez, U. M., R. M. C. Barros, and P. Sinnecker. 2005. "Antioxidant Activity of Chlorophylls and Their Derivatives." *Food Research International (Ottawa, Ont.)* 38 (8–9): 885–891. https://doi.org/10.1016/j.foodres.2005.02.012.

Latif, Amera Abd El, Doaa H. Assar, Ebtihal M. Elkaw, Hanafy A. Hamza, Dalal Hussien M. Alkhalifah, Wael N. Hozzein, and Ragaa A. Hamouda. 2021. "Protective Role of Chlorella Vulgaris with Thiamine against Paracetamol Induced Toxic Effects on Haematological, Biochemical, Oxidative Stress Parameters and Histopathological Changes in Wistar Rats." *Scientific Reports* 11 (1): 3911. https://doi.org/10.1038/s41598-021-83316-8.

Lee, J. H., Han, J. W., Ko, J. Y. et al. 2016. "Protective effect of a freshwater alga, Spirogyra sp., against lipid peroxidation in vivo zebrafish and purification of antioxidative compounds using preparative centrifugal partition chromatography." *Journal of Applied Phycology* 28: 181–189. https://doi.org/10.1007/s10811-015-0548-y.

Lee, Myoungsook, and Myung-Ae Bae. 2007. "Docosahexaenoic Acid Induces Apoptosis in CYP2E1-Containing HepG2 Cells by Activating the c-Jun N-Terminal Protein Kinase Related Mitochondrial Damage." *The Journal of Nutritional Biochemistry* 18 (5): 348–354. https://doi.org/10.1016/j.jnutbio.2006.06.003.

Lee, Sang Yoon, Jihyun Lee, HyoJung Lee, Bonglee Kim, Jaehwan Lew, Namin Baek, and Sung-Hoon Kim. 2016. "MicroRNA134 Mediated Upregulation of JNK and Downregulation of NFkB Signalings Are Critically Involved in Dieckol Induced Antihepatic Fibrosis." *Journal of Agricultural and Food Chemistry* 64 (27): 5508–5514. https://doi.org/10.1021/acs.jafc.6b01945.

Lee, Yoon Jin, Jeong Min Lee, Ji Sung Lee, Hwa Young Lee, Bo Hyun Park, Young Hoon Kim, Joon Koo Han, and Byung Ihn Choi. 2015. "Hepatocellular Carcinoma: Diagnostic Performance of Multidetector CT and MR Imaging—A Systematic Review and Meta-Analysis." *Radiology* 275 (1): 97–109. https://doi.org/10.1148/radiol.14140690.

Li, Wenjun, Hai-Nan Su, Yang Pu, Jun Chen, Lu-Ning Liu, Qi Liu, and Song Qin. 2019. "Phycobiliproteins: Molecular Structure, Production, Applications, and Prospects." *Biotechnology Advances* 37 (2): 340–353. https://doi.org/10.1016/j.biotechadv.2019.01.008.

Li, Yong-Xin, Isuru Wijesekara, Yong Li, and Se-Kwon Kim. 2011. "Phlorotannins as Bioactive Agents from Brown Algae." *Process Biochemistry* 46 (12): 2219–2224. https://doi.org/10.1016/j.procbio.2011.09.015.

Li, Yu-Ji, Ming Dong, Fan-Min Kong, and Jian-Ping Zhou. 2015. "Folate-Decorated Anticancer Drug and Magnetic Nanoparticles Encapsulated Polymeric Carrier for Liver Cancer Therapeutics." *International Journal of Pharmaceutics* 489 (1–2): 83–90. https://doi.org/10.1016/j.ijpharm.2015.04.028.

Lichtenthaler, Hartmut K. 1987. "[34] Chlorophylls and Carotenoids: Pigments of Photosynthetic Biomembranes." In *Methods in Enzymology*, 148:350–382. Plant Cell Membranes. Academic Press. https://doi.org/10.1016/0076-6879(87)48036-1.

Lin, Liyuan, Shengtao Yang, Zhenbang Xiao, Pengzhi Hong, Shengli Sun, Chunxia Zhou, and Zhong-Ji Qian. 2021. "The Inhibition Effect of the Seaweed Polyphenol, 7-Phloro-Eckol from Ecklonia Cava on Alcohol-Induced Oxidative Stress in HepG2/CYP2E1 Cells." *Marine Drugs* 19 (3): 158. https://doi.org/10.3390/md19030158.

Lin, Shibo, Katrin Hoffmann, and Peter Schemmer. 2012. "Treatment of Hepatocellular Carcinoma: A Systematic Review." *Liver Cancer* 1 (3–4): 144–158. https://doi.org/10.1159/000343828.

Lin, Yuan, Xingsi Qi, Hengjian Liu, Kuijin Xue, Shan Xu, and Zibin Tian. 2020. "The anticancer Effects of Fucoidan: A Review of Both in Vivo and in Vitro Investigations." *Cancer Cell International* 20: 154. https://doi.org/10.1186/s12935-020-01233-8.

Liu, Xijiao, Hanyu Jiang, Jie Chen, You Zhou, Zixing Huang, and Bin Song. 2017. "Gadoxetic Acid Disodium-Enhanced Magnetic Resonance Imaging Outperformed Multidetector Computed Tomography in Diagnosing Small Hepatocellular Carcinoma: A Meta-Analysis." *Liver Transplantation: Official Publication of the American Association for the Study of Liver Diseases and the International Liver Transplantation Society* 23 (12): 1505–1518. https://doi.org/10.1002/lt.24867.

Liu, Yixiang, Di Zhang, Guang-Ming Liu, Qingchou Chen, and Zhenhua Lu. 2019. "Ameliorative Effect of Dieckol-Enriched Extraction from Laminaria Japonica on Hepatic Steatosis Induced by a High-Fat Diet via β-Oxidation Pathway in ICR Mice." *Journal of Functional Foods*. https://10.1016/j.jff.2019.04.051.

Liu, Zonghua, Yanpeng Jiao, Yifei Wang, Changren Zhou, and Ziyong Zhang. 2008. "Polysaccharides-Based Nanoparticles as Drug Delivery Systems." *Advanced Drug Delivery Reviews* 60 (15): 1650–1662. https://doi.org/10.1016/j.addr.2008.09.001.

Lopes, Graciliana, Paula B. Andrade, and Patrícia Valentão. 2016. "Phlorotannins: Towards New Pharmacological Interventions for Diabetes Mellitus Type 2." *Molecules : A Journal of Synthetic Chemistry and Natural Product Chemistry* 22 (1): 56. https://doi.org/10.3390/molecules22010056.

Ma, Danhui, Jiayi Wei, Sinuo Chen, Heming Wang, Liuxin Ning, Shi-Hua Luo, Chieh-Lun Liu, Guangqi Song, and Qunyan Yao. 2021. "Fucoidan Inhibits the Progression of Hepatocellular Carcinoma via Causing LncRNA LINC00261 Overexpression." *Frontiers in Oncology* 11 (April): 653902. https://doi.org/10.3389/fonc.2021.653902.

Manivasagan, Panchanathan, Subramaniyan Bharathiraja, Madhappan Santha Moorthy, Sudip Mondal, Hansu Seo, Kang Dae Lee, and Junghwan Oh. 2018. "Marine Natural Pigments as Potential Sources for Therapeutic Applications." *Critical Reviews in Biotechnology* 38 (5): 745–761. https://doi.org/10.1080/07388551.2017.1398713.

Marrero, Jorge A., Laura M. Kulik, Claude B. Sirlin, Andrew X. Zhu, Richard S. Finn, Michael M. Abecassis, Lewis R. Roberts, and Julie K. Heimbach. 2018. "Diagnosis, Staging, and Management of Hepatocellular Carcinoma: 2018 Practice Guidance by the American Association for the Study of Liver Diseases." *Hepatology (Baltimore, Md.)* 68 (2): 723–750. https://doi.org/10.1002/hep.29913.

Martín-Algarra, Salvador, Enrique Espinosa, Jordi Rubió, Juan José López López, José Luis Manzano, Lorenzo Alonso Carrión, Arrate Plazaola, Adnan Tanovic, and Luis Paz-Ares. 2009. "Phase II Study of Weekly Kahalalide F in Patients with Advanced Malignant Melanoma." *European Journal of Cancer* 45 (5): 732–735. https://doi.org/10.1016/J.EJCA.2008.12.005.

Mohamed, Ayman S., Shimaa A. Sadek, Sarah S. Hassanein, and Amel M. Soliman. 2019. "Hepatoprotective Effect of Echinochrome Pigment in Septic Rats." *The Journal of Surgical Research* 234 (February): 317–324. https://doi.org/10.1016/j.jss.2018.10.004.

Munakata, Tsubasa, Yuqiong Liang, Seungtaek Kim, David R. McGivern, Jon Huibregtse, Akio Nomoto, and Stanley M. Lemon. 2007. "Hepatitis C Virus Induces E6AP-Dependent Degradation of the Retinoblastoma Protein." *PLOS Pathogens* 3 (9): e139. https://doi.org/10.1371/journal.ppat.0030139.

Nagamine, Takeaki, Kou Hayakawa, Takahiko Kusakabe, Hisashi Takada, Kyoumi Nakazato, Etsuko Hisanaga, and Masahiko Iha. 2009. "Inhibitory Effect of Fucoidan on Huh7 Hepatoma Cells through Downregulation of CXCL12." *Nutrition and Cancer* 61 (3): 340–347. https://doi.org/10.1080/01635580802567133.

Nagtegaal, Iris D, Robert D Odze, David Klimstra, Valerie Paradis, Massimo Rugge, Peter Schirmacher, Kay M Washington, Fatima Carneiro, and Ian A Cree. 2020. "The 2019 WHO Classification of Tumours of the Digestive System." *Histopathology* 76 (2): 182–188. https://doi.org/10.1111/his.13975.

Ohno, Tomohiko, Masahito Shimizu, Yohei Shirakami, Tsuneyuki Miyazaki, Takayasu Ideta, Takahiro Kochi, Masaya Kubota, Hiroyasu Sakai, Takuji Tanaka, and Hisataka Moriwaki. 2016. "Preventive Effects of Astaxanthin on Diethylnitrosamine-Induced Liver Tumorigenesis in C57/BL/KsJ-Db/Db Obese Mice." *Hepatology Research: The Official Journal of the Japan Society of Hepatology* 46 (3): E201–209. https://doi.org/10.1111/hepr.12550.

Omata, Masao, Ann-Lii Cheng, Norihiro Kokudo, Masatoshi Kudo, Jeong Min Lee, Jidong Jia, Ryosuke Tateishi, et al. 2017. "Asia-Pacific Clinical Practice Guidelines on the Management of Hepatocellular Carcinoma: A 2017 Update." *Hepatology International* 11 (4): 317–370. https://doi.org/10.1007/s12072-017-9799-9.

Ormond, Alexandra B., and Harold S. Freeman. 2013. "Dye Sensitizers for Photodynamic Therapy." *Materials (Basel, Switzerland)* 6 (3): 817–840. https://doi.org/10.3390/ma6030817.

Pangestuti, Ratih, and Se-Kwon Kim. 2011. "Biological Activities and Health Benefit Effects of Natural Pigments Derived from Marine Algae." *Journal of Functional Foods* 3 (4): 255–266. https://doi.org/10.1016/j.jff.2011.07.001.

Park, Eek Joong, Jun Hee Lee, Guann-Yi Yu, Guobin He, Syed Raza Ali, Ryan G. Holzer, Christoph H. Osterreicher, Hiroyuki Takahashi, and Michael Karin. 2010. "Dietary and Genetic Obesity Promote Liver Inflammation and Tumorigenesis by Enhancing IL-6 and TNF Expression." *Cell* 140 (2): 197–208. https://doi.org/10.1016/j.cell.2009.12.052.

Pereira, David M., Patrícia Valentão, and Paula B. Andrade. 2014. "Marine Natural Pigments: Chemistry, Distribution and Analysis." *Dyes and Pigments* 111 (December): 124–134. https://doi.org/10.1016/j.dyepig.2014.06.011.

Petrick, Jessica L., Andrea A. Florio, Ariana Znaor, David Ruggieri, Mathieu Laversanne, Christian S. Alvarez, Jacques Ferlay, Patricia C. Valery, Freddie Bray, and Katherine A. McGlynn. 2020. "International Trends in Hepatocellular Carcinoma Incidence, 1978–2012." *International Journal of Cancer* 147 (2): 317–330. https://doi.org/10.1002/ijc.32723.

Prueksapanich, Piyapan, Panida Piyachaturawat, Prapimphan Aumpansub, Wiriyaporn Ridtitid, Roongruedee Chaiteerakij, and Rungsun Rerknimitr. 2018. "Liver Fluke-Associated Biliary Tract Cancer." *Gut and Liver* 12 (3): 236–245. https://doi.org/10.5009/gnl17102.

Rademaker-Lakhai, Jeany M., Simon Horenblas, Willem Meinhardt, Ellen Stokvis, Theo M. De Reijke, José M. Jimeno, Luis Lopez-Lazaro, José A. Lopez Martin, Jos H. Beijnen, and Jan H. M. Schellens. 2005. "Phase I Clinical and Pharmacokinetic

Study of Kahalalide F in Patients with Advanced Androgen Refractory Prostate Cancer." *Clinical Cancer Research* 11 (5): 1854–1862. https://doi.org/10.1158/1078-0432.CCR-04-1534.

Rao, A. V., and L. G. Rao. 2007. "Carotenoids and Human Health." *Pharmacological Research* 55 (3): 207–216. https://doi.org/10.1016/j.phrs.2007.01.012.

Ravi, Mathangi, Shilpa Tentu, Ganga Baskar, Surabhi Rohan Prasad, Swetha Raghavan, Prajisha Jayaprakash, Jeyaraman Jeyakanthan, Suresh K. Rayala, and Ganesh Venkatraman. 2015. "Molecular Mechanism of anticancer Activity of Phycocyanin in Triple-Negative Breast Cancer Cells." *BMC Cancer* 15 (1): 1–13. https://doi.org/10.1186/S12885-015-1784-X.

Rawat, Diwan S., Mukesh C. Joshi, Penny Joshi, and Himanshu Atheaya. 2006. "Marine Peptides and Related Compounds in Clinical Trial." *Anticancer Agents in Medicinal Chemistry* 6 (1): 33–40. https://doi.org/10.2174/187152006774755519.

Ruocco, Nadia, Susan Costantini, Stefano Guariniello, and Maria Costantini. 2016. "Polysaccharides from the Marine Environment with Pharmacological, Cosmeceutical and Nutraceutical Potential." *Molecules (Basel, Switzerland)* 21 (5): E551. https://doi.org/10.3390/molecules21050551.

Sheu, J. H., G. H. Wang, P. J. Sung, and C. Y. Duh. 1999. "New Cytotoxic Oxygenated Fucosterols from the Brown Alga Turbinaria Conoides." *Journal of Natural Products* 62 (2): 224–227. https://doi.org/10.1021/np980233s.

Shoelson, Steven E., Laura Herrero, and Afia Naaz. 2007. "Obesity, Inflammation, and Insulin Resistance." *Gastroenterology* 132 (6): 2169–2180. https://doi.org/10.1053/j.gastro.2007.03.059.

Silva, Tiago H., Anabela Alves, Elena G. Popa, Lara L. Reys, Manuela E. Gomes, Rui A. Sousa, Simone S. Silva, João F. Mano, and Rui L. Reis. 2012. "Marine Algae Sulfated Polysaccharides for Tissue Engineering and Drug Delivery Approaches." *Biomatter* 2 (4): 278–289. https://doi.org/10.4161/biom.22947.

Soo Yeon Park, Young-Sang Kim, Chang-Bum Ahn, andJae-Young Je. 2016. "Partial Purification and Identification of Three Antioxidant Peptides with Hepatoprotective Effects from Blue Mussel (Mytilus Edulis) Hydrolysate by Peptic Hydrolysis—Google Search." *Journal of Functional Foods* 20 (January): 88–95.

Sprygin, V. G., N. F. Kushnerova, S. E. Fomenko, E. S. Drugova, L. N. Lesnikova, V. Yu. Merzlyakov, and T. V. Momot. 2017. "The Influence of an Extract from the Marine Brown Alga Sargassum Pallidum on the Metabolic Reactions in the Liver under Experimental Toxic Hepatitis." *Russian Journal of Marine Biology* 43 (6): 479–484. https://doi.org/10.1134/S1063074017060098.

Stadnichuk A. N. 1995. "Phycobiliproteins: Determination of Chromophore Composition and Content." *Phytochemical Analysis* 6: 281–288.

Stevenson, Christopher S., Elizabeth A. Capper, Amy K. Roshak, Brian Marquez, Chris Eichman, Jeffrey R. Jackson, Michael Mattern, William H. Gerwick, Robert S. Jacobs, and Lisa A. Marshall. 2002. "The Identification and Characterization of the Marine Natural Product Scytonemin as a Novel Antiproliferative Pharmacophore." *The Journal of Pharmacology and Experimental Therapeutics* 303 (2): 858–866. https://doi.org/10.1124/jpet.102.036350.

Sun, Si-Nan, Wei-Dong Jia, Hao Chen, Jin-Liang Ma, Yong-Sheng Ge, Ji-Hai Yu, and Jian-Sheng Li. 2013. "Docosahexaenoic Acid (DHA) Induces Apoptosis in Human Hepatocellular Carcinoma Cells." *International Journal of Clinical and Experimental Pathology* 6 (2): 281–289.

Sun, Xiao-Man, Lu-Jing Ren, Quan-Yu Zhao, Xiao-Jun Ji, and He Huang. 2018. "Microalgae for the Production of Lipid and Carotenoids: A Review with Focus on Stress Regulation and Adaptation." *Biotechnology for Biofuels* 11: 272. https://doi.org/10.1186/s13068-018-1275-9.

Sung, Hyuna, Jacques Ferlay, Rebecca L. Siegel, Mathieu Laversanne, Isabelle Soerjomataram, Ahmedin Jemal, and Freddie Bray. 2021. "Global Cancer Statistics 2020: GLOBOCAN Estimates of Incidence and Mortality Worldwide for 36 Cancers in 185 Countries." *CA: A Cancer Journal for Clinicians* 71 (3): 209–249. https://doi.org/10.3322/caac.21660.

Takatani, Naoki, Yuka Kono, Fumiaki Beppu, Yuko Okamatsu-Ogura, Yumiko Yamano, Kazuo Miyashita, and Masashi Hosokawa. 2020. "Fucoxanthin Inhibits Hepatic Oxidative Stress, Inflammation, and Fibrosis in Diet-Induced Nonalcoholic Steatohepatitis Model Mice." *Biochemical and Biophysical Research Communications* 528 (2): 305–310. https://doi.org/10.1016/j.bbrc.2020.05.050.

Tamura, Takashi, Asahi Hishida, and Kenji Wakai. 2019. "Coffee Consumption and Liver Cancer Risk in Japan: A Meta-Analysis of Six Prospective Cohort Studies." *Nagoya Journal of Medical Science* 81 (1): 143–150. https://doi.org/10.18999/nagjms.81.1.143.

Tang, Hai-Feng, Yi Yang-Hua, Xin-Sheng Yao, Qiang-Zhi Xu, Shu-Yu Zhang, and Hou-Wen Lin. 2002. "Bioactive Steroids from the Brown Alga Sargassum Carpophyllum." *Journal of Asian Natural Products Research* 4 (2): 95–101. https://doi.org/10.1080/10286020290027362.

Tokushige, Katsutoshi, Etsuko Hashimoto, Yoshinori Horie, Makiko Taniai, and Susumu Higuchi. 2011. "Hepatocellular Carcinoma in Japanese Patients with Nonalcoholic Fatty Liver Disease, Alcoholic Liver Disease, and Chronic Liver Disease of Unknown Etiology: Report of the Nationwide Survey." *Journal of Gastroenterology* 46 (10): 1230–1237. https://doi.org/10.1007/s00535-011-0431-9.

Trefts, Elijah, Maureen Gannon, and David H. Wasserman. 2017. "The Liver." *Current Biology: CB* 27 (21): R1147–1151. https://doi.org/10.1016/j.cub.2017.09.019.

Turnšek, Jernej, John K. Brunson, Thomas J. Deerinck, Miroslav Oborník, Aleš Horák, Vincent A. Bielinski, and Andrew E. Allen. 2019. "Phytotransferrin Endocytosis Mediates a Direct Cell Surface-to-Chloroplast Iron Trafficking Axis in Marine Diatoms." https://doi.org/10.1101/806539.

Vadiraja, Bhat B., Nilesh W. Gaikwad, and K. M. Madyastha. 1998. "Hepatoprotective Effect of C-Phycocyanin: Protection for Carbon Tetrachloride AndR-(+)-Pulegone-Mediated Hepatotoxicty in Rats." *Biochemical and Biophysical Research Communications* 249 (2): 428–431. https://doi.org/10.1006/BBRC.1998.9149.

Veereman, G., J. Robays, L. Verleye, R. Leroy, C. Rolfo, E. Van Cutsem, D. Bielen, et al. 2015. "Pooled Analysis of the Surgical Treatment for Colorectal Cancer Liver Metastases." *Critical Reviews in Oncology/Hematology* 94 (1): 122–135. https://doi.org/10.1016/j.critrevonc.2014.12.004.

Venugopal, Vazhiyil. 2008. "Marine Products for Healthcare." *Marine Products for Healthcare*, October. https://doi.org/10.1201/9781420052640.

Vervoort, Hélène, and William Fenical. 1969. "Tamandarins A and B: New Cytotoxic Depsipeptides from a Brazilian Ascidian of the Family Didemnidae." *Proceedings of the National Academy of Sciences of the United States of America* 106 (2): 782–792. https://doi.org/10.1021/jo991425a.

Viskari, Pertti J., and Christa L. Colyer. 2003. "Rapid Extraction of Phycobiliproteins from Cultured Cyanobacteria Samples." *Analytical Biochemistry* 319 (2): 263–271. https://doi.org/10.1016/s0003-2697(03)00294-x.

Wells, Mark L., Philippe Potin, James S. Craigie, John A. Raven, Sabeeha S. Merchant, Katherine E. Helliwell, Alison G. Smith, Mary Ellen Camire, and Susan H. Brawley. 2017. "Algae as Nutritional and Functional Food Sources: Revisiting Our Understanding." *Journal of Applied Phycology* 29 (2): 949–982. https://doi.org/10.1007/s10811-016-0974-5.

Welzel, Tania M., Lene Mellemkjaer, Gridley Gloria, Lori C. Sakoda, Ann W. Hsing, Laure El Ghormli, Jorgen H. Olsen, and Katherine A. McGlynn. 2007. "Risk

Factors for Intrahepatic Cholangiocarcinoma in a Low-Risk Population: A Nationwide Case-Control Study." *International Journal of Cancer* 120 (3): 638–641. https://doi.org/10.1002/ijc.22283.

Yasui, Kohichiroh, Etsuko Hashimoto, Yasuji Komorizono, Kazuhiko Koike, Shigeki Arii, Yasuharu Imai, Toshihide Shima, et al. 2011. "Characteristics of Patients with Nonalcoholic Steatohepatitis Who Develop Hepatocellular Carcinoma." *Clinical Gastroenterology and Hepatology: The Official Clinical Practice Journal of the American Gastroenterological Association* 9 (5): 428–433; quiz e50. https://doi.org/10.1016/j.cgh.2011.01.023.

Yoon, Jin-Soo, Anandam Kasin Yadunandam, Soon-Jin Kim, Hee-Chul Woo, Hyeung-Rak Kim, and Gun-Do Kim. 2013. "Dieckol, Isolated from Ecklonia Stolonifera, Induces Apoptosis in Human Hepatocellular Carcinoma Hep3B Cells." *Journal of Natural Medicines* 67 (3): 519–527. https://doi.org/10.1007/s11418-012-0709-0.

Yuen, Vincent Wai-Hin, and Carmen Chak-Lui Wong. 2020. "Hypoxia-Inducible Factors and Innate Immunity in Liver Cancer." *The Journal of Clinical Investigation* 130 (10): 5052–5062. https://doi.org/10.1172/JCI137553.

Zhou, Jian, Huichuan Sun, Zheng Wang, Wenming Cong, Jianhua Wang, Mengsu Zeng, Weiping Zhou, et al. 2020. "Guidelines for the Diagnosis and Treatment of Hepatocellular Carcinoma (2019 Edition)." *Liver Cancer* 9 (6): 682–720. https://doi.org/10.1159/000509424.

13 Applications of Marine Biochemical Pathways to Develop Bioactive and Functional Products

Toni-Ann Benjamin, Imran Ahmad, and Muhammad Bilal Sadiq

CONTENTS

DOI: 10.1201/9781003303909-13

13.1 INTRODUCTION

Marine waste, also known as marine debris/litter, is defined as "any persistent solid material that is manufactured or processed and directly or indirectly, intentionally or unintentionally, disposed of or abandoned into the marine environment or the Great Lakes" (NOAA, 2021). The most common materials that make up the debris are plastics, glass, metal, paper, cloth, rubber, and wood. More than 300 million tons of plastic are produced every year, and 8 million tons of it end up in our oceans. The other 20% are a collection of accidental/deliberate charges or lost/abandoned fishing gear and traps (EPA, 2021).

Plastic materials are cheap, lightweight, and adaptable in many food industry sectors, from storing frozen products to microwaveable meals. However, many of them are made from, or contain, petrochemicals (i.e., ethylene) that can be linked to causing acute adverse health effects, such as coughing, dizziness, or shortness of breath (Kongtip, 2013). In addition, the toxins from plastic can negatively impact the marine environment, such as contributing to the spread of invasive microorganisms and bacteria (IUCN, 2018). Furthermore, marine wildlife can suffer from mistaking the plastic debris for prey and either suffer from lacerations or die from starvation.

It is reported as of right now, there are 5.25 trillion macro and micro pieces of plastic in the ocean, which equates to roughly 269,000 tons (244,032,695 kg; Parker, 2015). National Geographic reports that the amount of plastic trash entering our oceans is expected to triple to 29 million metric tons by 2040 if our habits were to remain the same. Other studies estimate 600 million metric tons (Parker, 2020). Pew Charitable Trusts and SYSTEMIQ developed a model based on the projected 30 million metric tons and found that systemwide changes to our relationship with plastic would yield an 82% reduction in plastic trash and would cost US$600 billion to implement.

The Agricultural and food industry generates large amounts of food waste. However, one of the biggest contributors to plastic pollution is food packaging materials. Most food packaging is designed to be single-use and is nonrecyclable, or they are not. There are developments toward creating more eco-friendly packaging but based on the Pew and SYSTEMIQ model, and plastic trash would decrease by 38% and cost an additional US$140 billion. One strategy that the model predicted could be useful is "reduction and substitution." This strategy would yield a 52% decrease. Even though it would cost an additional US$250 billion, it seems as though food scientists are headed toward finding alternative methods to produce plastic. Scientists are looking into novel approaches to developing safer film materials that are also cost-effective. Some studies have looked into using components recovered from waste products and by-products as a suitable form of food packaging. Other studies have considered developing active packaging from seafood waste, such as biopolymers and bioactive agents (Debeaufort, 2021; Suleria et al., 2015).

13.2 BIOACTIVE COMPOUNDS/INGREDIENTS

Food industries produce food loss on a massive scale, generating approximately 38% during food processing (Helkar et al., 2016). The U.S. Department of Agriculture (USDA) defines food waste as "food discarded by retailers due to color or appearance and plate waste by consumers" (USDA, 2013). It is estimated that 125–160 billion pounds of edible food are wasted ever year and that 25% of that waste leads to freshwater pollution (FoodPrint, 2020). Additionally, single-use food packaging materials make up almost half of solid waste. The Environmental Protection Agency reported in 2017 that out of the 258 million tons of solid waste 63% consisted of packaging materials (EPA, 2017). All of this contributes to the global issue of plastic pollution and how to regulate it on an international level. Focusing solely on eliminating marine litter places limits on tackling the problem altogether because it doesn't address the hazardous chemical components produced by floating plastic, as well as how microplastics affect water, soil, and air quality. Carlini and Kleine (2018) suggest that by approaching marine litter as a "broader plastic pollution issue," a more effective, comprehensive approach can begin to develop. One novel approach is utilizing marine waste bioactive compounds to develop safer film materials or as ingredients for functional foods.

What are "functional foods"? While there is not a universal definition, functional foods can be classified as nutrient-rich or -enhanced foods that are beneficial to our health. Dietary items, such as fiber, protein, minerals, vitamins, and antioxidants, are considered functional foods ingredients and are recognized as Foods for Specified Health Use. The term was first coined in Japan in the late 1980s due to the food industries' response to the rise of noncommunicable diseases occurring in the elderly Japanese population (Fernandes et al., 2019). *Nutraceuticals* is another term for functional foods—it was coined in 1991 by the Foundation for Innovation in Medicine, but it is not particularly popular among consumers (Hasler, 2002). Studies are currently being conducted to identify how functional foods and food ingredients can improve human health and how food process residues can be used as an alternative source of proteins and bioactive compounds.

Bioactive compounds, or bioactive ingredients, are compounds that are present in foods, animals, or plants that influence the body once consumed (Fernandes et al., 2019). They are phytochemicals, which can be extracted from food or food by-products and are able to regulate metabolic functions leading to beneficial effects (Galanakis, 2017). Marine sources supply a vast array of bioactive molecules, such as collagen, peptides, polyunsaturated fatty acids, chitin, antioxidant compounds, and catalysts in biodiesel synthesis (Mutalipassi et al., 2021). Aquatic product processing industries produce huge amounts of marine waste through processing by-products that contain valuable bioactive compounds and proteins, which can be used to treat high-risk disorders and/or diseases (Grienke et al., 2014).

Bioactive compounds isolated from seafood by-products via chemical, enzymatic, and fermentation technologies contain antimicrobial, anticancer, antitumor, antihypertensive, and anticoagulant effects. Mussel lipids, for example, contain compounds that are known to treat rheumatoid arthritis (RA), and fish by-products contain compounds that can be used as an antioxidant or as an antimicrobial component in functional foods. Fish and shellfish waste contains significant levels of high-quality protein, which is used as a source for bio-functional peptide mining (Atef & Ojagh, 2017; Harnedy & Fitzgerald, 2012). Marine organisms are enriched with bioactive molecules extracted from different waste streams and formulated into clinically proven products to fortify human health (Suleria et al., 2016). Depending on which waste stream bioactive molecules are being extracted from, those methods differ.

This chapter explores the waste streams of various bioactive compounds (chitin, chitosan, astaxanthin, omega-3-fatty acids, collagen, glucosamine) and their properties and properties' impact on various commercial industries.

13.2.1 Chitin/Chitosan

Chitin is a homopolymer of N-acetylglucosamine. It is the second-most abundant and renewable natural resource. It is commonly found in invertebrates—crustaceans, insects, arthropod exoskeletons, and mollusks—and some mushrooms, fungi, and algae cell walls. Crustacean shells are rich in chitin and, once extracted from crustacean processing waste, can be used as a functional food ingredient. The conventional preparation of chitin from these waste materials involves demineralization and deproteinization. A further deacetylation step will produce chitosan, which can be utilized as a functional food ingredient. However, this can cause problems in disposing of the waste generated and inconsistent physiological properties of chitin (Santos et al., 2020; Doan et al., 2019).

A more cost-effective and eco-friendly solution was required to extract chitin, and so far, fermentation has shown to be just that. Chitin obtained from fermentation was demonstrated to have better quality than obtained by the chemical method (Liu et al., 2014). One study compared the physiological and structural characteristics of enzymatically produced chitin and commercial chitin, results were shown to be significantly favorable toward enzyme-assisted chitin extraction. Enzymatic chitins exhibited a higher degree of acetylation and had a smooth microfibrillar structure than commercial chitin (Marzieh et al., 2019).

Chitosan is a polysaccharide obtained from chitin deacetylation through acid hydrolysis and other alkaline treatments. This polymer can be found in seafood waste, as well as from fungi cell walls. Chitosan is commonly produced at a commercial scale using the conventional method; however, it can be produced alternatively. Recently, one study successfully performed microwave irradiation and found that this method reduced the extraction time by 1/16 and produced chitosan with a higher molecular weight and similar crystallinity to commercial chitosan (El Knidri et al., 2016). The degree of deacetylation and pH are the main factors that affect chitosan's absorption ability to remove oils, greases, and heavy metals from wastewater (Yadav et al., 2019). These alternative methods for producing chitin have been shown to have stable properties and compositions that could prove applicable in the biomedical or pharmaceutical industry.

13.2.2 Omega-3 Fatty Acids

Omega-3 fatty acids, along with omega-6 fatty acids, are polyunsaturated fatty acids (PUFAs) and consist of long, hydrocarbon chains. Omega-3 fatty acids contain significant amounts of eicosapentaenoic acid (EPA) and docosahexaenoic acid (DHA), which are strongly linked to the prevention of coronary heart disease, neural disorders, arthritis, asthma, and dermatosis (Swanson, 2012; Xu, 2015; Zhang et al., 2018). EPA and DHA are omega-3 fats and provide many health benefits, from anti-inflammatory processing and lipid mediators to reduction of cardiovascular diseases and fetal development (Swanson et al., 2012). DHA is a major structural component of the mammalian central nervous system and is abundantly found in brain and retina cells (Djuricic & Calder, 2021; Xu, 2015). It has also been shown to decrease inflammation and improve vascular endothelial functions, which have been shown to protect against heart disease.

Fish oil supplements have become popular in both Europe and the United States due to the many health benefits of EPA and DHA. Fish head and liver, as well as shellfish cephalothorax and carapace, predominantly contain omega-3 fatty acids. DHA is mainly produced from fish oil and has been known to improve metabolic diseases by lowering blood pressure and triacylglycerol concentrations (Kim & Mendis, 2006). Shrimp oil, for example, is produced through biowaste processing of northern shrimp and is not only rich in omega-3 PUFAs but astaxanthin as well (Suleria et al., 2016). However, approximately only 5% of global fish oil production is used to extract those contents for use as food ingredients and food supplements, with the rest being used for fish farming (Chu, 2017; Ciriminna, et al., 2017). A few companies have already collected and recycled fish and seafood leftovers to convert them into fish oil and other materials. However, there has been a push toward safer and greener extraction techniques instead of the conventional method.

One common conventional method used for extracting omega-3 fish oil is wet pressing. The extraction of fish oil involves fish cooking, pressing, decantation, and centrifugation, usually resulting in approximately 30% of omega-3 fatty acids and other compounds (Pateiro et al., 2021; Bonilla-Mendez & Hoyos-Concha, 2018). Another method is solvent extraction. It is used mostly on shellfish but is not commonly used in the fish industry (Sadighara et al., 2015). Pateiro (2021) reported that a 1:1 mixture of hexane–isopropanol (*v*/*v*) had a 25.44% increase of omega-3 content in the cephalothorax compared to other mixtures.

The conventional methods do provide satisfactory results, but the process can be lengthy, and the product can be toxic or degraded based on the chemicals used. Food industries are seeking viable and sustainable processes to ensure the stability and purity of those products. There are numerous methods developed to extract omega-3s from marine sources and process wastes. Alternative "green" extraction methods include supercritical fluid extraction (SFE), ultrasound-assisted extraction (UAE), and microwave-assisted extraction (MAE; Duarte, 2014; Chu, 2017). This technology can be used as a pretreatment combined with other treatments to improve extraction yields. SFE is the most common "green" extraction technology used to obtain oil from marine side streams, but the acidity of extracted oil can be high and has very low oxidative stability when in high temperatures or light (Rubio-Rodriguez et al., 2008). MAE, although not widely used for extraction, is suitable for large industries because minimal energy is required and can be combined with other treatments to increase extraction yields (Pateiro et al., 2021).

A vast amount of health benefits is linked to omega-3 PUFAs. They have antioxidant and anti-inflammatory properties that contribute to reducing the risk and severity of numerous diseases (Djuricic & Calder, 2021). Omega-3 supplements and fish oil are popular in the nutraceutical market, but there are studies focused on expanding this compound into other food products and technological applications. To conclude, marine side streams are a valuable source of omega-3 oils, and their application in the food industry would prove to be a sustainable strategy.

13.2.3 Astaxanthin

Astaxanthin is a ketocarotenoid that is part of a class consisting of more than 750 naturally occurring pigments synthesized from natural sources, such as plants, algae, and photosynthetic bacteria. The carotenoid pigments that are widely distributed are red, orange, and yellow colors (Ngamwonglumlert & Devahastin, 2019; Davinelli et al., 2018). Astaxanthin is a photosynthetic pigment present in marine food and contains bioactive components from macro/microalgae, industrial waste from fish and other marine animals (Hamed et al., 2015; Fernandes et al., 2019). The microalga *Haematococcus pluvialis* is considered to have the highest accumulation of astaxanthin and can act as a strong coloring agent and potent antioxidant.

Most marine fish and terrestrial animals cannot synthesize carotenoids, so they are supplemented into their diet to enhance muscle pigmentation and improve color and market value (Maoka, 2020; Nanda et al., 2021). Astaxanthin is commonly used in aquaculture feed, and as a functional food ingredient, in fact, it was first commercially used for pigmentation in the aquaculture industry to increase astaxanthin content in farmed salmon (Venugopal, 2016; Davinelli et al., 2018). When extracting bioactive compounds (chitosan or omega-3 fatty acids), astaxanthin ends up as a by-product due to its red-yellow pigmentation, and a discoloration step is applied to ensure pure extraction of those compounds. Nevertheless, astaxanthin is effective in adding value when combined with other bioactive compounds (Ali et al., 2021; Suleria et al., 2015).

Astaxanthin is associated with proteins found on the exoskeletons of crustacean shell waste, such as deepsea shrimp (*Acanthephyra purpurea*), is mainly used as a nutrient supplement and color additive, which is approved by the FDA (Vazquez et al., 2013; Suleria et al., 2015). Crabmeat contains a healthy amount of the water-soluble vitamin, folate (B_9), which is known to prevent cardiovascular diseases and brain disorders, like Alzheimer's and Parkinson's disease. Crab carapace also contains significant amounts of carotenoids, including astaxanthin and its esters, which have strong antioxidant capacity are currently being used in functional food formulations to improve human health (Nanda et al., 2021).

Carotenoids extracted from seafood processing waste are rich in astaxanthin and display significant antioxidant activity, which is higher than α-carotene and β-carotene, and lutein (Sadighara et al., 2015). Carotenoids are unstable due to their hydrophobicity and if organic solvents are used in the extraction process, which is environmentally hazardous and requires solvent recycling (Fernandes, 2016). New techniques have been developed that are environmentally safe and can increase the bioavailability and solubility of these bioactive compounds, such as liposomal encapsulation and solvent evaporation (Pateiro et al., 2019). Trypsin hydrolysis is used to extract carotenoids from shrimp and crab shell waste, which is used as an ingredient in aquafeed to improve the color of salmonid and crustacean species (Ozogul et al., 2021). Liquid chromatography/mass spectrometry tandem mass spectrometry (LC-MS/MS) is an essential tool for the characterization and identification of low-molecular compounds and has been used to isolate carotenoids (i.e., astaxanthin monoesters) from microalgae *N. oleoabundans* (Duarte et al., 2014). Ultimately, these different modern innovative extraction technologies are used to improve the extraction efficiency of astaxanthin and other bioactive compounds. Ongoing research has investigated the green extraction of these bioactive compounds that have vast applications in agriculture, environment, food, textile, pharmaceutical, and other biomedical fields.

13.2.4 Collagen/Gelatin

Collagen is a common fibrous protein found in all connective tissues (i.e., skin, bones, ligaments, tendons, and cartilage). The most abundant, cost-effective, and eco-friendly source of the bioactive compound is available as marine collagen obtained through marine waste streams (Cheung & Li-Chan, 2016; Suleria et al., 2015). Seafood processing by-products contain a rich content of functional molecules, such as proteins, bioactive peptides, collagen, polyunsaturated fatty acids, chitin, and fat-soluble vitamins (Lucarini et al., 2020). Collagen is characterized by a triple-helix structure made by three crosslinked alpha-amino acid chains, consisting of two homologous α1-chains and one α2-chain (Lionetto & Esposito Corcione, 2021). Gelatin is a protein derived from the partial hydrolysis of native collagen followed by thermal treatment. Further enzymatic hydrolysis can be used to extract collagen peptides from gelatin (Lionetto & Esposito Corcione, 2021).

Collagen is a major structural protein and, like gelatin, is obtained through fish skin, bones, and scales (Harnedy & FitzGerald, 2012). One of the main advantages of using marine gelatin sources is that they are not associated with the outbreak of bovine spongiform encephalopathy, also known as mad cow disease, due to the increasing demand for collagen from beef and pork by-products (Benjakul et al., 2012). Collagen and gelatin derived from fish waste are an ideal alternative source because the products derived would be safe from diseases and not limited to religious sentiments (i.e., Judaism, Islam–Judaism, Hinduism), which would make them halal or kosher (Lionetto & Esposito Corcione, 2021; Fernandes, 2016; Shavandi et al., 2019).

There has been an increase in the production of fish by-products, and the isolation of collagen from fish skin provides an opportunity for seafood industries. Verified Market Research reported that global marine collagen is expected to rise to US$1,055.2 million by 2026 (Ozogul et al., 2021). Marine collagen and gelatin contain significant amounts of bioactive compounds (Shavandi et al., 2019). Collagen makes up approximately 70–80% of fish skin as dry matter, and fish scales represent about 4% of the total weight, amounting to 18–130 million tons (Lionetto & Esposito Corcione, 2021).

Shark, ray, and skate contain up to 10% (w/w) connective tissue proteins (Harnedy & FitzGerald, 2012). Fishbone represents 30% of collagen, while 60–70% consists of calcium phosphate and hydroxyapatite, making them an alternative source of calcium and is widely applicable for tissue engineering (Atef & Ojagh, 2017; Senevirathne & Kim, 2012a, 2012b; Kim & Mendis, 2006). Gelatin obtained from fish skin contains a rich source of amino acids (mainly glycine), which has a positive impact on nail and hair growth, as well as on reducing arthritis symptoms (Shavandi et al., 2019).

While type I, II, and IV collagen are mainly extracted from fish-processing streams, it can also be extracted from jellyfish, sea urchin, starfish, octopus, or sea cucumber connective tissue (Lionetto & Esposito Corcione, 2021; Mutalipassi et al., 2021). Marine collagen and gelatin are unique proteins consisting mainly of nonpolar amino acids (glycine, alanine, valine, proline) but differ in composition based on their environment, which plays a role in their effectiveness (Senevirathne & Kim, 2012b). For instance, cold water fish gelatin has a lower thermal stability than warm water fish and mammalian gelatin. This is beneficial for the thermo-mechanical process because food industries can save energy consumption and cost, as well as increase commercial feasibility (Lionetto & Esposito Corcione, 2021). Fish gels with high thermal stability can be good candidates for use in sterilized products (Shavandi et al., 2019). Collagen and gelatin obtained through fish waste streams exhibit a multitude of properties, from film-forming abilities to high biodegradability, which could prove beneficial to cosmetic, pharmaceutical, biomedical, and food industries.

13.2.5 Chondroitin Sulfate/Glucosamine

Chondroitin sulfates, also known as glycosaminoglycan (GAG), are polysaccharides composed of repetitions of disaccharides linked by an oxygen atom (Al Khawli et al., 2020). Chitosan is the deacetylated form of chitin, which is composed primarily of glucosamine, also known as 2-amino-2-deoxy-β-D-glucose (Elieh-Ali-Komi & Hamblin, 2016). Glucosamine (GlcN) is a water-soluble amino monosaccharide that acts as a substrate for the constitution of glycosaminoglycan chains and is one of the most abundant monosaccharides in the human body (Sibi et al., 2013; Vasiliadis & Tsikopoulous, 2017).

GAG is commonly obtained from mammalian tissues. However, due to bovine spongiform encephalopathy and other food chain crisis, marine organisms have become a popular alternative source (Vazquez et al., 2013; Bellaaj-Ghorbel et al., 2013). GAGs can be extracted from numerous marine organisms such as shark cartilage, shrimp, sponges, sea urchins, sea cucumbers, and mussels (Suleria et al., 2015). Mussel *P. canaliculus* is rich in GAGs and was clinically proved to reduce joint pain and enhance joint mobility (Mutalipassi et al., 2021). GlcN can also be retrieved from marine waste streams and has significant properties such as anticancer, anti-inflammatory, and antibacterial effects (Brasky et al., 2011; Shavandi et al., 2019). Chondroitin sulfate plays a key role in different biological pathways (i.e., inflammation, cell proliferation, and differentiation), such as providing cartilage with resistance and elasticity to resist tensile stresses during strenuous conditions (Sibi et al., 2013; Vasiliadis & Tsikopoulous, 2017).

Chondroitin sulfate and glucosamine are sometimes consumed in combination, usually as nutraceuticals, for the treatment of osteoarthritis to eliminate pain and inflammation, as well as in food, cosmetic, and clinical industries (Ibanez-Sanz et al., 2020; Suleria et al., 2015). They are common, cost-effective nutritional supplements with chondroprotective properties used to improve joint health in elderly patients. Laboratory and animal studies have shown that glucosamine and chondroitin sulfate may be effective in preserving cartilage in the early stages of idiopathic osteoarthritis. However, those findings do not correlate with clinical level I studies, and more research in this area is needed (Vasiliadis & Tsikopoulous, 2017).

13.3 EXTRACTION METHODS IN SEAFOOD PROCESSING

The discovery and development of marine bioactive compounds are relatively new and the new environmentally friendly and sustainable extraction technologies. Environmentally friendly or "green" technology is regarded as technology proposed to lessen the impact of humans on the environment. Green technologies are innovative methods for bioactive compound extraction because of their enhanced extraction efficiency in a short amount of time and a considerable increase in antioxidant properties with minimal degradation of the product. Supercritical carbon dioxide (SC-CO_2) extraction, SFE, MAE, PLE, and enzyme-assisted extraction are considered green methods (Bonilla-Mendez & Hoyos-Concha, 2018; Ibanez et al., 2012; Ozogul et al., 2021).

These novel techniques are favorable at industrial facilities because of their high extraction efficiency, reduced time costs and temperature consumption, solvent selectivity, and are eco-friendly (Ozogul et al., 2021). Another name for this technology is nonthermal processing technology and some examples include High hydrostatic pressure (HHP), pulsed electric field (PEF), dense phase carbon dioxide (DPCD), and UAE, which also have a minimal impact on the environment (Duarte et al., 2014; Ali et al., 2021; Pal & Suresh, 2016).

13.3.1 CHITIN/CHITOSAN

Chitosan is a cationic polymer obtained by the deacetylation of chitin and can also be found in invertebrates and fungal species (Santos et al., 2020; Yadav et al., 2019; El Knidri et al., 2016). Chitin is the second-most abundant natural polysaccharide, after cellulose, on earth. The most popular and economical method of chitosan production is the deacetylation process of chitin, but industries are looking into alternative methods of chitosan extraction, such as obtaining the material through biowaste or fungal cell walls (Santos et al., 2020). Chitin is usually extracted from crustacean waste by demineralization and deproteinization however alkaline proteases from fish by-products can assist in the process (Venugopal, 2016; Ozogul et al., 2021).

For chitin recovery from seafood-processing wastes, different fermentation strategies are used. Fermentation using lactic acid bacteria, *Bacillus sp.*, and *Pseudomanas* species are commonly used to produce chitin from shrimp waste. In 2013, Ghorbel-Bellaaj et al. extracted chitin via microbial fermentation using *Bacillus pumilus* A1 and found it effective in producing a high-quality strain. Another extraction method is liquid fermentation. Doan et al. (2019) used an alkaline protease-producing strain (*Brevibacillus parabrevis*) and found that this method produced high deproteinization rates, and the supernatant had high growth-enhancing activity on lactic acid bacteria. This method would be difficult to apply at the industrial scale due to the risk of microbial contamination (Ozogul et al., 2021). Enzyme-assisted extraction is also a nonthermal processing technique and can be applied to chitosan via extraction from chitin via a chitin deacetylase enzyme. However, pretreatment is needed, such as grinding, sonication, and heating (Yadav et al., 2019).

13.3.2 Astaxanthin

Carotenoids is one the major pigments found in the by-products of the seafood process, typically crustacean waste. Enzymatic extraction is cost-effective and relies on the use of proteases (i.e., trypsin) to extract carotenoids in the form of carotenoproteins (Suresh & Prabhu, 2013). Hydrolytic enzymes are commonly used in shrimp wastes to extract carotenoids without disrupting the texture and color quality (Ali et al., 2021). Astaxanthin recovered displays significant antioxidant activity, which could improve the shelf life of fish products (Fernandes, 2016).

The solvent used is very important because it can affect astaxanthin extract quality. Astaxanthin is unstable when organic solvents are used, requiring solvent recycling because they are environmentally hazardous (Fernandes, 2016). Astaxanthin levels can vary depending on the waste stream the compound was recovered from. Extraction using organic solvents from crustacean waste streams exhibit higher extraction efficiency and antioxidant activity; acetone extraction exhibited the maximum level (48.64 μg/g) compared to other extracts (Ozogul et al., 2021). Solvents that are nonflammable, nontoxic, nonvolatile, and effective at low temperatures are the most effective for extracting astaxanthin: nonpolar solvents and vegetable oil commonly used for the industrial extraction of carotenoids from crustacean by-products. One popular vegetable oil used for astaxanthin extraction is flaxseed oil because it is stable at low temperatures (30–40°C) and may prove to be a healthier food options for consumers. Using this method to extract astaxanthin from crustacean processing streams are recommended to prevent atherosclerosis, cardiovascular, eye, and neurodegenerative diseases (Stachowiak & Szulc, 2021; Pu et al., 2010).

13.3.3 Omega-3 Fatty Acids

The fish oil omega-3 industry is considering the purity and stability of the products extracted. Enzyme-assisted extraction is a common method for recovering oils at a large scale because it requires less energy, is cost-effective, and is easy to operate. Enzymatic hydrolysis is applied to improve the recovery of oil and protein from marine by-products. Recovery depends on the type of protease enzyme used and other contributing factors (pH, concentration, temperature) (Bonilla-Mendez & Hoyos-Concha, 2018). When recovering oil from tuna (*Thunnus albacares*) heads the enzymatic method exhibited the highest yield of EPA and DHA (Ali et al., 2021).

13.3.4 Collagen/Gelatin

Fish waste streams contain organic (collagen, fat, lecithin, scleroprotein, various vitamins, etc.) and inorganic (hydroxyapatite, calcium phosphate, etc.)

components (Lionetto & Esposito Corcione, 2021). Collagen and gelatin have similar macromolecular structures, differing in amino acid composition and can be extracted and processed efficiently by chemical treatments along with mechanical methods, such as pH adjustment, homogenization, or sonication (Mutalipassi et al., 2021). Functional food ingredients can be recovered from fish silage through enzymatic extraction (Bonilla-Mendez & Hoyos-Concha, 2018). Acetic acid extraction method is a feasible and cost-effective method for extracting collagen from fish skin and scales. It is also a common method used to preserve the triple-helix structure of collagen (Kim & Mendis, 2006). Thermal treatments disrupt the stabilizing triple helical structure of collagen, resulting in a coiled conformation, known as gelatin. Hot water is commonly used for the extraction and isolation of gelatin from fish skin (Kim & Mendis, 2006). Acid treatments are also utilized for the extraction of gelatin from fish skin (Shavandi et al., 2019).

UAE is widely used to extract collagen from fish skin and scales. However, gelatin strength and melting point decreased due to the high temperatures, resulting in the degradation of the molecular structure. PEF extraction yields maximum collagen content from marine bones and when combined with UAE extraction methods, displayed less oxidation and hydrolysis (Ali et al., 2021).

Enzymatically hydrolyzed fish skin gelatin has better biological activities than peptides derived from fish muscle protein and obtains antioxidative peptides (Cheung & Li-Chan, 2016; Senevirathne & Kim, 2012b). Four different groups of fish proteases (acidic/aspartic, serine, thiol or cysteine, metalloproteases) are used to isolate specific compounds from different fisher products (Venugopal, 2016). Serine proteases would be used to extract myofibrillar proteins and collagen from fish muscle or collagenases (metalloprotease), along with trypsin or trypsin, would be used to extract collagen from fish bones, skin, scales, fins, and others. Collagenases are useful in the removal of supportive tissue, such as skinning squid or for the preparation of caviar (Svenning, 1993). Gu et al. (2011) reported that collagen extracted from Atlantic salmon (*Salmo salar L.*) via enzymatic hydrolysis had angiotensin-converting-enzyme ACE inhibitory activities of AP (alanine–proline) and VR (valine–arginine) that were approximately 20- and 4-fold higher, respectively, than the control salmon skin collagen peptides.

13.3.5 Chondroitin Sulfate/GlcN

Chondroitin sulfate can be synthesized and extracted using various methods. Chondroitin sulfate extraction is a cost-effective method for releasing important bioactive compounds and exhibiting antiviral properties against viruses, such as herpes simplex virus and HIV (Venugopal, 2016; Al Khawli et al., 2020). Protease treatments are used to isolate chondroitin sulfate from cartilage (Venugopal, 2016). PEF can be used to extract chondroitin sulfate from marine food bone using a high electric-field intensity (Ali et al., 2021). Acid hydrolysis is the preferred method for glucosamine release with concentrated HCl, but it can lead to the breakdown of GlcN and decreased recovery if left for extended periods (Sibi et al., 2013). Microbial production of chondroitin sulfate has been reported obtaining high extraction yields, 80% chondroitin with 90% purity (Vazquez et al., 2013).

13.4 APPLICATIONS IN TECHNOLOGY

13.4.1 Active Ingredient Carriers

13.4.1.1 Chitin/Chitosan

Chitosan-based nanoparticles are nontoxic, nonallergenic, biocompatible, and biodegradable, making them an effective "drug-loading vehicle" (Santos et al., 2020). Chitosan has biodegradable and biocompatible properties, which allows it to be a good carrier compound for active ingredients. Chitosan oligomers were found to possess antitumor activities, and chitin is applicable as an anti-aging cosmetic (Kim & Mendis, 2006). Nanotechnology can be applied in the pharmaceutical, cosmetic, food, biomedical, chemical, and textile industries. Chitin nanotechnology can increase the effectiveness of the active ingredient at a lower dosage, reducing the risk of environmental contamination (Yadav et al., 2019). When chitin is crosslinked with polylactic acid (PLA) it is an effective anti-HIV drug delivery application (Elieh-Ali-Komi & Hamblin, 2016).

13.4.1.2 Astaxanthin

Astaxanthin is a super, natural antioxidant, and microencapsulation has proved to be applicable in the food, pharmaceutical, cosmetic, biomedical, and other industries as a type of preservative (Ozogul et al., 2021). Astaxanthin extracted from *H. pluvalis* has been used in the cosmetology industry to improve skin conditions in both men and women, such as suppressing hyperpigmentation (Shavandi et al., 2019).

13.4.1.3 Omega-3 Fatty Acids

Omega-3 and omega-6 PUFAs are essential fatty acids and are key components for human health. PUFAs derived from fish oil is beneficial in treating various ailments such as cancer, depression, cardiovascular diseases, arthritis, and diabetes mellitus (Ashraf et al., 2020; Djuricic & Calder, 2021). Omega-3 (EPA and DHA) fatty acids exhibit anti-inflammatory properties that act against inflammatory diseases (Kim & Mendis, 2006). The FDA approved a prescription of omega-3 fatty acids to treat high triglyceride levels (McKenney & Sica, 2007).

13.4.1.4 Collagen/Gelatin

Marine collagen and gelatin have a wide array of applications, including health foods, cosmetics, and biomedicine as drug/delivery carriers or wound dressings (Lionetto & Esposito Corcione, 2021). Collagen sheets have been used as a drug carrier for cancer treatment (Kim & Mendis, 2006). Collagen/gelatin matrices are a common market supplement for the maintenance of bone integrity and can reduce osteoarthritis pain in patients.

13.4.1.5 Chondroitin Sulfate/GlcN

Chondroitin sulfate exhibit antiviral, antimetastatic, anti-inflammatory, and anticoagulant properties (Al Khawli et al., 2020; Mutalipassi et al., 2021). When incorporated with PLA materials, they can be applied in the biomedical field as implants, bone fixation, pins, and drug delivery systems (Karakut et al., 2019).

13.4.2 Tissue Engineering

13.4.2.1 Chitin/Chitosan

Numerous studies have observed how chitin/chitosan-and-alginate-based films perform when incorporated with active agents. Chitosan and chitin derivatives have wound healing properties through type IV collagen synthesis (Araujo et al., 2019). When chitin and chitosan are combined with composite biomaterials—PLA or collagen alginate—they can form scaffolds, films, fibers, and gels (Kumar et al., 2020).

Chitosan has a more rigid structure than chitin, but it can be compressed, and studies have shown that when it is incorporated with a gelatin-like material, it can be used for nerve repair. Chitin-based scaffolds incorporated with alginate have been shown as capable bone repair alternatives and have been shown to increase vascularization and assist in the "deposition of the calcified matrix and connective tissue" of scaffold structures (Santos et al., 2020; Yadav et al., 2019; Elieh-Ali-Komi & Hamblin, 2016).

13.4.2.2 Omega-3 Fatty Acids

Omega-3 fatty acids have multiple beneficial properties that could prove useful in the medical industry. Recent studies have been conducted, observing how marine sourced omega-3 fatty acids modulate white and brown tissue functions. EPA and DHA have anti-inflammatory effects, which could contribute to improving weight loss maintenance and metabolic complications in humans (Kalupahana et al., 2020). Coldwater fish are rich in omega-3 PUFAs, which have a high lipid content, and those properties can contribute to advanced wound care. Kotronoulas (2019) found that acellular Atlantic cod skin is uniquely rich in omega-3 fatty acids and can promote keratin migration *in vitro* models and can enhance wound healing properties.

Other studies involving adipose tissue include the use of stem cells. One such study studied the impact omega-3 fatty acids had on mediating cardioprotective effects using adipose-derived stem cells (Parshyna et al., 2017). The researchers found that DHA did not have a significant effect on cell viability, but adipose-derived stem cells could influence microvascular function in humans. However, omega-3 fatty acids are associated with a reversible tissue biomarker of breast cancer and could serve as an alternative treatment method for breast cancer (Hidaka et al., 2015).

13.4.2.3 Astaxanthin

Crustacean waste streams are a prominent source of carotenoids, crab wastes are a promising source of astaxanthin, and have strong antioxidant capacity.

Carotenoids are known to act as precursors of vitamin A, antioxidants, reproduction, and improve immunity (Maoka, 2020).

In vitro and *in vivo* studies have shown that astaxanthin has advantageous physiological effects that are beneficial to human health. Some of the properties include anti-inflammatory and anticancer activity, as well as prevention against atherosclerosis, cardiovascular and other degenerative diseases, which is why the nutraceutical market is in high demand for the encapsulated product (Hamed et al., 2015; Pateiro et al., 2019; Nanda et al., 2021; Higuera-Ciapara et al., 2006).

Astaxanthin is a derivative of chitosan and has applications in tissue engineering and regenerative medicine. Choi et al. (2019) observed astaxanthin's influence on the proliferation of adipose-derived mesenchymal stem cells in tissue-engineered constructs. They found that astaxanthin had a positive impact on mesenchymal stem cells. Based on the swelling ability of hydrogels infused with astaxanthin, it can be a useful tool for 3D culture systems and various tissue engineering applications.

13.4.2.4 Collagen/Gelatin

New technological approaches involved in the use of marine collagen/gelatin in the biomedical industry are currently in development to better human lives and the environment. Collagen and gelatin display a wide array of medical applications in the healthcare industry (plastic surgery, ophthalmology, dentistry, etc.). The use of fish gelatin hydrogel/composite sources obtained by protein-based biopolymers (i.e., collagen, gelatin, chiton/chitosan, cellulose) has been observed as an alternative source for medical textile applications (Atma, 2022). Marine gelatin exhibits properties that are beneficial for surface behavior involving emulsion, foam formation, stabilization, adhesion and cohesion, protective colloid functions, and film-forming capacities (Lionetto & Esposito Corcione, 2021).

Collagen-based polymers used as implants for cartilage, bone, and skin are biocompatible and exhibit low-toxicity behavior, allowing them to interact with host tissue without causing a mutagenic immunologic reaction (Shavandi et al., 2019). Hydroxyproline is important in bone tissue engineering because, when combined with collagen, it can mimic the natural composition of bone (Zamri et al., 2021). For example, a 3D collage scaffold was developed, consisting of hydroxyapatite and collagen from tilapia fish scales, for a corneal tissue engineering application. The scaffold was tested on rabbit corneal cells, with SIRC (Statens Seruminstitut Rabbit Cornea), and it had a positive effect on cell proliferation and exhibited gas permeability (Shavandi et al., 2019).

13.4.2.5 Chondroitin Sulfate/GlcN

Chondroitin sulfate also has wide biomedical applications, such as controlling tissue permeability and hydration, macromolecular transport between cells, and bacterial invasion control (Bilal et al., 2020). GAGs bind to proteins and form proteoglycans. Proteoglycans can capture growth factors that make GAGs suitable for tissue regeneration (Al Khawli et al., 2020). Marine collagen combined with chondroitin sulfate, derived from shark cartilage, can be used to simulate the human cartilage extracellular matrix when developing scaffolds (Xu et al., 2021).

13.4.3 Pharmaceutical Applications

13.4.3.1 Chitin/Chitosan

Chitin-based materials can induce immune responses (Araujo et al., 2019). Chitosan-based wound dressings are currently on the market for clinical use, and all products (ChitoFlex) are FDA approved (Santos et al., 2020). They are currently being used as anticoagulants to treat medical conditions, such as thrombosis.

13.4.3.2 Omega-3 Fatty Acids

Omega-3 fatty acids, especially EPA and DHA, are associated with fetal development, cardiovascular function, and neurological function (Swanson et al., 2012). DHA is the main component of omega-3 and omega-6 fatty acids and is responsible for ensuring the function of cell membranes, tissue metabolism, and hormonal pathways (Ashraf et al., 2020). Due to the positive attributes of omega-3 fatty acids, the FDA has approved the use of these products to treat high triglyceride levels. Epanova was the first omega-3 drug to be approved by the FDA to treat severe hypertriglyceridemia in adults. Epanova, Vascepa, and generic versions of this drug is derived from fish oil and EPA–DHA mixture (McKenney & Sica, 2007; FDA, 2019).

13.4.3.3 Astaxanthin

Carotenoids' antioxidant properties contribute to them being active agents in the prevention of cancer (Hamed et al., 2015). Due to astaxanthin's high antimicrobial activity, it can be applied in pharmaceutical industries for the treatment or prevention of pathogenic microbial infections (Dalei & Sahoo, 2015). Astaxanthin has been shown to benefit liver health and liver processes. AstaREAL, for example, is a product shown to "[improve] blood lipids and [increase] adiponectin, preventing fatty liver disease, reducing the risk for atherosclerotic plaque, and lowering hypertension by improving vascular tone" (Shavandi et al., 2019).

13.4.3.4 Collagen/Gelatin

Collagen and gelatin have a wide range of biological activities such as antioxidant, antihypertensive, antitumor, anticancer, antibacterial, and anti-HIV (Senevirathne & Kim, 2012a, 2012b). Marine gelatin is derived from the fibrous protein collagen and can be formulated into capsules or tablets or to encapsulate nutrients and vitamins. The bioactive compound can also be used in the fabrication of wound healing pads (Shavandi et al., 2019). Collagen production decreases with age and a bad diet. Therefore, collagen supplements are intended to preserve consumers' skin, hair, nails, and body tissues. There are already companies (Aquareneur, Copalis, etc.) selling collagen products and new marine ingredients made of collagen are currently in development (Mutalipassi et al., 2021).

Enzymatic hydrolysis of collagen from marine fish waste can be developed as antihypertensive components in functional foods or nutraceuticals, such as managing hypertension, high blood pressure, and diabetes (Kim et al., 2012; Cheung & Li-Chan, 2016).

13.4.3.5 Chondroitin Sulfate/GlcN

GlcN and chondroitin sulfate are commonly used to treat pain and inflammation associated with osteoarthritis. Chondroitin sulfate exhibits antiarthritic activity and has been shown to reduce joint pain as well as enhance joint mobility. Biolane is a GAG-metalloprotease product and is utilized for normal tissue remodeling (Mutalipassi et al., 2021).

Studies have reported a synergistic effect with the use of nonsteroidal anti-inflammatory drug use and the combination of glucosamine and chondroitin sulfate. Recent evidence suggests that glucosamine and chondroitin sulfate anti-inflammatory properties may contribute to the prevention of lung cancer and colorectal cancer (Brasky et al., 2011; Ibanez-Sanz et al., 2020).

13.4.4 Antimicrobial/Antibacterial Agents

13.4.4.1 Chitin/Chitosan

One of the most studied properties of chitosan and its derivatives have been its antimicrobial activity and its ability to bind to the surface of bacterial walls/plasma membranes. Studies have investigated the changes in cell permeability; if the cell has a higher degree of acetylation and molecular weight, antibacterial activity is mediated, blocking bacterial transport. However, antibacterial activity will occur if a lower degree of acetylation, molecular weight, and pH levels but only toward gram-negative bacteria. (Santos et al., 2020).

Chitin and chitosan derivatives act as inhibitors of ACE, associated with hypertension (Kim & Mendis, 2006). As a result, chitosan derivatives have been used in commercial disinfectants and topical antimicrobials (Elieh-Ali-Komi & Hamblin, 2016).

13.4.4.2 Omega-3 Fatty Acids

Omega-3 fatty acids possess antimicrobial, autoimmune, anticancer, and anti-inflammatory properties (Pateiro et al., 2019). Humans are incapable of synthesizing PUFAs with more than 18 carbons, which is why they are obtained through food. Most seafood contains PUFAs and contributes to the regulation of blood clots and blood pressure (Hamed et al., 2015). Marine algae are rich in PUFAs, which contribute to preventing cardiovascular diseases, osteoarthritis, and diabetes (Al Khawli et al., 2020).

13.4.4.3 Astaxanthin

Dalei and Sahoo (2015) studied the antimicrobial activity of astaxanthin extracted by six different polar, organic solvents and found that methanol extraction exhibited the highest zone of inhibition against *E. coli*, *Bacillus*, *Staphylococcus*, and *Pseudomonas* strains.

13.4.4.4 Collagen/Gelatin

Bioactive peptides that have been isolated from marine fish sources, especially from fish skin, are a good source of ACE inhibitory activity (Kim et al., 2012; Ozogul et al., 2021). Collagen extracted from unicorn leatherjacket skin and chitosan enhanced bacteriostatic capacity and fungistatic activity, which can also be used to improve the elasticity and/or brittleness of film materials (Lionetto & Esposito Corcione, 2021).

Enzyme-extracted gelatin shows potential as a multifunctional nutraceutical because it exhibits significant antimicrobial activity. Cheung and Li-Chan (2016) observed gelatin extracted from steelhead fish skin using pepsin, Corolase, and papain and found that gelatin hydrolysates displayed strong ACE and dipeptidyl peptidase IV (DPP-IV) inhibitory activity. Collagen extracted through enzymatic hydrolysis produces similar results. Enzymatically extracted collagen also exhibits high antibacterial activity when combined with other biopolymers. Wu et al. (2021) found that PLA combined with fish scale–derived hydroxyapatite and an antibacterial agent derived from eggshell composite displayed significant antibacterial and antimicrobial activity against *S. aureus* and *E. coli* compared to other composite forms.

13.4.4.5 Chondroitin Sulfate/GlcN

Chondroitin sulfate and GlcN are commonly used to together for therapeutic purposes because they naturally exhibit antibacterial, antifungal, and antiviral activity. Glucosamine and chondroitin sulfate immobilized on PLA-based films have been shown to improve their antibacterial activity against *E. coli* and *S. aureus* with a 99% growth inhibition (Karakut et al., 2019). Poly-N-acetyl-glucosamine nanofibers have been utilized for *in vivo* treatments of cutaneous wounds (Lindner et al., 2011). Chondroitin sulfate has shown promising results for topical applications in wound healing with silver nanoparticles as an antibacterial agent (Im et al., 2013).

13.4.5 Food Technology

13.4.5.1 Chitin/Chitosan

Chitosan is a well-studied biopolymer that has been reported to extend the shelf life of various food products (cheese, banana, dairy, meat, etc.) as well as improve food packaging matrices (Kumar et al., 2020). Chitosan exhibits antimicrobial activity and nontoxic properties, and when incorporated into food packaging materials, the compound can preserve the nutritional quality of food. Chitosan films and coatings are usually prepared using an acetic acid solution. While it can release an unpleasant flavor and odor, it has efficient microbiological and sensory qualities (Ozogul et al., 2021). Chitosan can also be used as a stabilizer for seafood-based products (Venugopal, 2016).

13.4.5.2 Omega-3 Fatty Acids

Several researchers have studied how microencapsulation of omega-3 fatty acids impacts food products. PUFAs are susceptible to lipid oxidation, but if they are utilized with other antioxidants they can extend the shelf life of other food products (Hamed et al., 2015; Ibanez et al., 2012). Omega-3s have been used to enrich and fortify different types of foods and beverages. Microencapsulated omega-3 fatty acids have been shown to increase the nutritional value of bread. For example, PUFAs in powder form improved the technological properties, such as reducing chewiness and gumminess (Pateiro et al., 2021).

13.4.5.3 Astaxanthin

Astaxanthin is a major carotenoid found in crustacean shell waste, and dietary carotenoids have nutritional and therapeutic importance for human health. β-carotene is a major natural colorant applied to food and drinks, and astaxanthin can be used as an ingredient in aquafeed to improve the color of salmonid and crustacean species (Venugopal, 2016; Hamed et al., 2015). Astaxanthin's significant antioxidant activity acts as a contributing factor in preventing microbial contamination and the oxidation of lipid compounds, which can be applied as an alternative method for active packaging or acting as a natural preservative (Fernandes, 2016).

13.4.5.4 Collagen/Gelatin

Fish scales are a nutritional food source and are a rich source of proteins (i.e., collagen) and calcium phosphate. Collagen extracted from fish waste streams is widely used in the food and beverage industry as an antioxidant, emulsifier, preservative, texturizer, and thickener, making it a good candidate for improving the rheological properties of food products (Fernandes, 2016). Fish gelatin, when combined with other thickeners or gelling agents (pectin), has been shown to improve bulk density, compressibility, elasticity, and firmness when used in lower ratios. Coldwater fish skin collagen has been reported to prevent texturization of food products, as well as in low-fat spreads and dairy products (Shavandi et al., 2019). Karthickeyan et al. (2007) investigated the effectiveness of immunostimulants derived from fish scale waste of *Nemipterus japonicus* in aquafeed. The study concluded that the formulated feed could prevent microbial disease, decrease mortality, and develop an immune response in the fish.

Marine collagen has multiple uses in the food packaging industry. Collagen films and coatings are important for maintaining, protecting, and extending the shelf life of foods. One such use is the industrial application of collagen in edible casings for the meat processing industries (Lionetto & Esposito Corcione, 2021). Recently fish gelatin has been recommended for the preparation of biodegradable films in active packaging. Incorporating other biopolymers can improve food preservation. For example, antioxidants from plant extracts (olive, orange, fruit berries, etc.) can improve food preservation (Lionetto & Esposito Corcione, 2021). Collagen incorporated with chitosan and chitin to improve the mechanical properties of biopolymers and increase shelf life by delaying/inhibiting microbial growth, reducing oxidation and spoilage, and improving thermal stability (Santos et al., 2020; Yadav et al., 2019; Lionetto & Esposito Corcione, 2021).

However, there are noticeable drawbacks if collagen and gelatin were used solely as the packaging material, such as poor water vapor barrier, mechanical strength, and low oxygen permeability. This is due to fish collagen films exhibiting low thermal stability (Sommer et al., 2021). Plasticizers are used to increase the flexibility and molecular mobility of biopolymer matrices. In combination with chemical and enzymatic treatments and other biopolymers, the final processing steps for both bioactive compounds are used to improve on those defects. Other treatments include utilizing soy protein isolate (Ahmad et al., 2016) to enhance collagen water vapor barrier, natural-based crosslinking agents (Liguori et al., 2019), and nanofillers (i.e., nano-SiO_2 particles, metal ions, chitosan nanoparticles) to improve the performance of food packaging systems (Hosseini et al., 2018). Fish stromal and myofibrillar proteins can be used to produce biodegradable, edible films with good barrier properties against gases, organic volatiles, and lipids, which can be beneficial for producing active packaging materials (Lionetto & Esposito Corcione, 2021).

13.4.5.5 Chondroitin Sulfate/GlcN

Nutritional supplements, functional foods, and beverage sales are the major components of US$2 billion glucosamine market (Barrow, 2010). GlcN is a common functional ingredient in food and beverages, next to omega-3 fish oil, and is primarily used as a dietary supplement, usually combined with chondroitin sulfate. Chondroitin sulfate and glucosamine exhibit anti-inflammatory, antioxidant, and anabolic properties which are why they were marketed as food supplements (Ashraf et al., 2020). They could be introduced into their diet to reduce pain caused by degenerative joint disease (Brasky et al., 2011). On the other hand, purified chondroitin sulfate can be used as a food preservative due to its emulsifying properties (Vazquez et al., 2013).

13.5 LEGAL REGULATIONS AND ECOLOGICAL IMPLICATIONS

Plastic polymers are hazardous to the environment, and as a result, biodegradable alternatives are in high demand. Polymers derived from marine waste processing streams are a potential alternative. They are eco-friendly, renewable, and consist of functional food ingredients that are beneficial to human health. The marketability of seafood-derived functional foods is on the rise. Industries are adopting new technologies to utilize compounds derived from marine processing waste streams. Green technologies are considerably safer and less harmful than other extraction techniques. They also utilize eco-friendly solvents (i.e., CO_2 and water) that do not negatively impact the environment.

Global climate change is a growing factor that needs to be considered because the feasibility of the bioactive compounds is influenced by biotic and abiotic factors (i.e., pH, salinity, temperature, nutrients; Meena et al., 2020). As mentioned previously, marine collagen has vastly different properties when extracted from cold-water and warm-water fish. Therefore, the marine environment is going to exhibit changes that will negatively impact marine animals and change the properties of those bioactive compounds. Therefore, more studies on bioactive compounds derived from marine viscera and processing by-products should be conducted to evaluate their nutritional and functional applications.

Major pharmaceutical companies have explored the use of marine compounds for decades, and their main target is cancer. However, there are some *in vitro* studies that have results that do not correlate with clinical trials. For example, there is a lack of case studies supporting the correlation between the intake of omega-3 PUFAs and their health benefits (Ozogul et al., 2021). This can lead to many doctors and clinicians dismissing the effects of the treatment as a kind of placebo effect. Efforts should be made to process marine functional foods responsibly since their consumption has the potential to reduce the occurrence of chronic illnesses. Legislation and safety issues for extraction methods should also be addressed. There is currently no restriction on the amount of collagen usage in food items (Pal & Suresh, 2016). This essentially leaves the responsibility up to individual companies to regulate how much bioactive compound they are distributing in their products and/or supplements. The FDA has approved the usage of some marine bioactive substances but only in low doses and are usually in adjunct with other commercially known drugs (Elieh-Ali-Komi & Hamblin, 2016). It is a strong recommendation that scientists across different academic disciplines, clinicians, and industrial partners collaborate and efficiently evaluate these marine-derived drugs to ensure they are sustainable for market distribution and consumers' health and safety.

There are other challenges in the commercialization of seafood-processing bioactive compounds, including sustainability of seafood by-product resources, consumer acceptance, safety profile, cost of products, reproducibility, and the economic viability of those products (Pal & Suresh, 2016). The requirements to get marine-derived drugs on the market is a lengthy and costly one—the approval period can be between 8–15 years and can cost an average of US$900 million to get the product on the market. Not to mention, these drugs have a very high risk of failure and, due to their toxicity, are not sustainable (Lindequist, 2016).

Consumer acceptance is important because their attitude towards food products can affect their overall marketability. Pal and Suresh (2016) suggest referring to disposed of seafood materials. They should be called "offal" instead of "waste" as a way to ensure consumer acceptance and dispel any negative connotations towards this alternative nutraceutical source. Labeling is important because it can help consumers choose products that align with their preferences. Government labels are habitually mandatory and would benefit the aquaculture industry economically. The labels would address consumer concerns and would dispel any concerns over the safety of the product. Policies that address the weaknesses of aquaculture should be required so that poor fisheries are not exploited, ensuring the availability of marine products for future applications (Gomez et al., 2019).

All food production procedures need appropriate regulatory standards, and in this case, the aquaculture industry would need to follow the same standards if marine bioactive compounds are to be implemented into food and dairy products. Regulations differ based on the country, but EU countries have a divested interest in the regulation of novel foods (Rotter et al., 2021). Marine bioactive compounds that are extracted through green technology would be regulated by the European Union because those governments have invested funds ensuring the quality, safety, and eco-friendly production of novel food ingredients (Gomez et al., 2019). Overall, the recent advancements in marine products (drugs, food, bone grafts, etc.) demonstrate the enormous potential of marine natural products. Regulation and legislation would ensure the public of its safety as well as its stability in the market.

13.6 CONCLUSION

This chapter discussed the possibility of utilizing marine viscera (i.e., shellfish and fish) as natural bioactive compounds with health benefits for the potential use in the various commercial industries—especially in the food industry as a functional food ingredient. This approach explored the various properties displayed by these marine bioactive compounds (chitosan, chitin, collagen, chondroitin sulfate, omega-3 fatty acid, glucosamine) and what was discovered is that they have a wide range of usage from acting as an antioxidant to being utilized in mammalian bone scaffolding. Additionally, eco-friendly extraction methods and solvents were discussed as an alternative method to recover pure, valuable compounds in a sustainable manner. The fish industry would benefit from marine bioactive compound extraction because many of the compounds are biodegradable and renewable, which happen to be the popular trend in the effort to reduce ocean pollution. Many industries have begun to utilize marine bioactive compounds in their products, such as the food, cosmetic, biomedical, and pharmaceutical industries.

The FDA has approved the use of some of these bioactive compounds to be used in adjunct with other commercially known products (i.e., omega-3 and chondroitin sulfate-glucosamine supplements). While the bioactive compounds due produce promising results pertaining to human health, they are conducted under laboratory settings, and more clinical studies are required to corroborate the results presented *in vitro* studies. Furthermore, many of these new green technologies are unable to be used at a large scale and require more study and investigation. For example, HHP is an efficient extraction method but can cause degradation in collagen compounds due to either intrinsic forces or denaturation during the extraction process. In other scenarios, pretreatment is required to assist in extraction. Discussions around the regulation of these compounds are being had, especially in the food and biomedical industry.

Utilizing marine bioactive compounds as a functional food ingredient is a fairly new approach but based on how rapidly these compounds have been incorporated into other industrial areas, it demonstrates the feasibility of this concept and commercialization.

REFERENCES

Ahmad, M., Nirmal, N.P., Danish, M., Chuprom, J., & Jafarzedeh, S. (2016). Characterisation of composite films fabricated from collagen/chitosan and collagen/soy protein isolate for food packaging applications. *Research Advances*, 6: 82191–82204.

Al Khawli, F., Martí-Quijal, F.J., Ferrer, E., Ruiz, M.J., Berrada, H., Gavahian, M., Barba, F.J., & de la Fuente, B. (2020). Aquaculture and its by-products as a source of nutrients and bioactive compounds. *Advances in Food and Nutrition Research*, 92: 1–33. DOI:10.1016/bs.afnr.2020.01.001

Ali, A., Wei, S., Liu, Z., Fan, X., Sun, Q., Xia, Q., Liu, S., Hao, J., & Deng, C. (2021). Non-thermal processing technologies for the recovery of bioactive compounds from marine by-products. *LWT—Food Science and Technology*, 147: 111549. DOI:10.1016/j.lwt.2021.111549

Araujo, D., Ferreira, I.C., Torres, C.A.V., Neves, L., & Freitas, F. (2019). Chitinous polymers: Extraction from fungal sources, characterization and processing towards value-added applications. *J Chem Tecnol Biotechnol*, 95: 1277–1289. DOI:10.1002/jctb.6325

Ashraf, S.A., Adnan, M., Patel, M., Siddiqui, A.J., Sachidanandan, M., Snoussi, M., & Hadi, S. (2020). Fish-based bioactives as potent nutraceuticals: Exploring the therapeutic perspective of sustainable food from the sea. *Marine Drugs*, 18(5): 265. https://doi.org/10.3390/md18050265

Atef, M., & Ojagh, S.M. (2017). Health benefits and food application of bioactive compounds from fish by-products: A review. *Journal of Functional Foods*, 35: 673–681. DOI:10.101016/j.jff.2017.06.034

Atma, Y. (2022). Synthesis and application of fish gelatin for hydrogels/composite hydrogels: A review. *Biointerface Research in Applied Chemistry*, 12(3): 3966–3976. DOI:10.33263/BRIAC123.39663976

Barrow, C.J. (2010). Marine nutraceuticals: Glucosamine and omega-3 fatty acids. New trends for established ingredients. *AgroFOOD Industry Hi-tech*, 21(2). https://www.teknoscienze.com/Contents/Riviste/PDF/AF2_2010_RGB_38-43.pdf

Bellaaj-Ghorbel, O., Hajji, S., Younes, I., Chaabouni, M., Nasri, M., & Jellouli, K. (2013). Optimization of chitin extraction from shrimp waste with *Bacillus pumilus* A1 using response surface methodology. *International Journal of Biological Macromolecules*, 61: 243–250. DOI:10.1016/j.ijbiomac.2013.07.001

Benjakul, S., Kittiphattanabawon, P., & Regenstein, J.M. (2012). Fish gelatin. Simpson, B.K., Nollet, L.M.L., Toldra, F., Benjakul, S., Paliyath, G., Hui, Y.H. (eds.), *Food Biochemistry and Food Processing*, (2nd ed.), John Wiley & Sons, 388–405.

Bilal, M., Qindeel, M., Nunes, L.V., Duarte, M., Ferreira, L., Soriano, R.N., & Iqbal, H. (2020). Marine-Derived Biologically Active Compounds for the Potential Treatment of Rheumatoid Arthritis. *Marine Drugs*, 19(1): 10. https://doi.org/10.3390/md19010010

Bonilla-Mendez, J.R., & Hoyos-Concha, J.L. (2018). Methods of extraction, refining and concentration of fish oil as a source of omega-3 fatty acid. *Corpoica Ciencia y Tecnologia Agropecuaria*, 19(3): 645–668. DOI:10.21930/recta.vol19_num2_art:684

Brasky, T.M., Lampe, J.W., Slatore, C.G., & White, E. (2011). Use of glucosamine and chondroitin and lung cancer risk in the VITamins And Lifestyle (VITAL) cohort. *Cancer Causes & Control: CCC*, 22(9): 1333–1342. DOI:10.1007/s10552-011-9806-8

Carlini, G., & Kleine, K. (2018). Advancing the international regulation of plastic pollution beyond the United Nations Environment Assembly resolution on marine litter and microplastics. *Review of European, Comparative & International Environmental Law (RECIEL)*, 27: 234–244. DOI:10.1111/reel.12258

Cheung, I.W.Y., & Li-Chan, E.C.Y. (2016). Enzymatic production of protein hydrolysates from steelhead (*Oncorhynchus mykiss*) skin gelatin as inhibitors of dipeptidyl-peptidase IV and angiotensin-I converting enzyme. *Journal of Functional Foods*, 28: 254–264: DOI:10.1016/j.jff.2016.10.030

Choi, B.Y., Chalissery, E.P., Kim, M.H., Kang, H.W., Choi, I.W., & Nam, S.Y. (2019). The influence of astaxanthin on the proliferation of adipose-derived mesenchymal stem cells in gelatin-methacryloyl (GelMA) hydrogels. *Materials*, 12(15): 2416. DOI:10.3390/ma12152416

Chu, W. (2017). How can firms enhance and improve omega-3 fish oil extraction? *NUTRAingredients*. https://www.nutraingredients.com/Article/2017/09/18/How-can-firms-enhance-and-improve-omega-3-fish-oil-extraction

Ciriminna, R., Meneguzzo, F., Delisi, R., & Pagliaro, M. (2017). Enhancing and improving the extraction of omega-3 from fish oil. *Sustainable Chemistry and Pharmacy*, 5: 54–59. DOI:10.1016/j.scp.2017.03.001

Dalei, J., & Sahoo, D. (2015). Extraction and characterization of astaxanthin from the crustacean shell waste from shrimp processing industries. *International Journal of Pharmaceutical Sciences and Research*, 6(6): 2532–2537. DOI:10.13040/IJPSR.0975-8232.6(6).2532-37

Davinelli, S., Nielson, M.E., & Scapagnini, G. (2018). Astaxanthin in skin health, repair, and disease: A comprehensive review. *Nutrients*, 10(4): 522. DOI:10.3390/nu10040522

Debeaufort, F. (2021). Active biopacking produced from by-products and waste from food and marine industries. *FEBS Open Bio*, 11(4): 984–998. DOI:10.1002/2211-5463.13121

Djuricic, I., & Calder, P.C. (2021). Beneficial outcomes of omega-6 and omega-3 polyunsaturated fatty acids on human health: An update for 2021. *Nutrients*, 13(2421). DOI:10.3390/nu13073421

Doan, C.T., Tran, T.N., Nguyen, V.B., Vo, T.P.K., Nguyen, A.D., & Wang, S.L. (2019). Chitin extraction from shrimp waste by liquid fermentation using an alkaline protease-producing strain, *Brevibacillus parabrevis*. *International Journal of Biological Macromolecules*, 131: 706–715. DOI:10.1016/j.ijbiomac.2019.03.117

Duarte, K., Justino, C.I.L., Pereira, R., Freitas, A.C., Gomes, A.M., Duarte, A.C., & Rocha-Santos, T.A.P. (2014). Green analytical methodologies for the discovery of bioactive compounds from marine sources. *Trend in Environmental Analytical Chemistry*, 3–4: 43–52. DOI:10.1016/j.teac.2014.11.001

Elieh-Ali-Komi, D., & Hamblin, M.R. (2016). Chitin and chitosan: Production and application of versatile biomedical nanomaterials. *International Journal of Advanced Research*, 4(3): 411–427.

El Knidri, H., El Khalfaouy, R., Laajeb, A., Addaou, A., & Lahsini, A. (2016). Eco-friendly extraction and characterization of chitin and chitosan from the shrimp shell waste via microwave irradiation. *Process Safety and Environmental Protection*, 104: 395–405. DOI:10.1016/j.psep.2016.09.020

Fernandes, P. (2016). Enzymes in fish and seafood processing. *Frontiers in Bioengineering and Biotechnology*, 4: 59. DOI:10.3389/fbioe.2016.00059

Fernandes, S.S., Coelho, M.S., & Salas-Mellado, M.M. (2019). Bioactive compounds as ingredients of functional foods: Polyphenols, carotenoids, peptides from animal and plant sources new. *Bioactive Compounds*, 129–142. DOI:10.1016/B978-0-12-814774-0.00007-4

FoodPrint. (2020). The problem of food waste. *The Environmental Impact of Food Packaging*. https://foodprint.org/issues/the-problem-of-food-waste/

Galanakis, C.M. (2017). Introduction. *Nutraceutical and Functional Food Components. Effects of Innovative Processing Techniques*, 1–14. DOI:10.1016/B978-0-12-805257-0.00001-6

Gomez, B., Pateiro, M., Barba, F.J., Marszalek, K., Puchalski, C., Lewandowski, W., Simal-Gandara, J., & Lorenzo, J.M. (2019). Legal regulations and consumer attitudes regarding the use of products obtained from aquaculture. *Advances in Food and Nutrition Research*, 92: 1043–4526. DOI:10.1016/bs.afnr.2019.11.002

Grienke, U., Silke, J., & Tasdemir, D. (2014). Bioactive compounds from marine mussles and their effects on human health. *Food Chemistry*, 48–60. DOI:10.1016/j.foodchem.2013.07.027

Gu, R.Z., Li, C.Y., Liu, W.Y., Yi, W.X., & Cai, M.Y. (2011). Angiotensin I-converting enzyme inhibitory activity of low-molecular-weight peptides from Atlantic salmon (*Salmo salar*) skin. *Food Research International*, 44: 1536–1540. DOI:10.1016/j.foodres.2011.04.006

Hamed, I., Ozogul, F., Ozogul, Y., & Regenstein, J.M. (2015). Marine bioactive compounds and their health benefits: A review. *Comprehensive Reviews in Food Science and Food Safety*, 14: 446–465. DOI:10.1111/1541-4337.12136

Harnedy, P.A., & FitzGerald, R.J. (2012). Bioactive peptides from marine processing waste and shellfish: A review. *Journal of Functional Foods*, 6–24. DOI:10.1016/j.jff.20111.0901

Hasler, C.M. (2002). Functional foods: Benefits, concerns and challenges – A position paper from the American Council on Science and Health. *Journal of Nutrition*, 132(12): 3772–3781. DOI:10.1093/jn/132.12.3772

Helkar, P.B., Sahoo, A.K., & Patil, N.J. (2016). Review: Food industry by-products used as a functional food ingredients. *Internal Journal of Waste Resources*, 6: 3. DOI:10.4172/2252-5211.1000248

Hidaka, B.H., Li, S., Harvey, K.E., Carlson, S.E., Sullivan, D.K., Kimler, B.F., Zalles, C.M., & Fabian, C.J. (2015). Omega-3 and omega-6 fatty acids in blood and breast tissue of high-risk women and association with atypical cytomorphology. *Cancer Prevention Research*, 8(5): 359–364. DOI:10.1158/1940-6207.CAPR-14-0351

Higuera-Ciapara, I., Felix-Valenzuela, L., & Goycoolea, F.M. (2006). Astaxanthin: A review of its chemistry and applications. *Critical Reviews in Food Science and Nutrition*, 46(2): 185–196. DOI:10.1080/10408690590957188

Hosseini, S.F., & Gomez-Guillen, M.C. (2018). A state-of-the-art review on the elaboration of fish gelatin as bioactive packaging: Special emphasis on nanotechnology-based approaches. *Trends in Food Science Technology*, 79: 125–135.

Ibanez, E., Herrero, M., Mendiola, J.A., & Castro-Puyana, M. (2012). Extraction and characterization of bioactive compounds with health benefits from marine resources: Macro and micro algae, cyanobacteria, and invertebrates. *Marine Bioactive Compounds: Sources, Characterizations and Applications*. DOI:10.1007/978-4614-1247-2_2

Ibanez-Sanz, G., Guino, E., Morros, R., Quijada-Manuitt, M.A., de la Pena-Negro, L.C., & Moreno, V. (2020). Chondroitin sulphate and glucosamine use depend on non-steroidal anti-inflammatory drug use to modify the risk for colorectal cancer. *Cancer Epidemiology, Biomarkers & Prevention*, 29(9): 1809–1816. DOI:10.1158/1055-9965

Im, A.R., Kim, J.Y., Kim, H.S., & Cho, S. (2013). Wound healing and anti-bacterial activities of chondroitin sulfate—and acharan sulfate—reduced silver nanoparticles. *Nanotechnology*, 24(39): 395102. DOI:10.1088/0957-4484/24/39/395102

International Union for Conservation of Nature (IUCN). (2018). Marine plastics. https://www.iucn.org/resources/issues-briefs/marine-plastics

Kalupahana, N.S., Goonapienuwala, B.L., & Moustaid-Moussa, N. (2020). Omega-3 fatty acids and adipose tissue: Inflammation and browning. *Annual Review of Nutrition*, 40(1): 25–49

Karakut, I., Ozaltin, K., Vesela, D., Lehocky, M., Humpolicek, P., & Mozetic, M. (2019). Anti-bacterial activity and cytotoxicity of immobilized glucosamine/chondroitin sulfate on polylactic acid films. *Polymer Biointerfaces*, 11(7): 1186. DOI:10.3390/polym11071186

Karthickeyan, A., Gayathri, V., Banumathy, P., & Hemalatha, K. (2007). Extracton of bio-compound from marine waste and its usage as a possible immune stimulant. *International Journal of Innovative Research in Science, Engineering and Technology*, 5(1): 3297

Kim, S.K., & Mendis, E. (2006). Bioactive compounds from marine processing by-products—A review. *Food Research International*, 39: 383–393. DOI:10.1016/j.foodres.2005.10.010

Kim, S.K., Ngo, D.H., & Vo, T.S. (2012). Marine fish-derived bioactive peptides as potential antihypertensive agents. *Advances in Food and Nutrition Research*, 65: 1043–4526. DOI:10.1016/B978-0-12-416003-3.00016-0

Kongtip, P., Singkaew, P., Yoosook, W., Chantanakul, S., & Sujiratat, D. (2013). Health effects of people living close to a petrochemical industrial estate in Thailand. *Journal of the Medical Association of Thailand*, 96(Suppl. 5): S64–72. PMID: 24851575

Kotronoulas, A., Jonasdottir, H.S., Siguroardottir, R.S., Halldorsson, S., Haraldsson, G.G., & Rolfsson, O. (2019). Wound healing grafts: Omega-3 fatty acid lipid content differentiates the lipid profiles of acellular Atlantic cod skin from traditional dermal substitutes. *Journal of Tissue Engineering and Regenerative Medicine*, 14: 441–451. DOI:10.1002/term.3005

Kumar, S., Mukherjee, A., & Dutta, J. (2020). Chitosan based nanocomposite films and coatings: Emerging antimicrobial food packaging alternatives. *Trends in Food Science & Technology*, 97: 196–209. DOI:10.1016/j.tfis.2020.01.002

Liguori, A., Uranga, J., Panzavolta, S., Guerrero, P., de la Caba, K., & Focarete, M.L. (2019). Electrospinning of fish gelatin solution containing citric acid: An environmentally friendly approach to prepare cross-linked gelatin fibers. *Materials*, 12: 2808.

Lindequist, U. (2016). Marine-derived pharmaceuticals—challenges and opportunities. *Biomolecules & Therapeutics*, 24(6): 561–571. DOI:10.4062/biomolther.2016.181

Lindner, H.B., Zhang, A., Eldridge, J., Demcheva, M., Tsichilis, P., Seth, A., Vournakis, J., & Muise-Helmericks, R.C. (2011). Anti-bacterial effects of Poly-N-Acetyl-Glucosamine nanofibers in cutaneous wound healing: Requirement for Akt1. *PLOS ONE*, 6(7). DOI:10.1371/annotation/3966fcd7-6127-45e0-80e6-55063875e826

Lionetto, F., & Esposito Corcione, C. (2021). Recent applications of biopolymers derived from fish industry waste in food packaging. *Polymers*, 13(14): 2337. https://doi.org/10.3390/polym13142337

Liu, P., Liu, S., Guo, N., Mao, X., Lin, H., Xue, C., & Wei, D. (2014). Cofermentation of *Bacillus licheniformis* and *Gluconobacter oxydans* for chitin extraction from shrimp waste. *Biochemical Engineering Journal*, 91: 10–15. DOI:10.1016/j.bej.2014.07.004

Lucarini, M., Zuorro, A., Di Lena, G., Lavecchia, R., Durazzo, A., Benedetti, B., & Lombardi-Boccia, G. (2020). Sustainable management of secondary raw materials from the marine food-chain: A case-study perspective. *Sustainability*, 12: 8997. DOI:10.3390/su12218997

Maoka, T. (2020). Carotenoids as natural functional pigments. *Journal of Natural Medicines*, 74(1): 1–16. DOI:10.1007/s11418-019-01364-x

Marzieh, M.N., Zahra, F., Tahereh, E., & Sara, K.N. (2019). Comparison of the physicochemical and structural characteristics of enzymatic produced chitin and commercial chitin. *International Journal of Biological Macromolecules*, 139: 270–276. DOI:10.1016/j.ijbiomac.2019.07.217

McKenney, J.M., & Sica, D. (2007). Role of prescription omega-3 fatty acids in the treatment of hypertriglyceridemia. *Pharmacotherapy*, 27(5): 715–728. DOI:10.1592/phco.27.5.715

Meena, F., Wijesinghe, P.A.U.I., Thiripuranathar, G., Uzair, B., Iqbal, H., Khan, B.A., & Meena, B. (2020). Ecological and industrial implications of dynamic seaweed-associated microbiota interactions. *Marine Drugs*, 18(12): 641. DOI:10.3390/md18120641

Mutalipassi, M., Esposito, R., Ruocco, N., Viel, T., Costantiini, M., & Zupo, V. (2021). Bioactive compounds of nutraceutical value from fishery and aquaculture discards. *Foods*, 10: 1495. DOI:10.3390/foods10071495

Nanda, P.K., Das, A.K., Dandapat, P., Dhar, P., Bandyopadhyay, S., Dib, A.L., Lorenzo, J.M., & Gagaouta, M. (2021). Nutritional aspects, flavor profile and health benefits of crab meat based novel food products and valorization of processing waste to wealth: A review. *Trends in Food Science & Technology*, 112: 252–267. DOI:10.1016/j.tifs.2021.03059

National Oceanic and Atmospheric Administration (NOAA). (2021). What is marine debris? *OR&R'S Marine Debris Division*. https://marinedebris.noaa.gov/discover-marine-debris/what-marine-debris

Ngamwonglumlert, L., & Devahastin, S. (2019). Carotenoids. *Encyclopedia of Food Chemistry*, 40–52. DOI:10.1016/B978-0-08-100596-5.21608-9

Ozogul, F., Cagalj, M., Simat, V., Ozogul, Y., Tkaczewska, J., Hassoun, A., Kaddour, A.A., Kuley, E., Rathod, N.B., & Phadke, G.G. (2021). Recent developments in valorisation of bioactive ingredients in discard/seafood processing by-products. *Trends in Food Science & Technology*, 116(1): 559–582. DOI:10.1016/j.tifs.2021.08.007

Pal, G.K., & Suresh, P.V. (2016). Sustainable valorisation of seafood by-products: Recovery of collaged and development of collagen-based functional food ingredients. *Innovative Food Science and Emerging Technologies*. DOI:10.1016/j.ifset.2016.03015

Parker, L. (2015). Ocean trash: 5.25 trillion pieces and counting, but big questions remain. *Science*. https://www.nationalgeographic.com/science/article/150109-oceans-plastic-sea-trash-science-marine-debris

Parker, L. (2020). Plastic trash flowing into the seas will nearly triple by 2040 without drastic action. *Science*. https://www.nationalgeographic.com/science/article/plastic-trash-in-seas-will-nearly-triple-by-2040-if-nothing-done

Parshyna, I., Lehmann, S., Grahl, K., Pahkle, C., Frenzel, A., Weidlich, H., & Morawietz, H. (2017). Impact of omega-3 fatty acids on expression of angiogenic cytokines and angiogenesis by adipose-derived stem cells. *Atherosclerosis Supplements*, 30: 303–310. DOI:10.1016/j.atherosclerosissup.2017.05.040

Pateiro, M., Dominguez, R., Varzakas, T., Munekata, P.E.S., Fierro, E.M., & Lorenzo, J.M. (2021). Omega-3-rich oils from marine side streams and their potential application in food. *Marine Drugs*, 19(233). DOI:10.3390/md19050233

Pateiro, M., Munekata, P.E.S., Tsatsanis, C., Dominguez, R., Zhang, W., Barba, F.J., & Lorenzo, J.M. (2019). Evaluation of the protein and bioactive compound bioaccessibility/bioavailability and cytotoxicity of the extracts obtained from aquaculture and fisheries by-products. *Advances in Food and Nutrition Research*, 92: 1043–4526. DOI:10.1016/bs.afnr.2019.12.002

Pu, J., Betchtel, P.J., & Sathivel, S. (2010). Extraction and shrimp astaxanthin with flaxseed oil: Effects on lipid oxidation and astaxanthin degradation rates. *Biosystem Engineering*, 107(4): 364–371. DOI:10.1016/j.biosystemseng.2010.10.001

Rotter, A., Barbier, M., Bertoni, F., Bones, A.M., Cancela, M.L., Carlsson, J., Carvalho, M.F., Cegłowska, M., Chirivella-Martorell, J., Conk, D.M., Cueto, M., Dailianis, T., Deniz, I., Díaz-Marrero, A.R., Drakulovic, D., Dubnika, A., Edwards, C., Einarsson, H., Erdoğan, A., Eroldoğan, O.T., Ezra, D., Fazi, S., FitzGerald, R.J., Gargan, L.M., Gaudêncio, S.P., Gligora, U.M., Ivošević, de N., Jónsdóttir, R., Kataržytė, M., Klun, K., Kotta, J., Ktari, L., Ljubešić, Z., Lukić, B.L., Mandalakis, M., Massa-Gallucci, A., Matijošytė, I., Mazur-Marzec, H., Mehiri, M., Nielsen, S.L., Novoveská, L., Overlingė, D., Perale, G., Ramasamy, P., Rebours, C., Reinsch, T., Reyes, F., Rinkevich, B., Robbens, J., Röttinger, E., Rudovica, V., Sabotič, J., Safarik, I., Talve, S., Tasdemir, D., Theodotou, S.X., Thomas, O.P., Toruńska-Sitarz, A., Varese, G.C., & Vasquez, M.I. (2021). The essentials of marine technology. *Frontiers in Marine Science*, 8: 158. DOI:10.3389/fmars.2021.629629

Rubio-Rodriguez, N., de Diego, S.M., Beltran, S., Jaime, I., Sanz, M.T., & Rovira, J. (2008). Supercritical fluid extraction of the omega-3 rich oil contained in hake (*Merluccius capensis-Merluccius paradoxus*) by-products: Study of the influence of process parameters on the extraction yield and oil quality. *Journal of Supercritical Fluids*, 47(2): 215–226

Sadighara, P., Ardabili, F.H., & Kazemi, V. (2015). Extraction of shrimp waste oil and its fortification with shrimp waste pigments. *Journal of Food Safety and Hygiene*, 1(2):69–71.

Santos, V.P., Marques, N.S.S., Maia, P.C.S.V., Lima, M.A.B., Franco, L.O., & Campos-Takaki, G.M. (2020). Seafood waste as attractive source of chitin and chitosan production and their applications. *International Journal of Molecular Sciences*, 21(4920): 1–17. DOI:10.3390/ijms21124290

Senevirathne, M., & Kim, S.K. (2012a). Development of bioactive peptides from fish proteins and their health promoting ability. *Advances in Food and Nutrition Research*, 65: 1043–4526. DOI:10.1016/B978-012-416003-3.00015-9

Senevirathne, M., & Kim, S.K. (2012b). Utilization of seafood processing by-products: Medicinal application. *Advances in Food and Nutrition Research*, 65: 1043–4526. DOI:10.1016/B978-012-416003-3.00032-9

Shavandi, A., Hou, Y., Carne, A., McConnell, M., & Bekhit, AEA (2019). Marine waste utilization as a source of functional and health compounds. *Advances in Food and Nutrition Research*, 87: 1043–4526. DOI:10.1016/bs.afnr.2018.08.001

Sibi, G., Dhananjaya, K., Ravikumar, K.R., Mallesha, H., Venkatesha, R.T., Trivedi, D., Bhusal, K.P., & Gowada, K. (2013). Preparation of glucosamine hydrochloride from crustacean shell waste and its quantification by RP-HPLC. *American-Eurasion Journal of Scientific Research*, 8(2): 63–67. DOI:10.5829/idosi.aejsr.2013.8.2.7381

Sommer, A., Dederko-Kantowicz, P., Staroszczyk, H., Sommer, S., & Michalec, M. (2021). Enzymatic and chemical cross-linking of bacterial cellulose/fish collagen composites—A comparative study. *International Journal of Molecular Science*, 22: 3346.

Stachowiak, B., & Szulc, P. (2021). Astaxanthin for the food industry. *Molecules (Basel, Switzerland)*, 26(9): 2666. DOI:10.3390/molecules26092666

Suleria, H.A.R., Masci, P., Gobe, G., & Osborne, S. (2016). Current and potential uses of bioactive molecules from marine processing waste. *Journal of the Science of Food and Agriculture*, 96: 1064–1067. DOI:10.1002/jsfa.7444

Suleria, H.A.R., Osborne, S., Masci, P., & Gobe, G. (2015). Marine-based nutraceuticals: An innovative trend in the food and supplement industries. *Marine Drugs*, 13(10): 6336–6351. DOI:10.3390/md13106336

Suresh, P.V. & Prabhu G.N. (2013). Seafood. Chandrasekaran, M. (editor), *Valorization of Food Processing By-Products*, CRC Press, Taylor & Francis, New York, 685–736.

Svenning, R. (1993). Biotechnological scaling of fish. *Infofish International*, 6: 30.

Swanson, D., Block, R., & Mousa, S.A. (2012). Omega-3 fatty acids EPA and DHA: Health benefits throughout life. *Advances in Nutrition*, 3(1): 1–7. DOI:10.3945/an.111.000893

United States Department of Agriculture (USDA), Office of the Chief Economist. (2013). Food waste FAQs. *USDA*. https://www.usda.gov/foodwaste/faqs

US Environmental Protection Agency (EPA). (2017). Advancing sustainable materials management: Facts and figures. *EPA*. https://www.epa.gov/facts-and-figures-about-materials-waste-and-recycling/advancing-sustainable-materials-management-0

US Environmental Protection Agency (EPA). (2021). Sources of aquatic trash. *EPA*. https://www.epa.gov/trash-free-waters/sources-aquatic-trash

US Food and Drug Administration (FDA). (2019). FDA approves use of drug to reduce risk of cardiovascular events in certain adult patient groups. https://www.fda.gov/news-events/press-announcements/fda-approves-use-drug-reduce-risk-cardiovascular-events-certain-adult-patient-groups

Vasiliadis, H.S., & Tsikopoulos, K. (2017). Glucosamine and chondroitin for the treatment of osteoarthritis. *World Journal of Orthopedics*, 8(1): 1–11. DOI:10.5312/wjo.v8.i1.1

Vazquez, J.A., Rodriquez-Amado, I., Montemayor, M.I., Fraguas, J., Gonzalez, M.P., & Murado, M.A. (2013). Chondroitin sulfate, hyaluronic acid and chitin/chitosan production using marine waste sources: Characteristics, applications and eco-friendly. *Marine Drugs*, 11: 747–774. DOI:10.3390/md1103070747

Venugopal, V. (2016). Enzymes from seafood processing waste and their applications in seafood processing. *Advances in Food and Nutrition Research*, 78: 1043–4526. DOI:10.1016/bs.afnr.2016.06.004

Wu, C.S., Wang, S.S., Wu, D.Y., & Shih, W.L. (2021). Novel composite 3D-printed filament made from fish scale-derived hydroxyapatite, eggshell and polylactic acid via a fused fabrication approach. *Additive Manufacturing*, 46: 102169. DOI:10.1016/j.addma.2021.102169

Xu, N., Peng, X.L., Li, H.R., Liu, J.X., Cheng, J.S.Y., Qi, X.Y., Ye, S.J., Gong, H.L., Zhao, X.H., Yu, J., Xu, G., & Wei, D.X. (2021). Marine-derived collagen as biomaterials for human health. *Frontiers in Nutrition*, 8: 493. DOI:10.3389/fnut.2021.702108

Xu, R. (2015). Importance bioactive properties of omega-3 fatty acids. *Italian Journal of Food Science*, 27(2): 129–135. DOI:10.14674/1120-1770/ijfs.v177

Yadav, M., Goswami, P., Paritosh, K., Kumar, M., Pareek, N., & Vivekanand, V. (2019). Seafood waste: A source for preparation of commercially employable chitin/chitosan materials. *Bioresources and Bioprocessing*, 6: 8. DOI:10.1186/s40643-019-0243-y

Zamri, M.F.M.A., Bahru, R., Amin, R., Khan, M.U.A., Razak, S.I.A., Hassan, S.A., Kadir, M.R.A., & Nayan, N.H.M. (2021). Waste to health: A review of waste derived materials for tissue engineering. *Journal of Cleaner Production*, 290: 125792. DOI:10.1016/j.jclepro.2021.125792

Zhang, Y., Ward, V., Dennis, D., Plechkova, N.V., Armenta, R., & Rehmann. (2018). Efficient extraction of a docosahexaenoic acid (DHA)-rich lipid fraction from *Thraustochytrium* sp. using ionic liquids. *Materials*, 11(1986). DOI:10.3390/ma11101986

14 Marine-Derived *Aspergillus*

A Source of Bioactive Compounds and Its Medical Applications

Janakiraman V, Monisha KG, Ramakrishnan V, and Shiek SSJ Ahmed

CONTENTS

14.1 MARINE WORLD

14.1.1 Introduction

Marine bioactive compounds play a key role in human life, mostly for the scope of biotechnology and pharmacology. Owing to their vast diversity and

DOI: 10.1201/9781003303909-14

their potential bioactivity, scientists face challenges to come up with novel techniques to isolate bioactive compounds for a more robust future. The seas are the major reservoirs of different organisms with vast diversity, producing various metabolites with diverse biological activities suitable for drug discovery. For centuries, we have been using natural products to produce foods, pigments, insecticides, medicines, fragrances, fertilizers, and so on. Traditionally, terrestrial plants have been used for conventional folk medicine because of their availability and accessibility (Nicoletti and Vinale 2018). Of total pharmaceutical drug sales, natural plant–derived products contribute 25%, while microbial products contribute 12%. The thermal range of the marine environment around the globe ranges from 350°C in underwater hydrothermal vents to deep-freezing temperatures in the Antarctic, with a pressure range of 1–1000 atm, covering photic and nonphotic zones (Joffe and Thomas 1989).

Indeed, the fact that the marine environment has the largest biodiversity, compared to other terrestrial environments, using marine-derived products as a source of therapeutic agents is still in its infancy. The limitation of marine-derived products is the hassle of collecting marine organisms. The marine environment has extensive speciation of almost every phylogenetic level from smaller microorganisms to larger mammals. Thanks to modern technology, we are able to get samples from deep seawater using innovative diving techniques and auto- or remote-operating machines. In the sea, at a depth of 900m, 5000 new compounds have been isolated in past decades (McCarthy and Pomponi 2004).

In particular, approximately 10,000 novel metabolites have been isolated from marine species that possess pharmacological effects (Jha and Zi-Rong 2004). Previously, it was thought that marine plants and animals were producers of drugs. For instance, the squid model has been used to study the transmission of nerves to obtain precise knowledge about the mesenteries of vision. Moreover, the physiology of large nerve axons was elucidated in eye models of horseshoe crabs, sharks, and skates. Likewise, the regulation of the cell cycle, cell reproduction, and its molecular mechanism were demonstrated using a surf clam model. In line with earlier studies, current research provides evidence on fungi as an increasing recognition of the symbiotic and adaptation ability to sustain in various environmental conditions. It is also reported, that in several cases, fungi are the real producers of drugs.

Marine biotechnology deals with marine species, which have been used to make or synthesize or discover new bioactive compounds for different specific uses in both the aquatic and terrestrial environment (Harvey 2000). The bioactive compounds isolated from marine organisms include the following: seaweed, cyanobacteria, sea slugs, soft corals, cnidarians, sea hares, sponges, tunicates, bryozoans, nudibranchs, mollusks, marine fungus, and other marine organisms (Donia and Hamann 2003). Marine species are the major sources, which starts from extracting fish liver oil from fish, which are omega-3 fatty acids, to isolation bioactive compounds from marine microorganisms. In the following sections, we cover some important species and their bioactive compounds.

14.2 MARINE SPECIES AND ACTIVE COMPOUNDS

14.2.1 Cyanobacteria

In terms of both general and marine species, one of the abundant sources of bioactive compounds is *Cyanobacteria.* Cyanobacteria are classified into five divisions of microalgae, of which much research interest is focused on dinoflagellates *pyrrophyta* and blue-green algae *Cyanophyta* (Moore et al. 1988). Marine cyanobacteria are the sources for the production of large-scale commercial vitamins such as vitamin E and vitamin B complex (Plavšić et al. 2002). Cyanobacteria are also used for the production of various commercial products and natural food coloring agents, such as cyanobacteria-derived carotenoids, phycobiliprotein pigments, feed additives, fertility, and color density improvement of the egg yolk.

Numerous metabolites have been extracted and isolated from cyanobacteria; most of them are freshwater species because of its facile nature compared to other marine species. Of 100 marine-isolated compounds, 50% exhibits potential bioactivity. In Hawaii, two different toxic strains of cyanobacteria have been isolated from *Lyngbya mausculata*: lyngbyatoxin-A and debromoaplysiatoxin. The toxic strains were found to be structurally unique, inflammatory in nature and were isolated anatoxin-a from *Anabae ciecinalis* (Cardellina et al. 1979).

Many marine metabolites have been evaluated and proved to play active role in cancer cell apoptosis by inducing the signaling enzyme, a member of kinase C protein (Fujiki and Sugimura 1987; Jha and Zi-Rong 2004). Novel sulfur derivatives, fusaperazines A and B, containing dioxopiperazine isolated from *Fusarium chlamydospore*, were cultured in Japanese marine red alga Capopeltis, and two known compounds isolated by the process of fermentation from the fungus *Tolypocladium* species (Asolkar et al. 2002). Chalcomycin-B has shown activity against microalgae and microorganisms (Lin et al. 2008). Fungus *Leptosphaeria sp.*, which belongs to *Sargassum tortile* (Japanese brown alga), from that isolated four novel epipolysulphanyldioxopiperazine (Yamada et al. 2002). Of the cytotoxic activity noted in the P388 cell line on each compound, one of the leptosins showed extraordinary cytotoxicity against 39 human cancer cell lines on the disease-oriented panel and precisely inhibited protein kinases and topoisomerase-II (Stevenson et al. 2002). From cyanobacterium, several species were isolated and exhibited biological activities, such as antifungal activity, antiproliferative, and anti-inflammatory activity, in human body system. The species *Stigonema* were isolated from a pigment of extracellular sheath proved and exhibit the antiproliferative and anti-inflammatory activity of scytonemin (Murakami et al. 1988). Whereas Goniodomin A, an isolate of dinoflagellate *Goniodoma pseudogoniaulax*, a polyether macrolide exhibits antifungal, has the potency to suppress angiogenesis by inhibiting the migration of endothelial cells and the formation of a fibroblast growth factor (bFGF)–induced tubes (Koehn et al. 1992). Furthermore, Microcolin A, a linear immunosuppressive peptide isolated from *Lyngbya majusculata*, inhibits the lymphocyte reaction, a mixture of two-way murine in nanomolar concentrations (Carte 1996). From *Lyngbya majusculata*, novel curacin A– and thiozoline-containing compounds have been extracted and purified resulting in the active involvement of antiproliferative agent with slight sensitivity for colon, breast cancer–derived, and renal cell lines. Moreover, marine-based novel antibiotic activity has been noted in dinoflagellates; *Gambierdiscus toxicus* showed antifungal activity, and *Ptychodiscus brevis* yielded brevitoxins (Cohen et al. 1990). From prorocentrum species-isolated Okadaic acid, a polyether fatty acid plays an important molecule in studying eukaryotic signaling pathways, a selective phosphatase inhibitor protein (Faulkner 2001). Therefore, the bioactive compounds, strains and metabolites of cyanobacteria are well suitable for comprehensive biotechnological applications due to their bioavailability of organic nutrients.

14.2.2 Seaweed

Another abundant marine source is seaweed found in clear tropical waters and intertidal zones with less bioassay attention. There are numerous seaweeds in the form of algae with tremendous economic potential in human biological activities, food, and cosmetics and in the extraction of fertilizers and industrial chemicals. Marine alga such as marine *sphaerococcus coronopifilius* (red alga), *Ulva lactuca* (green alga) and *Portieria hornemannii* exhibits the biological antibacterial activity (Donia and Hamann 2003) and anti-inflammatory activity and possess an antitumor compound (Ali et al. 2002). Moreover, *Stypodium zonale*, a brown tropical alga stypoldione, plays a predominant role in the inhibition of microtubule polymerization and mitotic spindle formation.

Few marine algae have been proved to possess steroid activity that modest antibacterial efficacy against some bacterial species. Similar to this, a green *Codium iyengarii* found in the Arabian Sea off the coast of Karachi is the source of iyengadione, iyengarosides A, and iyengarosides B, which exhibit their steroid nature against the bacterial species (Ali et al. 2002). Moreover, several clinical trials have elucidated the cytotoxic activity of marine seaweed in human tumor cell lines. Similarly, the National Cancer Institute (NCI) in the United States derived the compound halomon, a penta-halogenated monoterpene, from *Portieria hornemannii*. This compound elucidated its predominant cytotoxic activity in renal, brain, and colon tumor cell lines and was further elected for preclinical drug development. In addition, the species *Sargassum carpophyllum*–derived novel bioactive sterols have been shown a significant potential for inducing abnormal morphology in plant fungus (*Pyricularia oryzae*) and to withhold the cytotoxic activity against tumor cell lines (Blunt et al. 2004).

Some algae species possess a special ability to transform polyunsaturated fatty acids; for example, arachidonic acids are converted to eicosanoids and oxylipins. For the maintenance of homeostasis, arachidonic derivatives play a major role in mammal organization, along with metabolite production occurring in diseases such as ulcers, heart disease, psoriasis, asthma, cancer, and arteriosclerosis.

14.2.3 Sponges

Approximately there are 10,000 sponges noted in the marine organisms. Of those, 11 sponge genera have bioactive activity observed so far. *Discodemia*, *Petrosia*, and *Halicona* are specific sponges exhibiting excellent anticancer and anti-inflammatory activity (Bhimba et al. 2013). After the discovery of spongouridine from the Caribbean sponge *Cryptotethia crypta*, the dominant role of sponges metabolites were determined. The marine sponge spongouridine has proven its tumor-inhibiting activity in arabinosyl nucleoside. By suppressing the DNA polymerase, the converted bioactive compounds of arabinosyl cytosine were used in a bench-side clinical application to treat acute mylocytic leukemia and non-Hodgkin's lymphoma (Fusetani et al. 1992). Also, there are about 300 chemical analogs were found in compound manolide, which is derived from a Pacific sponge and has been clinically proved to exhibit anti-inflammatory activity. Marine sponge *Dercitus*-derived compounds dercitin and aminoacridine alkaloid possess cytotoxic activity and increase the life expectancy of mice by enduring P388 cancer. Similarly, Burres et al. has demonstrated the cytotoxic activity of dercitin and aminoacridine against melanoma B16 cells and Lewis lung cancer. A polyether macrolide, halichondrinB, isolated from sponge *Theonella* holds potential anticancer bioactivity (Sakemi et al. 1988). In addition, the theopederins obtained from *Mycale* species and onnamide obtained from *Theonella* species also exhibited in vitro cytotoxic and in vivo anticancer activity in solid tumor models and leukemia (Pettit et al. 1993). NCT (National Critical Technologies) isolated cribrostatin, a metabolite of *Isoquinolinequinone* sponge species, from *Cribrochalina* that showed antitumor activity against nine melanoma cells (Pettit et al. 1993). Similarly, few other compounds, spongstatin and varitriol, that are extreme chemo-resistant cancer types (cell lines of breast, renal, central nervous system) in the tumor panel of NCT were isolated from *Spongia* and *Emericella variecolor* species found in the Indian Ocean, (Shoji et al. 1992).

In addition to antitumor activity, there are several marine sponge compounds, including varixanthone and glycolipid caminoside A, that exhibit antibacterial and antimicrobial activity. Furthermore, the marine-derived compounds scalaradial and manolide were proved to inhibit phospholipase, which plays a crucial role in the inflammatory process. The marine compound xestobergsterol also suppresses the mast cell–released immunoglobulin E–mediated histamine and has been proved to have efficacy over an antiallergic drug (Chan et al. 1993; Lysek et al. 2002).

14.2.4 Bryozoans

Bryozoans are mostly used to derive alkaloids because they lack bioactive compounds (Blunt et al. 2004). *Flustra foliacae*–derived deformylflustrabromine has shown moderate cytotoxic activity against cell-line HCT-116 (Narkowicz et al. 2002). The reservoirs of convolutindole A and H are *Bryozoan Amathia* convolute, obtained from the east coast of Tasmania, showed activity against parasite nematode and *Haemonchus contortus* (Blackman and Matthews 1985). The compound *Watersipora suntorquata*–derived bryoanthrathiophene, obtained from Tsutsumi Island, Japan, has the potency to suppress angiogenesis during the proliferation of bovine aorta endothelial cells. From *Amathia wilsoni*, isolated asymmetric alkaloids syntheses, amathamide A and B derivatives were obtained from Tasmania (Zhang et al. 1994). A bryozoan *Bugula neritina* has extraordinary anticancer selectivity against renal, human leukemia, lung cancer, and melanoma cell lines. It also regulates protein kinase-C (PKC), which is a signal transduction enzyme (Prinsep et al. 1991). *Anithia convolute*–derived compound convolutamide A possess cytotoxic activity against rodent KB epidermoid cancer cells and L1210 leukemia cells. Alkaloid β-carboline isolated from *Cribricellina cribreria* possess antibacterial, antifungal, antiviral, and cytotoxic activity (Holst et al. 1994). *Flustra foliacae*–derived indole alkaloids exhibit potential antimicrobial activity (Imhoff 2016).

14.2.5 Marine Fungi

In recent years, marine-derived fungi have been recognized by many researchers and biotechnologists due to their abundant biologically active compounds. From marine fungi, numerous bioactive compounds have been isolated and are well known for their clinical significance. They are used in various fields of product development. Some drugs purified from the marine fungus are presently used in clinics and hospitals. Some are still in different stages of the clinical trial. Previously, researchers neglected using marine fungus as a source of novel drug development because of doubts like fungi species' diversity, availability, and clinical significance. Fungi species live in both terrestrial and marine environments, showing that they have the potential to adapt to the environment. They can live in conditions like extreme pressure and high salt content. Deep seawater fungus also exists. The most observed fungal species are *Penicillium* and *Aspergillus* (Lee et al. 2013).

The very first marine fungal–derived bioactive compound cephalosporin, isolated from *Acremonium chrysogenum* in the 1940s, is indispensable in antibacterial treatment. New bioactive compounds and products are derived from marine Fungi (Hwang et al. 2019). In drug development, fungal peptides are used to study bioactivity. One of the approved antifungal drugs is caspofungin acetate (CANCIDAS®), derived from marine *Glarea lozoyensis* (Liu et al. 2020). Countless marine-derived fungi are well-known terrestrial genera, especially *Penicillium*, *Aspergillus*, *Fusarium*, *Phoma*, and *Cladosporium* (Hwang et al. 2019).

Major sources of new bioactive molecules are marine-derived fungi, and from 2001–2017, more than 613 new products have been identified from corals, algae, sponges, and other marine organisms. From 2011–2017, over 613 new natural marine-derived products reported in the marine habitat of China. Of those, 170 new natural products belong to the *Aspergillus* genera (Park et al. 2017). The foremost abundant fungal filamentous species are *Aspergillus* and *Penicillium* (Nicoletti and Vinale 2018). The proportion percentage shows that the comparison of one bioactive terpenoid compared to the whole bioactivity of marine fungi–derived terpenoids.

14.3 MARINE FUNGUS *ASPERGILLUS*

14.3.1 Introduction

An Italian priest and biologist Pier Antonio named the *Aspergillus* genus because of its similarity to aspergillum (holy water sprinkler). *Aspergillus* reproduction is via asexual spores (conidia) that resemble multicellular conidiophores. The identification and classification of *Aspergillus* have been identified with the observation of conidiophores morphology based on their shape, size, arrangement, and color of nonsexual spores. Table 14.1 describes the classification of *Aspergillus* species.

Officially, 340 filamentous *Aspergillus* species have been noted (Wang et al. 2018). The rich sources of bioactive compounds present in the genus *Aspergillus* include polyketides, peptides, terpenoids, and others. Simple-structured diphenyl ethers belong to a group of polyketides derived from various species of *Aspergillus* that have diverse bioactivities, such as antiviral, antimicrobial, regulating actin function, cytotoxicity, phytocidal activities, radical scavenging, β-glucuronidase enzyme inhibition, and anti-Aβ 42 aggregations (Xu et al. 2020). Furthermore, more than 18 species of *Aspergillus* have been sequenced and analyzed to elucidate its distinguished role in identifying genomic to phenotype relationships and functional genomics studies in humans. The study of marine microbial diversity is the source for biotechnology and pharmacology to identify and explore new bioactive compounds. The new bioactive marine compounds play a potential role in human health. The seas and oceans are sources of species producing secondary structures along with exciting bioactivity (Nicoletti and Vinale 2018). Figure 14.1 shows the percentage of various bioactivities of *Aspergillus* species.

TABLE 14.1
Classification of *Aspergillus*

Kingdom	Fungi
Phylum	Ascomycota
Sub-Phylum	Pezizomycotina
Class	Eurotiomycetes
Order	Eurotiales
Family	Trichocomaceae
Genus	Aspergillus

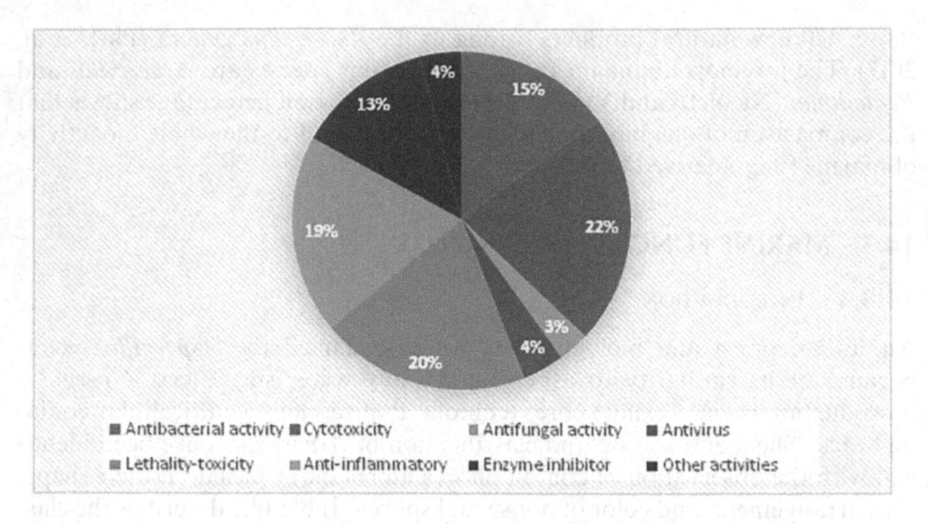

FIGURE 14.1 The proportion percentage shows that the comparison of one bioactive terpenoid compared to the whole bioactivity of marine fungi–derived terpenoids (Jiang et al. 2020).

Aspergillus metabolite with indole alkaloids serves as the active component in various clinical drugs, such as tadafil, rizatriptan, sumatriptan, and fluvastatin (Da Silva et al. 2017). Bioactive compounds of *Aspergillus terrus*, comprising terrain, butyrolactone V, and butyrolactone I, have been elucidated its antiparasitic and antibacterial activity of the isolated compounds (Hwang et al. 2019). After several clinical trials, plinabulinhas been proposed as a novel anticancer drug (Li et al. 2019). Also, *Aspergillus Fumigatus* and *Aspergillus flavus*, belonging to *Aspergillus* species, proved to cause health issues, including allergic bronchopulmonary aspergillosis.

Aspergillus is well known for both their positive and negative sides. The positive role that *Aspergillus* species plays is a predominant role in the production of enzymes, pharmaceuticals, biotechnology, food fermentation, and drug discovery. For instance, in East Asia, *Aspergillus oryzae* and *Aspergillus niger* are widely used for fermenting food and producing amylase, pectinases, and citric acid enzymes. Koji molds, which are nothing but *Aspergillus oryzae*, have been used for making alcoholic beverages, vinegar, soybean pastes, and soy sauce. On the other hand, the negative part of this species results from its extremely harmful toxicity if consumed; it may even cause liver failure. Toxin-secreting *Aspergillus* species, such as *Aspergillus flavus* and *Aspergillus parasiticus*, ooze mycotoxin aflatoxins that are well known for their contamination of plant-synthesized feeds and foods. The intake of mycotoxin aflatoxin–contaminated food causes liver necrosis, liver cancer, and aflatoxicosis (Wang et al. 2018).

14.3.2 Metabolites of *Aspergillus*

14.3.2.1 Terpenoids

Terpenoids, a vast and diverse class of natural organic compounds derived from the strains of marine-derived *Aspergillus*, exhibit various biological activities, such as antibacterial, antifungal, antitumor, antiparasitic, and antiviral. They are also known for their structural and chemical diversity in secondary metabolites and for possessing various bioactive compounds. Figure 14.2 explains the different forms of terpenes isolated from *Aspergillus* species in 2015–2019 period.

The fungus *Aspergillus Versicolor* obtained from underwater ocean sediment. Two new butyrolactone classes of monoterpenoids have been isolated from *Aspergillus versicolor* that include pestalotiolactones A, C, and D (Zhang et al. 2008). Figures 14.3 and 14.4 show the structures of pestalotiolactones C and D.

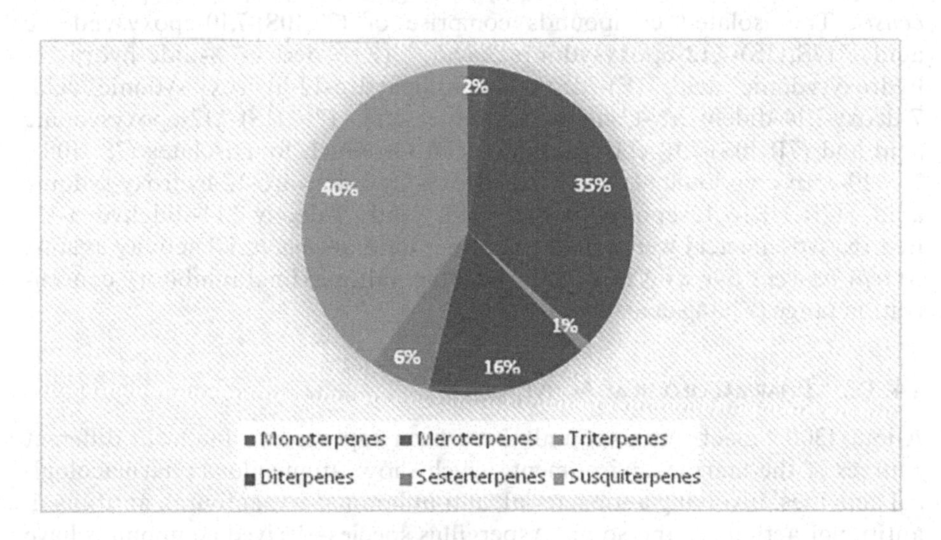

FIGURE 14.2 The proportion of various marine fungi–derived terpenes between 2015 and 2019 (Jiang et al. 2020).

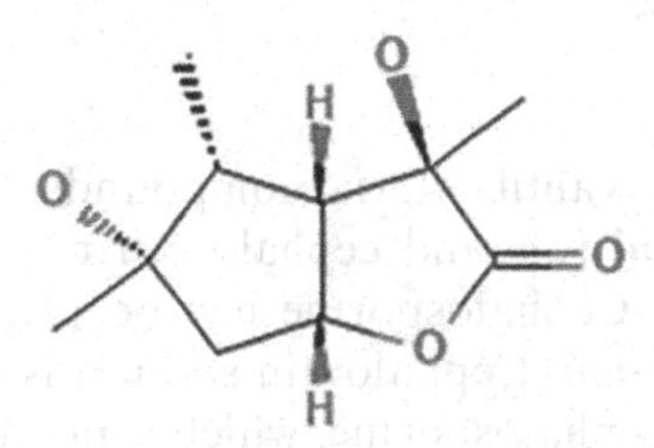

FIGURE 14.3 Chemical structure of pestalotiolactones C

(Source: PubChem SID–406871342)

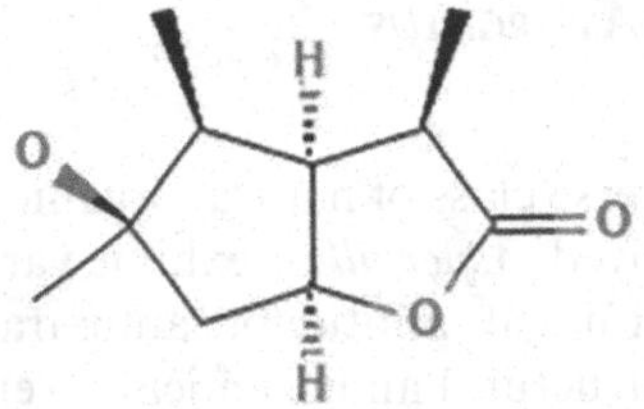

FIGURE 14.4 Chemical structure of pestalotiolactones D (Source: PubChem SID–406871343).

Seven novel phenolic bisabolane sesquiterpenoids have been isolated from *Aspergillus* species, which is endophytic fungus habitat in *Xylocarpus moluccensis.* The isolated compounds comprise of (7S,10S)-7,10-epoxysyndonic acid, (7S,11S)-7,12-epoxysydonic acid, (Z)-7-deoxy-7,8-dide-hydra-12-hydroxysydonic acid, (E)-7deoxy-7, 8-didehydro-12-hydroxy-sydonic acid, 7-deoxy-7,14-didehydro-12-hydroxysydonic acid, (7R,11S)-7,12-epoxysydonic acid and (7R,10S)-7,10-epoxysydonic acid. Of which four isolates (7S, 10S)-7, 10-epoxysyndonic acid, (E)-7deoxy-7,8-didehydro-12-hydroxy-sydonic acid, (7R,11S)-7,12-epoxysydonic acid, and 7-deoxy-7,14-didehydro-12-hydroxysydonic acid were found to prompt mild antibacterial activity against *Staphylococcus aureus* ATCC25923 with the half-maximal inhibitory concentration range (Zhang et al. 2008).

14.3.3 Pharmacological Activities

About 1300 bioactive compounds have been reported so far from different sources of the marine environment, which shows tremendous pharmacological activities, involving antibacterial, anti-inflammatory, antiviral, antifungal, antitumor activities, and so on. Aspergillus species–derived compounds have cytotoxic and antimicrobial potency (Hwang et al. 2019), of which 13 new diterpenoids from *Aspergillus wentii*SD-310, collected in the South China Sea at a depth of 2038 m, were isolated, demonstrating its nature toward cytotoxicity and antibacterial activity.

14.3.3.1 Antibacterial

In the development of new antibacterial compounds, the marine derived–fungal source plays a crucial role and cephalosporines are the classic example of antimicrobial agents. Cephalosporine has been derived from the fungus *Cephalosporium acremonium.* Cephalothin sodium is a semisynthetic-derived antibiotic compound of cephalosporine, which is potent against gram-positive and gram-negative bacteria (Hwang et al. 2019). Figures 14.5 and 14.6 show the chemical structure of cephalosporine and cephalothin sodium.

From *Aspergillus sydowi*, 17 pure compounds were extracted and purified by column chromatography on Si gel and semi-preparative high-performance

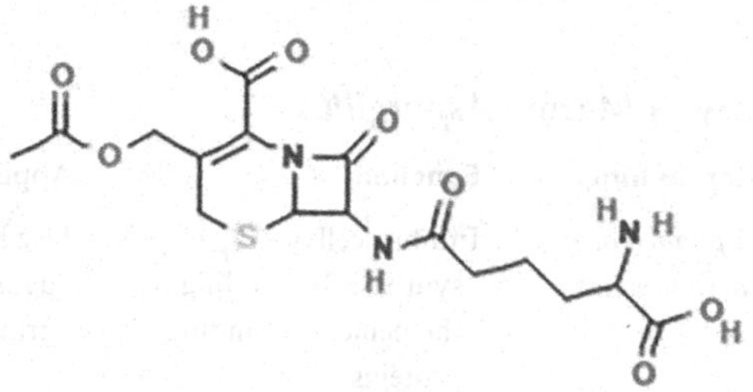

FIGURE 14.5 Chemical structure of cephalosporine

(Source: PubChem CID–2669).

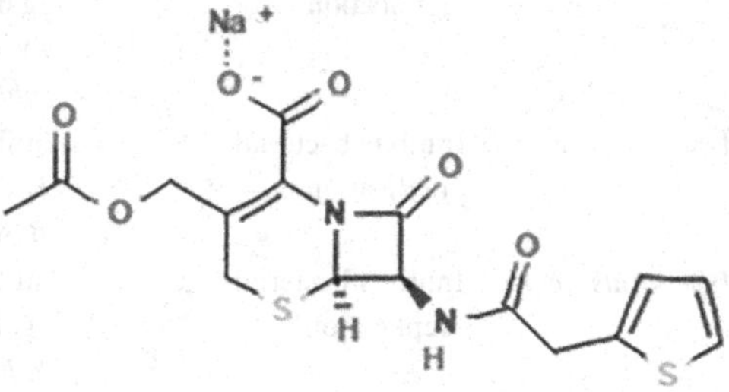

FIGURE 14.6 Chemical structure of cephalothin sodium

(Source: PubChem CID–23675321).

liquid chromatography. Of those, two new compounds 29-nordammarane triterpenoid and oxaspiro(4.4)lactam exhibits the antibacterial activity against *Escherichia coli*, *Lysoleikticus*, and *Bacillus subtilis* (Jiang et al. 2020). Also, compounds such as aspewentins I and J were isolated from *Aspergillus wentii*, and their inhibitory activities were established against marine bacteria such as *E. tarda*, *V. harveyi*, and *V. parahaemolyticus* (Sun et al. 2013). Likewise, the compound yicathin C, possessing the antibacterial activity against *E. coli*, *Colletotrichum lagenarium*, and *Staphylococcus aureus*, were isolated (Kim et al. 2017). Table 14.2 explains the *Aspergillus*–derived metabolites and their antibacterial activity. Figure 14.7 shows the chemical structure of yicathin C.

14.3.3.2 Anticancer

The biologically active molecules are abundant in marine-derived fungi that possess antitumor activity. The fungal metabolite 1386A, collected from the South China Sea, showed a half-maximal inhibitory concentration (IC50) of 4.79 µmol/L when 1386A incubated with DU-145 cells expressed the highest inhibition. The marine fungus *Aspergillus fumigates* is used to isolate secondary metabolite demethoxyfumitremorgin C, which has been reported to inhibit PC-3 cells (Liu et al. 2014). Apochalasin V, derived from *Aspergillus* species habitat in the gut of *Ligia oceanic*, exhibits moderate cytotoxic activity against prostate cancer (PC-3) cells with a half-maximal inhibitory concentration of

TABLE 14.2
Antibacterial Activity of Marine *Aspergillus*

Metabolite	Marine fungi	Function	Application
Cephalosporine	*Cephalosporium acremonium*	Inhibits cell wall synthesis by binding to the penicillin-binding proteins	Used as an antibiotic against gram-positive and gram-negative bacteria
Cephalothin sodium	*Cephalosporium acremonium*	Inactivate penicillin-binding protein, which leads to bacterial cell lysis	Antibacterial
29-nordammarane triterpenoid	*Aspergillus sydowi*	Inhibits bacterial replication	Antibactrial activity against *Escherichia coli*, *Lysoleikticus*, and *Bacillus subtilis*.
Oxaspiro(4.4)lactam	*Aspergillus sydowi*	Inhibits bacterial replication	Antibactrial activity against *E. coli*, *Lysoleikticus* and *B. subtilis*.
Aspewentins I and J	*Aspergillus wentii*	Inhibits bacterial replication	Antibacterial activity against *E. tarda*, *V. harveyi*, and *V. parahaemolyticus*
Yicathin C	*Aspergillus wentii*	Inhibits bacterial replication	Antibacterial activity against *E. coli*, *Colletotrichum lagenarium*, and *Staphylococcus aureus*

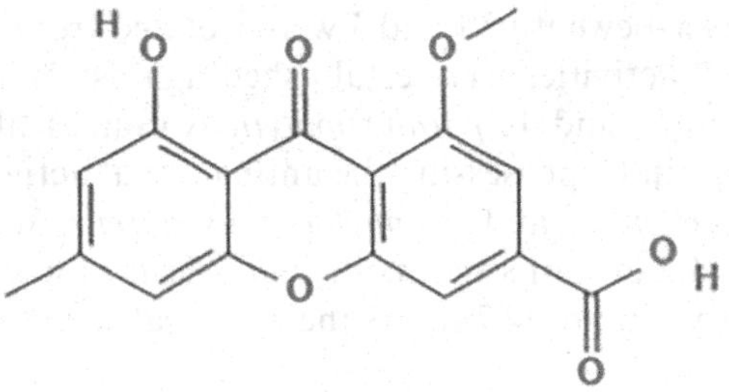

FIGURE 14.7 Structure of yicathin C (PubChem CID–102191487)

30.4 μM in MTT assay (Li et al. 2016). Also, the asperolides E isolated from *Aspergillus wentii*SD-330 showed cytotoxic activities against MCF-7, HeLa, and NCI-H446 with IC50 values of 10.0, 11.0, and 16.0μM, respectively (Du et al. 2007).

From marine *Aspergillus glaucus*–derived compound, aspergiolide A has cytotoxic activity against A-549, HL-60, P388, and BEL-7402 cell lines (Suja

TABLE 14.3
Antitumor Activity of Marine *Aspergillus*

Metabolite	Marine fungi	Function	Application
Demethoxyfumitremorgin C	*Aspergillusfumigates*	Inhibit cell growth	Anticancer activity against A-549 and HL-60 cells
Apochalasin V	*Aspergillus* species	Acts as a cytotoxicity compound	Anticancer
(+)-3,3,7,7,8,8-hexahydroxy-5,5-dimethylbianthraquinone	*Aspergillus candidus* and *Aspergillus terrus*	Inhibit topo I isomerase	Anticancer
Malformin A	*Aspergillus niger*	Cytotoxic activity	Anticancer activity against breast cancer cell line
Diketopiperazine	*Aspergillus sydowi*	A alkaloid expresses cytotoxic activity	Anticancer activity against A-549 and HL-60 cells
Aspergiolide A	*Aspergillus glaucus*	A novel anthraquinone derivative with naphtho[1,2,3-de] chromene-2,7-dione skeleton	Cytotoxic activity against A-549, HL-60, P388, and BEL-7402 cell lines.

et al. 2014). Correspondingly, the compound(+)-3, 3, 7, 7, 8, 8-hexahydroxy-5, 5-dimethylbianthraquinone, which is a topo I isomerase inhibitor extracted and isolated from *Aspergillus candidus* and *Aspergillus terrus*, exhibits in vitro anticancer and cytotoxic activity (Dobretsov et al. 2016). The pure compound of malformin A, isolated from *Aspergillus niger*, exhibits strong anticancer activity against breast cancer cell lines in dimethyl sulfoxide solvent (Rochfort et al. 2005). Likewise, the three marine-derived diketopiperazine alkaloids derived from *Aspergillus sydowi* exhibit the nature of cytotoxic activity against A-549 and HL-60 cells (Jiang et al. 2020).

Table 14.3 explains the *Aspergillus*–derived metabolites that exhibit antitumor activity.

Figures 14.8 through 14.12 show the chemical structure of demethoxyfumitremorgin C,(+)-3, 3, 7, 7, 8, 8-hexahydroxy-5, 5-dimethylbianthraquinone, malformin A, diketopiperazine, and aspergiolide A.

14.3.3.3 Antiviral

A marine microorganism, especially *Aspergillus* species compound, yields structurally potent diverse metabolites with antiviral activity in human viral infections. Aspochalasin L derived from *Aspergillus flavipes* exhibits antiviral

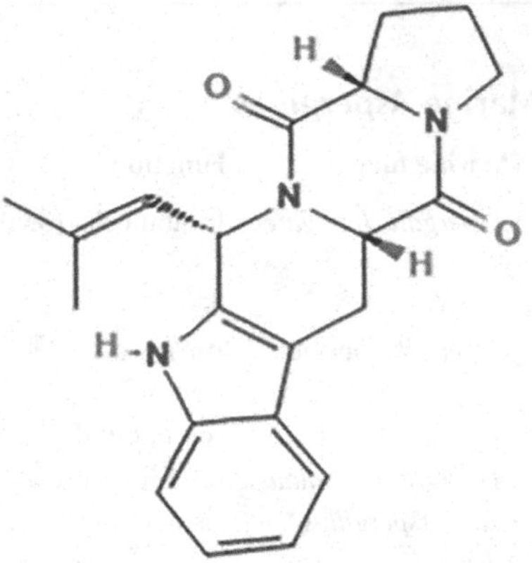

FIGURE 14.8 Chemical structure of demthoxyfumitremorgin C

(Source: PubChem CID–10337896)

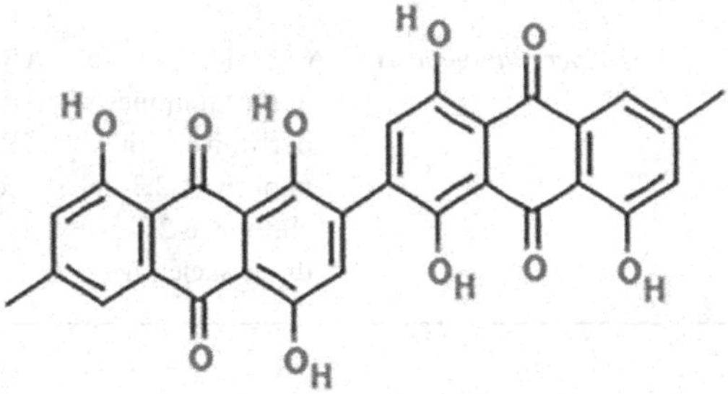

FIGURE 14.9 Chemical structure of (+)-3, 3, 7, 7, 8, 8-hexahydroxy-5, 5-dimethylbia nthraquinone

(Source: PubChem CID–25112177).

activity against HIV integrase with IC50 of 71.7 μM (Ma et al. 2017). As well as aspergillipeptides D and E derived from *Aspergillus* Species SCSIO 41501 explicit antiviral activity against herpes simplex virus type 1 (HSV-1) with a half inhibition concentration (IC50) value of 9.5 and 19.8 mM under noncytotoxic concentration in Vero cell line, confirmed by Marfey's method (Capon et al. 2003). Table 14.4 denotes *Aspergillus*–derived metabolites exhibit antiviral activity. Figures 14.13 through 14.15 show the chemical structure of aspochalasin L and aspergillipeptides D and E.

14.3.3.4 Antiparasitic

Marine-derived *Aspergillus* species play a predominant role in antiparasitic activity among human models. Da Silva et al. 2017 confirmed and examined the antiparasitic activity of three compounds (terrein [648.7μM], butyrolactone

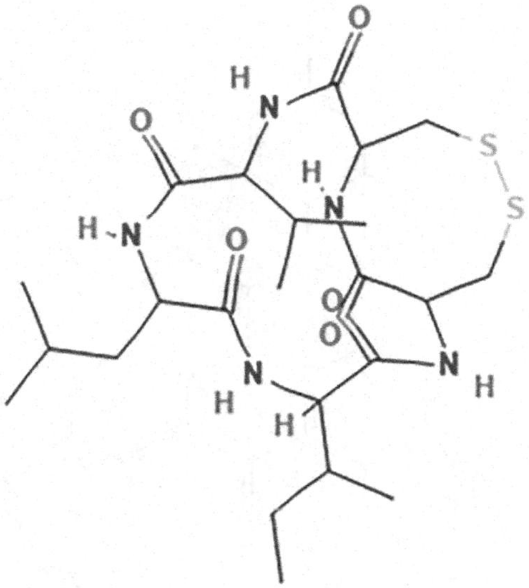

FIGURE 14.10 Structure of malformin A

(Source: PubChem CID–4005).

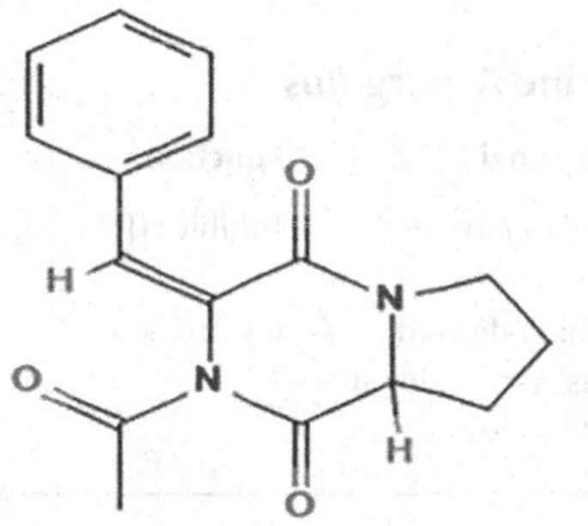

FIGURE 14.11 Chemical structure of diketopiperazine

(Source: PubChem CID–5363439).

I [235.6 μM], and butyrolactone V) derived from *Aspergillus terrus*, which reduced the motor activity of adult worm *S. mansori*. Furthermore, the results also stated that the terrein (129.3 μM) and butyrolactone I (471.2 μM) compounds have the efficacy to destroy 100% of all parasites in 48 hours by using the assay principle. In addition, a metabolite, marcfortine A, isolated from A*spergillus carneus*, was also expressed and demonstrated the inhibitory effect of the motility of parasite with respect to its significant antiparasitic activity against *Haemonchus contortus* (Capon et al. 2003). Table 14.5 explains the antiparasitic activity of marine *Aspergillus* species. Figures 14.16 to14.18 show the chemical structures of terrein, butyrolactone I, and marcfortine A.

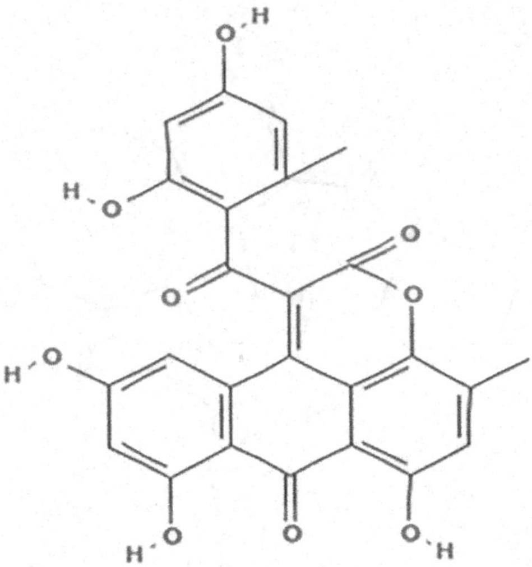

FIGURE 14.12 Chemical structure of aspergiolide A

(Source: PubChem CID–16105171).

TABLE 14.4
Antiviral Activity of Marine *Aspergillus*

Metabolite	Marine Fungi	Function	Application
Aspochalasin L	*Aspergillus flavipes*	Inhibit HIV integrase	Antiviral activity against HIV
Aspergillipeptides D and E	Gorgonian-derived fungus Aspergillus sp. SCSIO 41501	Cytotoxicity	Antiviral activity against herpes simplex virus type 1

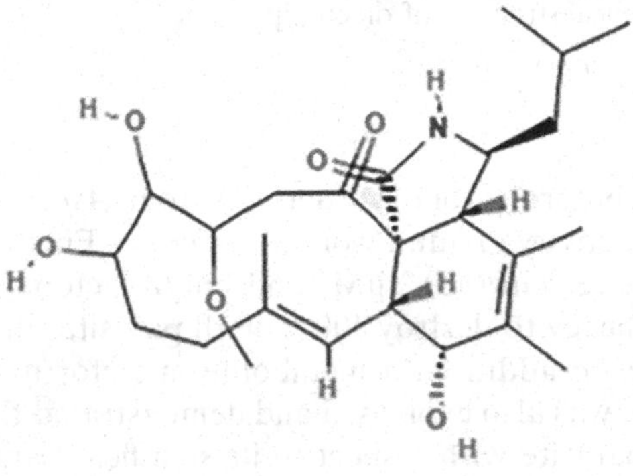

FIGURE 14.13 Chemical structure of aspochalasin L

(Source: PubChem CID–11396751).

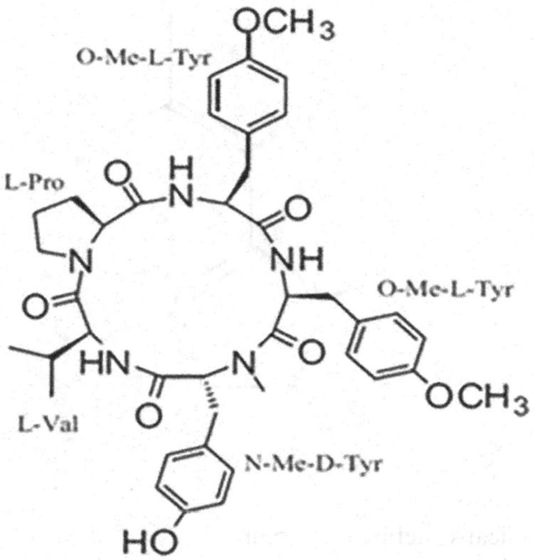

FIGURE 14.14 Chemical structure of aspergillipeptides D (Ma et al. 2017).

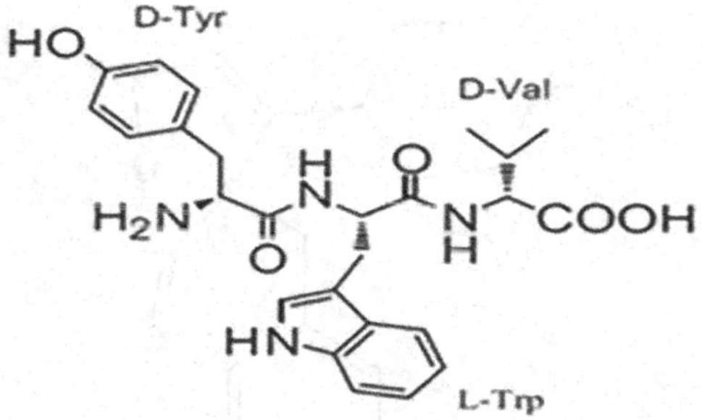

FIGURE 14.15 Chemical structure of aspergillipeptides E (Ma et al. 2017).

TABLE 14.5
Antiparasitic Activity of Marine *Aspergillus*

Metabolite	Marine Fungi	Function	Application
Terrein and butyrolactone I	*Aspergillus terrus*	Reduce motor activity or induce mortality	Antiparasitic activity against adult worm *S. mansori*
Marcfortine A	*Aspergillus carneus*	Inhibits the motility of parasite	Antiparasitic activity against *Haemonchus contortus*

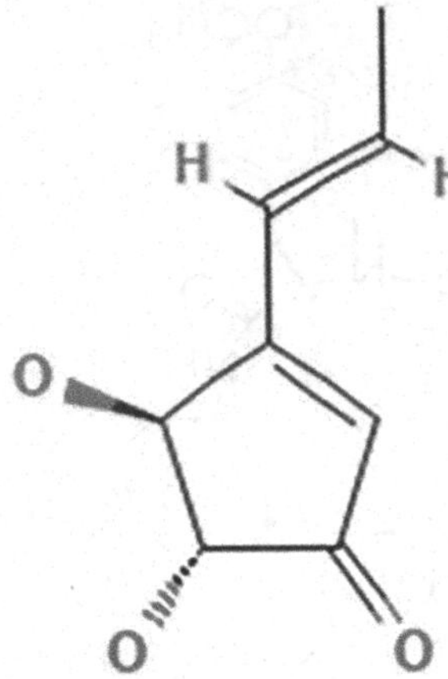

FIGURE 14.16 Chemical structure of terrein

(Source: PubChem CID–443834839).

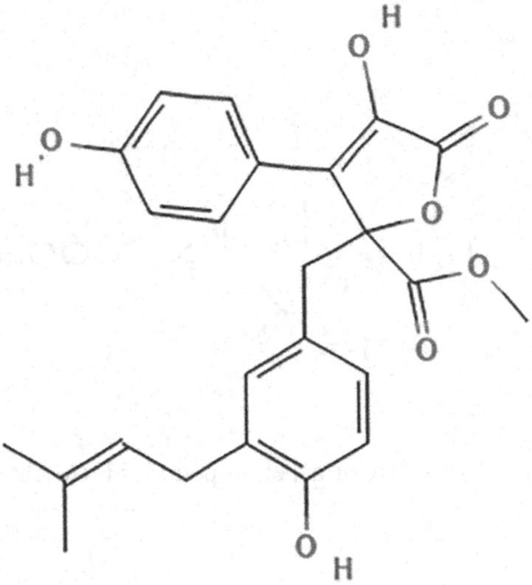

FIGURE 14.17 Chemical structure of butyrolactone I

(Source: PubChem CID–5235506).

14.4 CONCLUSION

Currently, many virulent viruses and bacteria are mutating rapidly. Pathogens are resistant to the drugs on the market, which results in serious health problems or even death. So, the need for new biologically active compounds is higher. Marine-derived species are abundant in nature, resulting in potential bioactive compounds and metabolites could that pave the way for the development of new drugs. The availability of marine fungi makes

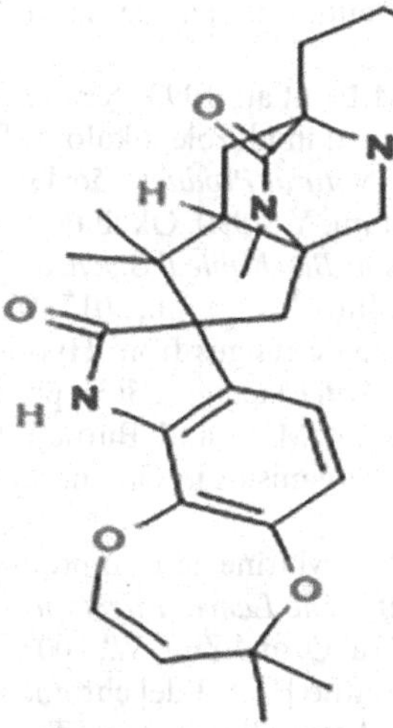

FIGURE 14.18 Chemical structure of marcfortine A

(Source: PubChem CID-433181).

them one of the better options for finding novel drugs. One of the abundant fungal species in the sea is *Aspergillus*. Marine *Aspergillus* is rich in bioactive compounds, and thus, thousands of bioactive compounds have been isolated and studied for clinical significance. A few drugs are currently in various stages of clinical trial, benefiting millions of people. Therefore, with abundant future technology and techniques, novel drugs can be developed using the resources available in the ocean, and the role of marine *Aspergillus* is indispensable.

REFERENCES

Ali, M.S., Saleem, M., Yamdagni, R. and Ali, M.A., 2002. Steroid and antibacterial steroidal glycosides from marine green alga Codium iyengarii Borgesen. *Natural Product Letters*, *16*(6), pp. 407–413.

Asolkar, R.N., Maskey, R.P., Helmke, E. and Laatsch, H., 2002. Chalcomycin B, a new macrolide antibiotic from the marine isolate Streptomyces sp. B7064. *The Journal of Antibiotics*, *55*(10), pp. 893–898.

Bhimba, V.C., Beulah, M.C. and Vinod, V., 2013. Efficacy of bioactive compounds extracted from marine sponge Haliclona exigua. *Asian J Pharm Clin Res*, *6*, pp. 347–354.

Blackman, A.J. and Matthews, D.J., 1985. Amathamide alkaloids from the marine bryozoan Amathia wilsoni Kirkpatrick. *Heterocycles (Sendai)*, *23*(11), pp. 2829–2833.

Blunt, J.W., Copp, B.R., Munro, M.H., Northcote, P.T. and Prinsep, M.R., 2004. Marine natural products. *Natural Product Reports*, *21*(1), pp. 1–49.

Capon, R.J., Skene, C., Stewart, M., et al., 2003. Aspergillicins A—E: Five novel depsipeptides from the marine-derived fungus Aspergillus carneus. *Organic & Biomolecular Chemistry*, *1*(11), pp. 1856–1862.

Cardellina, J.H., Marner, F.J. and Moore, R.E., 1979. Seaweed dermatitis: Structure of lyngbyatoxin A. *Science*, *204*(4389), pp. 193–195.

Carte, B.K., 1996. Biomedical potential of marine natural products. *Bioscience*, *46*(4), pp. 271–286.

Chan, G.W., Mong, S., Hemling, M.E., et al., 1993. New leukotriene B4 receptor antagonist: Leucettamine A and related imidazole alkaloids from the marine sponge Leucetta microraphis. *Journal of Natural Products*, *56*(1), pp. 116–121.

Cohen, P., Holmes, C.F. and Tsukitani, Y., 1990. Okadaic acid: A new probe for the study of cellular regulation. *Trends in Biochemical Sciences*, *15*(3), pp. 98–102.

Da Silva, I.P., Brissow, E., Kellner Filho, L.C., et al., 2017. Bioactive compounds of Aspergillus terreus—F7, an endophytic fungus from Hyptis suaveolens (L.) Poit. *World Journal of Microbiology and Biotechnology*, *33*(3), pp. 1–10.

Dobretsov, S., Tamimi, Y., Al-Kindi, M.A. and Burney, I., 2016. Screening for anticancer compounds in marine organisms in Oman. *Sultan Qaboos University Medical Journal*, *16*(2), p. e168.

Donia, M. and Hamann, M.T., 2003. Marine natural products and their potential applications as anti-infective agents. *The Lancet Infectious Diseases*, *3*(6), pp. 338–348.

Du, L., Zhu, T., Fang, Y., Liu, H., Gu, Q. and Zhu, W., 2007. Aspergiolide A, a novel anthraquinone derivative with naphtho [1, 2, 3-de] chromene-2, 7-dione skeleton isolated from a marine-derived fungus Aspergillus glaucus. *Tetrahedron*, *63*(5), pp. 1085–1088.

Faulkner, D.J., 2001. Marine natural products. *Natural Product Reports*, *18*(1), pp. 1R–49R.

Fujiki, H. and Sugimura, T., 1987. New classes of tumor promoters: Teleocidin, aplysiatoxin, and palytoxin. *Advances in Cancer Research*, *49*, pp. 223–264.

Fusetani, N., Sugawara, T. and Matsunaga, S., 1992. Bioactive marine metabolites. 41. Theopederins AE, potent antitumor metabolites from a marine sponge, Theonella sp. *The Journal of Organic Chemistry*, *57*(14), pp. 3828–3832.

Harvey, A., 2000. Strategies for discovering drugs from previously unexplored natural products. *Drug Discovery Today*, *5*(7), pp. 294–300.

Holst, P.B., Anthoni, U., Christophersen, C. and Nielsen, P.H., 1994. Marine alkaloids, 15. Two alkaloids, flustramine E and debromoflustramine B, from the marine bryozoan Flustra foliacea. *Journal of Natural Products*, *57*(7), pp. 997–1000.

Hwang, J.Y., Lee, J.H., Park, S.C., et al., 2019. New peptides from the marine-derived fungi Aspergillus allahabadii and Aspergillus ochraceopetaliformis. *Marine Drugs*, *17*(9), p. 488.

Imhoff, J.F., 2016. Natural products from marine fungi—Still an underrepresented resource. *Marine Drugs*, *14*(1), p. 19.

Jha, R.K. and Zi-Rong, X., 2004. Biomedical compounds from marine organisms. *Marine Drugs*, *2*(3), pp. 123–146.

Jiang, M., Wu, Z., Guo, H., Liu, L. and Chen, S., 2020. A review of terpenes from marine-derived fungi: 2015–2019. *Marine Drugs*, *18*(6), p. 321.

Joffe, S. and Thomas, R., 1989. Phytochemicals: A renewable global resource. *AgBiotech-News-and-Information (UK)*, *1*(5), pp. 697–700.

Kim, Y.S., Kim, S.K. and Park, S.J., 2017. Apoptotic effect of demethoxyfumitremorgin C from marine fungus Aspergillus fumigatus on PC3 human prostate cancer cells. *Chemico-biological Interactions*, *269*, pp. 18–24.

Koehn, F.E., Longley, R.E. and Reed, J.K., 1992. Microcolins A and B, new immunosuppressive peptides from the blue-green alga Lyngbya majuscula. *Journal of Natural Products*, *55*(5), pp. 613–619.

Lee, Y.M., Kim, M.J., Li, H., et al., 2013. Marine-derived Aspergillus species as a source of bioactive secondary metabolites. *Marine Biotechnology*, *15*(5), pp. 499–519.

Li, X.D., Li, X., Li, X.M., et al., 2016. Tetranorlabdane diterpenoids from the deep sea sediment-derived fungus Aspergillus wentii SD-310. *Planta Medica*, *82*(09/10), pp. 877–881.

Li, X.D., Li, X.M., Yin, X.L., Li, X. and Wang, B.G., 2019. Antimicrobial sesquiterpenoid derivatives and monoterpenoids from the deep-sea sediment-derived fungus Aspergillus versicolor SD-330. *Marine Drugs*, *17*(10), p. 563.

Lin, W., Fang, L.K., Liu, J.W., Cheng, W.Q., Yun, M. and Yang, H.L., 2008. Inhibitory effects of marine fungal metabolites from the South China Sea on prostate cancer cell line DU-145. *International Journal of Internal Medicine*, *35*, pp. 562–564.

Liu, Y., Zhao, S., Ding, W., Wang, P., Yang, X. and Xu, J., 2014. Methylthio-aspochalasins from a marine-derived fungus Aspergillus sp. *Marine Drugs*, *12*(10), pp. 5124–5131.

Liu, Z., Frank, M., Yu, X., et al., 2020. Secondary metabolites from marine-derived fungi from China. In *Progress in the Chemistry of Organic Natural Products 111* (pp. 81–153). Springer, Cham.

Lysek, N., Rachor, E. and Lindel, T., 2002. Isolation and structure elucidation of deformylflustrabromine from the North Sea bryozoan Flustra foliacea. *Zeitschrift für Naturforschung C*, *57*(11–12), pp. 1056–1061.

Ma, X., Nong, X.H., Ren, Z., et al., 2017. Antiviral peptides from marine gorgonian-derived fungus Aspergillus sp. SCSIO 41501. *Tetrahedron Letters*, *58*(12), pp. 1151–1155.

McCarthy, P.J., and Pomponi, S.A., 2004. A search for new pharmaceutical drugs from marine organisms. *Marine Biomedical Research*, *22*, pp. 1–2.

Moore, R.E., Patterson, M.L. and Carmichael, W.W., 1988. Biomedical importance of marine organisms. *Fautin, DG, Ed*, pp. 143–150.

Murakami, M., Makabe, K., Yamaguchi, K., Konosu, S. and Wälchli, M.R., 1988. Goniodomin A, a novel polyether macrolide from the dinoflagellate Goniodoma pseudogoniaulax. *Tetrahedron Letters*, *29*(10), pp. 1149–1152.

Narkowicz, C.K., Blackman, A.J., Lacey, E., Gill, J.H. and Heiland, K., 2002. Convolutindole A and convolutamine H, new nematocidal brominated alkaloids from the marine bryozoan Amathia convoluta. *Journal of Natural Products*, *65*(6), pp. 938–941.

Nicoletti, R. and Vinale, F., 2018. Bioactive compounds from marine-derived Aspergillus, Penicillium, Talaromyces and Trichoderma species. *Marine Drugs*, *16*(11), p. 408.

Park, H.S., Jun, S.C., Han, K.H., Hong, S.B. and Yu, J.H., 2017. Diversity, application, and synthetic biology of industrially important Aspergillus fungi. *Advances in Applied Microbiology*, *100*, pp. 161–202.

Pettit, G.R., Cichacz, Z.A., Gao, F., Herald, C.L. and Boyd, M.R., 1993. Isolation and structure of the remarkable human cancer cell growth inhibitors spongistatins 2 and 3 from an eastern Indian Ocean Spongia sp. *Journal of the Chemical Society, Chemical Communications*, (14), pp.1166–1168.

Plavšić, M., Terzic, S., Ahel, M. and Van Den Berg, C.M.G., 2002. Folic acid in coastal waters of the Adriatic Sea. *Marine and Freshwater Research*, *53*(8), pp. 1245–1252.

Prinsep, M.R., Blunt, J.W. and Munro, M.H., 1991. New cytotoxic β-carboline alkaloids from the marine bryozoan, Cribricellina cribraria. *Journal of Natural Products*, *54*(4), pp. 1068–1076.

Rochfort, S., Ford, J., Ovenden, S., et al., 2005. A novel aspochalasin with HIV-1 integrase inhibitory activity from Aspergillus flavipes. *The Journal of Antibiotics*, *58*(4), pp. 279–283.

Sakemi, S., Ichiba, T., Kohmoto, S., Saucy, G. and Higa, T., 1988. Isolation and structure elucidation of onnamide A, a new bioactive metabolite of a marine sponge, Theonella sp. *Journal of the American Chemical Society*, *110*(14), pp. 4851–4853.

Shoji, N., Umeyama, A., Shin, K., et al., 1992. Two unique pentacyclic steroids with cis C/D ring junction from Xestospongia bergquistia Fromont, powerful inhibitors of histamine release. *The Journal of Organic Chemistry*, *57*(11), pp. 2996–2997.

Stevenson, C.S., Capper, E.A., Roshak, A.K., et al., 2002. The identification and characterization of the marine natural product scytonemin as a novel antiproliferative pharmacophore. *Journal of Pharmacology and Experimental Therapeutics*, *303*(2), pp. 858–866.

Suja, M., Vasuki, S. and Sajitha, N., 2014. Anticancer activity of compounds isolated from marine endophytic fungus Aspergillus terreus. *World Journal of Pharmacy and Pharmaceutical Sciences*, *3*(6), pp. 661–672.

Sun, R.R., Miao, F.P., Zhang, J., Wang, G., Yin, X.L. and Ji, N.Y., 2013. Three new xanthone derivatives from an algicolous isolate of Aspergillus wentii. *Magnetic Resonance in Chemistry*, *51*(1), pp. 65–68.

Wang, Y.N., Mou, Y.H., Dong, Y., et al., 2018. Diphenyl ethers from a marine-derived Aspergillus sydowii. *Marine Drugs*, *16*(11), p. 451.

Xu, K., Yuan, X.L. and Li, C., 2020. Recent discovery of heterocyclic alkaloids from marine-derived Aspergillus species. *Marine Drugs*, *18*(1), p. 54.

Yamada, T., Iwamoto, C., Yamagaki, N., et al., 2002. Leptosins M—N1, cytotoxic metabolites from a Leptosphaeria species separated from a marine alga. Structure determination and biological activities. *Tetrahedron*, *58*(3), pp. 479–487.

Zhang, H.P., Shigemori, H., Ishibashi, M., et al., 1994. Convolutamides A~ F, novel γ-lactam alkaloids from the marine bryozoan Amathia convoluta. *Tetrahedron*, *50*(34), pp. 10201–10206.

Zhang, M., Wang, W.L., Fang, Y.C., Zhu, T.J., Gu, Q.Q. and Zhu, W.M., 2008. Cytotoxic alkaloids and antibiotic nordammarane triterpenoids from the marine-derived fungus Aspergillus sydowi. *Journal of Natural Products*, *71*(6), pp. 985–989.

15 Biocatalytic Reduction of Organic Compounds by Marine-Derived Fungi

Gabriel S. Baia, David E. Q. Jimenez, and André Luiz Meleiro Porto

CONTENTS

15.1 INTRODUCTION

Biocatalysis is an important field of research applied mainly to the synthesis of asymmetric molecules and is known as the connection between different areas of natural science and engineering science with the objective of making use of enzymes, organisms, cells or molecular equivalents to obtain products of academic and industrial interest [1].

Biocatalysis is linked to biotechnology which can be defined as an interdisciplinary area strongly linked to scientific and technological research, whose main objective is to develop processes and products using biological agents. For the application of biotechnological processes, knowledge is required in several areas, such as organic chemistry, physics, bioinformatics, mechanical engineering, catalysis, mathematics, materials science and biology (microbiology, molecular biology, biochemistry) [2].

Biotransformation involves the use of enzymes contained in living cells (fermentation processes), while biocatalysis usually employ enzymes that are in an isolated, immobilized form; that are not part of living cells; and that catalyze reactions in xenobiotic substrates. Biocatalysis involves processes in which biological catalysts are used to transform a xenobiotic substrate into a product via metabolic enzymatic reactions [3, 4].

DOI: 10.1201/9781003303909-15

Asymmetric methodologies are necessary for pharmaceutical medicaments that have one or more chiral centers [5]. Biocatalysis is a tool for the synthesis of enantiomer-enriched compounds employing enzymes as chiral catalysts.

One example of such an enzyme is *ene-reductases* (ERs; EC 1.6.99.1), which catalyze the asymmetric reduction of activated alkenes. ERs are important for the production of enantio-pure chemical compounds in a single step [6]. Four classes of ERs exist that have been used in biocatalytic protocols and have been observed to differ in reaction mechanism, substrate, stereoselectivity and enantioselectivity.

ERs from *old yellow enzyme* (OYE) are flavin mononucleotide (FMN)–containing ERs that catalyze the asymmetric reduction of activated C=C bonds. Brenna et al. (2012) reported the use of ER OYE3 expressed from *Saccharomyces cerevisiae* in the reduction of methyl (*Z*)-2-bromobut-2-enoate and methyl (*E*)-2-bromobut-2-enoate to methyl (*S*)-2-bromobutanoate. The reduction reaction conditions applied were 30°C, 130 rpm in buffer solution (Na_2HPO_4/KH_2PO_4) for 24 h, to produce the reduced product methyl (*S*)-2-bromobutanoate in >98% yield and 93 and 98% enantiomeric excess (*ee*) of the *S*-enantiomer showed in Figure 15.1 [7].

Enoate reductases (EnoRs) can catalyze the NADH-dependent reduction of C=C bonds of nonactive enolates such as cyclic ketones, methyl ketones and α,β-unsaturated aldehydes. Dobrijevic et al. [8] described the use of the EnoRs pQR1907, pQR1908, pQR1909, pQR1440, pQR1442, pQR1443 in the reduction of 2-methylcyclohex-2-en-1-one to 2-methylcyclohexan-1-one. The reduction reaction conditions were 30°C, 600 rpm in Tris-HCl for 24 h. The EnoRs produced the reduced product (*R*)-2-methylcyclohexan-1-one in 39–99% yields and 83–99% *ee* of the *R*-enantiomer observed in Figure 15.2 [8].

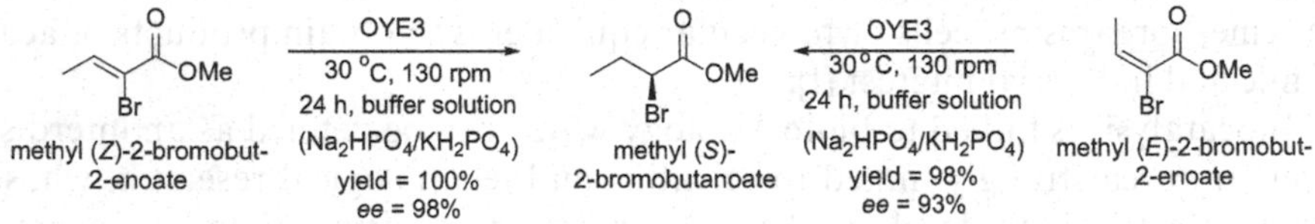

methyl (*Z*)-2-bromobut-2-enoate → OYE3, 30 °C, 130 rpm, 24 h, buffer solution (Na_2HPO_4/KH_2PO_4), yield = 100%, *ee* = 98% → methyl (*S*)-2-bromobutanoate ← OYE3, 30 °C, 130 rpm, 24 h, buffer solution (Na_2HPO_4/KH_2PO_4), yield = 98%, *ee* = 93% ← methyl (*E*)-2-bromobut-2-enoate

FIGURE 15.1 Bioreduction of methyl (*Z*)-2-bromobut-2-enoate and methyl (*E*)-2-bromobut-2-enoate using ER-OYE3 (based on Brenna et al. [7]).

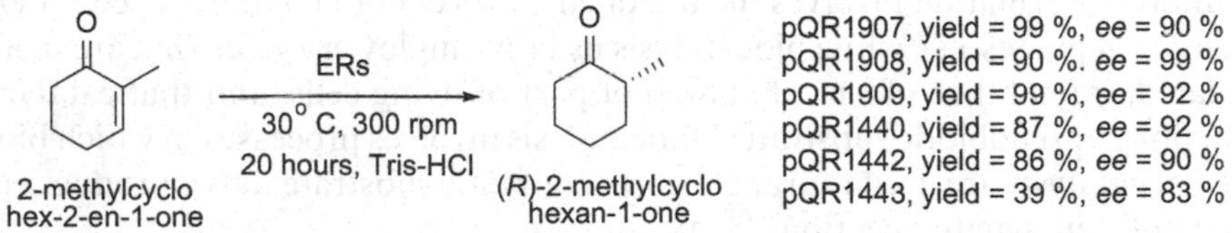

2-methylcyclohex-2-en-1-one → ERs, 30 °C, 300 rpm, 20 hours, Tris-HCl → (*R*)-2-methylcyclohexan-1-one

pQR1907, yield = 99 %, *ee* = 90 %
pQR1908, yield = 90 %. *ee* = 99 %
pQR1909, yield = 99 %, *ee* = 92 %
pQR1440, yield = 87 %, *ee* = 92 %
pQR1442, yield = 86 %, *ee* = 90 %
pQR1443, yield = 39 %, *ee* = 83 %

FIGURE 15.2 Bioreduction of 2-methylcyclohex-2-en-1-one using EnoRs (based on Dobrijevic et al. [8]).

Medium-chain dehydrogenase/reductase (MDR) oxidoreductases form a large enzyme superfamily of 1000 members [9]. These enzymes represent many different activities, for example, *alcohol dehydrogenases (ADHs)* [9]. The ADHs are enzymes that can transform ketones/aldehydes to alcohols and vice versa at the expense of a nicotinamide cofactor that acts as hydride donor and acceptor [9]. Gonzalo et al. [10] used ADHs enzymes to the reduction of ketones to alcohol derivatives as shown in Figure 15.3. The reduction reaction conditions were 30°C and 300 rpm in micro-aqueous hexane media for 40–42 h. The ADHs resulted in 67–96% conversion and >99% *ee* for all products [10].

Short-chain dehydrogenase/reductase (SDRs) are a large family of NAD(P)H-dependent oxidoreductases, sharing sequence motifs and exhibiting analogous mechanisms. SDR enzymes play critical roles in alcohols, sugars, steroids, lipids, amino acids and aromatic compounds [11]. Peng et al. [12] developed a stereoselective synthesis of (*R*)-phenylephrine using an SDR from *Serratia marcescens* BCRC 10948. The reduction reaction used 1-(3-hydroxyphenyl)-2-(methylamino)ethanone at 30°C and 120 rpm in a buffer solution (Na_2HPO_4/KH_2PO_4) for 24 h to give the alcohol in 51% yield and 99% *ee* of the *R*-configuration alcohol as shown in Figure 15.4 [12].

The ERs most used in research is the OYE family of flavin oxidoreductases (EC 1.6.99.1). These enzymes can catalyze the asymmetric reduction of α,β-unsaturated alkenes in a ping-pong bi-bi mechanism. The hydride transfer from the nicotinamide moiety of NAD(P)H to the enzyme-bound flavin occurs in a reductive half-reaction. When oxidized to $NAD(P)^+$, the transfer of the hydride ion from the reduced flavin to the alkene takes place in an oxidative

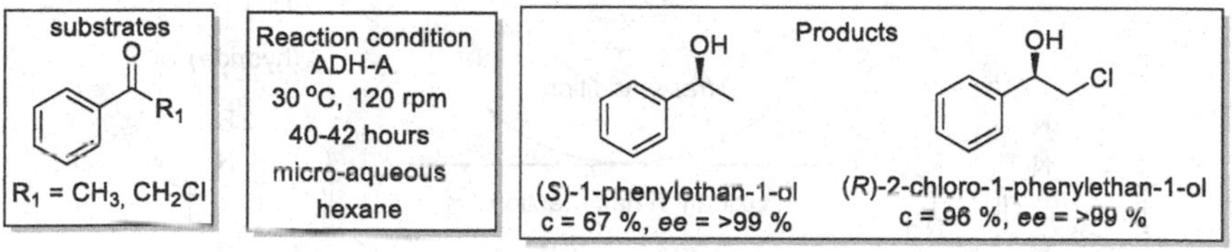

FIGURE 15.3 Bioreduction of ketones derivatives using ADH-A enzyme (based on Gonzalo et al. [10]).

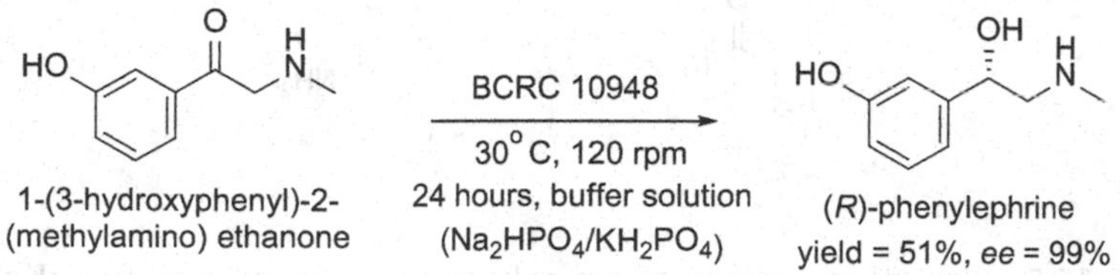

FIGURE 15.4 Bioreduction of 1-(3-hydroxyphenyl)-2-(methylamino)ethanone using SDRs from *Serratia marcescens* BCRC 10948 (based on Peng et al. [12]).

half-reaction [13]. The alkene is bound to the active site of the enzyme by hydrogen-bonding via His/His or His/Asn amino acid interaction, which enhances the C=C bond polarization and supports the hydride transfer from $FMNH_2$ to the C-β of the alkene [13]. Proton transfer occurs simultaneously from the Tyr amino acid residue of the opposite side and completes the reduction reaction in Figure 15.5.

The aldo-keto reductase (AKR) enzymes can catalyze asymmetric reductions of carbonyl compounds. The difference between ERs and AKRs is that the latter needs coenzymes, for example, nicotinamide adenine dinucleotide (NADH), nicotinamide adenine dinucleotide phosphate (NADPH) and flavin [14]. The carbonyl group reduction proceeds in two steps: (1) the hydroxyl group of the substrate is reduced, and the coenzyme is oxidized to $NAD(P)^+$, and (2) the $NAD(P)^+$ and alcohol product system are dissociated from the enzyme. Sheng et al. (2019) described the use of AKRs by perakine reductase in the reduction of (*E*)-4-phenylbut-3-en-2-one to (*S,E*)-4-phenylbut-3-en-2-ol. The reduction reaction conditions were 30°C and 150 rpm in Kpi buffer for 10 h. AKR affords the alcohol product in 60% yield and >99% *ee* of the *S*-enantiomer as shown in Figure 15.6 [15].

To our knowledge, there are no reports in the literature about the application of isolated ERs in biocatalytic reactions in marine microorganisms. Next, we present a series of carbonylated xenobiotic compounds and conjugated systems that have been reduced using whole cells from marine microorganisms.

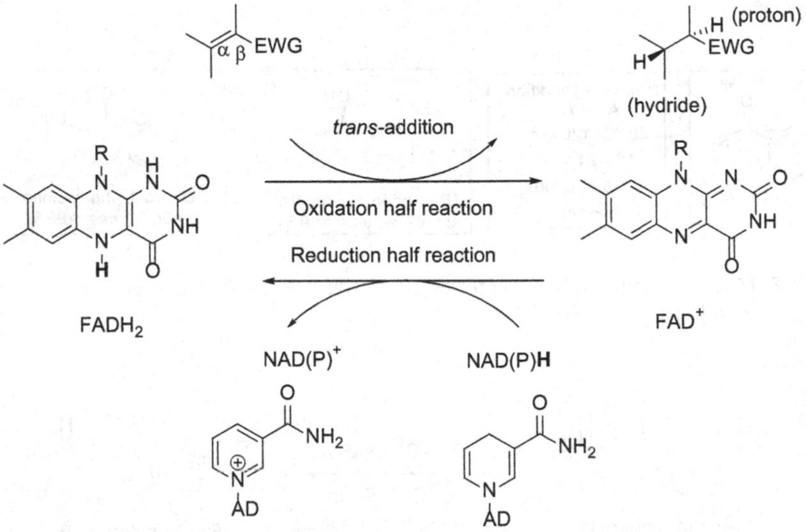

FIGURE 15.5 General mechanism of ene-reductase enzyme (adapted from Kohli and Massey [13])

Author: Kohli and Massey [13].

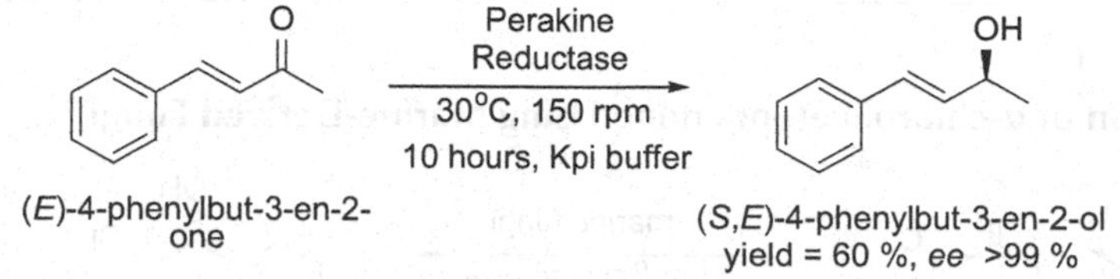

Perakine Reductase
30°C, 150 rpm
10 hours, Kpi buffer

(*E*)-4-phenylbut-3-en-2-one

(*S*,*E*)-4-phenylbut-3-en-2-ol
yield = 60 %, *ee* >99 %

FIGURE 15.6 Bioreduction of (*E*)-4-phenylbut-3-en-2-one using perakine reductase (based on Sheng et al. 2019).

15.2 APPLICATION OF BIOTRANSFORMATION REACTIONS USING MARINE-DERIVED MICROORGANISMS IN THE REDUCTION OF ORGANIC COMPOUNDS

Many marine microorganisms have been used in biotechnology and biocatalysis research [16]. Some examples are the fungi that can be associated with different organisms, such as porifers, fish, cnidarians, tunicates and algae, among others. Enzymatic reactions by fungi from the marine environment can be carried out in the laboratory. Many scientific papers have reported bioreduction reactions using marine-derived fungi. Herein are shown representative examples of the reduction of organic compounds using microorganisms: Rocha et al. [17] described the use of 2-chloro-1-phenylethan-1-one to bioreduction reaction using marine fungi *Trichoderma* sp. Gc1, *P. miczynskii* Gc5, *A. sydowii* Gc12, *Bionectria* sp. Ce5, *A. sydowii* Ce15, *P. raistrickii* Ce16 and *A. sydowii* Ce19. The bioreduction of α-chloroacetophenone to (*S*)-(-)-2-chloro-1-phenylethanol was optimized at 32°C and 150 rpm in a buffer solution (Na_2HPO_4/KH_2PO_4) and a time of 48–72 h to give the alcohol in up to 99% conversion and 17–66% *ee* from and only formed the alcohols with *S*-configuration as show in Table 15.1 [17]. To our knowledge, this is the first study of the reduction of α-chloroacetophenone with marine-derived fungi.

Rocha et al. [18] developed a protocol using the marine fungus *A. sydowii* Ce19 in the reduction of 2-bromo-1-phenylethan-1-one at 32°C and 150 rpm in a buffer solution (Na_2HPO_4/KH_2PO_4) and a time of 2–120 h. (*R*)-2-bromo-1-phenylethan-1-ol was obtained with 100% conversion and 54–55% *ee* of the *R*-enantiomer observed in Figure 15.7 [18].

Rocha et al. (2012) reported an asymmetric reduction of iodoacetophenone derivatives to iodophenylethanols using nine marine fungi: *A. sclerotiorum* CBMAI 849, *A. sydowii* Ce19, *B. felina* CBMAI 738, *M. racemosus* CBMAI 847, *P. citrinum* CBMAI 1186, *P. miczynskii* Ce16, *P. miczynskii* Gc5, *P. oxalicum* CBMAI 1185, and *Trichoderma* sp. Gc1 in buffer solution (Na_2HPO_4/KH_2PO_4), at 32°C and 120 rpm for 48 h [19]. All the fungi afforded (*S*)-*o*-iodophenylethanol and (*S*)-*m*-iodophenylethanol. However, the fungi *B. felina* CBMAI 738, *P. miczynskii* Gc5, *P. oxalicum* CBMAI 1185 and *Trichoderma* sp. Gc1 gave only (*R*)-*p*-iodophenylethanol in Figure 15.8.

Mouad et al. [20] reported the bioreduction of fluoroacetophenone derivatives using the marine-derived fungi *Botryosphaeria* sp. CBMAI 1197,

TABLE 15.1
Reduction of α-chloroacetophenone Using Marine-Derived Fungi

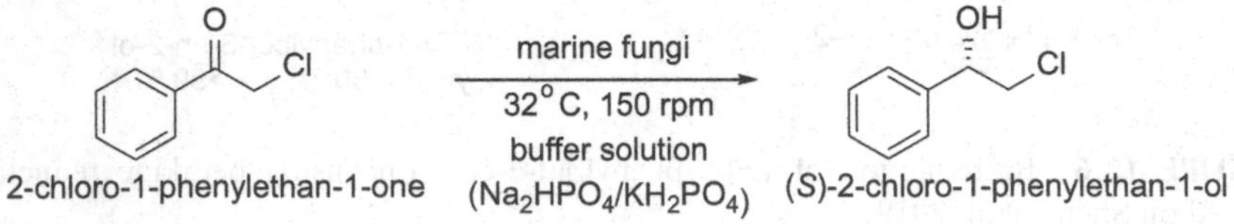

2-chloro-1-phenylethan-1-one → marine fungi, 32°C, 150 rpm, buffer solution (Na_2HPO_4/KH_2PO_4) → (*S*)-2-chloro-1-phenylethan-1-ol

Fungus	Time (h)	c (%) alcohol	ee (%, ac)
Penicillium miczynskii Ge5	48	95	50 (*S*)
Trichoderma sp. Ge1	48	30	66 (*S*)
Aspergillus sydowii Ge12	48	23	20 (*S*)
A. sydowii Ce19	72	78	20 (*S*)
A. sydowii Ce15	72	88	35 (*S*)
Bionectria sp. Ce5	72	99	22 (*S*)
P. miczynskii Ce16	72	99	17 (*S*)

c = conversion; ac = absolute configuration

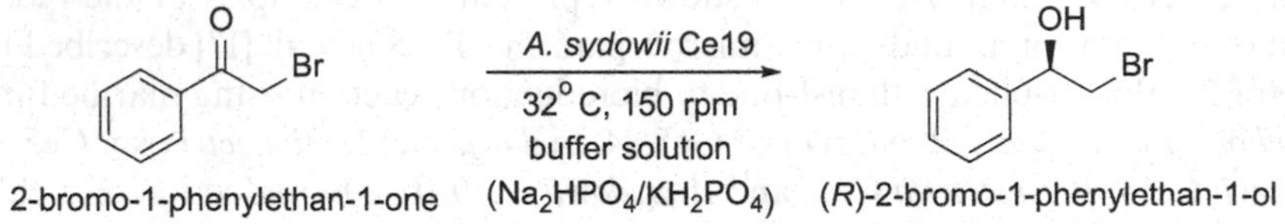

2-bromo-1-phenylethan-1-one → *A. sydowii* Ce19, 32°C, 150 rpm, buffer solution (Na_2HPO_4/KH_2PO_4) → (*R*)-2-bromo-1-phenylethan-1-ol

FIGURE 15.7 Bioreduction of α-bromoacetophenone using marine-derived fungus *A. sydowii* Ce19 (based on Rocha et al. [18]).

Eutypella sp. CBMAI 1196, *Hidropisphaera* sp. CBMAI 1194 and *Xylaria* sp. CBMAI 1195. The reduction of fluoroacetophenones furnished the asymmetric fluorophenylalcohols. The reduction reaction was optimized at 32°C and 130 rpm in a 2% malt extract culture medium for 7 days [20]. In general, all fungi reduced the 2,2,2-trifluoroacetophenone to (*S*)-2,2,2-trifluorol-phenylethanol with conversions of 5–100% and 55–99% *ee* of the *S*-enantiomer as shown in Figure 15.9.

Rocha et al. [21] carried out a study into the stereoselective bioreduction of α-azido ketones using whole cells of marine-derived fungi. The authors reported the synthesis of α-keto azides from α-bromoacetophenones and sodium azide in acetone at room temperature (rt) [21]. The α-keto azides were then subjected to bioreduction with four strains of marine-derived fungi: *A. sclerotiorum* CBMAI 849, *C. cladosporioides* CBMAI 587, *P. raistrickii* CBMAI 931 and *P. citrinum* CBMAI 1186. The fungi catalyzed the asymmetric reduction of the α-keto azides to β-azidophenylethanols. The reaction was optimized using 2-azido1-(4-methoxyphenyl)ethanone at 32°C and 130 rpm in buffer

Substrates
R = o-I, m-I, p-I

Reaction condition
marine fungi
32 °C, 120 rpm
48 h, buffer solution
(Na_2HPO_4/KH_2PO_4)

Products

All fungi
c = 6-86 %
ee = 74-99 %
(*S*)-1-(2-iodophenyl)ethan-1-ol

All fungi
c = 5-56 %
ee = 35-99 %
(*S*)-1-(3-iodophenyl)ethan-1-ol

B. felina CBMAI 738, c = 72 %, *ee* = 99 %
P. oxalicum CBMAI 1185, c = 59 %, *ee* = 85 %
Trichoderma sp. Ge1, c = 73 %, *ee* = 55 %
(*R*)-1-(4-iodophenyl)ethan-1-ol

FIGURE 15.8 Bioreduction of iodoacetophenones using marine-derived fungi (based on Rocha et al. 2012).

2,2,2-trifluoro-1-phenylethan-1-one → marine fungi, 30 °C, 130 rpm, 7 days, malt extract 2% → (*S*)-2,2,2-trifluoro-1-phenylethan-1-ol

Botryosphaeria sp. CBMAI 1197, c = 100 %, *ee* = 99 %
Eutypella sp. CBMAI 1196, c = 38 %, *ee* = 99 %
Hidropisphaera sp. CBMAI 1194, c = 6 %, *ee* = 99 %
Xylaria sp. CBMAI 1195, c = 6 %, *ee* = 99 %

FIGURE 15.9 Bioreduction of 2,2,2-trifluoroaacetophenone to (*S*)-2,2,2-trifluorol phenylethanol using marine-derived fungi (based on Mouad et al. [20]).

solution (Na_2HPO_4/KH_2PO_4) for 1 day to give an alcohol conversion of 75–89% and >99% *ee* of the *R*-enantiomer in Figure 15.10.

Jimenez et al. [22] studied the biocatalytic reduction of 2-arylidenemalononitrile compounds using the marine-derived fungus *P. citrinum* CBMAI 1186. The bioreduction conditions were 32°C and 130 rpm in buffer solution (Na_2HPO_4/KH_2PO_4) for 6 days to give the reduced product in up to 99% yield in Figure 15.11a [22]. Another interesting reaction observed was the hydrolysis of one cyanide group of 2-(furan-2-ylmethyl)malononitrile to 2-cyano-3-(furan-2-yl)propanamide in 30% yield in Figure 15.11b.

marine fungi
32 °C, 130 rpm
1 day, buffer solution
(Na_2HPO_4/KH_2PO_4)

2-azido-1-(4-methoxyphenyl)ethan-1-one

(*R*)-2-azido-1-(4-methoxy phenyl) ethan-1-ol

A. sclerotiorum CBMAI 849
c = 75 %, *ee* = 99 %
C. cladosporioides CBMAI 587
c = 81 %, *ee* = 99 %
P. raistrickii CBMAI 931
c = 89 %, *ee* = 99 %
P. citrinum CBMAI 1186
c = 76 %, *ee* = 99 %

FIGURE 15.10 Bioreduction of 2-azido-1-(4-methoxyphenyl)ethanone using marine-derived fungi (based on Rocha et al. [21]).

a)
P. citrinum
CBMAI 1186
32 °C, 130 rpm
6 days, buffer solution
(Na_2HPO_4/KH_2PO_4)
yields = 12-99 %

Ar = C_6H_5, 4-FC_6H_4, 4-ClC_6H_4, 4-BrC_6H_4, 4-$NO_2C_6H_4$
4-OHC_6H_4, 4-$OMeC_6H_4$, 4-OH,3-$OMeC_6H_3$,
pyridine, tiophene, furfural

b)
P. citrinum
CBMAI 1186
32 °C, 130 rpm
1 day, buffer solution
(Na_2HPO_4/KH_2PO_4)
yield = 70 %

P. citrinum
CBMAI 1186
32 °C, 130 rpm
5 days, buffer solution
(Na_2HPO_4/KH_2PO_4)
yield = 30 %

FIGURE 15.11 Bioreduction of Knoevenagel adducts using marine-derived fungus *P. citrinum* CBMAI 1186 (based on Jimenez et al. [22]).

De Matos et al. [23] described the chemoselective bioreduction of (*E*)-1-(2-hydroxyphenyl)-3-phenylprop-2-en-1-one using the marine-derived fungus *P. raistrickii* CBMAI 931. The conditions of the reduction reaction were 32°C and 130 rpm in buffer solution (Na_2HPO_4/KH_2PO_4) for 7 days; the reduced products were obtained in 47–74% conversion as shown in Figure 15.12 [23].

Jimenez et al. [24] reported a protocol to obtain *E*-2-cyano-3(furan-2-yl) acrylamide in the reaction between aromatic aldehydes and cyanoacetamide in yields of up to 95% [24]. The authors used marine-derived fungi to reduce *E*-2-cyano-3(furan-2-yl) acrylamide at 32°C and 130 rpm in buffer solution (Na_2HPO_4/KH_2PO_4) for 3 days to give 92–98% yields and 97 to 99% *ee* as shown in Figure 15.13. The 2-cyano-3-(furan-2-yl)propanamide undergoes a ketenimine tautomerization *in situ* to produce an achiral ketenimine in polar solvents.

Morais et al. [25] presented the synthesis of 2-chloro-1-phenylethan-1-one using acetophenone, oxone® and NH_4Cl in 30 min by conventional and microwave heating [25]. The 2-chloro-1-phenylethan-1-one was used in the reduction reaction using the marine-derived fungi *P. citrinum* CBMAI 1186, *M. racemosus* CBMAI 847, *A. sydowii* CBMAI 935, *P. raistrickii* CBMAI 931, and *P. oxalicum* CBMAI 1185 with the objective of producing (*S*)-2-chloro-1-phenylethan-1-ol in Figure 15.14. The reaction was optimized at 32°C and 130 rpm in buffer solution (Na_2HPO_4/KH_2PO_4) in 6 days to give 15–32% conversion and 12–60% *ee*.

(*E*)-1-(2-hydroxyphenyl)-3-phenylprop-2-en-1-one → 1-(2-hydroxyphenyl)-3-phenylpropan-1-one

P. raistrickii CBMAI 931, 32 °C, 130 rpm, 7 days, buffer solution (Na_2HPO_4/KH_2PO_4)

Mycelia in the enzymatic broth	c = 64 %
Mycelia in the phosphate buffer pH 6	c = 60 %
Mycelia in the phosphate buffer pH 7	c = 72 %
Mycelia in the phosphate buffer pH 8	c = 47 %

FIGURE 15.12 Chemoselective reduction of 2′-hydroxychalcone using marine-derived fungus *P. raistrickii* CBMAI 931 (based on De Matos et al. [23]).

(*E*)-2-cyano-3-(furan-2-yl)acrylamide → 2-cyano-3-(furan-2-yl) propanamide

marine fungi, 32 °C, 130 rpm, 3 days, buffer solution (Na_2HPO_4/KH_2PO_4)

P. citrinum CBMAI 1186
yield = 98 %, *ee* = 99 % (*S*)
A. sydowii CBMAI 935
yield = 92 %, *ee* = 97 % (*R*)

FIGURE 15.13 Bioreduction of *E*-2-cyano-3-(furan-2-yl)acrylamide with marine-derived fungi (based on Jimenez et al. [24]).

acetophenone → (NH_4Cl, Oxone, 70 °C, 30 min, MeOH, CH) → 2-chloro-1-phenylethan-1-one (yield = 70 %) → (marine fungi, 32 °C, 130 rpm, 6 days, buffer solution (Na_2HPO_4/KH_2PO_4)) → (*S*)-2-chloro-1-phenylethan-1-ol

P. citrinum CBMAI 1186
c = 32 %, *ee* = 15 %
P. oxalicum CBMAI 1185
c= 18 %, *ee* = 21 %
A. sydowii CBMAI 935
c = 21 %, *ee* = 50 %
P. raistrickii CBMAI 931
c = 15, *ee* = 12
M. racemosus CBMAI 847
c = 16, *ee* = 60 %

FIGURE 15.14 Synthesis and bioreduction of 2-chloro-1-phenylethan-1-one using marine-derived fungi (based on Morais et al. [25]).

Organic chemistry has been focused on the development of environmentally friendly methodologies and sustainable processes with good efficiency and low cost. In this context, the marine-derive fungi were used by Ferreira et al. [26, 27] in bioreduction reactions. Inspired by the previously mentioned scientific studies, Birolli et al. [28] reported the Knoevenagel condensation between cyanoacetamide and aromatic aldehydes in the synthesis of (*E*)-2-cyano-3-(phenyl)acrylamide compounds using triethylamine as catalyst under microwave irradiation at in 80°C for 30 min to give 93–99% yields [28]. Rapid bioreduction reactions were developed using the Knoevenagel adducts and marine-derived fungus *Cladosporium* sp. CBMAI 1237 to obtain the reduced 2-cyano-3-phenylpropanamide derivatives at 32°C and 130 rpm for 8 h in buffer solution (Na_2HPO_4/KH_2PO_4) with conversion rates of 76–100% in Figure 15.15.

Ferreira et al. [29] studied the bioreduction of (*E*)-2-metyl-2-phenylacrylaldehyde using the marine-derived fungus *P. citrinum* CBMAI 1186 immobilized on chitosan. Initially, the reaction was realized in free-mycelia conditions to give a 40% yield of (*S*)-(+)-2-methyl-3-phenylpropan-1-ol with 10% *ee* [29]. When the fungal mycelia were immobilized on (noncommercial) chitosan, a 49% yield of (*S*)-(+)-2-methyl-3-phenylpropan-1-ol with 40% *ee* and a 35% yield of (±)-2-methyl-3-phenylacrilic acid were produced. When the fungus was immobilized on commercial chitosan, the production of (*S*)-(+)-2-methyl-3-phenylpropan-1-ol in 48% yield and 10% *ee* was observed in Figure 15.16.

Alvarenga and Porto [30] investigated the use of the marine-derived fungi *A. sydowii* CBMAI 935 and *M. racemosus* CBMAI 847 in the biocatalytic reduction of 2-azido-1-phenylethanone to 2-azido-1-phenylethanol at 32°C and 130 rpm in buffer solution (Na_2HPO_4/KH_2PO_4) for 6 days [30]. The fungus *A. sydowii* CBMAI 935 gave a yield of 78% and 80% *ee* of the *S*-enantiomer. The fungus *M. racemosus* CBMAI 847 gave a yield of 72% and 81% *ee* of the *R*-enantiomer as shown in Figure 15.17.

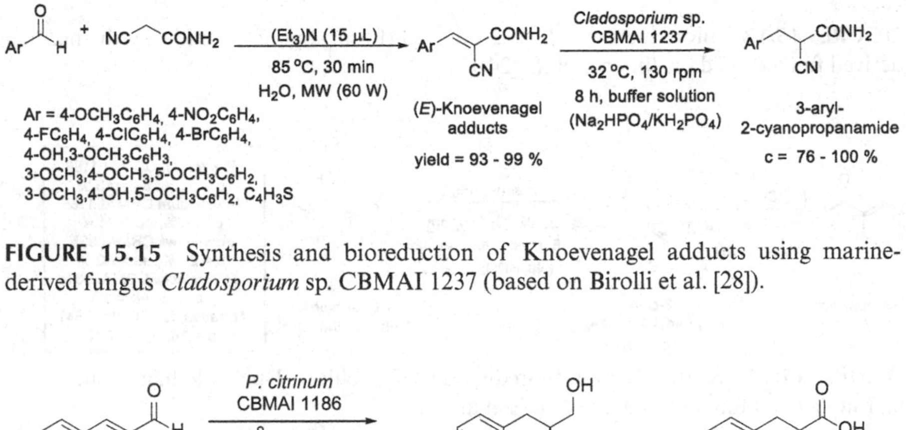

FIGURE 15.15 Synthesis and bioreduction of Knoevenagel adducts using marine-derived fungus *Cladosporium* sp. CBMAI 1237 (based on Birolli et al. [28]).

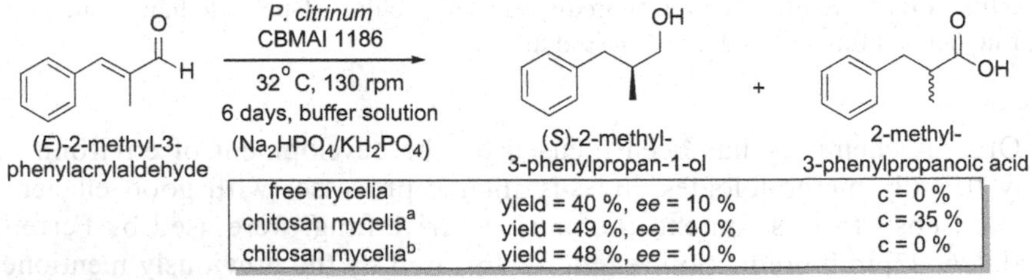

FIGURE 15.16 Bioreduction of (*E*)-3-methyl-3-phenylacryaldehyde using marine-derived fungus *P. citrinum* 1186 (based on Ferreira et al. [29]).

Notes: [a] noncommercial chitosan, [b] commercial chitosan.

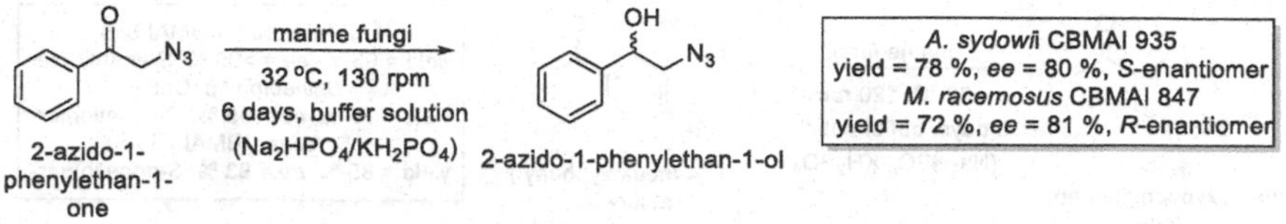

2-azido-1-phenylethan-1-one → marine fungi, 32 °C, 130 rpm, 6 days, buffer solution (Na_2HPO_4/KH_2PO_4) → 2-azido-1-phenylethan-1-ol

A. sydowii CBMAI 935
yield = 78 %, *ee* = 80 %, *S*-enantiomer
M. racemosus CBMAI 847
yield = 72 %, *ee* = 81 %, *R*-enantiomer

FIGURE 15.17 Bioreduction of 2-azido-phenylethanone using marine-derived fungi (based on Alvarenga and Porto [30]).

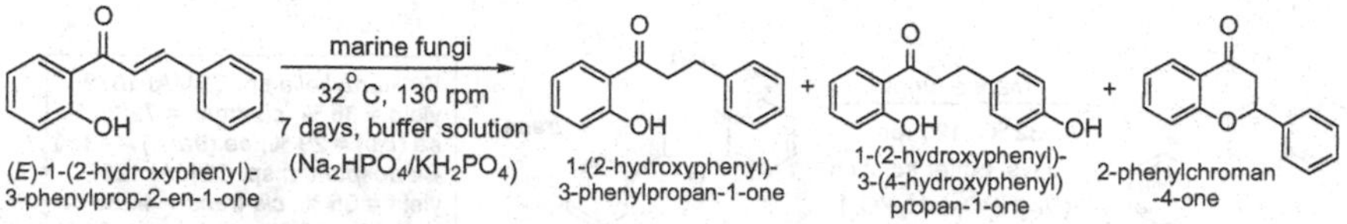

(*E*)-1-(2-hydroxyphenyl)-3-phenylprop-2-en-1-one → marine fungi, 32 °C, 130 rpm, 7 days, buffer solution (Na_2HPO_4/KH_2PO_4) → 1-(2-hydroxyphenyl)-3-phenylpropan-1-one + 1-(2-hydroxyphenyl)-3-(4-hydroxyphenyl) propan-1-one + 2-phenylchroman-4-one

Fungus	1-(2-hydroxyphenyl)-3-phenylpropan-1-one	1-(2-hydroxyphenyl)-3-(4-hydroxyphenyl) propan-1-one	2-phenylchroman-4-one
P. raistrickii CBMAI 931	c = 72 %	c = 0 %	c = 19 %
Cladosporium sp. CBMAI 1237	c = 18 %	c = 0 %	c = 7 %
P. oxalicum CBMAI 1996	c = 49 %	c = 0 %	c = 9 %
A. sydowii CBMAI 935	c = 10 %	c = 29 %	c = 9 %
Westerdykella sp. CBMAI 1679	c = 67 %	c = 0 %	c = 6 %
P. citrinum CBMAI 1186	c = 13 %	c = 0 %	c = 22 %
M. racemosus CBMAI 847	c = 8 %	c = 0 %	c = 20 %
A. sclerotiorum CBMAI 849	c = 81 %	c = 0 %	c = 7 %

FIGURE 15.18 Bioreduction of (*E*)-1-(2-hydroxyphenyl)-3-phenylprop-2-en-1-one using marine-derived fungi (based on De Matos et al. 2021).

De Matos et al. (2021) reported the use of the marine-derived fungi *P. raistrickii* CBMAI 931, *Cladosporium* sp. CBMAI 1237, *A. sydowii* CBMAI 935, *P. oxalicum* CBMAI 1996, *P. citrinum* CBMAI 1186, *M. racemosus* CBMAI 847, *Westerdykella* sp. CBMAI 1679, and *A. sclerotiorum* CBMAI 849 in the reduction of (*E*)-1-(2-hydroxyphenyl)-3-phenylprop-2-en-1-one at 32°C and 130 rpm, in buffer solution (Na_2HPO_4/KH_2PO_4) for 7 days. The authors observed three main products of the reduction reaction: 1-(2-hydroxyphenyl)-3-phenylpropan-1-one, 1-(2-hydroxyphenyl)-3-(4-hydroxyphenyl)propan-1-one and 2-phenylchroman-4-one in Figure 15.18 [31, 32].

Rocha et al. (2012) developed a reduction reaction using the marine-derived fungi *A. sclerotiorum* CBMAI 849, *Bionectria* sp. Ce5, and *B. felina* CBMAI 738. The fungi catalyzed the reduction of 1-(4-methoxyphenyl)ethan-1-one to 1-(4-methoxyphenyl)-ethanol at 32°C and 120 rpm in buffer solution (Na_2HPO_4/KH_2PO_4) for 3 days. The *Bionectria* sp. Ce5 afforded the reduced product (*R*)-1-(4-methoxyphenyl)-ethanol in 77% yield and 55% *ee* of the *R*-enantiomer as shown in Figure 15.19. *B. felina* CBMAI 738 formed (*S*)-1-(4-methoxyphenyl)-ethanol in 85% yield and 93% *ee* of the *S*-enantiomer. *A. sclerotiorum* CBMAI 849 afforded (*S*)-1-(4-methoxyphenyl)-ethanol in 95% yield and >99% *ee* of the *S*-enantiomer [33].

De Matos et al. (2021) reported the stereoselective reduction of flavonones using the marine-derived fungi *Westerdykella* sp. CBMAI 1679, *Cladosporium* sp. CBMAI 1237 and *Acremonium* sp. CBMAI 1676. The reduction of

1-(4-methoxyphenyl)ethan-1-one → marine fungi, 32 °C, 120 rpm, 3 days, buffer solution (Na_2HPO_4/KH_2PO_4) → 1-(4-methoxyphenyl) ethan-1-ol

A. sclerotiorum CBMAI 849
yield = 95 %, *ee* = >99 %. *S*-enantiomer
Bionectria sp. Ce5
yield = 77 %, *ee* = 55 %. *R*-enantiomer
B. felina CBMAI 738
yield = 85 %, *ee* = 93 %. *S*-enantiomer

FIGURE 15.19 Bioreduction of 1-(4-methoxyphenyl)ethan-1-one using marine-derived fungi (based on Rocha et al. 2012).

2-phenylchroman-4-one → marine fungi, 32 °C, 120 rpm, 7 days, buffer solution (Na_2HPO_4/KH_2PO_4) → (2*S*,4*S*)-2-phenylchroman-4-ol + *trans*

Westerdykella sp. CBMAI 1679
yield = 36 %, *cis-trans* = 72:28 %
ee (*cis*) = 29 %, *ee* (*trans*) = 43 %
Cladosporium sp. CBMAI 1237
yield = 35 %, *cis-trans* = 44:56 %
ee (*cis*) = 49 %, *ee* (*trans*) = 24 %
Acremonium sp. CBMAI 1676
yield = 22 %, *cis-trans* = 91:9 %
ee (*cis*) = 97 %, *ee* (*trans*) = 88 %

FIGURE 15.20 Bioreduction of 2-phenylchroman-4-one using marine-derived fungi (based on De Matos et al. 2021).

β-keto-1,2,3-triazole derivatives → *P. citrinum* CBMAI 1186, 32 °C, 130 rpm, 12 days, buffer solution (Na_2HPO_4/KH_2PO_4) → (*S*)-β-hydroxy-1,2,3-triazole derivatives

R = H, F, NO_2

R = H (yield = 60 %, *ee* = 99%)
F (yield = 49 %, *ee* = 98 %)
NO_2 (yield = 62 %, *ee* = 96 %)

FIGURE 15.21 Bioreduction of of β-keto-1,2,3-triazole using marine-derived fungus *P. citrinum* CBMAI 1186 (based on Alvarenga et al. [35]).

2-phenylchroman-4-one to 2-phenylchroman-4-ol was effected at 32°C and 130 rpm in buffer solution (Na_2HPO_4/KH_2PO_4) for 7 days [34]. *Westerdykella* sp. CBMAI 1679 resulted in the formation of 2-phenylchroman-4-ol in 36% yield and 72:28 *cis–trans* ratio. *Cladosporium* sp. CBMAI 1237 resulted in 35% yield and 44:56 *cis—trans*, while *Acremonium* sp. CBMAI 1676 afforded 2-phenylchroman-4-ol in 22% yield and 91:9 *cis–trans* as shown in Figure 15.20.

Alvarenga et al. [35] investigated the use of the marine-derived fungus *P. citrinum* CBMAI 1186 in the bioreduction of β-keto-1,2,3-triazole to β-hydroxy-1,2,3-triazole derivatives at 32°C and 130 rpm in buffer solution (Na_2HPO_4/KH_2PO_4) for 12 days [35]. The fungus gave a yield of 49–62% and 96–99% *ee* of the *S*-enantiomer as shown in Figure 15.21.

15.3 CONCLUSION AND PERSPECTIVES

The research progress in biocatalyst research shows increasing production rates with methodology optimization to improve the catalytic properties of fungi and enzymes in the chemical and pharmacological industries. Several fungi and enzymes have been applied effectively in organic synthesis to the production of high-value chemicals and pharmaceutical intermediates using biocatalytic methods to obtain compounds with enantioselectivity and can be seen as increasingly important tools in asymmetric organic synthesis.

Here, we presented the reduction of a series of carbonylated xenobiotic compounds and conjugated systems using whole cells from marine microorganisms.

However, we encourage researchers to isolate *ene*-reductase and dehydrogenase enzymes to enable proper characterization, as well as enzyme cloning and gene expression and site-directed enzymatic engineering studies to improve the performance of biocatalysts. Because enzymes in these classes are dependent on cofactors, the use of whole cells in fermenters shows great potential for large-scale applications.

ACKNOWLEDGMENTS

The authors are thankful for this study were financed by the Fundação de Amparo à Pesquisa do Estado de São Paulo (FAPESP, Projects 2016/20155-7 and 2019/07654-2) and Conselho Nacional de Desenvolvimento Científico e Tecnológico (CNPq, Project 302528/2017-2).

CONTRIBUTIONS

Gabriel S. Baia, David E. Q. Jimenez and André Luiz Meleiro Porto: conceptualization, methodology, software, formal analysis, investigation, data curation, writing—original draft, review and editing, visualization, conceptualization, supervision, project administration.

REFERENCES

[1] Grunwald, P. (2009). *Biocatalysis-biochemical fundamentals and applications*. Imperial College. 1052. https://doi.org/10.1142/p483

[2] Faber, K. (2004). *Biotransformations in organic synthesis*. Springer-Verlag. 454. https://www.springer.com/gp/book/9783642185373

[3] Semple, K. T., and Cain, R. B. (1996). Biodegradation of phenols by the alga *Ochoromonas danica*. *Applied and Environmental Microbiology*, n.4 (April): 1265–1273. https://www.ncbi.nlm.nih.gov/pmc/articles/PMC167892/

[4] Ke, T., and Klibnov, A. M. (1999). Enhancing enzymatic enantioselectivity in organic solvents by forming substrate salts. *Journal of the American Chemical Society*, n.14 (March): 3334–3340. https://pubs.acs.org/doi/abs/10.1021/ja984283v

[5] Carey, J. S., Laffan, D., Thomson, C., and Williams, M. T. (2006). Analysis of the reactions used for the preparation of drug candidate molecules. *Organic Biomolecular Chemistry*, n.4 (May): 2337–2347. https://pubs.rsc.org/en/content/articlelanding/2006/ob/b602413k

[6] Toogood, H., Mansell, D., Gardiner, J. M., and Scrutton, N. S. (2012). *Enantioselective Bioreduction of Carbon-Carbon Double Bonds*. Comprehensive Chirality: Ed. Elsevie. https://www.sciencedirect.com/science/article/pii/B9780080951676007138?via%3Dihub
[7] Brenna, E., Gatti, F. G., Mandredi, A., Monti, D., and Parmeggiani, F. (2011). Enoate reductase-mediated preparation of methyl (*S*)-2-bromobutanoate, a useful key intermediate for the synthesis of chiral pharmaceutical ingredients. *Organic Process Research & Development*, n.2 (July): 262–268. https://pubs.acs.org/doi/10.1021/op200086t
[8] Dobrijevic, D., Benhamou, L., Abil, E., Méndez, D., Dawson, N., Baud, D., Tappertzhofen, N., Moody, T. N., Orengo, C. A., Hailes, H. C., and Ward, J. M. (2019). Metagenomic ene-reductases for the bioreduction of sterically challenging enones. *RSC Advances*, n.9 (July): 36608–36614. https://pubs.acs.org/doi/10.1021/op200086t
[9] Rosas, H., Sanchez, A., Molina, R., Pardo, J., and Piñera, E. (2003). Diversity, taxonomy and evolution of medium-chain dehydrogenase/reductase superfamily. *European Journal of Biochemistry*, n.270 (August): 3309–3334. https://febs.onlinelibrary.wiley.com/doi/full/10.1046/j.1432-1033.2003.03704.x?sid=nlm%3Apubmed
[10] Gonzalo, G., Lavandera, I., Faber, K., and Kroutil, W. (2007). Enzymatic reduction of ketones in "micro-aqueous" media catalyzed by ADH-A from *Rhodococcus ruber*. *Organic Letters*, n.11 (March): 2163–2166. https://pubs.acs.org/doi/abs/10.1021/ol070679c
[11] Kavanagha, K., Jornvall, H., Persson, B., and Oppermann, U. (2008). Medium- and short-chain dehydrogenase/reductase gene and protein families. *Cellular and Molecular Life Sciences*, n.65 (November): 3895–3906. https://link.springer.com/article/10.1007%2Fs00018-008-8588-y
[12] Peng, G., Kuan, Y., Chou, H., Fu, T., Lin, S., Hsu, W., and Yang, M. (2014). Stereoselective synthesis of (*R*)-phenylephrine using recombinant *Escherichia coli* cells expressing a novel short-chain dehydrogenase/reductase gene from *Serratia marcescens* BCRC 10948. *Journal of Biotechnology*, n.170 (January): 6–9. https://www.sciencedirect.com/science/article/abs/pii/S0168165613004963?via%3Dihub
[13] Kohli, R. M., and Massey, V. (1998). The oxidative half-reaction of old yellow enzyme: The role of tyrosine 196. *The Journal of Biological Chemistry*, n.273 (December): 32763–32770. https://www.jbc.org/article/S0021-9258(19)58943-8/fulltext
[14] Eklund, H., and Ramaswamy, S. (2008). Three-dimensional structures of MDR alcohol dehydrogenases. *Cellular and Molecular Life Sciences*, n.65 (November): 3907–3917. https://link.springer.com/article/10.1007%2Fs00018-008-8589-x
[15] Sheng, C., Nana, S., Yuanyuan, C., Anbang. Li., Jie, P., Huajian, Z., Hongbin, Z., Su, Z., Lianli, S., and Jinhao, Z. (2019). Enantioselective reduction of α,β-unsaturated ketones and aryl ketones by perakine reductase. *Organic Letters*, n.21 (May): 4411–4414. https://pubs.acs.org/doi/abs/10.1021/acs.orglett.9b00950
[16] Rocha, L., De Souza, A. L., Rodrigues, U. P., Campana, S. P., Sette, L. D., and Porto, A. L. M. (2012). Immobilization of marine fungi on silica gel, silica zerogel and chitosan for biocatalytic reduction of ketones. *Journal of Molecular Catalysis B-Enzymatic*, n.84 (July): 160–165. https://www.sciencedirect.com/science/article/abs/pii/S1381117712001919?via%3Dihub
[17] Rocha, L. C., Ferreira, H. V., Pimenta, E. F., Berlinck, R. G. S., Seleghim, M. H. R., Darci, C. D. J., Sette, L. D., Bonugli, R. C., and Porto, A. L. M. (2009). Bioreduction of α-chloroacetophenone by whole cells of marine fungi. *Biotechnology Letters*, n.31 (June): 1559–1563. https://link.springer.com/article/10.1007%2Fs10529-009-0037-y
[18] Rocha, L. C., Ferreira, H. V., Pimenta, E. F., Berlinck, R. G. S., Rezende, M. O. O., Landgraf, M. D., Seleghim, M. H. R., Sette, L. D., and Porto, A. L. M. (2010). Biotransformation of α-bromoacetophenones by the marine fungus *Aspergillus*

sydowii. Marine Biotechnology, n.12 (November): 552–557. https://link.springer.com/article/10.1007/s10126-009-9241-y

[19] Rocha, L. C., Luiz, R. F., Rosset, I. G., Raminelli, C., Seleghim, M. H. R., Sette, L. D., and Porto, A. L. M. (2012). Bioconversion of iodoacetophenones by marine fungi. *Marine Biotechnology*, n.14 (June): 396–401. https://link.springer.com/article/10.1007%2Fs10126-012-9463-2

[20] Mouad, A. M., Oliveira, A. L. L., Debonsi, H. M., and Porto, A. L. M. (2015). Bioreduction of fluoroacetophenone derivatives by endophytic fungi isolated from the marine red alga *Bostrychia radicans. Biocatalysis*, n.1 (December): 141–147. https://www.degruyter.com/document/doi/10.1515/boca-2015-0011/html

[21] Rocha, L. C., Seleghim, M. H. R., Comasseto, J. V., Sette, L. D., and Porto, A. L. M. (2015). Stereoselective bioreduction of α-azido ketones by whole cells of marine-derived fungi. *Marine Biotechnology*, n.17 (August): 736–742. https://link.springer.com/article/10.1007%2Fs10126-015-9644-x

[22] Jimenez, D. E. Q., Ferreira, I. M., Birolli, W. G., Fonseca, L. P., and Porto, A. L. M. (2016). Synthesis and biocatalytic ene-reduction of Knoevenagel condensation compounds by the marine-derived fungus *Penicillium citrinum* CBMAI 1186. *Tetrahedron*, n.72 (November): 7317–7322. https://www.sciencedirect.com/science/article/abs/pii/S0040402016300771?via%3Dihub

[23] De Matos, I. L., Nitschke, M., and Porto, A. L. M. (2019). Hydrogenation of halogenated 2′-hydroxychalcones by mycelia of marine-derived fungus *Penicillium raistrickii. Marine Biotechnology*, v.21: 430–439. https://pubmed.ncbi.nlm.nih.gov/30895403/

[24] Jimenez, D. E. Q., Barreiro, J. C., Dos Santos, F. M., De Vasconcellos, S. P., Porto, A. L. M., and Batista, J. M. (2019). Enantioselective ene-reduction of *E*-2-cyano-3-(furan-2-yl) acrylamide by marine and terrestrial fungi and absolute configuration of (*R*)-2-cyano-3-(furan-2-yl) propanamide determined by calculations of electronic circular dichroism (ECD) spectra. *Chirality*, n.31 (July): 534–542. https://onlinelibrary.wiley.com/doi/10.1002/chir.23078

[25] Morais, A. T. B., Ferreira, I. M., Jimenez, D. E. Q., and Porto, A. L. M. (2018). Synthesis of α-chloroacetophenones with NH_4Cl/Oxone ® in situ followed by bioreduction with whole cells of marine-derived fungi. *Biocatalysis and Agricultural Biotechnology*, v.16 (October): 314–319. https://www.sciencedirect.com/science/article/abs/pii/S1878818117304899?via%3Dihub

[26] Ferreira, I. M., Rocha, L. R., Yoshioka, S. A., Nitschke, M., Jeller, A. H., Pizzuti, L., Seleghim, M. H. R., and Porto, A. L. M. (2014). Chemoselective reduction of chalcones by whole hyphae of marine fungus *Penicillium citrinum* CBMAI 1186, free and immobilized on biopolymers. *Biocatalysis and Agricultural Biotechnology*, n.3 (October): 358–364. https://www.sciencedirect.com/science/article/abs/pii/S1878818114000668?via%3Dihub

[27] Ferreira, I. M., Meira, E. B., Rosset, I. G., and Porto, A. L. M. (2015). Chemoselective biohydrogenation of α,β- and α,β,γ,δ-unsaturated ketones by the marine-derived fungus *Penicillium citrinum* CBMAI 1186 in a biphasic system. *Journal of Molecular Catalysis B: Enzymatic*, n.115 (May): 59–65. https://www.sciencedirect.com/science/article/abs/pii/S138111771500034X?via%3Dihub

[28] Birolli, W. G., Zanin, L. L., Jimenez, D. E. Q., and Porto, A. L. M. (2020). Synthesis of Knoevenagel adducts under microwave irradiation and biocatalytic ene-reduction by the marine-derived fungus *Cladosporium* sp. CBMAI 1237 for the production of 2-cyano-3-phenylpropanamide derivatives. *Marine Biotechnolohy*, n.22 (March): 317–330. https://link.springer.com/article/10.1007/s10126-020-09953-8

[29] Ferreira, I. M., Fiamingo, A., Campana-Filho, S. P., and Porto, A. L. M. (2020). Biotransformation of (*E*)-2-methyl-3-phenylacrylaldehyde using mycelia of

penicillium citrinum CBMAI 1186, both free and immobilized on chitosan. *Marine Biotechnology*, n.22 (February): 348–356. https://link.springer.com/article/10.1007%2Fs10126-020-09954-7

[30] Alvarenga, N., and Porto, A. L. M. (2017). Stereoselective reduction of 2-azido-1-phenylethanone derivatives by whole cells of marine-derived fungi applied to synthesis of enantioenriched β-hydroxy-1,2,3-triazoles. *Biocatalysis and Biotransformation*, n.35 (August): 388–396. https://www.tandfonline.com/doi/full/10.1080/10242422.2017.1352585

[31] De Matos, I., Nitschke, M., and Porto, A. L. P. (2021). Regioselective and chemoselective biotransformation of 2′-hydroxychalcone derivatives by marine derived fungi. *Biocatalysis and Biotransformation*, n.1 (July): 1–11. https://www.tandfonline.com/doi/abs/10.1080/10242422.2021.1956909?src=&journalCode=ibab20

[32] De Matos, I., Nitschke, M., Fonseca, L. J. P., and Porto, A. L. M. (1979). *Biotransformation of Flavonoids by Terrestrial and Marine Microorganisms*. Encyclopedia of Marine Biotechnology. https://onlinelibrary.wiley.com/doi/abs/10.1002/9781119143802.ch87

[33] Rocha, C. L., Ferreira, H. V., Luiz, R., Sette, L. D., and Porto, A. L. M. (2012). Stereoselective bioreduction of 1-(4-methoxyphenyl)ethanone by whole cells of marine-derived fungi. *Marine Biotechnology*, n. 14 (December): 358–362. https://link.springer.com/article/10.1007%2Fs10126-011-9419-y

[34] De Matos, I. L., Birolli, W. G., Santos. D., Nitschke, M., and Porto, A. L. M. (2021). Stereoselective reduction of flavanones by marine-derived fungi. *Molecular Catalysis*, n.513 (August): 111734–111744. https://www.sciencedirect.com/science/article/abs/pii/S2468823121003515

[35] Alvarenga, N., Porto, A. L. M., and Barreiro, J. C. (2018). Enantioselective separation of (±)-β-hydroxy-1,2,3-triazoles by supercritical fluid chromatography and high-performance liquid chromatography. *Chirality*, n.30 (July): 890–899. https://onlinelibrary.wiley.com/doi/10.1002/chir.22851

16 Biomacromolecules from Marine Organisms and Their Biomedical Application

Pitchiah Sivaperumal, Kannan Kamala, and Ganapathy Dhanraj

CONTENTS

16.1 INTRODUCTION

A large marine environment is composed of microorganisms, plants and animals, which are a wealthy resource of unique natural products like macromolecules (peptides, carbohydrates, proteins, lipids). The macromolecules derived from organisms and their associated microbes are biomacromolecules validated in pharmaceutical research. Generally, bio-macromolecules from marine have been chemically diverse with antimicrobial activity that used to develop drugs to control drug-resistant pathogens. Macromolecules are large molecules ranging from 100 to 10,000 mm in size and built by different types of subunits. The single subunits adhere to each other through covalent bonds to form large molecules. Generally, macromolecules are carbohydrates, proteins, lipids, extracellular polymeric substances and nucleic acids. Biomacromolecules from marine resource has a major impact on immunomodulatory drugs. The apoptosis of

DOI: 10.1201/9781003303909-16

malignant tumor cell growth induced by marine bio-macromolecules, and it has an advanced mechanism against multidrug-resistant microbes (Nikaido, 2009). Bio-macromolecules have improved bioactive properties, which is considered a pillar of chemotherapy (Yahya et al., 2019). In addition, biomacromolecules produced by marine organisms are being increasingly investigated for several biomedical applications (d'Ayala et al., 2008; Silva et al., 2012a, 2012b). The present chapter discusses the research on marine macromolecules and their biological importance carried out mostly during the last few years. Different aspects of biomedical applications and advanced technology for drug delivery applications are also discussed.

16.2 BIOMACROMOLECULES FROM MARINE MICROBES

Earlier studies have reported on the various biomacromolecules from the marine environment, which are categorized into polypeptides, steroids, alkaloids, terpenes, proteins and quinones. Several antimicrobial macromolecules have significant activity for developing commercial drugs, but none of the compounds is marketed yet.

Carbohydrate: A linear glycerol acid heptapeptide from *Paenibacillus profundus* sp., shows antimicrobial activity against *Bacillus subtilis*, *Staphylococcus epidermis*, *S. aureus* and *Enterococcus faecium*. The minimal zone of inhibition ranges from 20 to 24 mm (Kalinovskaya et al., 2013). The heptapeptide has an amphipathic nature because it holds both hydrophilic and hydrophobic amino acids in its cyclic structure. It has two protein amino acids and non-protein amino acids with two peptide bonds. Also, it has meristic fatty acid, which is bounded with therapeutic insulin peptide in humans (Kurtzhals et al., 1995; Jonassen et al., 2012; Zorzi et al., 2017). Similarly, a glycol hex depsipeptide compound, mollemycin A, has been derived from Streptomyces CMBM0244 that has antibacterial activity against various gram-negative and gram-positive bacteria at 2.5–40 µg/mL of concentration (Raju et al., 2014). A polysaccharide starch– and monosaccharide xylose–producing *Saccharophagus degradans* was reported by Gonzalez Garcia et al. (2015). A bioactive polysaccharide was isolated with the combination of glucose, mannose and galactose at 1.33:1:1.33 from *Fusarium oxysporum*. It has 61 kDa molecular weight, and it exhibited potential antioxidant activity (Chen et al., 2015). Likewise, antioxidant potential polysaccharide PS1–1 and PS1–2, PS2–1 also consisted by mannose, galactose and glucose from *Penicillium* sp. F23–2 (Sun et al., 2009). The *Phomaherbarum* strain was used to produce polysaccharide composed of glucan, glucoronic acid residues and glucopyranosyl that exhibits antitumor activity (Chen et al., 2009).

Lipids: Glycophingolipids and penicilliosides A and B were derived from the fungus *Pennicillium* sp. and isolated from the *Didemnum* sp. in the Red Sea. These compounds exhibited the antimicrobial activity against *Candida albicans*, *S. aureus* and *E. coli* (Murshid et al., 2016).

Proteins: A polyketide compound angucyclinone from marine *Streptomyces* sp., exhibited the antimicrobial activity (Abdalla et al., 2010). Macrolactin

W–producing *Bacillus* sp. showed antibacterial activity against *S. aureus, Pseudomonas aeruginosa, E. coli* and *Bacillus substilis* (Mondol et al., 2011). Similarly, *Clostridium difficile* was inhibited by Merochlorins A from *Streptomyces* sp. CNH189 (Sakoulas et al., 2012; Kaysser et al., 2012). The plant pathogenic bacterium *Xanthomonas campestris* and the dermatophytic fungi *Trichophyton rubrum* were inhibited by the polyketide compound malettinin from *Cladosporium* sp., isolated from Wadden Sea (Silber et al., 2014).

Extracellular Polymeric Substances (EPSs): Various marine microbes produced complex substances composed of carbohydrates, proteins, humic substances, lipids and nucleic acids (Manivasagan and Kim, 2014) are known as EPSs. It consists of organic and inorganic substances along with acetate, pyruvate, phosphate and succinate moieties (Priyanka and Ena, 2020). Enormous EPS-producing *Vibrio furnissi* VB0S3 was isolated from the coastal area of Goa (Bramhachari et al., 2007). Similarly, an *Enterobacter cloaca* was isolated from marine sediment, which has produced enormous acidic EPS exhibits significant emulsifying properties (Iyer et al., 2005). A gram-positive bacteria *Planococcus maitriensis* was isolated from the coastal waters of Bhavnagar in India produced an EPS used in oil recovery and bioremediation of oil (Priyanka and Ena, 2020). Marine bacterium *Alteronomas* sp., *Pseudoaltromonas* sp., and *Vibrio* sp., produced unique EPSs ranged from 0.5 to 4 g per liter of sugar base medium (Raza et al., 2011). The EPS from *Vibrio diabolicus* has composed with glucosamine and glucuronic acid (Arias et al., 2003).

16.3 BIOMEDICAL APPLICATION OF MARINE MICROBIAL METABOLITES

The marine environment represents a large source of bioactive compounds that are is chemically diverse and of low risk to humans (Silva et al., 2012a, 2012b). Microbes from extreme environments, like hydrothermal vents, hot springs, polar cold seeps and hyper-saline atolls, must adapt to survive in these extreme conditions, and they produce special active biomolecules, including macromolecules (Laurienzo, 2010). Generally, marine microbes have an osmotic tolerance capacity to produce polysaccharides that have elevated concentrations of sugar, which is economically important in commercial polysaccharide production (Mehta et al., 2014). Marine macromolecules components such as amino sugars, glucuronic acid, pyruvate and galacturonic acid have bioactive potentials, such as being antimicrobial, antiviral, antioxidant, antitumor, anticoagulant and other medicinal properties. In addition, these components are used as food additives and in the pharmaceuticals industry (Wang et al., 2012). Biomacromolecules are that prevent oxidative free radicals and their causes of coronary infirmity, enzyme inactivation, malignant cell growth, protein modification, cell membrane defect and DNA damage (Abdel Hamid et al., 2020; Asker and Shawky, 2010; Moskovitz et al., 2002). Sulfated EPSs have exhibited antiviral activity with minimal cytotoxic activity (Li et al., 2012). Also they have anticoagulant activities, such as inhibiting thrombin movement (Silva et al., 2010a, 2010b; De Jesus Raposo et al., 2015; Cao et al., 2019). Some of the biopolymers from marine sources are used as binders in tablet manufacturing due to their water-restricting limitation (Paolucci et al., 2015).

16.4 BIOLOGICAL ACTIVE MACROMOLECULES FROM MARINE PLANTS AND ALGAE

Marine habitats are affluent in biodiversity and 90% of biomass was established by marine flora (Nadia et al., 2016). It has significant therapeutically active substances, such as sulfated polysaccharides and polyphenols that have an implausible combination that is species-specific (Elsakhawy et al., 2017; Cao et al., 2019). Macromolecules from marine green, red and brown algae have high levels of multifunctional bioactivity, such as oxidative stress reduction, anticoagulants, and the inhibition of inflammation and microbiome modulation. In this section, we cover polysaccharides, agar, alginate and carrageenan and their biological properties.

Polysaccharides: The three main classes of polymers, polysaccharides, proteins and nucleic acids (McNaught and Wilkinson 1997), from marine organisms produce a variety of biological properties. Among them, polysaccharide-derived products of sulfated polymer glycosaminoglycans (GAG) have more biomedical applications. The marine polysaccharide rhamnan sulfate from green seaweed is primarily composed of L-rhamnose sugars linked with carbons (Harada et al., 1998). The rhamnan sulfate has been revealed as effective in fibrin polymerization and inhibiting thrombin (Li et al., 2012; Shammas et al., 2017). Likewise, the sulfated heteropolysaccharide ulvan is composed of L-iduronic acid/D-glucuronic acid with rhamnose residues and D-glucose and D-xylose residues (Wijesekara and Karunarathna, 2017). In addition, polysaccharides are biomacromolecules established by carbohydrate monomers linked by glycosidic bonds. The most typical polysaccharides in the marine environment are alginate, agar and carrageenan. All polysaccharides have parallel chemical structures, but the seemingly small differences are answerable for distinct properties of the molecules. The sulfated polymer GAGs are linear, multifaceted and polydisperse natural polysaccharides, classically bearing a repeating disaccharide unit organized by a hexose and a hexosamine. The presence of sulfated GAG is found in a diverse range of marine phyla, such as sponges. Porifera (Zierer and Mourao, 2000) have been well recognized, and increasing interest is being shown in different sectors of research, such as in the biochemical industry for biomedical and biopharmaceutical applications.

Moreover, the research on new natural sulfated polysaccharides with significant antithrombotic and anticoagulant action is an attractive alternative for the traditional heparin usage in medicine. There are many reports of other sulfated compounds from marine origins, such as heparin and fucans, also investigated (Brito et al., 2014; McLellan and Jurd 1992; Pavao et al., 1995). Venkata Rao and Sri Ramana (1991) reported structural studies of polysaccharide of the green seaweed *Chaetomorpha anteninna* of Indian waters. Antiviral polysaccharides have been reported from the red seaweed *Gracilaria corticate*, as well as brown seaweed *Stoechospermum marginatum* of Indian waters (Adhikari et al., 2005). Fucoidan is a polysaccharide derived from *Fucus vesiculosus*, which is composed of L-fucose, at 44%, and sulfate with a linear backbone, at 26%. Fucoidan is well known for the treatment of atherosclerosis (Wang et al., 2010). Similarly, laminarian polysaccharides consist of D-glucopyranose with a length of 20–25 disaccharides (Nelson et al., 1974). Generally, ulvan is extracted from the green algae *Ulva lactuca* and *U. pertusa* (Qi et al., 2012). Seaweed polysaccharides, including the sulfated ones, have been extensively reported to be exhibiting various bioactivities, for example, antiviral and anticoagulant properties (Siddhanta and Sai Krishnamurthy, 2001). Partially reduced sulphated alginic acid was reported to exhibit antithrombic activity (Shanmugam and Mody, 2000).

Agar and alginates: Agar generally extracted from red seaweeds *Gracilaria* sp. and *Gellidium* sp. in major quantities and from *Afeltia*, *Graciliaropsis* and *Pterocla* in lesser quantities (Kingsbury, 1984). Agar is composed of D-galactose and L-galactose connected with alpha and beta glycosidic bond (Li et al., 2014). Alginate is another important macromolecule, and it was discovered by Stanford in 1881, followed by a patent in which Stanford claimed the application of alginate as a pharmaceutical agent. Later on, from brown seaweed *Macrocystis pyrifera* the polysaccharide extracted by Kelco Co. in 1929. Furthermore, worldwide production of seaweed alginate has developed (d'Ayala et al., 2008). In addition, alginate is widely used as a gelling agent for different pharmaceutical, biomedical applications and personal care purposes (Hernandez-Carmona et al., 1998; Brownlee et al., 2005; Tonnesen and Karlsen, 2002). It is a polysaccharide from brown seaweed that has been used as a thickening agent and a colloidal substance (Gombotz and Wee, 1998). It is a linear-block copolymer composed of mannuronic acid and guluronic acid, which can be derived from bacteria too (Marguerite, 2008). Blood cholesterol reducing alginate was extracted from *Sargassum duplicatum* and *Turbinaria* sp. from a beach of Yogyakarta (Wahyu et al., 2015). A liquid form of alginate (2%) decreased cholesterol by 30% and body weight in 12 weeks in rats (81). Polyguluronate sulfate from alginate has anticoagulant activity and significantly stepping ahead in the treatment of cardiovascular disease (Zhao et al., 2007). In addition, Chattopadhyay et al. (2010) have reported that the alginate has exhibited potential 2,2-diphenyl-1-picrylhydrazyl (DPPH) radical reduction and ferric-reducing activity, reduction of hypertension and gut probiotic activity (Chattopadhyay et al., 2010). The achievement of the profitable development of alginate lies in its capability to retain water and in its gelling, viscosifying and stabilizing assets; in fact, it increases the viscosity of aqueous solutions and forms gels without temperature-dependence, in contrast with other polysaccharides of agar and carrageenan (Gomez et al., 2009). Other biotechnological applications can benefit from the specific biological effects of alginate, such as hypolipidemic and hypocholesterolemic effects (Smit, 2004).

Carrageenans: Carrageenan is a viscosifying sulfated polysaccharide from red seaweed that can reduce blood coagulation, lipids and atherosclerosis (Gates, 2014; Ledesma and Herrero, 2014). It consists of 15–40% of an ester sulfate group with a 100kDa molecular mass formed by the units of D-galactose, anhydrogalactose linked with glycosidic linkage. Generally, carrageenan is classified into different types based on the concentration of sulfate groups like s, λ, κ, ι, ε and μ, which differ by the position of ester sulfate groups (Necas and Bartosikova, 2013). A carrageenan-derivative gelatin is used as a cough remedy (Necas and Bartosikova, 2013).

Carrageenans have anti-inflammatory activity (Sini et al., 2010), and a higher concentration of carrageenans has been used for pathophysiological conditions (Porto et al., 2010; Silva et al., 2010a, 2010b). The blood coagulation was stopped for a long time by κ-carrageenan, and the activity was double with λ-carrageenan (Shanmugam and Mody, 2000). Also, it inhibited the human cytomegalovirus, herpes simplex virus and human rhinovirus (Stiles et al., 2008). Carrageenan prevents the viruses through inhibiting the initial binding of the virus to the cell (Grassauer et al., 2008). In addition, it exhibits antiproliferative activity in cancer cell lines and inhibits tumor growth in mice (Souza et al., 2007; Yuan et al., 2005; Zhou and Zheng, 1991). Rocha et al. (2005) reported that carrageenan from red algae possesses high antioxidant activity with potent inflammatory activity in mice. Also, lowering the level of blood cholesterol in humans (Panlasigui et al., 2003). Immobilized carrageenan was used to improve the degradation of chlorophenol and morpholine. Carrageenan is used as a stabilizer, emulsifier and colloidal gum in food industry (Tobacman et al., 2001). Alginate is widely used in the food industry to coat vegetable and fruits (Gundewadi et al., 2018; Hashemi et al., 2017), for its antiviral activity (Arroyo et al., 2020) and as a stabilizing and emulsifying agent (Severino et al., 2019; Jiang et al., 2020). Alginate was used to develop hydrogel dressing, which has good moisturizing, oxygen permeability and interfacial tension with soft tissue promotes as a wound-dressing material (Yasasvini et al., 2017; Raval et al., 2011; Siafaka et al., 2016). In addition, it has the potential to reduce obesity by lowering the absorption of glucose and cholesterol (Paxman et al., 2008).

16.5 BIOMACROMOLECULES FROM MARINE ANIMALS AND THEIR BIOLOGICAL IMPORTANCE

Chitin and chitosan: It is a common natural cationic polymer in crystal form from the exoskeleton of mollusks and crustaceans that is synthesized by the removal of acetyl group to soluble in dilute acid (Sudatta et al., 2020). Chitin production from shrimp waste by ammonium-based ionic liquid extraction was performed by Tolesa et al. (2019), and the yield was noted as 13.4% at 110°C for 24 h. Similarly the chitin from sea waste was extracted by fruit waste stream, which revealed the highest chitin production 0.89 g noted from potato peel (15 g/100 ml) fermentation (Tan et al., 2020). Chitosan has profound applications in the fields of biomedicines since it is having antibacterial, fungistatic, hemostatic, antitumoral and anticholesteremic properties (Krajewska, 2005). In economic sectors, chitin, chitosan and its by-products are being widely used in different pharmaceutical and veterinary medicine (Baldrick, 2010; Senel and McClure, 2004). Chitosan has been synthesized by removing an amine group through a deacetylation process (Sudatta et al., 2020). Chitosan also has the ability to prevent liver damage. Chitosan soluble in slight acidic solution due to the presence of glucosamine unit (Aranaz et al., 2009). Chitosan have been probably the most used marine-derived biopolymers in the preparation of drug delivery particles (Prabaharan and Mano 2005; Liew et al., 2006; Oh et al., 2008). Chitosan itself may be chemically modified to control the interaction with drug-loaded molecules. Moreover, Ramasamy et al., 2014 find out the protective effect of chitosan from the cuttlebone of *S. kobiensis* against carbon tetrachloride–induced hepatic damage in male Wistar rats. Different biological activities, such as antimicrobial activity, excellent chelation behavior, biocompatibility and nontoxicity activities, have been pointed out as main contributors to the performance of chitosan in the cited applications (Rinaudo, 2006; Silva et al., 2010a, 2010b; Honarkar and Barikani, 2009).

Collagen: Collagen is an important component for direct health-related applications. Animal-derivative collagens are widely used in the field of skin regeneration templates, dental composites, biodegradable matrices, plastic surgery and shields in ophthalmology, cardiovascular surgery, orthopedics, neurology and urology (Meena et al., 1999). Recently, marine collagen used to overcome the disease-related issues, as well as in innovative methodologies for tissue engineering, artificial organs and drug delivery applications (Song et al., 2006; Swatschek et al., 2002; Jeong et al., 2007).

16.6 MARINE BIOMACROMOLECULES IN BIOMEDICAL ENGINEERING APPLICATIONS

Overall biomedical engineering approach, matrices are developed to support cells, exciting their variation and proliferation to the development of novel new and functional tissue (Langer and Vacanti 1993). Such approaches are allowable for producing hybrid constructs that can be implanted in patients to induce the improvement of tissues or replace malfunctioning cells. Diverse materials have been recommended to be used in the processing of scaffolds, namely, biocompatible polymers derived from marine organisms. Naturally derived biomacromolecules are suggested to be similar to biological macromolecules wherever the biological environment is prepared to identify and easily deal with metabolic activities. Due to their resemblance to natural molecules and extracellular matrices, they may also evade the prompting immunological reactions, chronic inflammation, and toxicity.

Warm techniques have been extensively used to process marine-derived macromolecules that may be also attractive by the fact that cells or unstable proteins can be combined during the fabrication of the device. Freeze-drying has been extensively used to develop natural polymers, including marine-derived polysaccharides. A typical example is the production of scaffolds from chitosan (Madihally and Matthew, 1999). Crosslinking or the combination of other biomacromolecules could also be combined with different methods (Silva et al., 2008). In the case of orthopedic applications, scaffolds could be incorporated into stiffer prefabricated scaffolds (Prabaharan et al., 2007). Other marine-derived macromolecules of polysaccharides and collagen are also used in tissue engineering strategies. In addition, jellyfish collagen also been united with poly(lacticco-glycolic acid) and processed by freeze-drying and electrospinning to obtain tubular nano-scaffolds, in which smooth muscle cells and endothelial cells were shown to proliferate effect (Jeong et al., 2007). Electrospun scaffolds obtained from collagen (Rho *et al.,* 2006), poly (ethylene-co-vinyl alcohol) (Kenawy et al., 2003), collagen-PEO (poly-ethylene oxide) (Huang et al., 2001) and polyurethane (Khil et al., 2003) have been examined for potential application as wound healing. There also has been increasing concern about the combination of drugs in electrospun fibers in areas except skin care applications.

16.7 CONCLUSION

The marine environment has been confirmed to be an enormous source of bioactive materials, even though the available knowledge about marine biomaterials and their tool is quite at its beginning. In addition, marine ecosystem needs to protect so the diverse of natural component can be certainly used for biomedical purposes. In particular, marine bioactive macromolecules are important for several benefits in the search for new drugs. With their diverse chemical ecology, the marine organisms have an abundant capacity for providing effective, low-cost and safer therapeutic drugs that deserve broad exploration. In addition, using marine-origin macromolecules in biomedical engineering approaches, specifically their achievement in *in-vivo* studies, is still necessary to improve their potential in this precise area. Nevertheless, applied and basic research efforts regarding marine-derived macromolecules in the biomedical field need a close partnership between chemists and biologists, including marine microbiologists, marine biologists, biochemistry experts and computational scientists to accomplish screenings and detailed structural and biological studies.

ACKNOWLEDGMENT

The authors are grateful to Saveetha Dental College and Hospital, SIMATS, and the corresponding author acknowledges the SERB-DST for moral support through the TARE scheme (File no: TAR/2019/000143).

REFERENCES

Abdalla, M. A., E. Helmkeb and H. Laatscha, 2010. Fujianmycin C, a bioactive Angucyclinone from a marine derived Streptomyces sp. B6219. *Natural Product Communications* 5(12): 1917–1920.

Abdel Hamid, H. T., W. Wenlong and L. Qiaomin, 2020. Environmental sensitivity of flash flood hazards geospatial techniques. *Global Journal of Environmental Science and Management* 6(1): 31–46.

Adhikari, U., C. Mateu, E. B. Damonte and B. Ray, 2005. *Proceedings of CARBO XX Carbohydrate Conference*, Nov. 24–26th, Lucknow University, Lucknow. PP-12.

Aranaz, I., M. Mengibar, R. Harris, I. Panos, B. Miralles, N. Acosta, G. Galed and A. Heras, 2009. Functional characterization of chitin and chitosan. *Current Chemical Biology* 3: 203–230.

Arias, S., A. D. Moral, M. R. Ferrer, R. Tallon, E. Quesada and V. Bejar, 2003. Mauran, an exopolysaccharide produced by the halophilic bacterium Halomonas maura with a novel composition and interesting properties for biotechnology. *Extremophiles* 7: 319–326.

Arroyo, B., A. Bezerra, L. Oliveira, S. Arroyo, E. De Melo and A. Santos, 2020. Antimicrobial active edible coating of alginate and chitosan add ZnO nanoparticles applied in guavas (*Psidium guajava* L.). *Food Chemistry* 309: 125566.

Asker, M. and B. T. Shawky, 2010. Structural characterization and antioxidant activity of an extracellular polysaccharide isolated from Brevibacterium otitidis BTS 44. *Food Chemistry* 123: 315–320.

Baldrick, P., 2010. The safety of chitosan as a pharmaceutical excipient. *Regulatory Toxicology and Pharmacology* 56(3): 290–299.

Bramhachari, P. V., P. B. Kishor, R. Ramadevi, R. Kumar, B. R. Rao and S. K. Dubey, 2007. Isolationand characterization of mucous exopolysaccharide produced by *Vibrio furnissii* strain VB0S3. *Journal of Microbiology and Biotechnology* 17: 44–51.

Brito, A. S., R. S. Cavalcante, L. C. G. F. Palhares, A. J. Hughes, C. P. V. Andrae, E. A. Yates, M. A. Lima and S. F. Chavante, 2014. A non hemorrhagic hybrid heparin/heparan sulfate with anticoagulant potential. *Carbohydrate Polymers* 99: 372–378.

Brownlee, I. A., A. Allen, J. P. Pearson, P. W. Dettmar, M. E. Havler, M. R. Atherton and E. Onsoyen, 2005. Alginate as a source of dietary fiber. *Critical Review on Food Science and Nutrition* 45: 497–510.

Cao, S., X. He, L. Qin, M. He, Y. Yang, Z. Liu and W. Mao, 2019. Anticoagulant and antithrombotic properties in vitro and in vivo of a novel sulfated polysaccharide from marine green alga Monostroma nitidum. *Marine Drugs* 17: 247–268.

Chattopadhyay, N., T. Ghosh, S. Sinha, K. Chattopadhyay, P. Karmakar and B. Ray. 2010. Polysaccharides from Turbinaria conoides: Structural features and antioxidant capacity. *Food Chemistry* 118: 823–829.

Chen, S., D. K. Yin, W. B. Yao, Y. D. Wang, Y. R. Zhang and X. D. Gao, 2009. Macrophage receptors ofpolysaccharide isolated from a marine filamentous fungus *Phoma herbarum* YS4108. *Acta Pharmacologica Sinica* 30: 1008–1014.

Chen, Y. L., W. J. Mao, H. W. Tao, W. M. Zhu, M. X. Yan, X. Liu and T. Guo, 2015. Preparation and characterization of a novel extracellular polysaccharide with antioxidant activity from the mangrove-associated fungus *Fusarium oxysporum*. *Marine Biotechnol*ogy 17: 219–228.

D'Ayala, G. G., M. Malinconico and P. Laurienzo, 2008. Marine derived polysaccharides for biomedical applications: Chemical modification approaches. *Molecules* 13(9): 2069–2106.

De Jesus Raposo, M. F., A. M. B. De Moris and R. M. S. C. De Moris, 2015. Marine polysaccharide from algae with potential biomedical applications. *Marine Drugs* 13(5): 2967–3028.

Elsakhawy, T. A., A. S. Fatma and R. Y. Abd-EL-Kodoos, 2017. Marine microbial polysaccharides: Environmental role and applications (an overview). *Environment, Biodiversity and Soil Security* 1: 61–70.

Gates, K. W., 2014. Bioactive compounds from marine foods: Plant and animal sources. ed. B. H. Ledesma and M. Herrero. *Journal of Aquatic Food Product and Technology* 23: 313–317.

Gombotz, W. R., and S. Wee, 1998. Protein release from alginate matrices. *Advanced Drug Deliver Reviews* 31: 267–285.

Gomez, C. G., M. V. P. Lambrecht, J. E. Lozano, M. Rinaudoand and M. A. Villar, 2009. Influence of the extraction—purification conditions on final properties of alginates obtained from brown algae (*Macrocystispyrifera*). *International Journal of Biological Macromolecules* 44: 365–371.

GonzalezGarcia, Y., A. Heredia, J. C. Meza-Contreras, F. M. E. Escalante, R. M. Camacho-Ruiz and J. Cordova, 2015. Biosynthesis of extracellular polymeric substances by the marine bacterium *Saccharophagus degradans* under different nutritional conditions. *International Journal of Polymeric Science* 2015: 1–7.

Grassauer, A., R. Weinmuellner, C. Meier, A. Pretsch, E. P. Grassauer and H. Unger, 2008. Iota-carrageenan is a potent inhibitor of rhinovirus infection. *Virology Journal* 5: 1–13.

Gundewadi, G., S. Rudra, D. Sarkar and D. Singh, 2018. Nanoemulsion based alginate organic coating for shelf life extension of okra. *Food Package and Shelf Life* 18: 1–12.

Harada, N., and M. Maeda, 1998. Chemical structure of antithrombin-active rhamnan sulfate from Monostromnitidum. *Bioscience Biotechnology Biochemistry* 62: 1647–1652.

Hashemi, S., A. Khaneghah and M. Ghahfarrokhi, 2017. Basil-seed gum containing Origanum vulgare subsp. viride essential oilas edible coating for fresh cut apricots. *Postharvest Biology and Technology* 125: 26–34.

Hernandez-Carmona, G., D. J. McHugh, D. L. Arvizu-Higuera and Y. E. Rodriguez-Montesinos,1998. Pilot plant scale extraction of alginate from *Macrocystispyrifera*. 1. Effect of pre-extraction treatments on yield and quality of alginate. *Journal of Applied Phycology* 10(6): 507–513.

Honarkar, H. and M. Barikani, 2009. Applications of biopolymers I: Chitosan. *Chemical Monthly* 140(12): 1403–1420.

Huang, L., K. Nagapudi, R. P. Apkarian and E. L. Chaikof, 2001. Engineered collagen-PEO nanofibers and fabrics. *Journal of Biomaterials Science Polymer Edition* 12: 979–993.

Iyer, A., K. Mody and B. Jha, 2005. Characterization of an exopolysaccharide produced by a marine, *Enterobacter cloacae*. *Indian Journal of Experimental Biology* 43: 467–471.

Jeong, S., S. Y. Kim, S. K. Cho, M. S. Chong, K. S. Kim, H. Kim, S. B. Lee and Y. M. Lee, 2007. Tissue-engineered vascular grafts composed of marine collagen and PLGA fibers using pulsatile perfusion bioreactors. *Biomaterials* 28(6): 1115–1122.

Jiang, Y., G. Yu, Y. Zhou, Y. Liu, Y. Feng and J. Li, 2020. Effects of sodium alginate on microstructural and properties of bacterial cellulose nanocrystal stabilized emulsions. *Colloids and Surfaces A* 607: 1–28.

Jonassen, I., S. Havelund, T. H. Jensen, D. B. Steensgaard, P. O. Wahlund and U. Ribel, 2012. Design of the novel protraction mechanism of insulin degludec, an ultra-long-acting basal insulin. *Pharmacological Research* 29: 2104–2114.

Kalinovskaya, N. I., L. Romanenko, A. Kalinovsky, P. S. Dmitrenok and S. A. Dyshlovoy, 2013. A new antimicrobial and anticancer peptide producing by the marine deep sediment strain "Paenibacillus profundus" sp. nov. Sl 79. *Natural Product Communications* 8(3): 381–384.

Kaysser, L., P. Bernhardt, S. J. Nam, S. Loesgen, J. G. Ruby, P. Skewes-Cox, P. R. Jensen, W. Fenical and B. S. Moore, 2012. Merochlorins A—D, cyclic meroterpenoid antibiotics biosynthesized in divergent pathways with vanadium-dependent chloroperoxidases. *Journal of the American Chemical Society* 134(29): 11988–11991.

Kenawy, R., J. M. Layman, J. R. Watkins, G. L. Bowlin, J. A. Matthews, D. G. Simpson and G. E. Wnek, 2003. Electro spinning of poly (ethylene-co-vinyl alcohol) fibers. *Biomaterials* 24: 907–913.

Khil, M. S., D. I. Cha, H. Y. Kim, I. S. Kim and N. Bhattara, 2003. Electrospunnanofibrous polyurethane membrane as wound dressing. *Journal of Biomedical Material Research B* 67B: 675–679.

Kingsbury, J. M., 1984. The biology of Seaweed—Lobban, C. S., M. J. Wynne. *Bioscience* 34: 334. doi: 10.2307/1309407.

Krajewska, B., 2005. Membrane based process performed with use of chitin/chitosan materials. *Separation and Purification Technology* 41: 305–312.

Kurtzhals, P., S. Havelund, I. Jonassen, et al., 1995. Albumin binding of insulins acylated with fatty acids: Characterization of the ligand-protein interaction and correlation between binding affinity and timing of the insulin effect *in vivo*. *Biochemical Journal* 312: 725–731.

Langer, R. and J. P. Vacanti, 1993. Tissue engineering. *Science* 260: 920–926.

Laurienzo, P., 2010. Marine polysaccharides in pharmaceutical applications: An overview. *Marine Drugs* 8: 2435–2465.

Ledesma, H. B. and M. Herrero, 2014. *Bioactive Compounds from Marine Foods: Plant and Animal Sources*. 1st ed. Hoboken, NJ: Wiley-Blackwell, 464pp.

Li, H., W. Mao, Y. Hou, Y. Gao, X. Qi, C. Zhao, et al., 2012. Preparation, structure and anticoagulant activity of a low molecular weight fraction produced by mild acid hydrolysis of sulfated rhamnan fromMonostroma latissimum. *Bioresource Technology* 114: 414–418.

Li, M., G. Li, L. Zhu, Y. Yin, X. Zhao, C. Xiang, et al., 2014. Isolation and characterization of an agaro-oligosaccharide (AO)-hydrolyzing bacterium from the gut microflora of Chinese individuals. *PLoS ONE* 9(3): 1–9.

Liew, C. V., L. W. Chan, A. L. Ching and P. W. S. Heng, 2006. Evaluation of sodium alginate as drug release modifier in matrix tablets. *International Journal of Pharmaceutics* 309: 25–37.

Madihally, S. V. and H. W. T. Matthew, 1999. Porous chitosan scaffolds for tissue engineering. *Biomaterials* 20(12): 1133–1142.

Manivasagan, P. and S. K. Kim, 2014. Extracellular polysaccharides produced by marine bacteria. *Advanced Food Nutritional Res*earch 72: 79–94.

Marguerite, R., 2008. Main properties and current applications of some polysaccharides as biomaterials. *Polymer International* 57: 397–430.

McLellan, D. S. and K. M. Jurd, 1992. Anticoagulants from marine algae. *Blood Coagulation & Fibrinolysis* 3(1): 69–77.

McNaught, A. D. and A. Wilkinson, 1997. *Compodium of Chemical Terminology*. IUPAC. Oxford: Black Well Scientific Publications.

Meena, C., S. Mengi and S. Deshpande, 1999. Biomedical and industrial applications of collagen. *Journal of Chemical Science* 111(2): 319–329.

Mehta, A., C. Sidhu, A. K. Pinnaka and A. R. Choudhury, 2014. Extracellular polysaccharide production by a novel osmotolerant marine strain of *Alteromonas macleodii* and its application towards biomineralization of silver. *PLoS ONE* 9(6): 1–7.

Mondol, M., J. H. Kim, H. S. Lee, Y. J. Lee and H. J. Shin, 2011. Macrolactin W, a new antibacterial macrolide from a marine Bacillus sp, Bioorg. *Medicinal Chemistry Letters* 21(12): 3832–3835.

Moskovitz, J., M. B. Yim and P. B. Chock, 2002. Free radicals and disease. *Archives of Biochemistry and Biophysics* 397(2): 354–359.

Murshid, S. A., J. M. Badr and D. T. A. Youssef, 2016. Penicillosides A and B: New cerebrosides from the marine-derived fungus Penicillium species. *Revista Brasileria* 26: 29–33.

Nadia, R., C. Susan, G. Stefano and C. Maria, 2016. Polysaccharides from the marine environment with pharmacological, cosmeceutical and nutraceutical potential. *Molecules* 21(551): 1–16.

Necas, J. and L. Bartosikova, 2013. Carrageenan: A review. *Veterinarni Medicina* 58: 187–205.
Nelson, T. E. and B. A. Lewis, 1974. Separation and characterization of the soluble and insoluble components of insoluble laminaran. *Carbohydrate Research* 33: 63–74.
Nikaido, H., 2009. Multidrug resistant bacteria. *Annual Review of Biochemistry* 78: 119–146.
Oh, J. K., R. Drumright, D. J. Siegwart and K. Matyjaszewski, 2008. The development of microgels/nanogels for drug delivery applications. *Progress in Polymer Science* 33(4): 448–477.
Panlasigui, L. N., O. Q. Baello, J. M. Dimatangal and B. D. Dumelod, 2003. Blood cholesterol and luipid lowering effects of carrageenan on human volunteers. *Asia Pacific Journal of Clinical Nutrition* 12(2): 209–214.
Paolucci, M., G. Fasulo and M. G. Volpe, 2015. Employment of marine polysaccharides to manufacture functional biocomposites for aquaculture feeding applications. *Marine Drugs* 13(5): 2680–2693.
Pavao, M. S. G., P. A. S. Mourao, B. Mulloy and D. M. Tollefsen, 1995. A unique dermatan sulfate-like glycosaminoglycan from ascidian. Its structure and the effect of its unusual sulfation pattern on anticoagulant activity. *Journal of Biological Chemistry* 270: 31027–31036.
Paxman, J. R., J. C. Richardson, P. W. Dettnar and B. M. Corfe, 2008. Alginate reduces the increased uptake of cholesterol and glucose in overnight male subjects: A pilot study. *Nutrition Research* 28(8): 501–505.
Porto, G. G., B. C. Vasconcelos, V. A. Silva-Junior and E. S. Souza Andrade, 2010. The use of carrageenan for limiting the mandibular movement in rats: A preliminary experimental study. *Medicina Oral, Patologia Oral, Cirugia Bucal* 15: 653–657.
Prabaharan, M., and J. F. Mano, 2005. Chitosan-based particles as controlled drug delivery systems. *Drug Delivery* 12(1): 41–57.
Prabaharan, M., M. A. Rodriguez-Perez, J. A. de Saja and J. F. Mano, 2007. Preparation and characterization of poly (L-lactic acid)-chitosan hybrid scaffolds with drug release capability. *Journal of Biomedical Material Research Part B: Applied Biomaterials* 81B(2): 427–434.
Priyanka, S. and G. Ena, 2020. Role of marine microbial polysaccharides in sustainable environmental security. *Environmental Challenges and Issues in Present Scenario* 1: 16–23.
Qi, H., X. Liu, J. Zhang, Y. Duan, X. Wang and Q. Zhang, 2012. Synthesis and antihyperlipidemic activity of acetylated derivative of ulvan from Ulva pertusa. *International Journal of Biological Macromolecule* 50: 270–272.
Raju, R., Z. G. Khalil, A. M. Piggott, A. Blumenthal, D. L. Gardiner, T. S. S. Adams, and R. J. Capon, 2014. Mollemycin A: An antimalarial and antibacterial glycohexadepsipeptide- polyketide from an Australian marine-derived Streptomyces sp. (CMB-M0244). *Organic Letters* 16(6): 1716–1719.
Ramasamy, P., N. Subhapradha, S. Vairamani, and A. Shanmugam, 2014. Protective effect of chitosan from *Sepia kobiensis* (Hoyle 1885) cuttlebone against CCl_4 induced hepatic injury. *International Journal of Biological Macromolecules* 65: 559–563.
Raval, J. P., D. R. Naik and P. S. Patel, 2011. Preparation and evaluation of gatifloxacin dermal patches as wound dressing. *International Journal of Drug Development and Research* 247–259.
Raza, W., K. Makeen, Y. Wang, Y. Xu and S. Qirong, 2011. Optimization, purification, characterization and antioxidant activity of an extracellular polysaccharide produced by Paenibacillus polymyxa SQR-21. *Bioresource Technology* 102(10): 6095–6103.
Rho, S. K., L. Jeong, G. Lee, B. M. Seo, Y. J. Park, S. D. Hong, S. Roh, J. J. Cho, W. H. Park and B. M. Min, 2006. Electrospinning of collagen nanofibers: Effects on the

behavior of normal human keratinocytes and early stage wound healing. *Biomaterials* 27(8): 1452–1461.
Rinaudo, M. 2006. Chitin and chitosan: Properties and applications. *Progresses Polymeric Science* 31(7): 603–632.
Rocha, H. A. O., F. A. Moraes, E. S. Trindade, C. R. C. Franco, R. J. S. Torquato, S. S. Veiga, A. P. Valente, P. A. S. Mourao, E. L. Leite, H. B. Nader and C. P. Dietrich, 2005. Structural and haemostatic activities of a sulfated galactofucan from the brown alga *Spatoglossum schroederi*. An ideal antithrombotic agent?. *Journal of Biological Chemistry* 280: 41278–41288.
Sakoulas, G. S., J. Nam, S. Loesgen, W. Fenical, P. R. Jensen, V. Nizet and M. Hensler, 2012. Novel bacterial metabolites merochlorin A demonstrates in vitro activity against multidrug resistant methicillin-resistant Staphylococcus aureus. *PLoS One* 7(1): e29439.
Senel, S. and S. J. McClure, 2004. Potential applications of chitosan in veterinary medicine. *Advanced Drug Delivery Reviews* 56(10): 1467–1480.
Severino, P. C. Da Silva, L. Andrade, D. Oliveira, J. Campos and E. Souto, 2019. Alginate nanoparticles for drug delivery and targeting. *Current Pharmaceutical Design* 25: 1312–1314.
Shammas, A. N., H. Jeon-Slaughter, S. Tsai, H. Khalili, et al., 2017. Major limb outcomes following lower extremity endovascular revascularization in patients with and without diabetes mellitus. *Journal of Endovascular Therapy* 24: 376–382.
Shanmugam, M., and K. H. Mody, 2000. Heparinoid-active sulphated polysaccharides from marine algae as potential blood anticoagulant agents. *Current Science* 79: 1672–1683.
Siafaka, P. I., P. Barmbalexis and D. N. Bikiaris, 2016. Novel electrospun nanofibrous matrices prepared from poly(lactic acid)/poly(butylene adipate) blends for controlled release formulations of an anti-rheumatoid agent. *European Journal of Pharmaceutical Sciences* 88: 12–25.
Siddhanta, A. K., and A. Sai Krishnamurthy, 2001. Sterols from marine green algae of Indian waters. *Journal of Indian Chemical Society* 78: 431–437.
Silber, J., B. Ohlendorf, A. Labes, A. Wenzel-Storjohann, C. Näther and F. I. Johannes, 2014. Malettinin E, an antibacterial and antifungal tropolone produced by a marine Cladosporium strain. *Frontier Marine Science* 1: 1–6.
Silva, F. R. F., C. M. P. G. Dore, C. T. Marques, M. S. Nascimento, N. M. B. Benevides, H. A. O. Rocha, S. F. Chavante and E. L. Leite, 2010a. Anticoagulant activity, paw edema and pleurisy induced carrageenan: Action of major types of commercial carrageenans. *Carbohydrate Polymers*, 79: 26–33.
Silva, L. J. G., C. M. Lino, L. M. Meisel and A. Pena, 2012a. Selective serotonin re uptake inhibitors (SSRI)in the aquatic environment: An ecopharmacovigilance approach. *Science of the Total Environment* 437: 185–195.
Silva, L. P., D. Britto, M. H. R. Seleghim and O. B. G. Assis, 2010b. In vitro activity of water soluble quaternary chitosan chloride salt against *E. coli*. *World Journal of Microbiology and Biotechnology* 26: 2089–2092.
Silva, S. S., A. Motta, M. T. Rodrigues, A. F. M. Pinheiro, M. E. Gomes, J. F. Mano, R. L. Reis and C. Migliaresi, 2008. Novel genipin cross-linked chitosan/silk fibroin sponges for cartilage engineering strategies. *Biomacromolecules* 9(10): 2764–2774.
Silva, T. H., A. Alves, E. G. Popa, L. L. Reys, M. E. Gomes, R. Sousa and A. R. L. Reis, 2012b. Marine algae sulfated polysaccharides for tissue engineering and drug delivery approaches. *Biomaterials* 2: 278–289.
Sini, J. M., A. H. Yaro, L. O. Ayanwuyi, O. M. Aiyelero, S. M. Mallum and K. S. Gamaniel, 2010. Antinociceptive and antiinflammatory activities of the aqueous extract of the root bark of Combretumsericeum in rodents. *African Journal Biotechnology* 9: 8872–8876.

Smit, A. J., 2004. Medicinal and pharmaceutical uses of seaweed natural products: A review. *Journal of Applied Phycology* 16(4): 245–262.

Song, E., S. Y. Kim, T. Chun, H. J. Byun and Y. M. Lee, 2006. Collagen scaffolds derived from a marine source and their biocompatibility. *Biomaterials* 27(15): 2951–2961.

Souza, M. C. R., C. T. Marques, C. M. G. Dore, F. R. F. Silva, H. A. O. Rocha and E. L. leite, 2007. Antuioxidant activities of sulfated polysaccharide from brown and red seaweeds. *Journal of Applied Phycology* 19: 153–160.

Stiles, J., L. Guptill-Yoran, G. E. Moore and R. M. Pogranichniy, 2008. Effects of κ-carrageenan on in vitro replication of feline herpesvirus and on experimentally induced herpetic conjunctivitis in cats. *Investigative Ophthalmology & Visual Science* 49: 1496–1501.

Sudatta, B. P., V. Sugumar, R. Varma and P. Nigariga, 2020. Extraction, Characterization and Antimicrobial Activity of Chitosan from Pen Shell, Pinna Bicolor. *International Journal of Biological Macromolecule* 163: 423–430.

Sun, H. H., W. J. Ma, Y. Chen, S. D. Guo, H. Y. Li, X. H. Qi, Y. L. Chen and J. Xu, 2009. Isolation, chemical characteristics and antioxidant properties of the polysaccharides from marine fungus *Penicillium sp*. F23–2. *Carbohydrate Polymers* 78: 117–124.

Swatschek, D., W. Schatton, W. E. G. Muller and J. Kreuter, 2002. Micro particles derived from marine sponge collagen (SCMPs): Preparation, characterization and suitability for dermal delivery of all-trans retinol. *European Journal of Pharmaceutics and Biopharmaceutics* 54(2): 125–133.

Tan, Y. N., P. P. Lee and W. N. Chen, 2020. Microbial extraction of chitin from sea food waste using sugar derived from fruit waste stream. *AMB Express* 10: 1–11.

Tobacman, J. K., R. B. Wallace and M. B. Zimmerman, 2001. Consumption of carrageenan and other water soluble polymers used as food additives and incidence of mammary carcinoma. *Medical Hypotheses* 56(5): 589–598.

Tolesa, L. D., B. Gupta and M. J. Lee, 2019. Chitin and chitosan production from shrimp shells using ammonium based ionic liquids. *International Journal of Biological Macromolecules* 130: 818–826.

Tonnesen, H. H., and J. Karlsen, 2002. Alginate in drug delivery systems. *Drug Development and Industrial Pharmacy* 28(6): 621–630.

Venkata Rao, E., and K. Sri Ramana, 1991. Structural studies of a polysaccharide isolated from the green seaweed *Chaetomorphaanteninna. Carbohydrate Research* 217: 163–170.

Wahyu, M., S. Nonok and R. Endang, 2015. Decreasing blood cholesterol levels in rats induced by alginate of *Sargassum duplicatum* and *Turbinaria* sp. derived from Yogyakarta. *Asian Journal of Agriculture and Food Science* 3: 321–326.

Wang, W., S. X. Wang and H. S. Guan, 2012. The antiviral activities and mechanisms of marinepolysaccharides: An overview. *Marine Drugs* 10: 2795–2816.

Wang, Z., M. Ly, F. Zhang, W. Zhong, A. Suen, A. M. Hickey, J. S. Dordick and R. J. Linardt, 2010. E. coli K5 fermentation and the preparation of heparson a bioengineered heparin precursor. *Biotechnology and Bioengineering* 107(6): 964–973.

Wijesekara, I. and W. K. D. S. Karunarathna, 2017. Usage of seaweed polysaccharides as nutraceuticals. Nugegoda: Elsevier, 341–348. doi: 10.1016/B978-0-12-809816-5.00018-9

Yahya, E. B., A. A. Amirul, H. P. S. Abdul Khalil, N. G. Olaiya, M. O. Iqbal, F. Jummaat, A. K. Atty sofea and A. S. Adnan, 2019. Insights in to the role of biopolymer aerogel scaffolds in tissue engineering and regenerative medicine. *Polymer* 13(10): 1612.

Yasasvini, S., R. S. Anusa, B. N. VedhaHari, P. C. Prabhu and D. RamyaDevi, 2017. Topical hydrogel matrix loaded with Simvastatin microparticles for enhanced wound healing activity. *Materials Science and Engineering: C* 72: 160–167.

Yuan, H., W. Zhang, X. Li, X. Lü, N. Li, X. Gao and J. Song, 2005. Preparation and in vitro antioxidant activity of κ-carrageenan oligosaccharides and their oversulfated, acetylated, and phosphorylated derivatives. *Carbohydrate Research* 340: 685–690.

Zhao, X., G. Yu, H. Guan, N. Yue, Z. Zhang and H. Li, 2007. Preparation of low-molecular-weight polyguluronate sulfate and its anticoagulant and anti-inflammatory activities. *Carbohydrate Polymer* 69: 272–279.

Zhou, Y. C. and R. L. Zheng, 1991. Phenolic compounds and analog as superoxide anion scavengers and antioxidants. *Biochemical Pharmacology* 42: 1177–1179.

Zierer, M. S. and P. A. Mourao, 2000. A wide diversity of sulfated polysaccharides are synthesized by different species of marine sponges. *Carbohydrate Research* 328: 209–216.

Zorzi, A., S. J. Middendrop, J. Wilbs, K. Deyle and C. Heinis, 2017. Acylated heptapeptide binds albumin with high affinity and application as tag furnishes long acting peptides. *Nature Communications* 8: 16092.

17 Pharmaceutical Applications of Major Marine Nutraceuticals

Astaxanthin, Fucoxanthin, Ulvan, and Polyphenols

Madan Kumar P, Janani R, Priya S, Naveen J, and Baskaran V

CONTENTS

ABBREVIATIONS

ASX—Astaxanthin, CVD—Cardiovascular diseases, EPA—Eicosapentaenoic acid, FUC—Fucoxanthin, FDA—Food and Drug Administration, HMGB1 - High-mobility group box 1, HFD—High fat diet, IL—Interleukin, LDL—low-density lipoprotein, MAPK—Mitogen associated protein kinases, NF-κB—Nuclear factor- κB, PI3K/Akt—Phosphoinositide 3-kinase/protein

DOI: 10.1201/9781003303909-17

kinase B, TNF-α—Tumor necrosis factor-alpha, VEGF—Vascular endothelial growth factor, WAT—White adipose tissue

17.1 INTRODUCTION

Food plays a decisive role in one's health mainly in the prevention or onset and progression of lifestyle-associated diseases, such as diabetes, obesity, hypertension, atherosclerosis and cancer. Lately, the consumption of excess sugars, fats and calories, alongside sedentary lifestyles, have contributed to the severity of diseases. To treat or prevent these lifestyle-related diseases, pharmacological intervention alone is not adequate; hence, nutrition plays a vital role (Kuipers et al., 2011; Naveen and Baskaran, 2018; Naveen et al., 2021; Vani et al., 2021; Venkateish et al., 2021). Hence, researchers have started looking for safe and effective food ingredients to prevent or treat lifestyle-related diseases (Beppu et al., 2009a). In recent years, there has been a great awareness among people about the association among food, health and disease management, which consequently has directed research on the development of functional foods. In this context, great attention was bestowed on marine nutraceuticals in the last few decades mainly because of their extensive pharmacological significance. Marine nutraceuticals typically refer to those compounds, which are obtained from the sea. Marine nutraceuticals are mostly composed of lipids/fatty acids, polysaccharides, carotenoids, polyphenols, proteins, enzymes, minerals and vitamins. A vast number of *in vitro* and *in vivo* studies have demonstrated the health-promoting effects of marine nutraceuticals including antioxidant, antidiabetic, anti-inflammatory, anti-obesity, anticancer, cardioprotective and immunomodulatory, among others (Jayapal et al., 2019). In this chapter, the pharmacological significance of the major marine nutraceuticals, such as astaxanthin (ASX), fucoxanthin (FUC), ulvan and polyphenols are discussed in detail.

17.2 ASX

ASX is a lipid-soluble keto-carotenoid that is characterized under a group of carotenoids known as xanthophylls (Chang and Xiong, 2020; Kumar et al., 2021). It is widely seen in the marine organisms such as salmon, shrimp, krill, trout and crayfish (Hussein et al., 2006). ASX synthesis chiefly occurs in phytoplankton/microalgae, which then accumulates in the zooplankton and eventually enters crustaceans and the human food chain. ASX is also synthesized by algae, yeast and plants and through chemical synthesis (Kumar et al., 2021). The major source of natural ASX includes *Haematococcus pluvialis* (microalgae) and *Xanthophylomyces dendrorhous* (red yeast). Since humans cannot synthesize carotenoids, they are essentially consumed in through diet. The U.S. Food and Drug Administration (FDA) has approved ASX for application as a feed supplement for aquaculture as well as nutraceutical for humans (Guerin et al., 2003).

17.2.1 Chemical Structure of ASX

ASX has a polyene chain with conjugated double bonds and two β-ionone rings that has hydroxyl and keto groups. Depending on the hydroxyl group configuration in β-ionone rings, three stereoisomers exist for ASX namely (3S, 3′S), (3R, 3′R) and (3S, 3′R; Figure 17.1). The stereoisomer content of ASX varies between the organisms. For instance, esterified 3S, 3′S stereoisomer is found predominantly in *H. pluvialis* whereas *X. dendrorhous* contain nonesterified 3R, 3′R form of ASX (Ambati et al., 2014). While all the stereoisomers ([3S, 3′S], [3R, 3′S] and [3R, 3′R]) are found in the chemically synthesized ASX in 1:2:1 ratio, respectively (Higuera-Ciapara et al., 2006; Kumar et al., 2021). Besides, based on the configuration of double bonds in polyene chain, ASX has two geometric isomeric forms viz., trans (E) and cis (Z) isomers while all-trans-ASX (all-E-ASX) stands to be a prominent form in nature (Higuera-Ciapara et al., 2006). ASX occurs either in free, esterified with fatty acids or a protein-conjugated form. The monoester (esterified with one fatty acid), diester (esterified with two fatty acids) and the conjugated forms of ASX are predominant in natural sources since it has the advantage of high stability compared to the free form, which is susceptible to oxidation (Hussein et al., 2006).

17.2.2 Bioavailability of ASX

Like most other carotenoids, ASX is hydrophilic in nature and has low bioavailability, limiting its use in pharmaceutical applications. Methods such as incorporating lipids have been shown to enhance the bioavailability of ASX (Odeberg et al., 2003). Similarly, Rao et al. (2013) reported the enhancement of ASX bioavailability in experimental animals on dispersing the *H. pluvialis* biomass in olive oil. Also, strategies including incorporating ASX into liposomes, emulsification, microencapsulation, nanoemulsion and nanoparticle delivery have been shown to enhance ASX bioavailability (Peng et al., 2010; Affandi et al., 2011; Bustos-Garza et al., 2013; Wang et al., 2017).

17.2.3 Pharmaceutical Activities of ASX

ASX has gained significant interest in the recent years due to its potential pharmaceutical effects including antioxidant, antidiabetic, anti-inflammatory, anticancer, cardioprotective, hepatoprotective, neuroprotective, ocular protective and skin protective (Kumar et al., 2021). In this chapter, we aim to summarize the recent studies focused on the biological effects and underlying mechanisms of ASX in treating pathological conditions like inflammation, diabetes, cancer and cardiovascular diseases.

Numerous studies demonstrated the anti-inflammatory potential of ASX, where ASX administration significantly modulated the inflammation in several diseases including arthritis, diabetes, cancer, cardiovascular, eye, neurological and skin. In a dry eye model under hyperosmolarity,

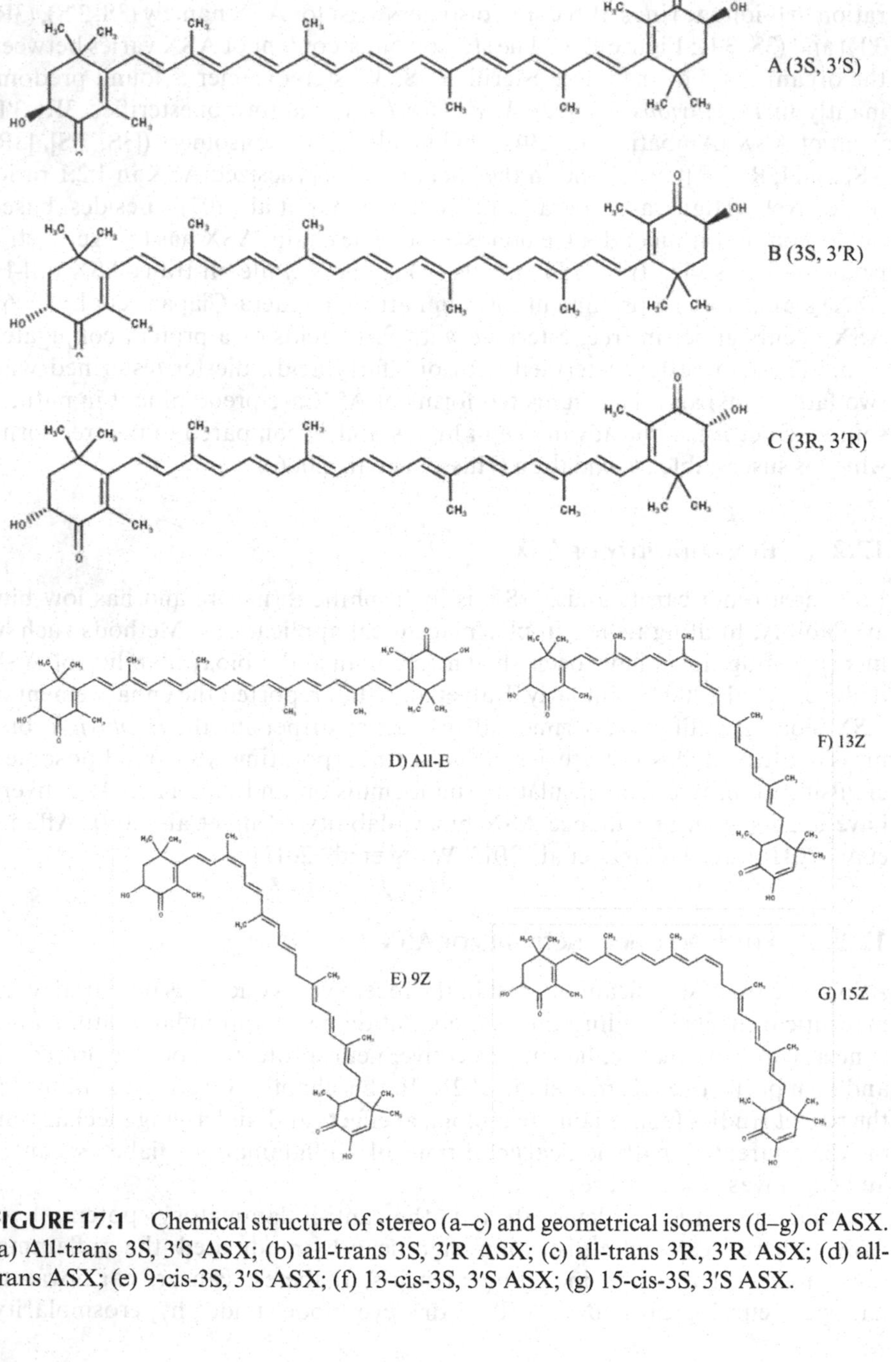

FIGURE 17.1 Chemical structure of stereo (a–c) and geometrical isomers (d–g) of ASX. (a) All-trans 3S, 3′S ASX; (b) all-trans 3S, 3′R ASX; (c) all-trans 3R, 3′R ASX; (d) all-trans ASX; (e) 9-cis-3S, 3′S ASX; (f) 13-cis-3S, 3′S ASX; (g) 15-cis-3S, 3′S ASX.

preconditioning with ASX significantly downregulated the expression of inflammatory markers, such as tumor necrosis factor-α (TNF-α), interleukin (IL)-1β, high-mobility group box 1 (HMGB1) and phosphoinositide 3-kinase/protein kinase B (PI3K/Akt) signaling (Li et al., 2020). ASX offered protection against arthritis and cardiovascular disease by attenuating oxidative stress and inflammatory response (Pereira et al., 2021). The effect of ASX in inflammatory-associated diseases has been reviewed in detail by Chang and Xiong (2020). In a randomized, double-blind placebo-controlled study, the supplementation of curcumin, eicosapentaenoic acid (EPA), ASX and gamma-linoleic acid reduced the inflammatory markers and improved the endothelial function, EPA levels and fatty acid indices, suggesting the clinical significance of nutraceuticals in treating inflammatory-associated diseases (Birudaraju et al., 2020).

ASX is well studied for the management of diabetes and diabetic complications. Oral administration of ASX (8 mg/day) for 8 weeks markedly improved the glucose metabolism in db/db mice, obese model for type II diabetes, as well as in type II diabetic patients (Sila et al., 2015; Mashhadi et al., 2018). ASX also been shown to modulate the diabetes-related complications, such as retinopathy, nephropathy and neuropathy, mainly by reducing the levels inflammatory mediators, including nuclear factor-κB (NF-κB), IL-1β, IL-6, TNF-α and vascular endothelial growth factor (VEGF). A study by Landon et al. (2020) demonstrated ASX's modulation on key pathways, such as mitogen-associated protein kinases (MAPK) and PI3K/Akt, pathways that are involved in diabetic complications. In diabetic retinopathy conditions, ASX treatment protected retinal cells by downregulating VEGF and increased the antioxidant status through PI3K/Akt/Nrf2 pathway (Lai et al., 2020; Janani et al., 2021). Similarly, ASX alleviated cognitive dysfunction and diabetic nephropathy by reducing oxidative stress and inflammation through Nrf2 and NF-κB pathways (Chen et al., 2018; Feng et al., 2018).

ASX exhibited anticancer effect by targeting various molecular mechanisms involved in the cancer cell progression and migration. ASX inhibited oxidative stress and inflammation, thereby promoting apoptosis of cancer cells (Song et al., 2011; Franceschelli et al., 2014). ASX treatment induced apoptosis of cancer cells by downregulating anti-apoptotic proteins (Bcl-2, p-Bad and survivin) and promoting the pro-apoptotic proteins (Bax/Bad and PARP) (Faraone et al., 2020). ASX treatment suppressed the proliferation of several cancer cells by targeting the NF-κB, STAT3, PI3K/AKT, MAPKs, PPAR-γ pathways (Sun et al., 2020). A systemic review conducted by Faraone et al. (2020) highlighted the detailed mechanism of anticancer activity of ASX in experimental models. In breast cancer cells, ASX treatment enhanced the antiproliferative effect of anticancer drugs (carbendazim and doxorubicin; Atalay et al., 2019; Fouad et al., 2021).

Cardiovascular diseases (CVDs) are a major cause of morbidity worldwide and typically associated with vascular damage. ASX treatment was shown to suppress oxidative stress and inflammation, which are the initial events of CVD (Pereira et al., 2021). Administration of ASX (10 mg/kg) to diabetic rats reduced the oxidation of low-density lipoprotein (LDL) and restored the expression of endothelial nitric oxide synthase (Zhao et al., 2011). In hyperlipidemic rats, ASX reduced coagulation and platelet aggregation and increased fibrinolytic activity (Deng et al., 2017). ASX also attenuated the thrombotic risk factors, reduced inflammatory markers and reversed the diabetic mediated changes in the plasma lipid profile of type II diabetic changes (Chan et al., 2019). In a recent study by Chen et al. (2020), ASX inhibited oxidative stress–induced mitochondrial dysfunction in vascular smooth muscle cells thereby prevented hypertensive mediated vascular remodeling. A pilot study by Kato et al. (2020) determined the effect of ASX the cardiac function in heart failure patients. The study further demonstrated that ASX supplementation reduced oxidative stress and improved cardiac contractility.

17.3 FUC

FUC, a marine carotenoid, is considered the most abundant natural carotenoid, accounting for ≥10% of the estimated total natural production of carotenoids. FUC is present in the macroalgae and microalgae, such as *Undaria pinnatifida*, *Laminaria japonica*, *Phaeodactylum tricornutum* and *Cylindrotheca closterium* (D'Orazio et al., 2012; Kim et al., 2012). In Southeast Asian countries, some seaweeds containing FUC are consumed in human diet. In addition, the safety of FUC has been extensively studied in several animal experiments (Beppu et al., 2009b; Tsukui et al., 2009; Hitoe et al., 2017). Based on remarkable biological properties and safety, FUC as a nutraceutical is highly regarded in the food and pharma industries.

17.3.1 Chemical Structure of FUC

The chemical name of FUC is (3S,5R,6S,3′S,5′R,6′R)-5,6-Epoxy-3′-ethanoyloxy-3,5′- dihydroxy-6′,7′-didehydro-5,6,7,8,5′,6′-hexahydro-beta,-beta-caroten-8-one with the molecular formula $C_{42}H_{58}O_6$ and a molecular weight of 658.91 (Yan et al., 1999). FUC has a unique chemical structure containing allene bonds, 5, 6-monoepoxide groups and nine conjugated double bonds. The conjugated double-bond system in FUC readily quenches singlet oxygen ($_1O^2$), and the electron-rich status of FUC makes it more suitable to react with free radicals. FUC is easily affected by heat, oxygen tension, pH and light. Nevertheless, the unique molecular structure and chirality of FUC are unstable. Because of its chiral structure and the allenic bond, FUC possesses higher antioxidant activity.

17.3.2 Bioavailability of FUC

It is essential to understand the metabolic process and methods to improve the bioavailability of FUC. FUC is metabolized into two main molecules, namely, fucoxanthinol and amarouciaxanthin (Figure 17.2). In an animal study, dietary FUC was rapidly hydrolyzed to fucoxanthinol in the gastrointestinal tract within 2h after FUC administration, and no unchanged FUC was detected in

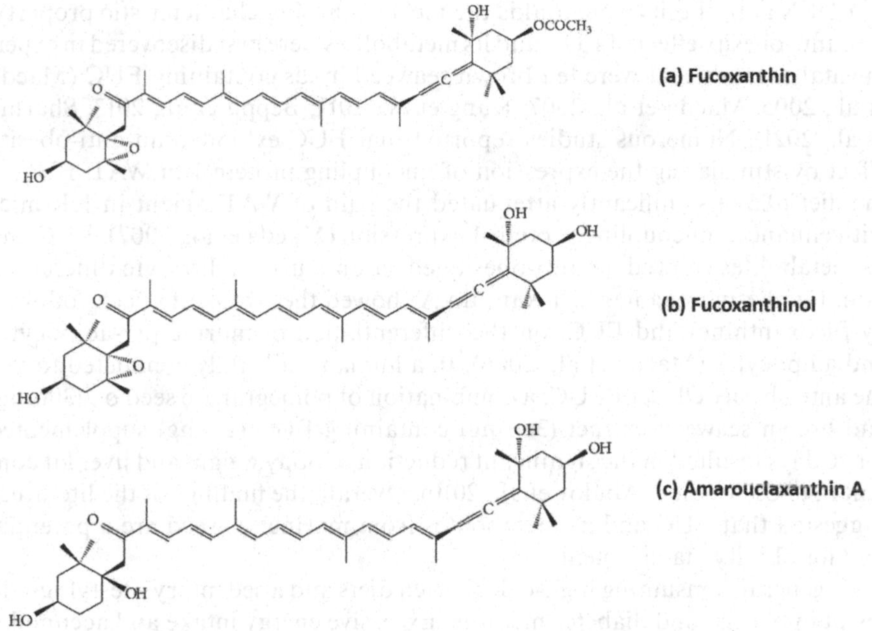

(a) Fucoxanthin

(b) Fucoxanthinol

(c) Amarouclaxanthin A

FIGURE 17.2 Chemical structures of (a) FUC, (b) fucoxanthinol, and (c) amarouciaxanthin A.

the plasma or liver (Asai et al., 2004; Matsumoto et al., 2010). In another study, FUC was further converted into amarouciaxanthin A through isomerization in mice liver microsomes and HepG2 cells (Matsumoto et al., 2010). An *in vivo* study demonstrated that dietary FUC is detected in the heart and liver as fucoxanthinol while in visceral adipose tissue and white adipose tissue (WAT) it is detected as amarouciaxanthin A (Airanthi et al., 2011). Many studies have debated the bioavailability of FUC in cellular, animal and human models. The bioavailability of FUC in Caco-2 cells was reported to be the lowest out of 11 carotenoids tested (Sugawara et al., 2001). Animal studies have shown the ratio of absorbed FUC to metabolite, FUC and amarouciaxanthin A exhibited higher absorption than ASX (Hashimoto et al., 2009). To increase the bioavailability and stability of FUC, approaches such as encapsulation of FUC into nanoparticles, nanoemulsions and other spray-dried powders are investigated.

17.3.3 Pharmaceutical Activities of FUC

FUC, a superior bioactive from marine seaweed, exhibits various pharmacological activities. Numerous animal studies have shown that FUC has the potential to prevent and treat lifestyle-related diseases, such as diabetes, obesity, CVD, cancer and other chronic diseases (Zhang et al., 2015; Jayapal et al., 2019; Sharma et al., 2021).

FUC's anti-obesity potential is the most promising characteristic property. The anti-obesity effect of FUC and its metabolites were first discovered in experimental animals that were fed brown seaweed lipids containing FUC (Maeda et al., 2005; Maeda et al., 2007; Kang et al., 2011; Beppu et al., 2013; Sharma et al., 2021). Numerous studies reported that FUC exhibited an anti-obesity effect by stimulating the expression of uncoupling protein-1 in WAT. FUC in the diet (0.2%) significantly attenuated the gain of WAT weight in KK-mice with enhanced uncoupling protein-1 expression (Maeda et al., 2007). FUC and its metabolites showed an anti-obesity effect on murine adipocyte differentiation. Furthermore, amarouciaxanthin A showed the strongest effect, followed by fucoxanthinol and FUC, on the differentiation of murine pre-adipocytes and adipocytes (Maeda et al., 2006). In a human trail study conducted to test the anti-obesity effect of FUC, a combination of pomegranate seed oil (300 mg) and brown seaweed extract (300 mg) containing FUC (2.4 mg) supplemented for 16 days resulted in the significant reduction of body weight and liver fat content in obese women (Abidov et al., 2010). Overall, the findings of the literature suggested that FUC and its metabolites from marine seaweed are a potential tool for obesity management.

In general, consuming high-calorie-rich diets and a sedentary lifestyle could result in obesity and diabetes mellitus. Excessive energy intake and accumulation of lipids in obesity can uplift the insulin resistance that leads to diabetes mellitus (Campfield et al., 1999). FUC has been shown to play an important role in reducing insulin resistance and elevated blood glucose levels. FUC administration to high-fat diet (HFD)–fed C57BL/6J mice resulted in significant reduction in blood glucose levels. However, FUC supplementation to C57BL/6J mice fed with normal chow diet did not affect the blood glucose levels, indicating the specificity of the blood glucose-lowering effect of FUC in diabetic conditions (Maeda et al., 2009; Hosokawa et al., 2010). TNF-α is involved in the development of type II diabetes, and its expression is positively correlated with insulin resistance (Hotamisligil et al., 1993). The administration of FUC (0.2%) to KK-mice significantly downregulated TNF-α mRNA levels and reduced the blood glucose and plasma insulin (Maeda et al., 2007). An FUC-rich diet ameliorated insulin resistance by promoting the expression of glucose transporter 4 mRNA in skeletal muscle tissues (Maeda et al., 2009).

In comparison with all other carotenoids extracted from marine seaweeds, FUC and its metabolites exhibited greater antiproliferative potential in several cancer types (Kumar et al., 2013; Zorofchian et al., 2014; Takahashi et al., 2015). The effect of different carotenoids (phytoene, phytofluene, ξ-carotene, lycopene, α-carotene, β-carotene, β-cryptoxanthin, canthaxanthin, ASX, capsanthin, lutein, zeaxanthin, vioaxanthin, neoxanthin, FUC) on the growth of human prostate cancer cells (PC-3, DU 145 and LNCap) were examined. Among the carotenoids, neoxanthin and FUC exhibited a higher antiproliferative effect (Kotake-Nara et al., 2001). FUC has also been reported to suppress the growth and number of tumours in animal models (Wang et al., 2012; Kim et al., 2013). In an animal colon cancer model, brown seaweed extract containing FUC exhibited chemopreventive activity against the preneoplastic marker (Das et al., 2006). The anticancer potential of FUC was reported to interfere with various pathways involved in cell cycle arrest, apoptosis or metastasis suppression. FUC treatment induced G0/Gl- and G2/M-phase cell-cycle arrest by altering the expression of various genes including GADD45, p21, p27, cyclin Dl, cyclin D2, CDK4 and survivin. The pro-apoptotic effect of FUC is well studied, and FUC-mediated apoptosis targets different molecular pathways, including Bcl-2, caspases, MAPK and NF-κB (D'Orazio et al., 2012; Kumar et al., 2013; Zhang et al., 2015). FUC administration to experimental animals significantly inhibited tumor development in a xenograft colorectal cancer model (Terasaki et al., 2017). In a similar study, FUC administration significantly reduced the numbers of colorectal cancer stem cells in colonic mucosa compared to control mice (Terasaki et al., 2019).

17.4 ULVAN

Ulvan is a major cell wall–derived sulfated polysaccharide present exclusively in green seaweed of the genus *Ulva*. The name "Ulvan" was derived from the term *Ulvin*, and Ulvacin introduced by Kylin for the different water-soluble polysaccharide fractions of *Ulva lactuca* (Kylin, 1946). Ulvan contributes 9–36% dry weight of the marine biomass seaweed and is chiefly formed of uronic acids (iduronic acid and glucuronic acid), sulfated rhamnose and xylose (Kidgell et al., 2019). Extracting ulvan can be achieved at about 80–90°C of water (Abdel-Fattah and Edress, 1972). Ulvan is generally composed of uronic acids (iduronic acid and glucuronic acid), sulfated rhamnose and xylose (Figure 17.3; Lahaye and Robic, 2007), and ulvan is the only polysaccharide consisting of both iduronic acid and rhamnose (Quemener et al., 1997). Ulvans are polyanionic heteropolysaccharides with sugar compositions with a range of rhamnose (5.0–92.2 mol), glucuronic acid (2.6–52.0 mol), xylose (0.0–38.0 mol) and iduronic acid (0.6–15.3 mol; Kidgell et al., 2019). The content of the sugars varies with the source and eco-physiological conditions and, importantly, depends on the extraction procedures and analytical techniques.

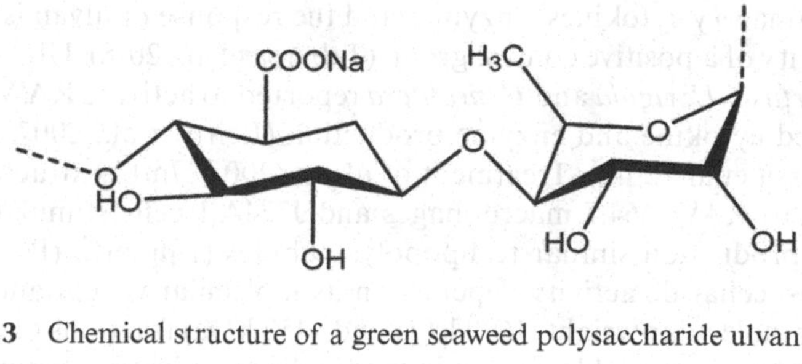

FIGURE 17.3 Chemical structure of a green seaweed polysaccharide ulvan.

17.4.1 Chemical Structure of Ulvan

The chemical structure of ulvan is composed of repeated sulfated disaccharide of neutral sugar linked with uronic acid (i.e., like gycosaminoglycans [GAGs]). Hence, ulvan is considered a candidate for the modulation of function and processes carried out by mammalian polysaccharides (Quemener et al., 1997; Lahaye et al., 1999). Also, ulvan is considered a potential biomolecule that has lot of applications in biomaterial science (biofilm prevention, tissue engineering, wound dressings) and as a nutraceutical (antioxidant, antiviral, immunomodulatory, antihyperlipidemic, anticancer) and has a major role in the agriculture and food industries (Wijesekara et al., 2011; Venkatesan et al., 2015). The health benefits of ulvan are directly related to its chemical structure. Hence, researchers need to understand the extraction and purification protocol, because extraction protocols affect the chemical structure of the polysaccharide and meanwhile its biological activity.

17.4.2 Pharmaceutical Activities of Ulvan

Li et al. (2018a) demonstrated the antioxidant activity of ulvan extracted from *Ulva pertusa* in the liver of hyperlipidemic Kunming mice model. The antioxidant activity of ulvan depends on its molecular weight and sulfation. The lower molecular weight and sulfated ulvan showed more antioxidant activity than its native form (Kidgell et al., 2019). Ulvan-rich polysaccharides extracted from *Ulva pertusa* showed antioxidant activity both *in vitro* and *in vivo* (Qi and Sun, 2015). Ulvan treatment to RAW 264.7 cells protected the hydrogen peroxide–induced oxidative damage by increasing the expression of catalase and superoxide dismutase (Le et al., 2019). Ulvan (1%) administration to hens significantly reduced serum MDA levels and increased catalase and superoxide dismutase, which correlated well with improved quality of eggs and egg-laying hens (Li et al., 2018b).

Ulvan from *U. rigida* increased nitric oxide generation and helped fight against pathogenic organisms (Leiro et al., 2007). Sulfated ulvan from *U. intestinalis* activated the immune system by macrophage stimulation (Rahimi et al., 2016). The effect of ulvan on inflammation has been studied using various molecular markers, active metabolites, enzymes, transcription-related molecules and immunoglobulins (Kidgell et al., 2019). Ulvan (50 μg/mL) extracted from *U. intestinalis* induced the production of pro-inflammatory cytokines, anti-inflammatory cytokines, enzymes and the response of ulvan is equivalent to the activity of a positive control group (Tabarsa et al., 2018). Ulvan extracted from *U. pertusa*, *U. rigida* and *U. prolifera* reported to activate RAW 264.7 cells and induced cytokine and enzyme production (Leiro et al., 2007; Cho et al., 2010; Tabarsa et al., 2012). Treatment of ulvan (200 μg/mL) extracted from *U. intestinalis* to RAW 264.7 macrophages and J774A.1 cells stimulated TNF-α and IL-1β production similar to lipopolysacchides (1 μg/mL) (Peasura et al., 2016). Polysaccharide activity depends on its molecular weight and sulfation. The higher molecular-weight (1690 kDa and 1450 kDa) fractions of ulvan of *U. pertusais* showed a twofold increase in macrophage activation compared to the lower molecular-weight fraction (365 kDa; Tabarsa et al., 2012). Later, Tabarsa et al. (2018) reported that the lower molecular-weight (28.7 kDa) fraction of ulvan exhibited higher immunomodulatory effect than the higher molecular-weight (87.2 kDa) fraction and concluded that the higher biological activity of the lower molecular-weight fraction was due to sulphation.

Oral treatment using ulvan extracted from *U. pertusa* to hypercholesterolemic rats for 21 days resulted in significant increase in the levels of high-density lipoproteins and cholesterol and a decrease in the triglyceride level compared to control rats (Pengzhan et al., 2003). Ulvan from *U. fasciata* when administered to hypercholesterolemic rats for 4 weeks showed a significant decrease in the level of lipid (30%), triglycerides (46%) and total cholesterol (69%) (Nikita et al., 2018). Experimental rats that were fed with *U. pertusa* polysaccharides showed a marked increase in bile acid production (Qi and Sun, 2015). Administering *U. fasciata* aqueous polysaccharide extracts at doses 200 and 400 mg/kg to diabetic rats reduced blood glucose levels (Abirami and Kowsalya, 2013). Ulvan (151.6 kDa) extracted from *U. pertusa* and used to treat Wistar rats showed a significant reduction in total cholesterol (45%) and LDL cholesterol (54%) (Pengzhan et al., 2003). Likewise, starch diets, along with lower molecular-weight ulvan (28.2 kDa) administered to rats showed a significant reduction in triglycerides (78%) and increased high-density lipoprotein (HDL) cholesterol (61%) (Li et al., 2018c). Qi and Sheng (2015) performed a comparative study of native and sulfated ulvan in hyperlipidemic rats and observed 28% reduction in serum total cholesterol in native Ulvan and 44% in sulfated ulvan group. Ulvan from *Monostroma nitidum* and treated with lipid-loaded hepatocytes showed a reduction in the cholesterol synthesis gene and induced cholesterol catabolism gene expression and LDL-gene expression (Hoang et al., 2015).

Ulvan species were reported to exhibit anticancer activity, that is, by apoptosis of the cancer cell through anti-inflammation properties (Thanh et al., 2016; Hu et al., 2018). Several studies demonstrated the anticancer effect of ulvan using cancer cells and animals (Shao et al., 2013; Thanh et al., 2016; Abd-Ellatef et al., 2017). Ulvan isolated from *U. prolifera* inhibited cell proliferation of human colon cancer cells and human gastric carcinoma cells in a dose-dependent manner (Cho et al., 2010). Ulvan extracted from *U. intestinalis* reduced tumor weight by 61–71% in mice model; that is, it was dose-dependent and the dose range was 100–400 mg/kg (Jiao et al., 2009). The anticancer activity of ulvan also depends on the molecular weight of ulvan and the degree of sulphation (Matloub et al., 2016). A recent study by Zhao et al. (2020) revealed that *U. lactuca* polysaccharides acted as an antioxidant, anti-inflammatory and antitumor molecule. A *U. lactuca* polysaccharide treatment inhibited tumor growth by enhancing p53 and IKKα and inhibiting the expression of p65, P13K/Akt and mTOR and CD31/VEGF expression. In addition, exciting potential applications of ulvan in cancer therapy as a nanoparticle for drug delivery systems (Li et al. 2018c), polysaccharide-protein complex for cancer treatment (Sun et al., 2017) and pH-responsive polysaccharide nanosystems to angiogenesis inhibition (Yang et al., 2017) have been reported. There is a need to ascertain the ulvan bioavailability and whether ulvan influence the efficacy of traditional chemotherapy drugs in combination therapies.

17.5 POLYPHENOLS

Marine polyphenols typically consist of flavonoids, phenolic terpenoids, bromophenols, phlorotannins and mycosporine-like amino acids. Seaweeds are a rich source of polyphenols. Green and red seaweeds are composed of bromophenols, phenolic acids and mycosporine-like amino acids (Corona et al., 2017). While phlorotannins are the unique class of polyphenols found exclusively in the brown seaweeds, constituting up to 15% of their dry weight (Targett and Arnold, 1998). Phlorotannins are polyphenolic compounds biosynthesized by polyketide pathway and typically formed by a group of complex polymers of phloroglucinol (1, 3, 5-trihydroxybenzene) (Arnold et al., 2002; Li et al., 2017). Bromophenols are another marine algae–derived polyphenolic compound composed of one or more benzene rings, a varying degree of bromine and hydroxyl-substituents.

Phlorotannins are reported for numerous pharmacological activities including antioxidant, antimicrobial, antifungal, antidiabetic, anti-inflammatory, anticancer, neuroprotective, and more (Vo et al., 2019). Phlorotannins extracted from the brown seaweed *C. trinodis* exhibited antioxidant activity (Sathya et al., 2017). The antifungal activity of phlorotannins isolated *C. nodicaulis*, *C. usneoides* and *F. spiralis* was studied against pathogenic yeast and filamentous fungi (Lopes et al., 2013). A study by Manandhar et al. (2019) highlighted the neuroprotective effect of phlorotannins against oxidative stress and inflammation. In lipopolysacchide-stimulated primary and RAW 264.7 macrophages, phlorotannin 6,6′-bieckol that are extracted from *E. cava*–inhibited NO and PGE2 production by suppressing the expression of iNOS and COX-2 (Yang et al., 2012). Eckol, a phlorotannin, showed antiproliferation and pro-apoptotic effects, thereby inhibiting tumor development in sarcoma 180 xenograft-bearing animals (Zhang et al., 2019).

Bromophenols can be isolated from many marine algae, including the red algae *Rhodomela larix* (Katsui et al., 1967). Bromophenols have been reported to possess a variety of biological activities, including antioxidant, antimicrobial, antidiabetic, anticancer and antithrombotic effects. Duan et al. (2007) showed the antioxidant effect of bromophenols isolated from *Symphyocladia latiuscula* by free radical scavenging assay. In a study conducted with eight strains of gram-positive and gram-negative bacteria, bromophenol derivatives from *Rhodomela confervoides* exhibited antibacterial activity (Xu et al., 2003). Bromophenols extracted from the *Leathesia nana* exhibited a cytotoxic effect against human cancer cell lines (Shi et al., 2009). A series of studies from Guo et al. demonstrated the anticancer effect of bromophenol derivatives (BOS-93 and BOS-102) in human lung cancer cells. In this study, applying a bromophenol-derivative treatment to A549 cells inhibited cell proliferation and induced G0/G1 phase arrest and apoptosis (Guo et al., 2018; Guo et al., 2019).

Mycosporine-like amino acids are a group of secondary metabolites that are present exclusively in red seaweeds. A study conducted with HeLa and U-937 cells highlighted the antiproliferative effect of mycosporine-like amino acids from methanolic extracts from edible wild-harvested (*C. crispus*, *M. stellatus*, *P. palmata*) and cultivated (*C. crispus*) marine red macroalgae (Athukorala et al., 2016). Recently,

Orfanoudaki et al. (2020) studied the effect of different mycosporine-like amino acids in HaCaT keratinocytes. Mycosporine-like amino acids treatment increased the proliferation and migration thereby exhibiting its wound-healing ability.

17.6 CONCLUSION AND FUTURE LINE OF WORK

Marine nutraceuticals are a great source of bioactives and known for greater pharmacological importance (Figure 17.4). The evidence supporting the notion of marine nutraceuticals as efficient antioxidant and pharmacological agents is

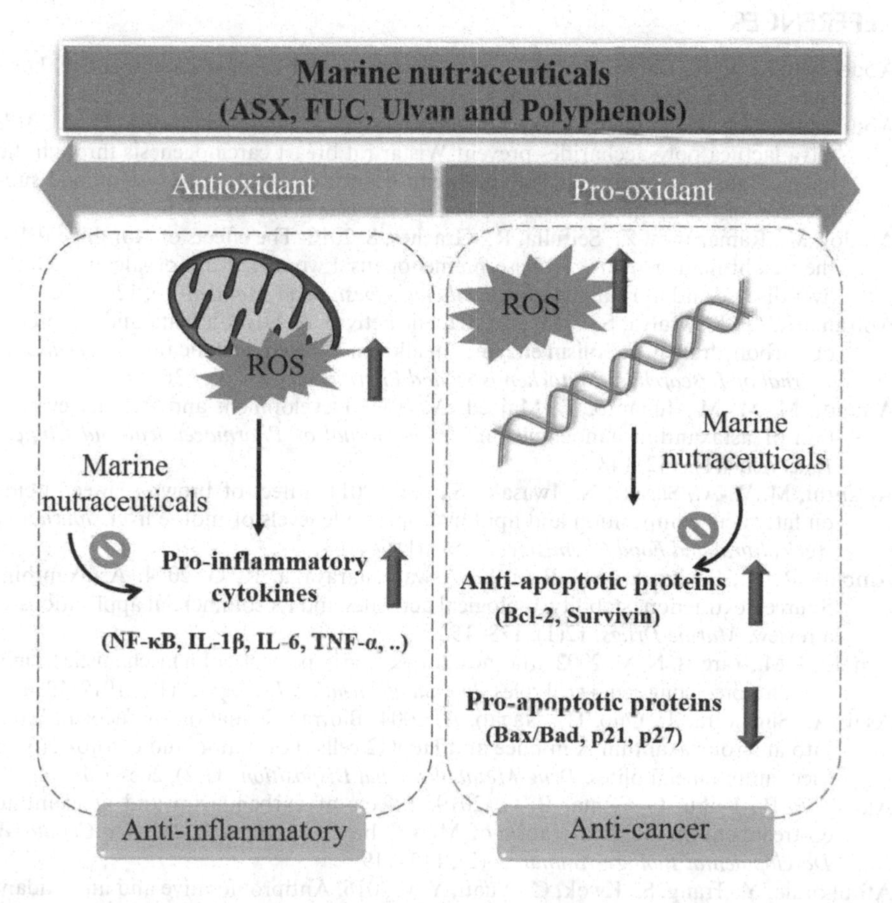

FIGURE 17.4 The possible mechanism of action by which the marine nutraceuticals (ASX, FUC, ulvan, polyphenols) inhibits the progression of diseases. In normal cells, the nutraceuticals act as antioxidants and suppress the stress-induced damage to cells, whereas the existence of high oxidative stress in cancer cells predisposes the nutraceuticals to act as pro-oxidants and leads to the apoptosis of cancer cells. Bax, Bcl-2-associated X protein; Bcl-2, B-cell lymphoma 2, Bad, Bcl-2-associated agonist of cell death; IL-1β, interleukin-1β; IL-6, interleukin-6; NF-κB, Nuclear factor kappa B; ROS, reactive oxygen species; TNF-α- tumor necrosis factor-α; VEGF, vascular endothelial growth factor.

very clear from the pre-clinical and clinical studies. Although it clearly indicated the promising utilization of marine nutraceuticals in health care, the knowledge gaps in the research of marine nutraceuticals must be addressed to position them in treating the diseases. First, the bioavailability and tissue distribution of marine carotenoids and polyphenols must be characterized in humans. Next, the biological effects of different marine nutraceuticals in humans should be studied in detail to understand the impact of product composition. Additionally, the optimal dose of marine nutraceuticals must be fixed based on evidence-based approach for various diseases.

REFERENCES

Abdel-Fattah, A. F., Edrees, M. 1972. A study on the polysaccharide content of Ulva lactuca L. *Qualitas Plantarum et Materiae Vegetabiles*, 22, 15–22

Abd-Ellatef, G. E. F., Ahmed, O. M., Abdel-Reheim, E. S., Abdel-Hamid, A. H. Z. 2017. Ulva lactuca polysaccharides prevent Wistar rat breast carcinogenesis through the augmentation of apoptosis, enhancement of antioxidant defense system, and suppression of inflammation. *Breast Cancer: Targets and Therapy*, 9, 67–83

Abidov, M., Ramazanov, Z., Seifulla, R., Grachev, S. 2010. The effects of Xanthigen™ in the weight management of obese premenopausal women with non-alcoholic fatty liver disease and normal liver fat. *Diabetes, Obesity and Metabolism*, 12(1), 72–81

Abirami, R. G., Kowsalya, S. 2013. Antidiabetic activity of Ulva fasciata and its impact on carbohydrate metabolism enzymes in alloxan induced diabetic rats. *International Journal of Research in Phytochemistry and Pharmacology,* 3(3), 136–141

Affandi, M. M. M., Julianto, T., Majeed, A. 2011. Development and stability evaluation of astaxanthin nanoemulsion. *Asian Journal of Pharmaceutical and Clinical Research*, 4(1), 142–148

Airanthi, M. W.-A., Sasaki, N., Iwasaki, S., et al. 2011. Effect of brown seaweed lipids on fatty acid composition and lipid hydroperoxide levels of mouse liver. *Journal of Agricultural and Food Chemistry*, 59(8), 4156–4163

Ambati, R. R., Phang, S. -M., Ravi, S., Aswathanarayana, R. G. 2014. Astaxanthin: Sources, extraction, stability, biological activities and its commercial applications—a review. *Marine Drugs*, 12(1), 128–152

Arnold, T. M., Targett, N. M. 2002. Marine tannins: The importance of a mechanistic framework for predicting ecological roles. *Journal of Chemical Ecology,* 28(10), 1919–1934

Asai, A., Sugawara, T., Ono, H., Nagao, A. 2004. Biotransformation of fucoxanthinol into amarouciaxanthin A in mice and HepG2 cells: Formation and cytotoxicity of fucoxanthin metabolites. *Drug Metabolism and Disposition*, 32(2), 205–211

Atalay, P. B., Kuku, G., Tuna, B. G. 2019. Effects of carbendazim and astaxanthin co-treatment on the proliferation of MCF-7 breast cancer cells. *In Vitro Cellular & Developmental Biology-Animal*, 55(2), 113–119

Athukorala, Y., Trang, S., Kwok, C., Yuan, Y. V. 2016. Antiproliferative and antioxidant activities and mycosporine-like amino acid profiles of wild-harvested and cultivated edible Canadian marine red macroalgae. *Molecules*, 21(1), E119

Beppu, F., Hosokawa, M., Yim, M. J., Shinoda, T., Miyashita, K. 2013. Down-regulation of hepatic stearoyl-CoA desaturase-1 expression by fucoxanthin via leptin signaling in diabetic/obese KK-A(y) mice. *Lipids*, 48(5), 449–455

Beppu, F., Niwano, Y., Sato, E., et al. 2009a. In vitro and in vivo evaluation of mutagenicity of fucoxanthin (FX) and its metabolite fucoxanthinol (FXOH). *The Journal of Toxicological Sciences*, 34(6), 693–698

Beppu, F., Niwano, Y., Tsukui, T., Hosokawa, M., Miyashita, K. 2009b. Single and repeated oral dose toxicity study of fucoxanthin (FX), a marine carotenoid, in mice. *The Journal of Toxicological Sciences*, 34(5), 501–510

Birudaraju, D., Cherukuri, L., Kinninger, A., et al. 2020. A combined effect of Cavacurcumin, Eicosapentaenoic acid (Omega-3s), Astaxanthin and Gamma—linoleic acid (Omega-6)(CEAG) in healthy volunteers-a randomized, double-blind, placebo-controlled study. *Clinical Nutrition ESPEN*, 35, 174–179

Bustos-Garza, C., Yáñez-Fernández, J., Barragán-Huerta, B. E. 2013. Thermal and pH stability of spray-dried encapsulated astaxanthin oleoresin from Haematococcus pluvialis using several encapsulation wall materials. *Food Research International*, 54(1), 641–649

Campfield, L. A., Smith, F. J. 1999. The pathogenesis of obesity. *Baillieres Best Practice and Research: Clinical Endocrinology & Metabolism*, 13(1), 13–30

Chan, K. C., Chen, S. C., Chen, P. C. 2019. Astaxanthin attenuated thrombotic risk factors in type 2 diabetic patients. *Journal of Functional Foods*, 53, 22–27

Chang, M. X., Xiong, F. 2020. Astaxanthin and its effects in inflammatory responses and inflammation-associated diseases: Recent advances and future directions. *Molecules*, 25(22), 5342

Chen, Q., Tao, J., Xie, X. 2018. Astaxanthin promotes Nrf2/ARE signaling to inhibit HG-induced renal fibrosis in GMCs. *Marine Drugs,* 16(4), 117

Chen, Y., Li, S., Guo, Y., et al. 2020. Astaxanthin attenuates hypertensive vascular remodeling by protecting vascular smooth muscle cells from oxidative stress-induced mitochondrial dysfunction. *Oxidative Medicine and Cellular Longevity*, 2020: 4629189

Cho, M., Yang, C., Kim, S. M., You, S. 2010. Molecular characterization and biological activities of water-soluble sulfated polysaccharides from Enteromorpha prolifera. *Food Science and Biotechnology,* 19, 525–533

Corona, G., Coman, M. M., Guo, Y., et al. 2017. Effect of simulated gastrointestinal digestion and fermentation on polyphenolic content and bioactivity of brown seaweed phlorotannin-rich extracts. *Molecular Nutrition and Food Research,* 61(11). doi: 10.1002/mnfr.201700223

Das, S. K., Hashimoto, T., Baba, M., Nishino, H., Komoto, A., & Kanazawa, K. 2006. Japanese kelp (Kombu) extract suppressed the formation of aberrant crypt foci in azoxymethane challenged mouse colon. *Journal of Clinical Biochemistry and Nutrition*, 38(2), 119–125

Deng, Z. Y., Shan, W. G., Wang, S. F., Hu, M. M., Chen, Y. 2017. Effects of astaxanthin on blood coagulation, fibrinolysis and platelet aggregation in hyperlipidemic rats. *Pharmaceutical Biology*, 55(1), 663–672

D'Orazio, N., Gemello, E., Gammone, M. A., De Girolamo, M., Ficoneri, C., Riccioni, G. 2012. Fucoxantin: A treasure from the sea. *Marine Drugs*, 10(3), 604–616

Duan, X. J., Li, X. M., Wang, B. G. 2007. Highly brominated mono- and bis-phenols from the marine red alga Symphyocladia latiuscula with radical-scavenging activity. *Journal of Natural Products,* 70, 1210–1213

Faraone, I., Sinisgalli, C., Ostuni, A., Armentano, M. F., Carmosino, M., Milella, L., Russo, D., Labanca, F., Khan, H. 2020. Astaxanthin anticancer effects are mediated through multiple molecular mechanisms: A systematic review. *Pharmacological Research*, 155, 104689

Feng, Y., Chu, A., Luo, Q., Wu, M., Shi, X., Chen, Y. 2018. The protective effect of astaxanthin on cognitive function via inhibition of oxidative stress and inflammation in the brains of chronic T2DM rats. *Frontiers in Pharmacology*, 9, 748

Fouad, M. A., Sayed-Ahmed, M. M., Huwait, E. A., Hafez, H. F., Osman, A. M. M. 2021. Epigenetic immunomodulatory effect of eugenol and astaxanthin on doxorubicin

cytotoxicity in hormonal positive breast Cancer cells. *BMC Pharmacology and Toxicology*, 22(1), 1–15
Franceschelli, S., Pesce, M., Ferrone, A., et al. 2014. Astaxanthin treatment confers protection against oxidative stress in U937 cells stimulated with lipopolysaccharide reducing O2− production. *PLoS One*, 9(2), e88359
Guerin, M., Huntley, M. E., Olaizola, M. 2003. Haematococcus astaxanthin: Applications for human health and nutrition. *Trends in Biotechnology*, 21(5), 210–216
Guo, C. L., Wang, L. J., Zhao, Y., et al. 2018. A novel bromophenol derivative BOS-102 induces cell cycle arrest and apoptosis in human A549 lung cancer cells via ROS-mediated PI3K/Akt and the MAPK signaling pathway. *Marine Drugs*, 16(2), 43
Guo, C. L., Wang, L., Zhao, Y., Jiang, B., Luo, J., Shi, D. 2019. BOS-93, a novel bromophenol derivative, induces apoptosis and autophagy in human A549 lung cancer cells via PI3K/Akt/mTOR and MAPK signaling pathway. *Experimental and Therapeutic Medicine,* 17(5), 3848–3858
Hashimoto, T., Ozaki, Y., Taminato, M., Das, S. K., Mizuno, M., Yoshimura, K., Maoka, T., Kanazawa, K. 2009. The distribution and accumulation of fucoxanthin and its metabolites after oral administration in mice. *British Journal of Nutrition*, 102(2), 242–248
Higuera-Ciapara, I., Felix-Valenzuela, L., Goycoolea, F. M. 2006. Astaxanthin: A review of its chemistry and applications. *Critical Reviews in Food Science and Nutrition,* 46(2), 185–196
Hitoe, S., Shimoda, H. 2017. Seaweed fucoxanthin supplementation improves obesity parameters in mild obese Japanese subjects. *Functional Foods in Health and Disease,* 7(4), 246–262
Hoang, M. H., Kim, J.-Y., Lee, J. H., You, S., Lee, S.-J. 2015. Antioxidative, hypolipidemic, and anti-inflammatory activities of sulfated polysaccharides from Monostroma nitidum. *Food Science and Biotechnology,* 24, 99–205
Hosokawa, M., Miyashita, T., Nishikawa, S., Emi, S., Tsukui, T., Beppu, F., Okada, T., Miyashita, K. 2010. Fucoxanthin regulates adipocytokine mRNA expression in white adipose tissue of diabetic/obese KK-Ay mice. *Archives of Biochemistry and Biophysics*, 504(1), 17–25
Hotamisligil, G. S., Shargill, N. S., Spiegelman, B. M. 1993. Adipose expression of tumor necrosis factor-alpha: Direct role in obesity-linked insulin resistance. *Science*, 259(5091), 87–91
Hu, Z., Hong, P., Cheng, Y., Liao, M., Li, S. 2018. Polysaccharides from Enteromorpha tubulosa: Optimization of extraction and cytotoxicity. *Journal of Food Processing and Preservation*, 42, e13373
Hussein, G., Sankawa, U., Goto, H., Matsumoto, K., Watanabe, H. 2006. Astaxanthin, a carotenoid with potential in human health and nutrition. *Journal of Natural Products*, 69(3), 443–449
Janani, R., Anitha, R. E., Perumal, M. K., Divya, P., Baskaran, V. 2021. Astaxanthin mediated regulation of VEGF through HIF1α and XBP1 signaling pathway: An insight from ARPE-19 cell and streptozotocin mediated diabetic rat model. *Experimental Eye Research*, 206, 108555
Jayapal, N., Perumal, M. K., Vallikannan, B. 2019. Biological activities and safety aspects of fucoxanthin. In *Handbook of Algal Technologies and Phytochemicals* (pp. 245–258)
Jiao, L. L., Li, X., Li, T. B., et al. 2009. Characterization and anti-tumor activity of alkali-extracted polysaccharide from Enteromorpha intestinalis. *International Immunopharmacology*, 9, 324–329

Kang, S.-I., Ko, H.-C., Shin, H.-S., Kim, H.-M., Hong, Y.-S., Lee, N.-H., Kim, S.-J. 2011. Fucoxanthin exerts differing effects on 3T3-L1 cells according to differentiation stage and inhibits glucose uptake in mature adipocytes. *Biochemical and Biophysical Research Communications,* 409(4), 769–774

Kato, T., Kasai, T., Sato, A., et al. 2020. Effects of 3-month astaxanthin supplementation on cardiac function in heart failure patients with left ventricular systolic dysfunction-a pilot study. *Nutrients,* 12(6), 1896

Katsui, N., Suzuki, Y., Kitamura, S., Irie, T. 1967. 5,6-dibromoprotocatechualdehyde and 2,3-dibromo-4,5-dihydroxybenzyl methyl ether: New dibromophenols from Rhodomela larix. *Tetrahedron,* 23, 1185–1188

Kidgell, J. T., Magnusson, M., de Nys, R., Glasson, C. R. K. 2019. Ulvan: A systematic review of extraction, composition and function. *Algal Research,* 39, 101422

Kim, Kil-Nam, Ahn, G., Heo, S.-J., Kang, S.-M., et al. 2013. Inhibition of tumor growth in vitro and in vivo by fucoxanthin against melanoma B16F10 cells. *Environmental Toxicology and Pharmacology,* 35(1), 39–46

Kim, S. M., Jung, Y.-J., Kwon, O.-N., et al. 2012. A potential commercial source of fucoxanthin extracted from the microalga Phaeodactylum tricornutum. *Applied Biochemistry and Biotechnology,* 166(7), 1843–1855

Kotake-Nara, E., Kushiro, M., Zhang, H., Sugawara, T., Miyashita, K., Nagao, A. 2001. Carotenoids affect proliferation of human prostate cancer cells. *The Journal of Nutrition,* 131(12), 3303–3306

Kuipers, R., De Graaf, D., Luxwolda, M., Muskiet, M., Dijck-Brouwer, D., Muskiet, F. 2011. Saturated fat, carbohydrates and cardiovascular. *Complex Acute Medicine: The Internist in the Lead,* 353, 372

Kumar, P. M., Naveen, J., Janani, R., Baskaran, V. 2021. Safety assessment and pharmaceutical effects of astaxanthin: An overview. *Global Perspectives on Astaxanthin,* 569–591

Kumar, S. R., Hosokawa, M., Miyashita, K. 2013. Fucoxanthin: A marine carotenoid exerting anti-cancer effects by affecting multiple mechanisms. *Marine Drugs,* 11(12), 5130–5147

Kylin, H. 1946. Fysiografiska Sällskapets i Lund Dorhanlingar. 16, pp. 102–105

Lahaye, M., Cimadevilla, E. A. C., Kuhlenkamp, R., Quemener, B., Lognoné, V., Dion, P. 1999. Chemical composition and 13C NMR spectroscopic characterisation of ulvans from Ulva (Ulvales, Chlorophyta). *Journal of Applied Phycology,* 11, 1–7

Lahaye, M., Robic, A. 2007. Structure and functional properties of ulvan, a polysaccharide from green seaweeds. *Biomacromolecules,* 8, 1765–1774

Lai, T. T., Yang, C. M., Yang, C. H. 2020. Astaxanthin protects retinal photoreceptor cells against high glucose-induced oxidative stress by induction of antioxidant enzymes via the PI3K/Akt/Nrf2 pathway. *Antioxidants,* 9(8), 729

Landon, R., Gueguen, V., Petite, H., Letourneur, D., Pavon-Djavid, G., Anagnostou, F. 2020. Impact of astaxanthin on diabetes pathogenesis and chronic complications. *Marine Drugs,* 18(7), 357

Le, B., Golokhvast, K. S., Yang, S. H., Sun, S. 2019. Optimization of microwave-assisted extraction of polysaccharides from ulva pertusa and evaluation of their antioxidant activity. *Antioxidants,* 8(5), 129

Leiro, J. M., Castro, R., Arranz, J. A., Lamas, J. 2007. Immunomodulating activities of acidic sul-phated polysaccharides obtained from the seaweed Ulva rigida C. Agardh. *International Immunopharmacology,* 7(7), 879–888

Li, H., Li, J., Hou, C., Li, J., Peng, H., Wang, Q. 2020. The effect of astaxanthin on inflammation in hyperosmolarity of experimental dry eye model in vitro and in vivo. *Experimental Eye Research,* 197, 108113

Li, J., Jiang, F., Chi, Z., Han, D., Yu, L., Liu, C. 2018c. Development of Enteromorpha prolifera polysaccharide-based nanoparticles for delivery of curcumin to cancer cells. *International Journal of Biological Macromolecules,* 112, 413–421

Li, Q., Luo, J., Wang, C., et al. 2018b. Ulvan extracted from green seaweeds as new natural additives in diets for laying hens. *Journal of Applied Phycology*, 30, 2017–2027

Li, W., Jiang, N., Li, B., Wan, M., Chang, X., Liu, H., Zhang, L., Yin, S., Qi, H., Liu, S. 2018a. Antioxidant activity of purified ulvan in hyperlipidemic mice. *International Journal of Biological Macromolecules,* 113, 971–975

Li, Y., Fu, X., Duan, D., Liu, X., Xu, J., Gao, X. 2017. Extraction and identification of phlorotannins from the brown alga, sargassum fusiforme (Harvey) setchell. *Marine Drugs,* 15(2), 49

Lopes, G., Pinto, E., Andrade, P. B., Valentão, P. 2013. Antifungal activity of phlorotannins against dermatophytes and yeasts: Approaches to the mechanism of action and influence on candida albicans virulence factor. *PLoS ONE*, 8, e72203

Maeda, H., Hosokawa, M., Sashima, T., Funayama, K., Miyashita, K. 2005. Fucoxanthin from edible seaweed, Undaria pinnatifida, shows antiobesity effect through UCP1 expression in white adipose tissues. *Biochemical and Biophysical Research Communications*, 332(2), 392–397

Maeda, H., Hosokawa, M., Sashima, T., Miyashita, K. 2007. Dietary combination of fucoxanthin and fish oil attenuates the weight gain of white adipose tissue and decreases blood glucose in obese/diabetic KK-Ay mice. *Journal of Agricultural and Food Chemistry*, 55(19), 7701–7706

Maeda, H., Hosokawa, M., Sashima, T., Murakami-Funayama, K., Miyashita, K. 2009. Anti-obesity and anti-diabetic effects of fucoxanthin on diet-induced obesity conditions in a murine model. *Molecular Medicine Reports*, 2(6), 897–902

Maeda, H., Hosokawa, M., Sashima, T., Takahashi, N., Kawada, T., Miyashita, K. 2006. Fucoxanthin and its metabolite, fucoxanthinol, suppress adipocyte differentiation in 3T3-L1 cells. *International Journal of Molecular Medicine*, 18(1), 147–152

Manandhar, B., Wagle, A., Seong, S. H., et al. 2019. Phlorotannins with potential anti-tyrosinase and antioxidant activity isolated from the marine seaweed Ecklonia stolonifera. *Antioxidants*, 8, 240

Mashhadi, N. S., Zakerkish, M., Mohammadiasl, J., Zarei, M., Mohammadshahi, M., Haghighizadeh, M. H. 2018. Astaxanthin improves glucose metabolism and reduces blood pressure in patients with type 2 diabetes mellitus. *Asia Pacific Journal of Clinical Nutrition,* 27(2), 341–346

Matloub, A. A., Aglan, H. A., Mohamed El Souda, S. S., Aboutabl, M. E., Maghraby, A. S., Ahmed, H. H. 2016. Influence of bioactive sulfated polysaccharide-protein complexes on hepatocarcinogenesis, angiogenesis and immunomodulatory activities. *Asian Pacific Journal of Tropical Medicine,* 9(12), 1200–1211

Matsumoto, M., Hosokawa, M., Matsukawa, N., et al. 2010. Suppressive effects of the marine carotenoids, fucoxanthin and fucoxanthinol on triglyceride absorption in lymph duct-cannulated rats. *European Journal of Nutrition,* 49(4), 243–249

Morelli, A., Puppi, D., Chiellini, F. 2017. *Perspectives on Biomedical Applications of Ulvan*. Elsevier Inc.: Amsterdam, The Netherlands; ISBN 9780128098172

Naveen, J., Baskaran, V. 2018. Antidiabetic plant-derived nutraceuticals: A critical review. *European Journal of Nutrition*, 57(4), 1275–1299

Naveen, J., Veeresh, B. T., Ramaprasad, T. R., Baskaran, V. 2021. Metabolomics of dietary fatty acids: Implications on life style diseases. In *Fats and Associated Compounds* (pp. 286–307)

Nikita, P., Victoria, L., Andrew, D. S., Lei, M., Daniel, C., Emily, Y., Aaron, B. 2018. Algal polysaccharides as therapeutic agents for atherosclerosis. *Frontiers in Cardiovascular Medicine*, 5, 153

Odeberg, J. M., Lignell, Å., Pettersson, A., Höglund, P. 2003. Oral bioavailability of the antioxidant astaxanthin in humans is enhanced by incorporation of lipid based formulations. *European Journal of Pharmaceutical Sciences*, 19(4), 299–304

Orfanoudaki, M., Hartmann, A., Alilou, M., et al. 2020. Absolute configuration of mycosporine-like amino acids, their wound healing properties and *in vitro* anti-aging effects. *Marine Drugs,* 18, 35

Peasura, N., Laohakunjit, N., Kerdchoechuen, O., Vongsawasdi, P., Chao, L. K. 2016. Assessment of biochemical and immunomodulatory activity of sulphated polysaccharides from Ulva intestinalis. *International Journal of Biological Macromolecules,* 91, 269–277

Peng, C. H., Chang, C. H., Peng, R. Y., Chyau, C. C. 2010. Improved membrane transport of astaxanthin by liposomal encapsulation. *European Journal of Pharmaceutics and Biopharmaceutics*, 75(2), 154–161

Pengzhan, Y., Ning, L., Xiguang, L., Gefei, Z., Quanbin, Z., Pengcheng, L. 2003. Antihyperlipidemic effects of different molecular weight sulfated polysaccharides from Ulva pertusa (Chlorophyta). *Pharmacological Research*, 48(6), 543–549

Pereira, C. P. M., Souza, A. C. R., Vasconcelos, A. R., Prado, P. S. 2021. Antioxidant and anti inflammatory mechanisms of action of astaxanthin in cardiovascular diseases. *International Journal of Molecular Medicine*, 47(1), 37–48

Qi, H., Sheng, J. 2015. The antihyperlipidemic mechanism of high sulfate content ulvan in rats. *Marine Drugs,* 13, 3407–3421

Qi, H., Sun, Y. 2015. Antioxidant activity of high sulfate content derivative of ulvan in hyperlipidemic rats. *International Journal of Biological Macromolecules,* 76, 326–329

Quemener, B., Lahaye, M., Bobin-Dubigeon, C. 1997. Sugar determination in ulvans by a chemical-enzymatic method coupled to high performance anion exchange chromatography. *Journal of Applied Phycology*, 9, 179–188

Rahimi, F., Tabarsa, M., Rezaei, M. 2016. Ulvan from green algae Ulva intestinalis: Optimization of ultrasound-assisted extraction and antioxidant activity. *Journal of Applied Phycology,* 28, 1–12

Rao, A. R., Baskaran, V., Sarada, R., Ravishankar, G. A. 2013. In vivo bioavailability and antioxidant activity of carotenoids from microalgal biomass—A repeated dose study. *Food Research International*, 54(1), 711–717

Sathya, R., Kanaga, N., Sankar, P., Jeeva, S. 2017. Antioxidant properties of phlorotannins from brown seaweed Cystoseira trinodis (Forsskål) C. Agardh. *Arabian Journal of Chemistry*, 10, S2608–S2614

Shao, P., Chen, X., Sun, P. 2013. In vitro antioxidant and antitumor activities of different sulfated polysaccharides isolated from three algae. *International Journal of Biological Macromolecules,* 62, 155–161

Sharma, P. P., Baskaran, V. 2021. Polysaccharide (laminaran and fucoidan), fucoxanthin and lipids as functional components from brown algae (Padina tetrastromatica) modulates adipogenesis and thermogenesis in diet-induced obesity in C57BL6 mice. *Algal Research,* 54, 102187

Shi, D., Li, J., Guo, S., Su, H., Fan, X. 2009. The antitumor effect of bromophenol derivatives in vitro and Leathesia nana extract in vivo. *Chinese Journal of Oceanology and Limnology*, 27(2), 277–282

Sila, A., Ghlissi, Z., Kamoun, Z., Makni, M., Nasri, M., Bougatef, A., Sahnoun, Z. 2015. Astaxanthin from shrimp by-products ameliorates nephropathy in diabetic rats. *European Journal of Nutrition*, 54(2), 301–307

Song, X. D., Zhang, J. J., Wang, M. R., Liu, W. B., Gu, X. B., Lv, C. J. 2011. Astaxanthin induces mitochondria-mediated apoptosis in rat hepatocellular carcinoma CBRH-7919 cells. *Biological and Pharmaceutical Bulletin,* 34(6), 839–844

Sugawara, T., Kushiro, M., Zhang, H., Nara, E., Ono, H., Nagao, A. 2001. Lysophosphatidylcholine enhances carotenoid uptake from mixed micelles by Caco-2 human intestinal cells. *The Journal of Nutrition*, 131(11), 2921–2927

Sun, S. Q., Zhao, Y. X., Li, S. Y., Qiang, J. W., Ji, Y. Z. 2020. Anti-tumor effects of astaxanthin by inhibition of the expression of STAT3 in prostate cancer. *Marine Drugs*, 18(8), 415

Sun, X., Zhong, Y., Luo, H., Yang, Y. 2017. Selenium-containing polysaccharide-protein complex in Se-enriched Ulva fasciata induces mitochondria-mediated apoptosis in A549 human lung cancer cells. *Marine Drugs,* 15

Tabarsa, M., Han, J. H., Kim, C. Y., You, S. G. 2012. Molecular characteristics and im-munomodulatory activities of water-soluble sulfated polysaccharides from Ulva pertusa. *Journal of Medicinal Food,* 15, 135–144

Tabarsa, M., You, S., Dabaghian, E. H., Surayot, U. 2018. Water-soluble polysaccharides from Ulva intestinalis: Molecular properties, structural elucidation and immunomodulatory activities. *Journal of Food and Drug Analysis,* 26(2), 599–608

Takahashi, K., Hosokawa, M., Kasajima, H., Hatanaka, K., Kudo, K., Shimoyama, N., Miyashita, K. 2015. Anticancer effects of fucoxanthin and fucoxanthinol on colorectal cancer cell lines and colorectal cancer tissues. *Oncology Letters,* 10(3), 1463–1467

Targett, N. M., Arnold, T. M. 1998. Mini review: Predicting the effects of brown algal phlorotannins on marine herbivores in tropical and temperate oceans. *Journal of Phycology*, 34(2), 195–205

Terasaki, M., Maeda, H., Miyashita, K., Tanaka, T., Miyamoto, S., Mutoh, M. 2017. A marine bio-functional lipid, fucoxanthinol, attenuates human colorectal cancer stem-like cell tumorigenicity and sphere formation. *Journal of Clinical Biochemistry and Nutrition*, 16–112

Terasaki, M., Masaka, S., Fukada, C., Houzaki, M., Endo, T., Tanaka, T., Maeda, H., Miyashita, K., Mutoh, M. 2019. Salivary glycine is a significant predictor for the attenuation of polyp and tumor microenvironment formation by fucoxanthin in AOM/DSS mice. *In Vivo,* 33(2), 365–374

Thanh, T. T. T., Quach, T. M. T., Nguyen, T. N., et al. 2016. Structure and cytotoxic activity of ulvan extracted from green seaweed Ulva lactuca. *International Journal of Biological Macromolecules,* 93, 695–702

Tsukui, T., Baba, N., Hosokawa, M., Sashima, T., Miyashita, K. 2009. Enhancement of hepatic docosahexaenoic acid and arachidonic acid contents in C57BL/6J mice by dietary fucoxanthin. *Fisheries Science,* 75(1), 261–263

Vani, V., Venkateish, V. P., Nivya, V., Baskaran, V., Madan Kumar, P. 2021. 8 Nutritional and anti-cancer effects of carotenoids from *Lactuca sativa. Advances in Health and Disease*, 33

Venkateish, V. P., Vani, V., Nivya, V., Baskaran, V., Madan Kumar, P. 2021. 2 Bioactives of *Lactuca sativa*: Nutritional and clinical importance. *Advances in Health and Disease*, 33

Venkatesan, J., Lowe, B., Anil, S., Manivasagan, P., Kheraif, A. A. A., Kang, K. -H., Kim, S. K. 2015. Seaweed polysaccharides and their potential biomedical applications. *Starch—Stärke,* 67, 381–390

Vo, T. S., Ngo, D. H., Kim, S. K. 2019. Pharmaceutical properties of marine polyphenols: An overview. *Acta Pharmaceutica Sciencia*, 57, 2

Wang, J., Chen, S., Xu, S., Yu, X., Ma, D., Hu, X., Cao, X. 2012. In vivo induction of apoptosis by fucoxanthin, a marine carotenoid, associated with down-regulating STAT3/EGFR signaling in sarcoma 180 (S180) xenografts-bearing mice. *Marine Drugs,* 10(9), 2055–2068

Wang, Q., Zhao, Y., Guan, L., Zhang, Y., Dang, Q., Dong, P., Li, J., Liang, X. 2017. Preparation of astaxanthin-loaded DNA/chitosan nanoparticles for improved cellular uptake and antioxidation capability. *Food Chemistry*, 227, 9–15

Wijesekara, I., Pangestuti, R., Kim, S. K. 2011. Biological activities and potential health benefits of sulfated polysaccharides derived from marine algae. *Carbohydrate Polym*ers, 84, 14–21

Xu, N., Fan, X., Yan, X., Li, X., Niu, R., Tseng, C. K. 2003. Antibacterial bromophenols from the marine red alga Rhodomela confervoides. *Phytochemistry,* 62, 1221–1224

Yan, X., Chuda, Y., Suzuki, M., Nagata, T. 1999. Fucoxanthin as the major antioxidant in Hijikia fusiformis, a common edible seaweed. *Bioscience, Biotechnology, and Biochemistry*, 63(3), 605–607

Yang, F., Fang, X., Jiang, W., Chen, T. 2017. Bioresponsive cancer-targeted polysaccharide nanosystem to inhibit angiogenesis. *International Journal of Nanomedicine,* 12, 7419–7431

Yang, Y. I., Shin, H. C., Kim, S. H., Park, W. Y., Lee, K. T., Choi, J. H. 2012. 6,6′-Bieckol, isolated from marine alga Ecklonia cava, suppressed LPS-induced nitric oxide and PGE_2 production and inflammatory cytokine expression in macrophages: The inhibition of NFκB. *International Immunopharmacology,* 12, 510–517

Zhang, H., Tang, Y., Zhang, Y., Zhang, S., Qu, J., Wang, X., Kong, R., Han, C., Liu, Z. 2015. Fucoxanthin: A promising medicinal and nutritional ingredient. *Evidence-Based Complementary and Alternative Medicine*, 2015, 1–10

Zhang, M. Y., Guo, J., Hu, X. M., et al. 2019. An in vivo anti-tumor effect of eckol from marine brown algae by improving the immune response. *Food & Function*, 10(7), 4361–4371

Zhao, C., Lin, G., Wu, D., et al. 2020. The algal polysaccharide ulvan suppresses growth of hepatoma cells. *Food Frontiers,* 1, 83–101

Zhao, Z. W., Cai, W., Lin, Y. L., et al. 2011. Ameliorative effect of astaxanthin on endothelial dysfunction in streptozotocin-induced diabetes in male rats. *Arzneimittelforschung,* 61(04), 239–246

Zorofchian, Moghadamtousi, S., Karimian, H., Khanabdali, R., Razavi, M., Firoozinia, M., Zandi, K., Abdul Kadir, H. 2014. Anticancer and antitumor potential of fucoidan and fucoxanthin, two main metabolites isolated from brown algae. *The Scientific World Journal,* 2014, 1–10

18 Exopolysaccharide Production from Marine Bacteria and Its Applications

Prashakha J. Shukla, Shivang B. Vhora, Ankita G. Murnal, Unnati B. Yagnik, and Maheshwari Patadiya

CONTENTS

DOI: 10.1201/9781003303909-18

18.1 INTRODUCTION: MARINE ECOSYSTEM

Approximately 70% of the earth's surface is covered with the world's oceans. The oceans are the principal continuous ecosystem on Earth. They comprise various habitats that maintain a wealth of marine biodiversity. Marine biodiversity and biogeochemical cycles are the irreplaceable part of the ocean for uptake of carbon dioxide that causes global warming. Life on earth began in the ocean, and the smooth functioning of marine ecosystem is essential for life to continue on earth. The marine ecosystem is the foundation of life, and it is significantly important for the sustainability and habitability of biota (Kennedy et al., 2002; Munn, 2011).

The marine ecosystem is affected by climate changes that can create new challenges. Changes in temperature in coastal and marine ecosystems will alter ecological cycles, such as productivity and species interactions. As temperature changes, it creates new combinations of species that will interact in unpredictable ways. Species unable to migrate or compete with other species for resources may face local or global extinction. A change in marine ecosystem is mainly dependent on precipitation and seawater level. An increase or decrease in precipitation enhances the risk of coastal flooding and an imbalance in the water level, and it forces the marine biota to migrate. The upwelling and downwelling of seawater may influence the vertical movement of ocean. These changes result in increase or decrease in essential nutrients and oxygen to the marine ecosystem (Kennedy et al., 2002). These changes are responsible for the evolution and extinction of species. Microbial diversity of marine ecosystem changes with alteration in physical, chemical and biotic variables. Due to these changes, marine biota has developed unique survival strategies. The marine ecosystem has thus emerged as a "gold mine" for research (Sobhana et al., 2015).

18.2 MARINE BACTERIA

Marine biota is critically important for the maintenance of a healthy and competitive environment in the ecosystem. Bacterial diversity plays a crucial role in each biogeochemical cycles, fluxes and processes occurring in the marine ecosystem. Microbes are extremely abundant and diverse, producing carbon products that are key in the regulation of Earth's climate, particularly CO_2 and CH_4. Marine bacteria also provide an untapped source of genetic information and biomolecules (Glöckner et al., 2012).

Research in marine microbial biodiversity mainly aims to understand the community structure and distribution pattern of diverse organisms. The unique metabolism of marine bacteria allows it to carry out many steps of the biogeochemical cycles that other organisms are unable to complete (Sobhana et al., 2015). Marine bacterial diversity varies with conditions such as the following:

- Presence of other organisms (symbiotic, free living, biofilm)
- Ocean surface and sediments
- Concentration of nutrients and required growth substrates (oligotrophic, mesotrophic, eutrophic)

Habitats having previously mentioned conditions are hot spots for the study of marine biodiversity and its biological activity. A majority of marine bacterial diversity is composed of gram-negative bacteria as their cell walls are better adapted for survival in the marine environment as compared to gram-positive bacteria (Das et al., 2006). Marine bacteria can withstand various extreme conditions, such as alkaline, hypersaline, fluctuation in temperature, pH, pressure and nutrient availability (Rampelotto, 2013). Their biological and physico-chemical systems undergo alterations with changes in the previously mentioned conditions (MacLeod, 1965; DeLong and Yayanos, 1987; Wai et al., 1999; Bowers et al., 2009). Hence, the marine environment exerts a driving force on bacteria for selecting new adaptive strategies for survival and adaptation.

18.3 EXOPOLYSACCHARIDES

As a survival strategy, marine bacteria secrete several secondary metabolites (De Carvalho and Fernandes, 2010). These secondary metabolites help in their protection and survival (Jensen and Fenical, 1996). The first compound was an "antibiotic" produced by a marine actinomycete *Chainiapurpurogena* in 1975 (Okazaki et al., 1975). Bioactive natural products with unique characteristics ("anti-tumor and anti-cancer" properties) were obtained from marine organisms, such as sponges and algae. From 1997 to 2008, 659 compounds have been isolated from five major marine bacterial phyla, namely, *Bacteroidetes, Firmicutes, Proteobacteria*, *Cyanobacteria* and *Actinobacteria* (Proksch et al., 2002; Williams, 2009).

Extracellular polymeric substances are one of the most important secondary metabolites secreted to the cell surface and produced by many species of bacteria, algae, fungi and some archaea (Sutherland, 1972). These substances include macromolecules, such as polysaccharides consisting 40–95%, the rest being proteins, nucleic acids and lipids (Wingender et al., 1999; Flemming and Wingender, 2001). Sutherland, in 1972, for the very first time used the term *exopolysaccharides* (EPSs) for microbial polysaccharides.

EPSs are the major fraction of the dissolved organic matter reservoir in the ocean. EPS-producing bacteria have been isolated from different zones of the marine environment, such as the sediment, the deep sea, hydrothermal vents, surface water and mangroves. Some marine bacteria belonging to genera *Bacillus, Rhodococcus, Halomonas, Alcanivorax, Pseudomonas, Marinobacter, Pseudoalteromonas* and *Alteromonas* have been isolated for EPS production (Chakraborty et al., 2016). EPSs are high-molecular-weight (HMW) carbohydrate polymers having heteropolymeric composition. EPSs represent an easily available reduced-carbon reservoir for marine organisms in the ocean and help the microbial communities survive under extreme conditions such as high or low temperature, salinity, nutrient availability, pH, pressure and presence of CO_2 or O_2 (Poli et al., 2010).

18.4 FUNCTIONAL DIVERSITY OF EPSs IN MARINE HABITAT

EPSs have a wide variety of structural diversity reflecting their diverse functions. The structure of EPSs differs from linear to highly branch. Based on their monomeric sugar unit, EPSs in bacteria serve various functions. They serve as structural polysaccharides (provide support), capsular polysaccharides (CPSs; help in cellular interaction and virulence) and biofilm-associated polysaccharides (help in the formation and stabilization of biofilm matrix and cell-to-surface interactions; Chakraborty et al., 2016).

18.4.1 Structural Polysaccharides

Structural polysaccharides provide structure and shape to the cells in marine plants, bacteria, animals, algae and fungi (Dumitriu, 2004). It serves as a significant communication factor in marine organisms for interactions with the surrounding microenvironment and works in the primary framework of the cell wall makeup of bacteria. The unique elasticity of these polysaccharides contain in their cell wall protects the shape and integrity of the cells from lysis (Rogers et al., 1980; Harz et al., 1990; Matias et al., 2003).

The cell wall polysaccharides of aquatic, terrestrial plants and seaweeds are made up of hemicellulose. It contains D-glucose, D-xylose, L-arabinose, D-glucuronic acid, D-mannose, D-galactose, D-galacturonic acid, L-fucose, 4-O-methyl, L-rhamnose and O-methylated neutral sugars. Moreover, arabinoxylans, chitin, cellulose and pectins are also included.

18.4.2 CPSs

CPSs are present in most bacteria. In the marine environment, CPSs are present either as heteropolysaccharides or homopolysaccharides with long repeating units of monosaccharides linked by glycosidic bonds (Roberts, 1996). They are closely attached to the surface of the cell or are present as slime that is not strongly attached to the cell (Whitfield, 1988). CPSs have a variety of functions; namely, they serve as a protecting layer toward the host community (specific and nonspecific; Roberts, 1996). CPSs form a hydrated gel structure around the bacterial cell to protect them against desiccation. They also promote the attachment of cell to the surface forming a biofilm (Costerton et al., 1987; Jenkinson, 1994). CPSs are more protective for organisms against toxic substances in the water column (Nichols et al., 2005). EPSs with antibacterial characteristic was produced by *Pseudoalteromonas tunicata*, which formed a biofilm on the surface of eukaryotic organisms (Egan et al., 2000; Debnath et al., 2007).

18.4.3 Biofilm-Associated Polysaccharides

Biofilm is a colonization of diverse bacterial species in a matrix containing various types of extracellular polymeric substances. The secretion of various types of polysaccharides creates a cementing material with novel composition of chemical and physical properties. EPSs are mainly polyanionic in nature because of the presence of ketal-linked pyruvate, phosphate, uronic acids and sulfate; some other EPSs are neutral in nature (Sutherland, 1990). Mack et al., 1996 reported the unique polycationic nature of EPSs as adhesive material obtained from *Staphylococcus epidermidis*.

Biofilm formation is the most important irreplaceable requirement of organisms for their survival. The amount and complexity of EPSs mainly depend on the type of environmental conditions, such as different levels of nitrogen and oxygen, temperature, availability of nutrients, pH and desiccation (Mayer et al., 1999). Fluctuations in environmental conditions result in changes in the structure of EPSs and the capacity of their attachment to the surface, forming a biofilm. The type of surface is a significant parameter because as the roughness of the surface increases, the production of EPSs also increases. Cell surface hydrophobicity, which is an important factor for the uptake of hydrocarbons, also depends on EPSs production (Flemming and Wingender, 2001).

The first step of biofilm formation is the adsorption of nutrients to the surface. Initial attachment starts with the diverse group of specific and non-specific microbial communities that adheres to the primary surface. These adhered microbial communities initiate cell differentiation, proliferation and EPSs synthesis, which results in mature biofilm formation (Chakraborty et al., 2016). Initial attachment can be reversible, but at the final stage of EPSs synthesis, the attachment is irreversible. Bacteria may reversibly attach to a surface using surface-associated nutrients (Wolfaardt et al., 1999). A biochemical interaction between the organisms and the surrounding surface environment is made by the production of EPSs. Secretions of exoenzymes are important in the cycling of organic and inorganic materials. This EPS-containing hydrated layer increases the cellular uptake of small molecules for energy and biomass (Decho et al., 1995; Manca et al., 1996). Biofilm formation helps the organisms with nutrient transportation and horizontal gene transfer (Caron, 1987). Changes in physical environmental conditions affect the stability of EPSs (Boyle and Reade, 1983).

Several studies have concluded that EPSs mediated biofilm formation is closely associated with quorum-sensing (QS) systems for the production of autoinducers (Di Donato et al., 2016). These QS mechanisms have been very well studied in the marine strain of *Vibrio fischeri* with two QS systems, *ain* and *lux* (Lupp and Ruby, 2005).

18.5 BIOSYNTHESIS AND GENETIC REGULATION OF EPSs

The biosynthesis of EPSs depends on the growth phase, pH, nutrients, temperature, salinity, mode of growth and more. EPSs are synthesized intracellularly throughout the growth phase or in different phases of growth under nutrient-limiting conditions and are then exported to the cell surface (Harder and Dijkhuizen, 1983). The biosynthesis of EPSs involves a large number of genes, enzymes and regulatory proteins.

The biosynthesis of EPSs occurs in four steps: (1) transport of sugar into the cell, (2) phosphorylation of sugar, (3) polymerization of sugar units and (4) export of EPSs to the cell surface (Madhuri and Prabhakar, 2014). The biosynthesis of EPSs is initiated when carbon substrate is available as the precursor to the cell. Synthesis is initiated by the transport of monosaccharides into the cell by diffusion, active transport or group translocation. In the cell, the sugar molecule is phosphorylated and converted into sugar-6-phosphate, which is further converted to sugar-1-phosphate using the enzyme phosphoglucomutase (Jolly et al., 2002). Sugar nucleotides (UDP-glu) are then formed from the sugar-1-phosphate with the help of enzyme pyrophosphorylase (Lieberman and Markovitz, 1970; Jolly et al., 2002). UDP-glu serves as an intermediate for other sugar moieties involved in polysaccharide assembly and synthesis. Bacterial polysaccharides consist of repeating units of sugars that are synthesized by glycosyltransferases (GTs) that transfer sugar to a glycosyl carrier lipid in the cytoplasmic membrane. In the last step, monosaccharide units of assembled polysaccharides are further modified by different enzymatic activities like acylation, acetylation, methylation and sulfation. They are then exuded from the cell in the form of loose slime or a capsule with the help of flippase, permease or ABC transporters (energy-dependent efflux transporter protein).

Transport of bacterial EPSs occurs through four general mechanisms: (1) Wzx/Wzy-dependent pathways, (2) ABC transporter–dependent pathways, (3) synthase-dependent pathways and (4) extracellular synthesis by use of a single sucrase protein (Schmid et al., 2015; Parkar et al., 2016).

18.5.1 Wzx/Wzy-Dependent Pathway

In this mechanism, each individual repeating unit of sugar is linked by a diphosphate anchor at the inner membrane, which is assembled by several GTs and translocated across the cytoplasmic membrane by the protein Wzx, which is also called as flippase. Polymerization occurs in the periplasmic space with the help of Wzy protein before they are exported to the surface of the cells. Polymerized repeating units are then transported from the periplasm to the surface of the cell by polysaccharide co-polymerase (PCP) and the outer membrane polysaccharide export (OPX) proteins.

Most of the carbohydrate polymers are assembled by the Wzx/Wzy-dependent pathway due to their highly diverse sugar pattern (4–5 sugars) and therefore are called heteropolymers. This pathway uses two types of enzymes: flippase (Wzx) and polymerase (Wzy; Figure 18.1; Schmid et al., 2015; Parkar et al., 2016).

18.5.2 ABC Transporter–Dependent Pathway

In this mechanism, EPSs are synthesized through the ABC transporter–dependent pathway, where the CPS is assembled by the action of GTs located in the inner cytoplasmic membrane.

This mechanism represents tripartite efflux pump–like complex composed of ABC transporters at the inner membrane, and periplasmatic proteins, PCP

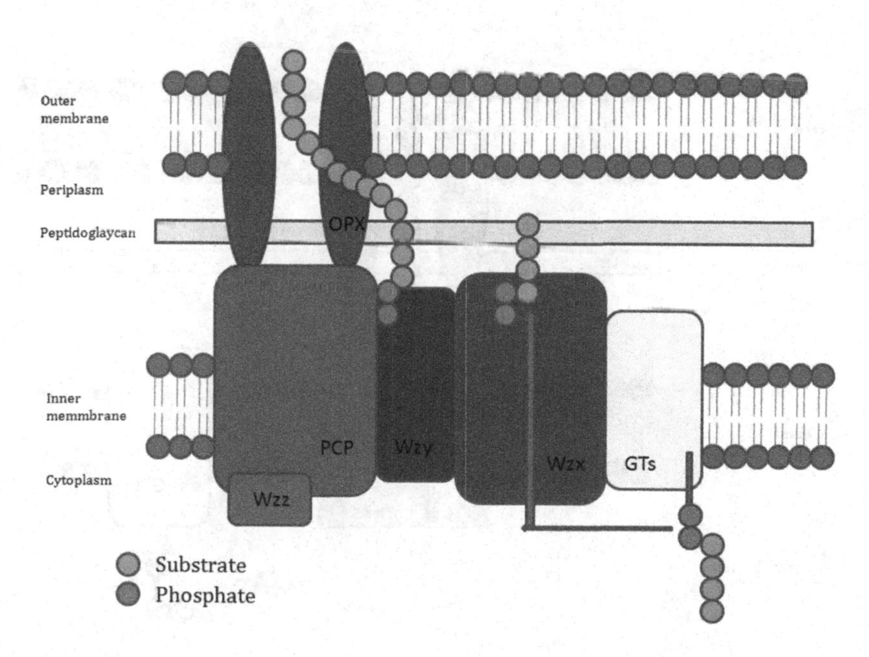

FIGURE 18.1 Wzy dependent pathway (modified from Schmid et al., 2015).

and OPX. These two proteins are closely similar to OPX and PCP proteins involved in the Wzx/Wzy-dependent pathway. CPSs produced via these mechanisms contain conserved glycolipid at the reducing terminus composed of phosphatidylglycerol and a poly-2-keto-3-deoxyoctulosonic acid (Kdo) linker. This is one of the major differences between of the Wzx/Wzy and the ABC transporter–dependent pathways (Figure 18.2; Schmid et al., 2015; Parkar et al., 2016).

18.5.3 Synthase-Dependent Pathway

A synthase-dependent pathway secretes complete polymer strands across the membrane and the cell wall and is independent of a flippase for translocating repeat units. In this pathway, polymerization and transport are carried out by the synthase complex, which is a complex protein envelope made up of multiple subunits and the tetratricopeptide repeat (TPR) proteins. Polymerization, as well as the translocation process, is performed by a single synthase protein (Figure 18.3; Schmid et al., 2015; Parkar et al., 2016). This pathway is often utilized for the assembly of homopolymers requiring only one type of sugar precursor.

18.5.4 Extracellular Synthesis by Use of a Single Sucrase Protein

Dextran and levan are extracellularly synthesized polysaccharides. The synthesis of dextran, levan and their derivatives is directly induced in the presence of

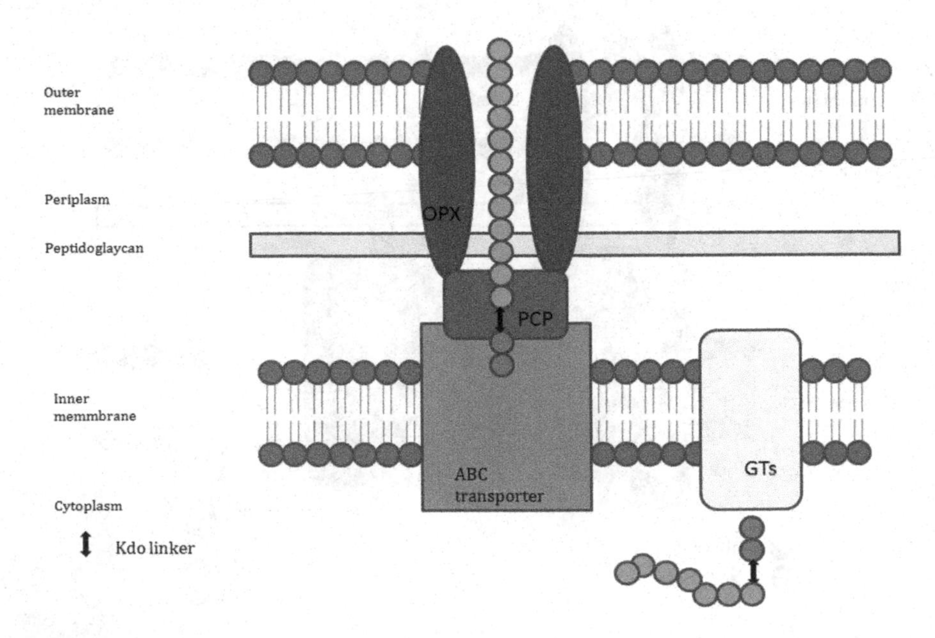

FIGURE 18.2 ABC transporter–dependent pathway (modified from Schmid et al., 2015).

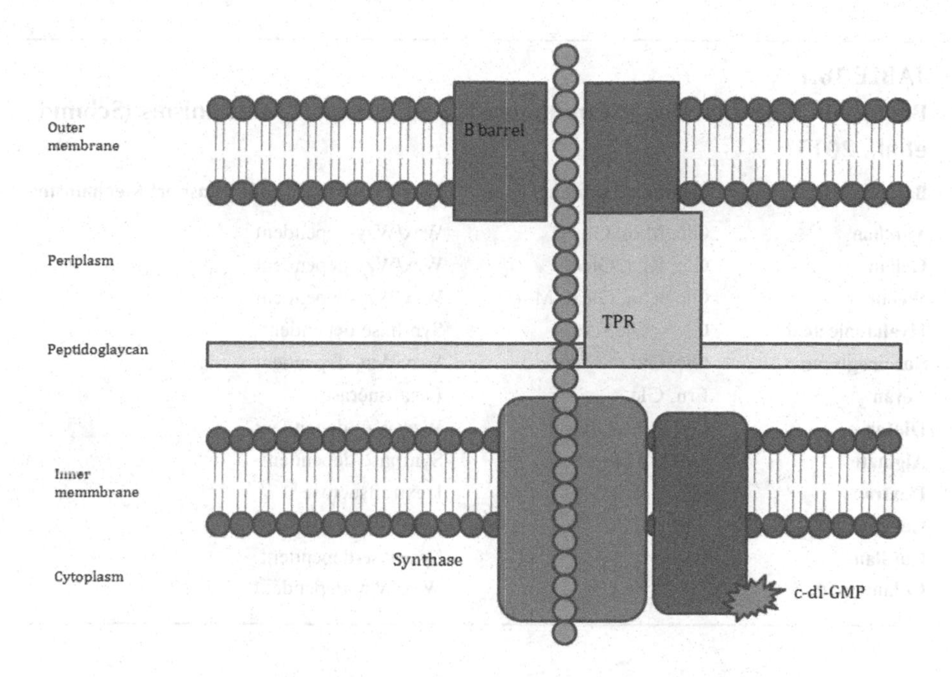

FIGURE 18.3 Synthase-dependent pathway (modified from Schmid et al., 2015).

sucrose (Schmid et al., 2015). The most common sucrase activity–based polymer is dextran, which mainly consists of α-(1–6) linked glucose. Dextran-based polymer is released outside the cell by the dextransucrase, the key enzyme for dextran synthesis (Schmid et al., 2015). Levan, which is produced by levansucrases, can also act as fructose transferases producing polyfructan (levan). Table 18.1 shows various bacterial EPSs, their sugar components and the transport mechanism.

18.6 GENETIC REGULATION OF EPSs

Genes involved in the EPSs biosynthetic pathways, encode various types of GTs, polymerizing and branching enzymes. These enzymes are also involved in addition of substituents and modification of sugar moieties. The function and mode of action of the genes and proteins are not totally understood. The genes encoding the enzymes can be found in most of the EPSs producing microbes clustered within the genome or on large plasmids. The transfer of complete gene clusters toward alternative host strains has been reported to result in altered compositions of the repeat units. On the basis of several GTs and secreting enzymes (1 to more than 23), EPS operons have been partially sequenced.

Operons for welan and diutan have been well studied in *Sphingomonas*sp. EPSs from the sphingan family are gellan, welan, diutan and rhamsan. Chemical structures of EPSs are different because of differently composed gene

TABLE 18.1
Bacterial EPSs, Their Sugar Components, and Transport Mechanisms (Schmid et al., 2015).

Bacterial EPSs	Sugar Components	EPSs Biosynthesis and Transport Mechanisms
Xanthan	Glc, Man, GluA	Wzx/Wzy-dependent
Gellan	Glc, Rha, GlcA	Wzx/Wzy-dependent
Welan	Glc, Rha, GlcA, Man	Wzx/Wzy-dependent
Hyaluronic acid	GlcA, GlcNAc	Synthase-dependent
Succinoglycan	Glc, Gal	Wzx/Wzy-dependent
Levan	Fru, Glc	Levansucrase
Diutan	Glc, Rha, GlcA	Wzx/Wzy-dependent
Alginate	GulA, ManA	Synthase-dependent
Dextran	Glc	Dextransucrase
Cellulose	Glc	Synthase-dependent
Curdlan	Glc	Synthase-dependent
Colanic acid	Glc, Fuc, GlcA, Gal	Wzx/Wzy-dependent

operons. The genes involved in the synthesis of the rhamnose are *rml ABCD*, respectively. In the case of the three sphingans, gellan, welan and diutan, the genes involved in the biosynthesis are named according to the corresponding polymer, *gel* for gellan, *wel* for welan and *dsp* for diutan. *spn* is the gene involved in sphingan biosynthesis. Succinoglycan (SG) is an acidic polymer produced by several *Rhizobium*, *Agrobacterium*, *Alcaligenes* and *Pseudomonas* strains. The different gene clusters involved in the operon of EPS biosynthesis and functions of various encoded proteins (Schmid et al., 2015). EPS synthesis thus involves multiple clusters of genes that are still unknown and a hot spot for research (Nilsson et al., 2011).

The genetic regulation of the synthesis of EPS mauran has been studied in the marine bacterium *Halomonasmaura*. The organization of genes *eps ABCDJ*, which are involved in mauran synthesis (Bouchotroch et al., 2001; Arias et al., 2003). Mauran is mainly synthesized through the Wzy pathway and the genes *epsA*, *epsB*, *epsC*, *epsD* and *epsJ* are responsible for regulation of autotyrosine kinase, sodium-sulfate symporter, outer membrane porin, phosphotyrosine phosphatase and flippase, respectively, which polymerizes the monosaccharides and forms mauran (Arco et al., 2005). *epsA* gene encodes an outer membrane protein, which is involved in the translocation of sugar moities across the membrane. The gene *epsB* participates in the polymerization cycle and dephosphorylates, the gene product phosphotyrosine phosphatase and undecaprenyl pyrophosphate. *epsC* is a membrane-specific located protein, which is involved in maintaining the length of polysaccharide chain. e*psD* (sodium-sulfate symporter) is a co-transporter of sodium sulfate/carboxylate

(Markovich and Murer, 2004). *epsJ* (flippase) transfers the polysaccharides across the membrane (Arco et al., 2005; Llamas et al., 2006).

18.7 STRUCTURE OF EPSs

Marine microbes are being highly explored because of their high diversity, complex physiological and metabolic systems. EPSs, which are significantly important biomolecules, are often produced in response to environmental disturbances such as a change in salinity, pH, nutrient requirement, osmotic stress and temperature (Junge et al., 2004; Krembs et al., 2002). The composition and structure of EPSs generally depend on the environmental conditions and the ecological niche. Structural changes in EPSs have been reported in *Pseudomonas* NCIB 11264, when grown in different N, P and C sources and in the stationary phase with suboptimal growth conditions (Williams and Wimpenny, 1978).

Most bacteria use carbohydrates as their sole source of carbon and energy and amino acids or other ammonium salts as their sole source of nitrogen. The structure and production of EPSs rely on the type of carbon and nitrogen substrates used (Poli et al., 2011).

EPSs mainly occur in two forms: homopolysaccharides and heteropolysaccharides. The main monosaccharide component commonly found in marine EPSs are (1) pentoses, such as D-arabinose, D-ribose and D-xylose; (2) hexoses, such as D-glucose, D-galactose, D-mannose, D-allose, L-rhamnose and L-fucose; (3) amino sugars, such as D-glucosamine and D-galactosamine; (4) uronic acids, such as D-glucuronic acids and D-galacturonic acids; and (5) other organic or inorganic components, such as sulfate, phosphate and acetic acid. The molecular weight of EPSs ranges from 1×10^5–3×10^5 Da. As per a report of Heymann et al. (2016), marine EPSs are of high molecular weight ranging from 4–15 KDa. The presence of uronic acids, ketal-linked pyruvate and inorganic ions, such as phosphate or sulfate, endows a polyanionic nature to EPSs whereas some EPSs are neutral in nature (Poli et al., 2010).

The monosaccharides have either 1,4-β- or 1,3-β-linkages which gives strong rigidity to these EPSs. The physical properties of EPSs are deeply affected by the way monosaccharides are arranged. The marine bacterium *Vibrio diabolicus* isolated from a Pompei worm tube secreted EPSs that had similar amounts of glucuronic acid and hexosamine. It is a hyaluronic acid–like polymer (Figure 18.4; Rougeaux et al., 1999).

A marine bacterium *Alteromonas infernus* was reported to secrete branched acidic heteropolysaccharide EPS having an HMW. Its monosaccharide repeating units are composed of galacturonic, glucuronic acid, galactose and glucose substituted with a sulfate group (Roger et al., 2004). *Alteromonas* strain 1644 possessed the ability to produce two different kinds of EPSs that differ in their viscosity and concentration of ions, making their separation difficult due to their gelling nature (Poli et al., 2010). The tertiary structure and characteristics of EPSs mainly depend on the presence of hydroxyl and

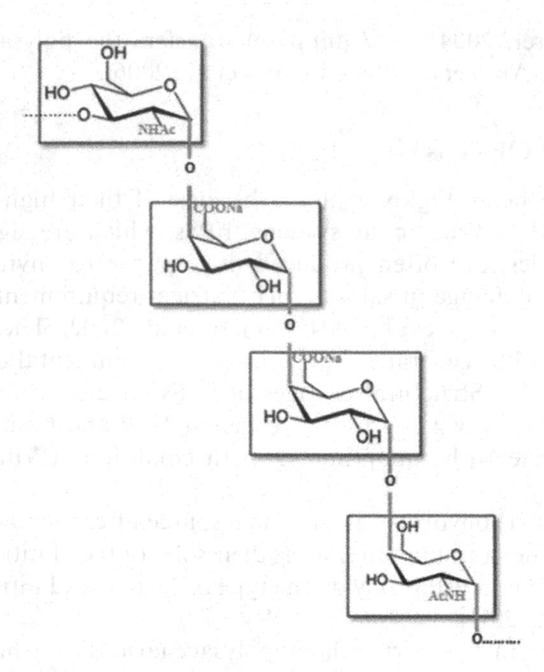

FIGURE 18.4 Chemical structure of EPSs produced by marine *Vibrio diabolicus* (modified from Rougeaux et al., 1999).

carboxyl groups. Microbial exopolymers are available either in dissolved form or as aggregates in a gel-like slime matrix. Figure 18.5 indicates the EPSs structure of psychrotolerant *Pseudoalteromonas* sp. SM9913 isolated from the deep sea of Antarctica at a depth of 1855 m. It contains a linear simple arrangement of α-1,6 glucose with an elevated degree of acetylation (Qin et al., 2007).

18.8 DISTRIBUTION OF EPS-PRODUCING MARINE BACTERIA

With due advancement of time, increasing concern is laid upon the isolation of new EPS-producing bacteria from extreme environments like deep-sea hydrothermal vents, mangroves, hypersaline environments, cold water, and sediments.

18.8.1 Hydrothermal Vents

Thermophilic bacteria from marine habitats grow in hot and salty environments with huge temperature fluctuations. They have developed structurally diverse lipids, enzymes, and biopolymers for their adaptation (Nicolaus et al., 2000, 2002, 2010). EPS-producing thermophilic bacteria *Bacillus thermantarticus*, *Alteromonas* sp., *Thermococcus litoralis* and *T. maritima* have been

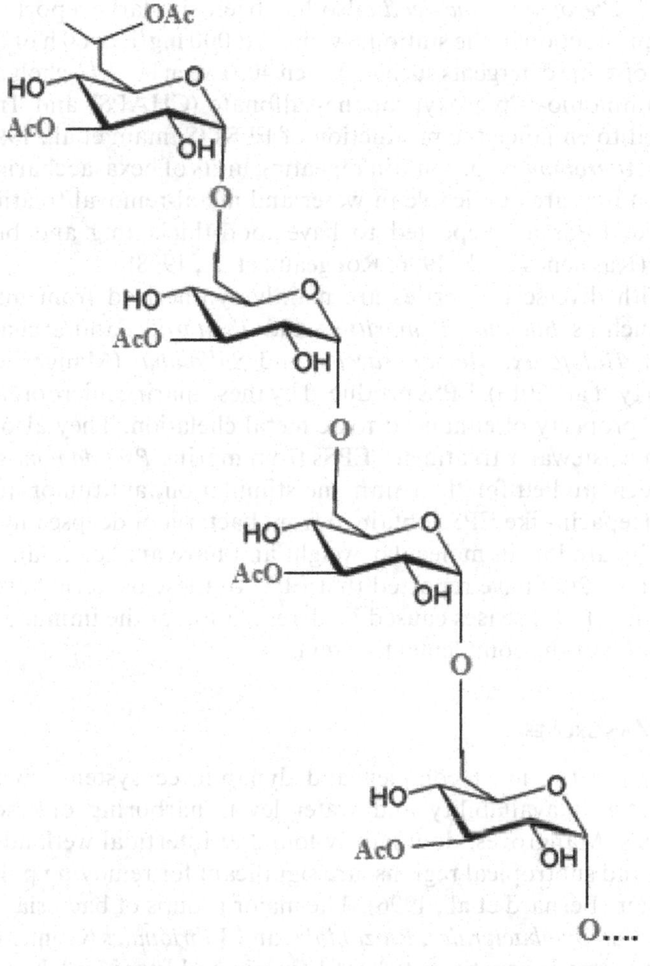

FIGURE 18.5 Chemical structure of EPSs produced by marine *Pseudoalteromonas* sp. SM9913 (modified from Qin et al., 2007).

reported on for their ability to grow beyond normal boundaries of life (Manca et al., 1996; Rinker and Kelly, 2000; Lelchat et al., 2015).

EPSs synthesized from this origin have a novel chemical composition with a wide range of biological and metal-binding properties (Guezennec, 2002). *Alteromonas macleodii* is an aerobic mesophilic bacterium isolated from the hydrothermal vent (depth of 2600 m) from the North Fiji Basin. The maximum production of EPSs was observed during the stationary phase in marine *Alteromonas* 1644 isolated from deepsea hydrothermal vents (Samain et al., 1997), strain 4004 belonging to genus *Geobacillus* cultured from sea sand in Ischia island, Italy (Moriello et al., 2003) and a new strain of *Bacillus licheniformis,* isolated from a marine hot-spring volcano island (Maugeri et al., 2002).

A strain of *Alteromonas macleodii*also has been similarly reported for maximum EPS production in the stationary phase (6000 mg/L at 60 h of incubation).

The use of some detergents such as Tween 40, Tween 80, 3-(3-cholamidopropyl) dimethyl ammonio-1-hydroxypropane-sulfonate (CHAPS) and Triton X has been proved to enhance the production of EPSs (Samain et al., 1997). Mostly, EPSs from *Alteromonas* sp. contain repeating units of hexasaccharide (3 uronosyl residues) that are applicable in water and metal-removal treatment. These xanthan-like EPSs are reported to have food-thickening and bone-healing properties (Raguenes et al., 1996; Rougeaux et al., 1998).

EPSs with diverse properties are mainly synthesized from many marine bacteria, such as *Bacillus*, *T. maritime* and *T. litoralis*, and archaea, such as *Haloarcula*, *Haloferax*, *Methanosarcina* and *Sulfolobus* (Maugeri et al., 2002; Chakraborty et al., 2016). EPSs produced by these marine microorganisms have the unique property of enhanced toxic metal chelation. They also act as biosorbents in wastewater treatment. EPSs from marine *Pseudomonas* and *Vibrio* sp. have been studied for their immune stimulation, antitumor and antiviral activities. Heparin-like EPSs obtained from bacteria of deepsea hydrothermal vents usually are low in molecular weight and have anticoagulant properties. Courtois et al., 2014 have reported that EPSs of these bacteria have been used in the treatment of diseases caused by deregulation of the immune system and overactivation of the complement system.

18.8.2 Mangroves

Mangroves are the most complex and dynamic ecosystems, with different salinity, nutrient availability and water levels harboring diverse microbial communities. Mangroves, dominantly found at intertidal wetlands coastlines in tropical and subtropical regions, are significant for removing pollution from marine water (Bernard et al., 1996). The major groups of bacteria in this habitat include *Campylobacterales*, *Rhizobiales* and *Vibrionales* (Gomes et al., 2010). Tropical mangroves constitute 91% of bacteria and fungi and 7% of algae and protozoa of the total microbial mass.

Pantoeaagglomerans strain KFS-9 isolated from mangrove forests have been reported to produce EPSs with high water solubility and antioxidant activity (Wang et al., 2007). *Pseudomonas* and *Azotobacter* spp. isolated from Pichavaram mangrove sediments synthesized alginate EPSs (Lakshmipriya, 2012).

18.8.3 Hypersaline Marine Environment

Hypersaline systems are harsh environments that have salt concentrations much greater than that of seawater, often close to or exceeding salt saturation (Oren, 2010; Biswas and Paul, 2014; Oren, 2016). Usually, hypersaline environments have high salinity and alkaline pH. The factors affecting biodiversity in such an environment include pressure, low nutrient availability, solar radiation and the presence of heavy metals and toxic compounds (Rodriguez-Valera,

1988). The EPS-producing microorganisms inhabiting such environments are members of archaea of the order *Halobacteriales*, namely, *Halobacterium, Halomonasalmeriensis, Haloarcula, Halorubrum, Haloferax, Natronobacterium, Natronococcus* and *Salinibacter ruber. Haloferax mediterranei* has been isolated from Mediterranean Sea that synthesizes high amounts of EPSs, which has an application in oil recovery (especially in oil deposits with high salinity; Anton et al., 1988; Parolis et al., 1996; Llamas et al., 2012).

Marine bacterium *Hahella chejuensis* belonging to phylum *Proteobacteria*, isolated from the sediment sample of Marado, Cheju Island, Korea, has been reported to synthesize copious amounts of glucose- and galactose-rich EPSs when grown in a sucrose-containing medium. These EPSs are used as biosurfactants for detoxifying polluted environments, such as those contaminated with petrochemical oils (Ko et al., 2000; Lee et al., 2001). EPSs from *Bacillus* and *Microbacterium* sp., having significant concentrations of hexosamines and uronic acids, possess biosurfactant activity. The unique anionic nature of these EPSs can chelate cations, which is applicable in bioremediation processes. EPSs from halophilic *Bacillus* sp. have tissue regeneration property (Ortega-Morales et al., 2005). EPSs from three novel halophilic species, *Idiomarina fontislapidosi, I. ramblicola* and *Alteromonas hispanica*, have been reported for emulsification and metal-chelating activities (Mata et al., 2008). *Halomonas alkaliantarctica* strain CRSS isolated from the Salt Lake in Cape Russell in Antarctica synthesized EPSs with high viscosity, primarily constituting glucose and fructose (Poli et al., 2007).

18.8.4 Cold Marine Environment

The deep sea is a major component of our planet's biosphere and represents 75 % of the total volume of the oceans. These environments are influenced by high pressure, low temperature and low nutrient concentration. Psychrotrophic bacteria possess the capability of growing above 20°C as well as tolerating low temperatures, whereas psychrophilic bacteria prefer to grow in temperatures as up to 20°C (0–20°C; Gounot, 1986). Few other examples of EPS-producing marine microbes include the psychrotolerant strain of *Pseudoalteromonas* sp. Bsi20310 from the sea ice of Antarctica (Bozal et al., 1997; Ma et al., 2012). *Bacillus licheniform is* strain B3–15, originating from a shallow marine vent, produced a novel EPSs, which is used as a counteractive agent for immune disorders caused by the herpes virus (Spanò and Arena, 2016), whereas *Pseudoalteromonas antarctica* strain NF3, isolated from a glacial marine sludge of the South Shetland Island in Antarctica, produced maximum EPSs during its exponential growth phase. *Pseudoalteromonas* strain SM9913, isolated from deepsea sediment from the Bohai Gulf, Yellow Sea, China, synthesized EPSs having linear arrangement of α (1 → 6) linkage of glucose with a high degree of acetylation, showing flocculation behavior and high biosorption capacity. EPSs produced by this strain are inversely proportional to the temperature fluctuation. *Pseudoalteromonas* strains CAM025 and CAM036, isolated from the Southern Ocean, have sulfate and uronic acids as galacturonic acid along with acetyl groups. In addition, strain CAM036 has been reported to possess a succinyl group. These features confer a polyanioinic nature to EPSs in a marine environment as at an alkaline pH, the acidic group on EPSs are ionized. This property makes EPSs bind cations, such as dissolved metals (Mancuso Nichols et al., 2004).

18.8.5 Sediments and Surface Water

The water from streams, rivers, lakes, wetlands, and the ocean constitutes the surface water. The surface water of the ocean is enriched with the primary microflora such as *Alteromonas, Colwellia* and *Pseudoalteromonas*. Many other species of marine bacteria dwelling in water and sediments produce diverse types of EPSs, which are a growing interest in industrial sectors. EPS-producing *Exiguobacterium* sp. was isolated and characterized from coastal (Mandapam) area of Pak Bay, Tamil Nadu, India, which showed the presence of amino sugars. EPSs from *Enterobacter cloacae*, emulsified xylene, hexane, paraffin oil, benzene, coconut oil, kerosene, cottonseed oil, groundnut oil, jojoba oil, castor oil and sunflower oil (Iyer et al., 2006). Another bioflocculant-producing marine bacterium *Virgibacillus* sp. was isolated from the marine sediment of Algoa Bay (Cosa et al., 2011). *Halomonas* sp. was isolated from surface water of Arabian Sea as a potent biofilm former synthesizing EPSs (Nisha and Thangavel, 2014). EPS-producing marine bacteria, namely, *Vibrio parahaemolyticus* and *Pseudomonas* sp. strain NCMB 2021 were isolated from marine water (Su and Liu, 2007) of Halifax, Nova Scotia (Wrangstadh et al., 1986). *Pseudoalteromonas* sp. strain S9 was isolated from the marine sediment; *Pseudoalteromonas atlantica* and *Hyphomonas* strain MHS-3 have been isolated from shallow-water sediments in Puget Sound (northwestern coast of the U.S. state of Washington, an inlet of the Pacific Ocean, and part of the Salish Sea) that have homogenous polysaccharides with high amounts of N-acetylgalactosamine (Quintero and Weiner, 1995).

18.9 BIOTECHNOLOGICAL POTENTIAL OF MARINE EPSS

The marine ecosystem harbors enormous microbial biodiversity which produces structurally diverse EPSs. The isolation and identification of new EPSs producing marine bacteria provide wider opportunities for novel industrial applications as discussed in the following sections.

18.9.1 Hydrocarbon Remediation

The marine environment is polluted by a variety of organic compounds, from both terrestrial and atmospheric sources. The contamination of the ocean principally with crude oil remains a major threat. Every year, about 1.3 million tons of petroleum enter into the marine environment (Board et al., 2003; Hassanshahian and Cappello, 2013). The components of crude oil (particularly aromatic hydrocarbons fractions) are toxic to marine organisms (Carls et al.,

1999; Heintz, 2007). However, the microbial community residing at polluted sites plays a vital role in the degradation of these compounds. Microbes have adapted several mechanisms, such as an increase in population size, genetic modification, biofilm formation and secretion of biosurfactants, to degrade organic contaminants (Singh et al., 2006).

Many marine bacteria produce EPSs for enhanced utilization of organic compounds as a sole carbon source. EPSs extracted from marine bacteria can significantly influence the overall performance of bioremediation. It also helps in attaching microbial cells to the surfaces of xenobiotics and subsequently helps in their degradation (Ta-Chen et al., 2008). *Alcanivorax borkumens* is one of the leading oil-degrading marine bacteria as it can form biofilm at the oil–water interface to increase the bioavailability and degradation of oil (Schneiker et al., 2006). Baelum et al. (2012) reported that genus *Colwellia* isolated from deepsea water at an extremely low temperature (5°C) can synthesize EPSs and degrade the components of crude oil. Ta-Chen et al., 2008 reported that EPSs are natural emulsifying agents and are important for remediating hydrocarbons. Bacteria such as *Halomonas eurihalina* and *Enterobacter cloacae* produced EPSs with emulsifying properties (Calvo et al., 2002; Iyer et al., 2006). EPSs can enhance the solubility of hydrophobic substrates (e.g., polycyclic aromatic hydrocarbon [PAHs], biphenyl) by numerous hydrophobic interactions (Pan et al., 2010). EPSs produced by the marine bacterium *Halomonas* sp. exhibited amphiphilic properties and high emulsifying qualities. The EPSs produced by this bacterium increase the solubilization and degradation of many hydrophobic organic pollutants, such as phenanthrene, fluorene, pyrene and biphenyl (Gutierrez et al., 2013).

Iwabuchi et al., 2000, 2002 reported that EPSs from *Rhodococcus rhodochrous* (named as S-2 EPS) led to 50% increase in degradation of multiple aromatic components in crude oil by native consortia of sea-water. These EPSs have also been used for *in situ* bioremediation (Ron and Rosenberg, 2002; Venosa and Zhu, 2003; Cappello et al., 2012). Afrouzossadat et al., 2012 reported that EPSs associated with toluene utilizing marine bacterium *Sporosarcina halophila* contain peroxidase enzymes, such as laccase and catalase (present in EPSs), which considerably affected the degradation of toluene. So extracellular enzymes immobilized in EPSs play a beneficial role in biodegradation. Table 18.2 illustrates the list of EPS-producing marine bacteria with their hydrocarbon degradation ability.

18.9.2 Metal Remediation

EPSs are composed of various negatively charged functional groups and vary in their interaction with ionic compounds (Zhang et al., 2010). Thus, EPSs are able to form organometallic complexes by electrostatic attraction with heavy metals (Beech and Sunner, 2004). This adds a promising application of EPSs for toxic heavy metal bioremediation (Shah et al., 2000; Lloyd and Renshaw, 2005). Extracellular polymeric substances have nonsugar components, such as proteins and nucleic acids. These components are important for metal absorption,

TABLE 18.2
EPS-Producing Marine Bacteria with Hydrocarbon Degradation Ability

Genus	Source	Hydrocarbon Degradation	References
Enterobacter	Sediment	*n*-Hexadecane	Hua et al., 2010
Halomonas	Seawater	Alkane, biphenyl	Gutierrez et al., 2013
Pseudomonas	Seawater	PAHs	Mangwani et al., 2014
Colwellia	Deepsea water	Crude oil	Marx et al., 2009
Alcanivorax	Seawater	Alkanes	Schneiker et al., 2006
Pseudoalteromonas	Deepsea water	PAHs, alkane	Chronopoulou et al., 2015
Rhodococcus	Seawater	PAHs	Iwabuchi et al., 2002

TABLE 18.3
Metal Biosorption by EPS-Producing Marine Bacteria

Genus	Source	Metal Biosorption	References
Enterobacter	Sediment	Cd, Cu, Co, Cr	Iyer et al., 2005, 2006
Alteromonas	Deepsea vent	Zn, Cd, Pb	Loaec et al., 1997
Marinobacter	Marine habitat	Pb and Cu	Bhaskar and Bhosle, 2006
Pseudoalteromonas	Deepsea sediments	Zn, Cu, Co	Qin et al., 2007
Pseudomonas	Sediments	Cd	Chakraborty and Das, 2014
Halomonas	Estuary	Cu, Zn, Pb, Cd, U	Gutierrez et al., 2012
Sulfate-reducing bacteria	Corroded surface	Cr, Ni, Mo, Cu, Zn	Beech and Cheung, 1995; Pal and Paul, 2008
Rhodobium	Seawater	Cd, Cu, Pb, Zn	Panwichian et al., 2011
Rhodovulum	Sediments	Cd	Watanabe et al., 2003
Bacillus	Seawater	Hg, Zn	Maugeri et al., 2002

because it enhances the rigidity of the tertiary structure and physical properties of EPSs. The occurrence of these nonsugar residues imparts acidic nature to EPSs, and the presence of uronic acid confers a negative surface charge (Decho and Lopez, 1993; Iyer et al., 2005). Table 18.3 indicates EPSs producing marine bacteria used in metal biosorption.

In another study, Gutierrez et al., 2012 reported that *Halomonas* sp. TG39 produced EPSs, rich in uronic acid that exhibited the ability to bind trace (e.g., Fe, Zn, Mn, and Si) and toxic metals (e.g., Cd and Pb). Maugeri et al., 2002 characterized EPSs from *Bacillus licheniformis*, isolated from marine hot springs, that could immobilize (or resist) Cd^{2+}, Zn^{2+}, As^{2+} and Hg^{2+}. EPSs produced by marine sulfate-reducing bacteria (SRB) can bind with heavy metals, such as Cr, Ni, and Mo. The binding efficiency of Mo to SRB-EPS is considerably greater as compared to Cr and Ni (Beech and Cheung, 1995). EPSs from *Rhodobium marinum* can bind with Pb and Cd (Panwichian et al., 2011). *Rhodovulum* sp. PS88 a photosynthetic bacterium has been reported for the removal of Cd (Watanabe et al., 2003). Marine EPSs of *Pseudomonas aeruginosa* JP-11 were used for synthesis of cadmium sulfide (CdS) nanoparticles (NPs) that could remove Cd from an aqueous solution (Raj et al., 2016).

18.9.3 Nanobiotechnology

Nanobiotechnology is receiving global attention due to its various applications. The synthesis of NPs with different sizes, morphologies and chemical composition is an important area of research in nanobiotechnology. EPSs have been recently focused on for NPs synthesis, because of their biodegradable, hydrophilic and biocompatible nature (Sarmento et al., 2006). Generally, NPs are easily produced from plant extracts and microbes, but the mechanisms for production are ill defined, and the final product is impure. EPSs from microbial origins seem to be a promising alternative as they can effectively act as strong reducing as well as stabilizing agents for metal NP production (Mehta et al., 2014).

There are few reports available on production of silver NPs (AgNPs) using polysaccharides, such as heparin, hyaluronic acid, cellulose, starch, alginic acid and others. However, most of them are unable to synthesize monodisperse colloidal suspension of AgNPs. *Alteromonas macleodii*, a marine osmotolerant strain, and *Bacillus subtilis* MSBN17 produced AgNPs used for biomineralization of silver (Sathiyanarayanan et al., 2013). Polyelectrolyte NPs are produced when two oppositely charged natural polymers interact with each other reversibly (Hamman, 2010). As reported by Deepak et al., 2016, *Lactobacillus acidophilus* is a strong candidate for producing polyelectrolyte NPs, which are useful in delivering the drugs through various routes of administration, such as in cancer treatments and insulin absorption (Fernández-Urrusuno et al., 1999; Lemarchand et al., 2003).

18.9.4 Control of Biofouling

Biological fouling is the attachment of organisms to submerged or wetted artificial surfaces. Biofouling can occur in two different ways: microbial fouling (adhesion of microorganisms to the surface) and macrofouling (adhesion of seaweed, mollusks, mussels or other organisms; Delauney et al., 2010). It is an ongoing and crucial problem for surfaces that are directly in contact with water, causing huge economic losses (Yebra et al., 2004; Cao et al., 2011). The first step of biofouling is the adsorption of organic and inorganic compounds to the primary film. Furthermore, the diverse group of specific and nonspecific microbial communities and multicellular organisms, such as microalgae and seaweeds, adhere to the primary film, accumulate and produce a polymeric matrix. At the last stage, large marine organisms, such as macroalgae, barnacles and mussels, attach to the preexisting microbial film (Delauney et al., 2010).

There are many techniques used to prevent fouling of artificial surfaces such as the use of biocide-coated surfaces (Nakayama et al., 1998; Morris and Walsh, 2000; Bearinger et al., 2003). However, natural antifouling chemicals produced by aquatic organisms or plants are the most promising methods for biofouling control. Coating surfaces with biological polymers or secondary metabolites can also prevent the formation of biofilm and, subsequently, biofouling. Guezennec et al. (2012) used EPSs from *Alteromonas, Pseudomonas*, and *Vibrio* spp. for an antibiofouling coating. They reported that EPSs can inhibit the primary colonization of bacteria, thereby minimizing successive biofouling. The presence of a polysaccharide film changes the hydrophobic/hydrophilic balance, which is important for adhering cells to surfaces (Yaskovich, 1998; Guezennec et al., 2012). *Vibrio alginolyticus, V. proteolyticus*, and *V. vulnificus* also are producers of antibiofouling EPSs (Qian et al., 2006; Kim et al., 2011).

EPSs extracted from *V. alginolyticus* prevented larval attachment and metamorphosis of marine invertebrates (e.g., *Hydroides elegans*, *Balanus amphitrite*, and *Bugula neritina*) causing fouling (Qian et al., 2006). Marine bacterial EPSs are nontoxic and do not cause any harm to native organisms residing near the EPS-coated surface. Thus, EPSs from marine bacteria can act as antibiofouling mediators.

18.9.5 Medical Field

EPSs have been used widely for medical purposes because of their unique and superior physical properties relative to other biomaterials. They have been used for diverse applications in biomedical field as ophthalmology, tissue engineering, antitumor, immune-stimulant activities, fibrinolytic agent, implantation of medical devices and artificial organs, prostheses, dentistry, bone repair and more (Shih, 2010). They are significantly applied in the field of glycochemistry and glycobiology in the form of glycosaminoglycans (GAGs) and are also being extensively used for the design and preparation of therapeutic drugs (Delbarre-Ladrat et al., 2014). EPSs from marine thermophilic bacilli were used for inducing a Th1 cytokine profile in humans (Arena et al., 2004). EPSs is also applied in antimetastatic treatment as reported by Heymann et al. (2016). Other reports for medicinal uses assigned to EPSs are antiproliferative, anticoagulant, antiviral and antiangiogenic activities (Arena et al., 2009; Llamas et al., 2010). EPSs from the genus *Bacillus* have been reported to possess strongest fibrinolytic activity (Lee and Kim, 2012).

The antiviral activity of sulfated EPSs has been attributed to their ability to inhibit virus particle adsorption in host cells (Llamas et al., 2010). *Pseudoalteromonas* sp., AM, *B. amyloliquefaciens* MJ7–66 and *B. subtilis* have been reported as promising candidates for the production of EPSs with interesting novel fibrinolytic activities (Omura et al., 2005; Al-Nahas et al., 2011; Lee and Kim, 2012). HMW mannans from yeast also possess fibrinolytic properties (Elinov et al., 1988).

18.9.6 Bioflocculant Production

Rapid development in industrialization and anthropogenic activities has led to an increase in the discharge of waste and wastewater containing organic and inorganic pollutants. Bioflocculant is a kind of biodegradable macromolecular flocculant produced by microorganisms. Because of their biodegradability, harmlessness and inability to produce secondary pollutants, bioflocculants have gained much wider attention in research (Gong et al., 2008). Flocculation is an essential process in the treatment of wastewater and dye effluents (Fujita, 2000).

EPSs act as natural bioflocculant and well replace chemical flocculants. *B. cereus* SK has been reported for its significant bioflocculant activity (Busi et al., 2017). EPSs of *Corynebacterium glutamicum* and *Bacillus* sp. As-101 also exhibited noticeable flocculation activity (Salehizadeh et al., 2000; He et al., 2004). EPS-producing *Pseudomonas* and *Staphylococcus* spp. were found to possess bioflocculant properties, and a novel bioflocculant MM1 was reported from the EPSs produced by these bacteria. Similarly, a consortia of *Rhizobium radiobacter* F2 and *B. sphaeicus* F6 and of *Oceanobacillus* and *Halobacillus* spp. have been applied in the industrial wastewater and river water treatments due to the bioflocculant properties of EPSs produced by them (Zhang et al., 2007; Wang et al., 2011; Cosa et al., 2014; Vijayendra and Shamala, 2014).

18.9.7 Emulsifying Agents

Numerous studies have been conducted for developing and implementing innovative technology to clean up contamination with petroleum hydrocarbons (King et al., 1997). Bioemulsifiers are able to emulsify these pollutants much more effectively than chemical surfactants. Halophilic EPSs producers are considered an interesting source for microbial-enhanced oil recovery, where polymers act as emulsifiers. The emulsification of petroleum has been noted with some strains, such as *Halobacterium salinarium*, *Haloferaxvolcanii* and *Halobacterium distributum* (Kulichevskaya et al., 1991). Martínez-Checa et al., 2007 studied the characteristics of the bioemulsifier V2–7 synthesized by strain F2–7 of *Halomonas eurihalin a*and found that it has the ability of emulsify a wide range of hydrocarbons, for example, n-tetradecane, n-hexadecane, n-octane, xylene mineral light and heavy oils, petrol and crude oil. EPSs of *Acinetobacter colcoaceticus* and *Pseudomonas* sp. produce EPSs emulsan (Abbasi et al., 2008; Carrión et al., 2015). EPSs from *Sphingomonaspausimobilis* GS-1, *Enterobacter cloaceae* and *Heurihalina* can emulsify various hydrocarbons and oils (Ashtaputre et al., 1995; Abbasi et al., 2008).

18.10 CONCLUSION

The marine environment is one of those niches that shows dynamic fluxes that encompass half the world's population. In the marine environment, the association and communication of macro- and microorganisms are the prominent features for the continuation of biogeochemical cycle and food web. The marine microbial community is critically essential for the maintenance of healthy and competitive environment in the ocean ecosystem. In recent times, research on marine bacterial diversity is increasing because of its ability to survive in extreme conditions, such as alkaline, hypersaline, fluctuation in temperature, pH, pressure and nutrient availability. Changes in environmental conditions exert a driving force on bacteria for selecting new adaptive strategies, leading to the synthesis of novel secondary metabolites. One of the important secondary metabolites are EPSs, which are of interest to researchers due to their significant role; marine bacterial EPSs with unique physical and chemical properties may find a wide range of applications in pharmaceutical, medical, food, chemical and other fields.

CONFLICTS OF INTEREST

The authors declare no conflicts of interest.

ACKNOWLEDGEMENT

None declare.

REFERENCES

Abbasi, A., Amiri, S., 2008. Emulsifying behavior of an exopolysaccharide produced by *Enterobacter cloacae*. *Afr. J. Biotechnol.* 7(10), pp. 1574–1576.

Afrouzossadat, H.A., Giti, E., SeyedMahdi, G., 2012. The role of exopolysaccharide, biosurfactant and peroxidase enzymes on toluene degradation by bacteria isolated from marine and wastewater environments. *Jundishapur J. Microbiol.* 2012(3), pp. 479–485.

Al-Nahas, M.O., Darwish, M.M., Ali, A.E., Amin, M.A., 2011. Characterization of an exopolysaccharide-producing marine bacterium, isolate *Pseudoalteromonas* sp. *AM. Afr. J. Microbiol. Res.* 5(22), pp. 3823–3831.

Anton, J., Meseguer, I., Rodriguez-Valera, F., 1988. Production of an extracellular polysaccharide by *Haloferax mediterranei*. *Appl. Environ. Microbiol.* 54(10), pp. 2381–2386.

Arco, Y., Llamas, I., Martínez-Checa, F., Argandona, M., Quesada, E., del Moral, A., 2005. epsABCJ genes are involved in the biosynthesis of the exopolysaccharide mauran produced by *Halomonas maura*. *Microbiol.* 151(9), pp. 2841–2851.

Arena, A., Gugliandolo, C., Stassi, G., Pavone, B., Iannello, D., Bisignano, G., Maugeri, T.L., 2009. An exopolysaccharide produced by Geobacillus thermodenitrificans strain B3-72: Antiviral activity on immunocompetent cells. *Immunol. Lett.* 123(2), pp. 132–137.

Arena, A., Pavone, B., Gugliandolo, C., Maugeri, T.L., Bisignano, G., 2004. Exopolysaccharides from marine thermophilic bacilli induce a Th1 cytokine profile in human PBMC. *Clin. Microbiol. Infect.* 10(2), pp. 366–375.

Arias, S., Del Moral, A., Ferrer, M.R., Tallon, R., Quesada, E., Bejar, V., 2003. Mauran, an exopolysaccharide produced by the halophilic bacterium *Halomonas maura*, with a novel composition and interesting properties for biotechnology. *Extremophiles.* 7(4), pp. 319–326.

Ashtaputre, A.A., Shah, A.K., 1995. Emulsifying property of a viscous exopolysaccharide from *Sphingomonas paucimobilis*. *World J. Microbiol. Biotechnol.* 11(2), pp. 219–222.

Baelum, J., Borglin, S., Chakraborty, R., Fortney, J.L., Lamendella, R., Mason, O.U., Malfatti, S.A., 2012. Deep-sea bacteria enriched by oil and dispersant from the Deepwater Horizon spill. *Environ. Microbiol.* 14(9), pp. 2405–2416.

Bearinger, J.P., Terrettaz, S., Michel, R., Tirelli, N., Vogel, H., Textor, M., Hubbell, J.A., 2003. Chemisorbed poly (propylene sulphide)-based copolymers resist biomolecular interactions. *Nat. Mater*. 2(4), pp. 259–264.

Beech, I.B., Cheung, C.S., 1995. Interactions of exopolymers produced by sulphate-reducing bacteria with metal ions. *Int. Biodeterior. Biodegradation*. 35(1–3), pp. 59–72.

Beech, I.B., Sunner, J., 2004. Biocorrosion: Towards understanding interactions between biofilms and metals. *Curr. Opin. Biotechnol.* 15(3), pp. 181–186.

Bernard, D., Pascaline, H., Jeremie, J.J., 1996. Distribution and origin of hydrocarbons in sediments from lagoons with fringing mangrove communities. *Mar. Poll. Bull.* 32(10), pp. 734–739.

Bhaskar, P.V., Bhosle, N.B., 2006. Bacterial extracellular polymeric substance (EPS): A carrier of heavy metals in the marine food-chain. *Environ. Int.* 32(2), pp. 191–198.

Biswas, J., Paul, A.K., 2014. Production of extracellular polymeric substances by halophilic bacteria of solar salterns. *Chin. J. Biol.* 2014, pp. 1–12.

Board, M., Board, O.S., National Research Council, 2003. *Oil in the Sea III: Inputs, Fates, and Effects*. National Academies Press, Washington, DC, USA.

Bouchotroch, S., Quesada, E., del Moral, A., Llamas, I., Béjar, V., 2001. *Halomonas maura* sp. nov., a novel moderately halophilic, exopolysaccharide-producing bacterium. *Int. J. Syst. Evol. Microbiol.* 51(5), pp. 1625–1632.

Bowers, K.J., Mesbah, N.M., Wiegel, J., 2009. Biodiversity of poly-extremophilic bacteria: Does combining the extremes of high salt, alkaline pH and elevated temperature approach a physico-chemical boundary for life? *Saline Syst*. 5(9), pp. 1–8.

Boyle, C.D., Reade, A.E., 1983. Characterization of two extracellular polysaccharides from marine bacteria. *J. Appl. Environ. Microbiol.* 46(2), pp. 392–399.

Bozal, N., Tudela, E., Rossello-Mora, R., Lalucat, J., Guinea, J., 1997. *Pseudoalteromonas antarctica* sp. nov., isolated from an Antarctic coastal environment. *Int. J. Syst. Evol. Microbiol.* 47(2), pp. 345–351.

Busi, S., Karuganti, S., Rajkumari, J., Paramanandham, P., Pattnaik, S., 2017. Sludge settling and algal flocculating activity of extracellular polymeric substance (EPS) derived from *Bacillus cereus* SK. *Water Environ. J.* 31(1), pp. 97–104.

Calvo, C., Martínez-Checa, F., Toledo, F., Porcel, J., Quesada, E., 2002. Characteristics of bioemulsifiers synthesised in crude oil media by *Halomonas eurihalina* and their effectiveness in the isolation of bacteria able to grow in the presence of hydrocarbons. *Appl. Microbiol. Biotechnol.* 60(3), pp. 347–351.

Cao, S., Wang, J., Chen, H., Chen, D., 2011. Progress of marine biofouling and antifouling technologies. *Chin. Sci. Bull.* 56(7), pp. 598–612.

Cappello, S., Genovese, M., Della Torre, C., Crisari, A., Hassanshahian, M., Santisi, S., Yakimov, M.M., 2012. Effect of bioemulsificant exopolysaccharide (EPS2003) on microbial community dynamics during assays of oil spill bioremediation: A microcosm study. *Mar. Poll. Bull.* 64(12), pp. 2820–2828.

Carls, M.G., Rice, S.D., Hose, J.E., 1999. Sensitivity of fish embryos to weathered crude oil: Part I. Low-level exposure during incubation causes malformations, genetic damage, and mortality in larval pacific herring (Clupea pallasi). *Environ. Toxicol. Chem.* 18(3), pp. 481–493.

Caron, D.A., 1987. Grazing of attached bacteria by heterotrophic microflagellates. *Microbiol. Ecol.* 13(3), pp. 203–218.

Carrión, O., Delgado, L., Mercade, E., 2015. New emulsifying and cryoprotective exopolysaccharide from Antarctic *Pseudomonas* sp. ID1. *Carbohydr. Polym.* 117, pp. 1028–1034.

Chakraborty, J., Das, S., 2014. Characterization and cadmium-resistant gene expression of biofilm-forming marine bacterium *Pseudomonas aeruginosa* JP-11. *Environ. Sci. Pollut. Res.* 21(24), pp. 14188–14201.

Chakraborty, J., Mangwani, N., Dash, H.R., Kumari, S., Kumar, H., Das, S., 2016. Marine bacterial exopolysaccharides. In: Se-Kwon, K. (Ed.). *Marine Glycobiology: Principles and Applications*. Busan, South Korea, pp. 235–257. DOI: 10.1201/9781315371399-18.

Chronopoulou, P.M., Sanni, G.O., Silas-Olu, D.I., Meer, J.R., Timmis, K.N., Brussaard, C.P., McGenity, T.J., 2015. Generalist hydrocarbon-degrading bacterial communities in the oil-polluted water column of the North Sea. *Microb. Biotechnol.* 8(3), pp. 434–447.

Cosa, S., Mabinya, L.V., Olaniran, A.O., Okoh, O.O., Bernard, K., Deyzel, S., Okoh, A.I., 2011. Bioflocculant production by *Virgibacillus* sp. Rob isolated from the bottom sediment of Algoa Bay in the Eastern Cape, South Africa. *Molecules.* 16(3), pp. 2431–2442.

Cosa, S., Okoh, A., 2014. Bioflocculant production by a consortium of two bacterial species and its potential application in industrial wastewater and river water treatment. *Pol. J. Environ. Stud.* 23(3), pp. 689–696.

Costerton, J.W., Cheng, K.J., Geesey, G.G., Ladd, T.I., Nickel, J.C., Dasgupta, M., Marrie, T.J., 1987. Bacterial biofilms in nature and disease. *Annu. Rev. Microbiol.* 41(1), pp. 435–464.

Courtois, A., Berthou, C., Guézennec, J., Boisset, C., Bordron, A., 2014. Exopolysaccharides isolated from hydrothermal vent bacteria can modulate the complement system. *PloS One.* 9(4), pp. 1–11.

Das, S., Lyla, P.S., Khan, S.A., 2006. Marine microbial diversity and ecology: Importance and future perspectives. *Curr. Sci.* 90(10), pp. 1325–1335.

Debnath, M., Paul, A.K., Bisen, P.S., 2007. Natural bioactive compounds and biotechnological potential of marine bacteria. *Curr. Pharm. Biotechnol.* 8(5), pp. 253–260.

De Carvalho, C.C., Fernandes, P., 2010. Production of metabolites as bacterial responses to the marine environment. *Mar. Drugs.* 8(3), pp. 705–727.

Decho, A.W., Herndl, G.J., 1995. Microbial activities and the transformation of organic matter within mucilaginous material. *Sci. Total Environ.* 165(1–3), pp. 33–42.

Decho, A.W., Lopez, G.R., 1993. Exopolymer microenvironments of microbial flora: Multiple and interactive effects on trophic relationships. *Limnol. Oceanogr.* 38(8), pp. 1633–1645.

Deepak, V., Ram K.P.S., Sivasubramaniam, S.D., Nellaiah, H., Sundar, K., 2016. Optimization of anticancer exopolysaccharide production from probiotic *Lactobacillus acidophilus* by response surface methodology. *Prep. Biochem. Biotechnol.* 46(3), pp. 288–297.

Delauney, L., Compère, C., Lehaitre, M., 2010. Biofouling protection for marine environmental sensors. *Ocean Sci.* 6(4), pp. 503–511.

Delbarre-Ladrat, C., Sinquin, C., Lebellenger, L., Zykwinska, A., Colliec-Jouault, S., 2014. Exopolysaccharides produced by marine bacteria and their applications as glycosaminoglycan-like molecules. *Front. Chem.* 2(85), pp. 1–15.

DeLong, E.F., Yayanos, A.A., 1987. Properties of the glucose transport system in some deep-sea bacteria. *J. Appl. Environ. Microbiol.* 53(3), pp. 527–532.

Di Donato, P., Poli, A., Taurisano, V., Abbamondi, G.R., Nicolaus, B., Tommonaro, G., 2016. Recent advances in the study of marine microbial biofilm: From the

involvement of quorum sensing in its production up to biotechnological application of the polysaccharide fractions. *J. Mar. Sci. Eng*. 4(2), pp. 34–47.
Dumitriu, S. (Ed.), 2004. *Polysaccharides: Structural Diversity and Functional Versatility*. CRC Press, Marcel Dekker, New York, pp. 1–1224.
Egan, S., Thomas, T., Holmström, C., Kjelleberg, S., 2000. Phylogenetic relationship and antifouling activity of bacterial epiphytes from the marine alga *Ulva lactuca*. *Environ. Microbiol*. 2(3), pp. 343–347.
Elinov, N., Ananyeva, E., Vitovskaya, G., Chlenov, M., Trushina, O., Gurina, S., Karavaeva, A., 1988. Investigation of fractions and fibrinolytic discussion activity of Yeast mannan. *Antibiot. Hymioter*. 33, pp. 359–362.
Fernández-Urrusuno, R., Calvo, P., Remuñán-López, C., Vila-Jato, J.L., Alonso, M.J., 1999. Enhancement of nasal absorption of insulin using chitosan nanoparticles. *J. Pharm. Res*. 16(10), pp. 1576–1581.
Flemming, H.C., Wingender, J., 2001. Relevance of microbial extracellular polymeric substances (EPSs)-Part I: Structural and ecological aspects. *Wat. Sci. Tech*. 43(6), pp. 1–8.
Fujita, M., Ike, M., Tachibana, S., Kitada, G., Kim, S.M., Inoue, Z., 2000. Characterization of a bioflocculant produced by *Citrobacter* sp. TKF04 from acetic and propionic acids. *J. Biosci. Bioeng*. 89(1), pp. 40–46.
Glöckner, F.O., Stal, L.J., Sandaa, R.A., Gasol, J.M., O'Gara, F., Hernandez, F., Pitta, P., 2012. Marine microbial diversity and its role in ecosystem functioning and environmental change. *Marine Board Position Paper*. 17, pp. 13–25.
Gomes, N.C., Cleary, D.F., Pinto, F.N., Egas, C., Almeida, A., Cunha, A., Smalla, K., 2010. Taking root: Enduring effect of rhizosphere bacterial colonization in mangroves. *PLoS One*. 5(11), pp. 1–11.
Gong, W.X., Wang, S.G., Sun, X.F., Liu, X.W., Yue, Q.Y., Gao, B.Y., 2008. Bioflocculant production by culture of *Serratia ficaria* and its application in wastewater treatment. *Bioresour. Technol*. 99(11), pp. 4668–4674.
Gounot, A.M., 1986. Psychrophilic and psychrotrophic microorganisms. *Cell. Mol. Life Sci*. 42(11), pp. 1192–1197.
Guezennec, J., 2002. Deep-sea hydrothermal vents: A new source of innovative bacterial exopolysaccharides of biotechnological interest? *J. Ind. Microbiol. Biotechnol*. 29(4), pp. 204–208.
Guezennec, J., Herry, J.M., Kouzayha, A., Bachere, E., Mittelman, M.W., Fontaine, M.N.B., 2012. Exopolysaccharides from unusual marine environments inhibit early stages of biofouling. *Int. Biodeterior. Biodegradation*. 66(1), pp. 1–7.
Gutierrez, T., Berry, D., Yang, T., Mishamandani, S., McKay, L., Teske, A., Aitken, M.D., 2013. Role of bacterial exopolysaccharides (EPS) in the fate of the oil released during the Deepwater Horizon oil spill. *PloS One*. 8(6), pp. 1–12.
Gutierrez, T., Biller, D.V., Shimmield, T., Green, D.H., 2012. Metal binding properties of the EPS produced by *Halomonas* sp. TG39 and its potential in enhancing trace element bioavailability to eukaryotic phytoplankton. *Biometals*. 25(6), pp. 1185–1194.
Hamman, J.H., 2010. Chitosan based polyelectrolyte complexes as potential carrier materials in drug delivery systems. *Mar. drugs*. 8(4), pp. 1305–1322.
Harder, W., Dijkhuizen, L., 1983. Physiological responses to nutrient limitation. *Annu. Rev. Microbiol*. 37(1), pp. 1–23.
Harz, H., Burgdorf, K., Höltje, J.V., 1990. Isolation and separation of the glycan strands from murein of *Escherichia coli* by reversed-phase high-performance liquid chromatography. *Anal. Biochem*. 190(1), pp. 120–128.
Hassanshahian, M., Cappello, S., 2013. Crude oil biodegradation in the marine environments. In: Chamy, R. and Rosenkranz, F. (Eds.). *Biodegradation-Engineering and Technology*. InTech Open. DOI: 10.5772/60894.

He, N., Li, Y., Chen, J., 2004. Production of a novel polygalacturonic acid bioflocculant REA-11 by *Corynebacterium glutamicum*. *Bioresour. Technol.* 94(1), pp. 99–105.

Heintz, R.A., 2007. Chronic exposure to polynuclear aromatic hydrocarbons in natal habitats leads to decreased equilibrium size, growth, and stability of pink salmon populations. *Integr. Environ. Assess. Manag*. 3(3), pp. 351–363.

Heymann, D., Ruiz-Velasco, C., Chesneau, J., Ratiskol, J., Sinquin, C., Colliec-Jouault, S., 2016. Antimetastatic properties of a marine bacterial exopolysaccharide-based derivative designed to mimic glycosaminoglycans. *Molecules*. 21(3), p. 309.

Hua, X., Wu, Z., Zhang, H., Lu, D., Wang, M., Liu, Y., Liu, Z., 2010. Degradation of hexadecane by *Enterobacter cloacae* strain TU that secretes an exopolysaccharide as a bioemulsifier. *Chemosphere*. 80(8), pp. 951–956.

Iwabuchi, N., Sunairi, M., Anzai, H., Nakajima, M., Harayama, S., 2000. Relationships between colony morphotypes and oil tolerance in *Rhodococcus rhodochrous*. *J. Appl. Environ. Microbiol*. 66(11), pp. 5073–5077.

Iwabuchi, N., Sunairi, M., Urai, M., Itoh, C., Anzai, H., Nakajima, M., Harayama, S., 2002. Extracellular polysaccharides of *Rhodococcus rhodochrous* S-2 stimulate the degradation of aromatic components in crude oil by indigenous marine bacteria. *J. Appl. Environ. Microbiol*. 68(5), pp. 2337–2343.

Iyer, A., Mody, K., Jha, B., 2005. Biosorption of heavy metals by a marine bacterium. *Mar. Poll. Bull*. 50(3), pp. 340–343.

Iyer, A., Mody, K., Jha, B., 2006. Emulsifying properties of a marine bacterial exopolysaccharide. *Enzyme Microb. Technol*. 38(1–2), pp. 220–222.

Jenkinson, H.F., 1994. Adherence and accumulation of oral *Streptococci. Trends Microbiol*. 2(6), pp. 209–212.

Jensen, P.R., Fenical, W., 1996. Marine bacterial diversity as a resource for novel microbial products. *J. Ind. Microbiol*. 17(5–6), pp. 346–351.

Jolly, L., Vincent, S.J., Duboc, P., Neeser, J.R., 2002. Exploiting exopolysaccharides from lactic acid bacteria. In: Siezen, R.J., Kok, J., Abee, T., Schasfsma, G. (Eds.). *Lactic Acid Bacteria: Genetics, Metabolism and Applications*. Springer, Dordrecht, pp. 367–374.

Junge, K., Eicken, H., Deming, J.W., 2004. Bacterial activity at -2 to -20°C in Arctic winter time sea ice. *J. Appl. Environ. Microbiol*. 70(1), pp. 550–557.

Kennedy, V.S., Twilley, R.R., Kleypas, J.A., Cowan Jr, J.H., Hare, S.R., 2002. *Coastal and Marine Ecosystems and Global Climate Change*. Pew Center on Global Climate Change, Arlington, VA.

Kim, M., Park, J.M., Um, H.J., Lee, K.H., Kim, H., Min, J., Kim, Y.H., 2011. The antifouling potentiality of galactosamine characterized from *Vibrio vulnificus* exopolysaccharide. *Biofouling*. 27(8), pp. 851–857.

King, R.B., Sheldon, J.K., Long, G.M., 1997. *Practical Environmental Bioremediation: The Field Guide*. CRC Press, Washington, DC, pp. 1–208.

Ko, S.H., Lee, H.S., Park, S.H., Lee, H.K., 2000. Optimal conditions for the production of exopolysaccharide by marine microorganism *Hahella chejuensis*. Biotechnol. *Bioprocess Eng*. 5(3), pp. 181–185.

Krembs, C.E., Eicken, H., Junge, K., Deming, J.W., 2002. High concentrations of exopolymeric substances in Arctic winter sea ice: Implications for the polar ocean carbon cycle and cryoprotection of diatoms. *Deep Sea Res. Part 1 Oceanogr. Res. Pap*. 49(12), pp. 2163–2181.

Kulichevskaya, I.S., Milekhina, E.I., Borzenkov, I.A., Zvyagintseva, I.S., Belyaev, S.S., 1991. Oxidation of petroleum-hydrocarbons by extremely halophilic archaebacteria. *Microbiol*. 60(5), pp. 596–601.

Lakshmipriya, V.P., 2012. Isolation and characterization of total heterotrophic bacteria and exopolysaccharide produced from mangrove ecosystem. *Int. J. Pharm. Biol. Sci. Arch*. 3(3), pp. 679–684.

Lee, H.A., Kim, J.H., 2012. Isolation of *Bacillus amyloliquefaciens* strains with antifungal activities from Meju. *Prev. Nutr. Food Sci.* 17(1), pp. 64–70.

Lee, H.K., Chun, J., Moon, E.Y., Ko, S.H., Lee, D.S., Lee, H.S., Bae, K.S., 2001. *Hahella chejuensis* gen. nov., sp. nov., an extracellular-polysaccharide-producing marine bacterium. *Int. J. Syst. Evol. Microbiol.* 51(2), pp. 661–666.

Lelchat, F., Cozien, J., Le Costaouec, T., Brandilly, C., Schmitt, S., Baudoux, A.C., Boisset, C., 2015. Exopolysaccharide biosynthesis and biodegradation by a marine hydrothermal *Alteromonas* sp. strain. *Appl. Microbiol. Biotechnol.* 99(6), pp. 2637–2647.

Lemarchand, C., Couvreur, P., Besnard, M., Costantini, D., Gref, R., 2003. Novel polyester-polysaccharide nanoparticles. *J. Pharma. Res.* 20(8), pp. 1284–1292.

Lieberman, M.M., Markovitz, A., 1970. Depression of guanosine diphosphate-mannose pyrophosphorylase by mutations in two different regulator genes involved in capsular polysaccharide synthesis in *Escherichia coli* K-12. *J. Bacteriol.* 101(3), pp. 965–972.

Llamas, I., Amjres, H., Mata, J.A., Quesada, E., Béjar, V., 2012. The potential biotechnological applications of the exopolysaccharide produced by the halophilic bacterium *Halomonas almeriensis*. *Molecules*. 17(6), pp. 7103–7120.

Llamas, I., Del Moral, A., Martínez-Checa, F., Arco, Y., Arias, S., Quesada, E., 2006. *Halomonas maura* is a physiologically versatile bacterium of both ecological and biotechnological interest. *Antonie Van Leeuwenhoek*. 89(3–4), pp. 395–403.

Llamas, I., Mata, J.A., Tallon, R., Bressollier, P., Urdaci, M.C., Quesada, E., Béjar, V., 2010. Characterization of the exopolysaccharide produced by *Salipiger mucosus* A3T, a halophilic species belonging to the Alphaproteobacteria, isolated on the Spanish mediterranean seaboard. *Mar. Drugs*. 8(8), pp. 2240–2251.

Lloyd, J.R., Renshaw, J.C., 2005. Bioremediation of radioactive waste: Radionuclide—microbe interactions in laboratory and field-scale studies. *Curr. Opin. Biotechnol.* 16(3), pp. 254–260.

Loaec, M., Olier, R., Guezennec, J., 1997. Uptake of lead, cadmium and zinc by a novel bacterial exopolysaccharide. *Water Res.* 31(5), pp. 1171–1179.

Lupp, C., Ruby, E.G., 2005. *Vibrio fischeri* uses two quorum-sensing systems for the regulation of early and late colonization factors. *J. Bacteriol.* 87(11), pp. 3620–3629.

Ma, Y., Shen, B., Sun, R., Zhou, W., Zhang, Y., 2012. Lead (II) biosorption of an Antarctic sea-ice bacterial exopolysaccharide. *Desalin. Water Treat.* 42(1–3), pp. 202–209.

Mack, D., Fischer, W., Krokotsch, A., Leopold, K., Hartmann, R., Egge, H., Laufs, R., 1996. The intercellular adhesin involved in biofilm accumulation of *Staphylococcus epidermidis* is a linear beta-1, 6-linked glucosaminoglycan: Purification and structural analysis. *J. Bacteriol.* 178(1), pp. 175–183.

MacLeod, R.A., 1965. The question of the existence of specific marine bacteria. *Bacteriol. Rev.* 29(1), pp. 9–23.

Madhuri, K.V., Prabhakar, K.V., 2014. Microbial exopolysaccharides: Biosynthesis and potential applications. *Orient. J. Chem.* 30(3), pp. 1401–1410.

Manca, M.C., Lama, L., Improta, R., Esposito, E., Gambacorta, A., Nicolaus, B., 1996. Chemical composition of two exopolysaccharides from *Bacillus thermoantarcticus*. *J. Appl. Environ. Microbiol.* 62(9), pp. 3265–3269.

Mancuso Nichols, C.A., Garon, S., Bowman, J.P., Raguénès, G., Guezennec, J., 2004. Production of exopolysaccharides by Antarctic marine bacterial isolates. *J. Appl. Microbiol.* 96(5), pp. 1057–1066.

Mangwani, N., Shukla, S.K., Rao, T.S., Das, S., 2014. Calcium-mediated modulation of *Pseudomonas mendocina* NR802 biofilm influences the phenanthrene degradation. *Colloids Surf. B Biointerfaces*. 114, pp. 301–309.

Markovich, D., Murer, H., 2004. The SLC13 gene family of sodium sulphate/carboxylate cotransporters. *Pflugers Arch.* 447(5), pp. 594–602.

Martínez-Checa, F., Toledo, F.L., El Mabrouki, K., Quesada, E., Calvo, C., 2007. Characteristics of bioemulsifier V2–7 synthesized in culture media added of hydrocarbons: Chemical composition, emulsifying activity and rheological properties. *Bioresour. Technol.* 98(16), pp. 3130–3135.

Marx, J.G., Carpenter, S.D., Deming, J.W., 2009. Production of cryoprotectant extracellular polysaccharide substances (EPS) by the marine psychrophilic bacterium *Colwellia psychrerythraea* strain 34H under extreme conditions. *Can. J. Microbiol.* 55(1), pp. 63–72.

Mata, J.A., Béjar, V., Bressollier, P., Tallon, R., Urdaci, M.C., Quesada, E., Llamas, I., 2008. Characterization of exopolysaccharides produced by three moderately halophilic bacteria belonging to the family *Alteromonadaceae*. *J. Appl. Microbiol.* 105(2), pp. 521–528.

Matias, V.R., Al-Amoudi, A., Dubochet, J., Beveridge, T.J., 2003. Cryo-transmission electron microscopy of frozen-hydrated sections of *Escherichia coli* and *Pseudomonas aeruginosa*. *J. Bacteriol.* 185(20), pp. 6112–6118.

Maugeri, T.L., Gugliandolo, C., Caccamo, D., Panico, A., Lama, L., Gambacorta, A., Nicolaus, B., 2002. A halophilic thermotolerant *Bacillus* isolated from a marine hot spring able to produce a new exopolysaccharide. *Biotechnol. Lett.* 24(7), pp. 515–519.

Mayer, C., Moritz, R., Kirschner, C., Borchard, W., Maibaum, R., Wingender, J., Flemming, H.C., 1999. The role of intermolecular interactions: Studies on model systems for bacterial biofilms. *Int. J. Biol. Macromol.* 26(1), pp. 3–16.

Mehta, A., Sidhu, C., Pinnaka, A.K., Choudhury, A.R., 2014. Extracellular polysaccharide production by a novel osmotolerant marine strain of *Alteromonas macleodii* and its application towards biomineralization of silver. *PloS One*. 9(6), pp. 1–11.

Moriello, V.S., Lama, L., Poli, A., Gugliandolo, C., Maugeri, T.L., Gambacorta, A., Nicolaus, B., 2003. Production of exopolysaccharides from a thermophilic microorganism isolated from a marine hot spring in flegrean areas. *J. Ind. Microbiol. Biotechnol.* 30(2), pp. 95–101.

Morris, R.S., Walsh IV, M.A., 2000. U.S. Patent No. 6,063,849. U.S. Patent and Trademark Office, Washington, DC, USA.

Munn, C., 2011. *Marine Microbiology, Ecology and Applications*, 2nd Edition. Garland Science. London.

Nakayama, T., Wake, H., Ozawa, K., Kodama, H., Nakamura, N., Matsunaga, T., 1998. Use of a titanium nitride for electrochemical inactivation of marine bacteria. *Environ. Sci. Technol.* 32(6), pp. 798–801.

Nichols, C.M., Guezennec, J., Bowman, J.P., 2005. Bacterial exopolysaccharides from extreme marine environments with special consideration of the southern ocean, sea ice, and deep-sea hydrothermal vents: A review. *Mar. Biotechnol.* 7(4), pp. 253–271.

Nicolaus, B., Kambourova, M., Oner, E.T., 2010. Exopolysaccharides from extremophiles: From fundamentals to biotechnology. *Environ. Technol.* 31(10), pp. 1145–1158.

Nicolaus, B., Lama, L., Panico, A., Moriello, V.S., Romano, I., Gambacorta, A., 2002. Production and characterization of exopolysaccharides excreted by thermophilic bacteria from shallow, marine hydrothermal vents of Flegrean Ares (Italy). *Syst. Appl. Microbiol.* 25(3), pp. 319–325.

Nicolaus, B., Panico, A., Manca, M.C., Lama, L., Gambacorta, A., Maugeri, T., Caccamo, D., 2000. A thermophilic *Bacillus* isolated from an eolian shallow hydrothermal vent able to produce exopolysaccharides. *Syst. Appl. Microbiol.* 23(3), pp. 426–432.

Nilsson, M., Chiang, W.C., Fazli, M., Gjermansen, M., Givskov, M., Tolker-Nielsen, T., 2011. Influence of putative exopolysaccharide genes on *Pseudomonas putida* KT2440 biofilm stability. *Environ. Microbiol.* 13(5), pp. 1357–1369.

Nisha, P., Thangavel, M., 2014. Isolation and characterization of biofilm producing bacteria from Arabian Sea. *Res. J. Recent Sci.* 3, pp. 132–136.

Okazaki, T., Kitahara, T., Okami, Y., 1975. Studies on marine microorganisms. V. A new antibiotic SS-228 Y produced by *Chainia* sp. isolated from shallow sea mud. *J. Antibiot.* 29(10), pp. 176–184.

Omura, K., Hitosugi, M., Zhu, X., Ikeda, M., Maeda, H., Tokudome, S., 2005. A newly derived protein from *Bacillus subtilis* natto with both antithrombotic and fibrinolytic effects. *J. Pharmacol. Sci.* 99(3), pp. 247–251.

Oren, A., 2010. Industrial and environmental applications of halophilic microorganisms. *Environ. Technol.* 31(8–9), pp. 825–834.

Oren, A., 2016. Life in hypersaline environments. In: Hurst, C. (Ed.). *Their World: A Diversity of Microbial Environments.* Springer, Cham, pp. 301–339.

Ortega-Morales, B.O., Gaylarde, C.C., Englert, G.E., Gaylarde, P.M., 2005. Analysis of salt-containing biofilms on limestone buildings of the Mayan culture at Edzna, Mexico. *Geomicrobiol. J.* 22(6), pp. 261–268.

Pal, A., Paul, A.K., 2008. Microbial extracellular polymeric substances: Central elements in heavy metal bioremediation. *Ind. J. Microbiol.* 48(1), pp. 49–64.

Pan, X., Liu, J., Zhang, D., 2010. Binding of phenanthrene to extracellular polymeric substances (EPS) from aerobic activated sludge: A fluorescence study. *Colloids Surf. B Biointerfaces.* 80(1), pp. 103–106.

Panwichian, S., Kantachote, D., Wittayaweerasak, B., Mallavarapu, M., 2011. Removal of heavy metals by exopolymeric substances produced by resistant purple nonsulfur bacteria isolated from contaminated shrimp ponds. *Electron. J. Biotechnol.* 14(4), pp. 2–2.

Parkar, D., Jadhav, R., Pimpliskar, M., 2016. Marine bacterial extracellular polysaccharides: A review. *J. Coast. Life Med.* 5(1), pp. 29–35.

Parolis, H., Parolis, L.A., Boán, I.F., Rodríguez-Valera, F., Widmalm, G., Manca, M.C., Sutherland, I.W., 1996. The structure of the exopolysaccharide produced by the halophilic Archaeon *Haloferaxmediterranei* strain R4 (ATCC 33500). *Carbohydr. Res.* 295, pp. 147–156.

Poli, A., Anzelmo, G., Nicolaus, B., 2010. Bacterial exopolysaccharides from extreme marine habitats: Production, characterization and biological activities. *Mar. Drugs.* 8(6), pp. 1779–1802.

Poli, A., Di Donato, P., Abbamondi, G.R., Nicolaus, B., 2011. Synthesis, production, and biotechnological applications of exopolysaccharides and polyhydroxyalkanoates by archaea. *Archaea.* 2011, pp. 1–13.

Poli, A., Esposito, E., Orlando, P., Lama, L., Giordano, A., de Appolonia, F., Gambacorta, A., 2007. *Halomonas alkaliantarctica* sp. nov., isolated from saline lake Cape Russell in Antarctica, an alkalophilic moderately halophilic, exopolysaccharide-producing bacterium. *Syst. Appl. Microbiol.* 30(1), pp. 31–38.

Proksch, P., Edrada, R., Ebel, R., 2002. Drugs from the seas—current status and microbiological implications. *Appl. Microbiol. Biotechnol.* 59(2–3), pp. 125–134.

Qian, P.Y., Dobretsov, S., Harder, T., Lau, C.K.S., 2006. Hong Kong University of Science, Anti-fouling exopolysaccharides isolated from cultures of *Vibrio alginolyticus* and *Vibrio proteolyticus.* U.S. Patent 7,090,856. US Patent and Trademark Office, Washington, DC, USA.

Qin, G., Zhu, L., Chen, X., Wang, P.G., Zhang, Y., 2007. Structural characterization and ecological roles of a novel exopolysaccharide from the deep-sea psychrotolerant bacterium *Pseudoalteromonas* sp. SM9913. *Microbiol.* 153(5), pp. 1566–1572.

Quintero, E.J., Weiner, R.M., 1995. Physical and chemical characterization of the polysaccharide capsule of the marine bacterium, *Hyphomonas* strain MHS-3. *J. Ind. Microbiol.* 15(4), pp. 347–351.

Raguenes, G., Pignet, P., Gauthier, G., Peres, A., Christen, R., Rougeaux, H., Guezennec, J., 1996. Description of a new polymer-secreting bacterium from a deep-sea hydrothermal vent, *Alteromonas macleodii* subsp. fijiensis, and preliminary characterization of the polymer. *Appl. Environ. Microbiol.* 62(1), pp. 67–73.

Raj, R., Dalei, K., Chakraborty, J., Das, S., 2016. Extracellular polymeric substances of a marine bacterium mediated synthesis of CdS nanoparticles for removal of cadmium from aqueous solution. *J. Colloid Interface Sci.* 462, pp. 166–175.

Rampelotto, P.H., 2013. Extremophiles and extreme environments. *Life*. 3(3), pp. 482–485.

Rinker, K.D., Kelly, R.M., 2000. Effect of carbon and nitrogen sources on growth dynamics and exopolysaccharide production for the hyperthermophilic archaeon *Thermococcus litoralis* and bacterium *Thermotoga maritima*. *Biotechnol. Bioeng.* 69(5), pp. 537–547.

Roberts, I.S., 1996. The biochemistry and genetics of capsular polysaccharide production in bacteria. *Annu. Rev. Microbiol.* 50(1), pp. 285–315.

Rodriguez-Valera, F., 1988. Characteristics and microbial ecology of hypersaline environments. In: Rodriguez-Valera, F. (Ed.). *Halophilic Bacteria*. CRC Press, Boca Raton, FL, pp. 3–30.

Roger, O., Kervarec, N., Ratiskol, J., Colliec-Jouault, S., Chevolot, L., 2004. Structural studies of the main exopolysaccharide produced by the deep-sea bacterium *Alteromonas infernus*. *Carbohydr. Res.* 339(14), pp. 2371–2380.

Rogers, H.J., Perkins, H.R., Ward, J.B., 1980. *Microbial Cell Walls and Membranes*. Chapman and Hall, London, UK, pp. 437–460.

Ron, E.Z., Rosenberg, E., 2002. Biosurfactants and oil bioremediation. *Curr. Opin. Biotechnol.* 13(3), pp. 249–252.

Rougeaux, H., Kervarec, N., Pichon, R., Guezennec, J., 1999. Structure of the exopolysaccharide of *Vibriodiabolicus* isolated from a deep-sea hydrothermal vent. *Carbohydr. Res.* 322(1–2), pp. 40–45.

Rougeaux, H., Talaga, P., Carlson, R.W., Guezennec, J., 1998. Structural studies of an exopolysaccharide produced by *Alteromonas macleodii* subsp. fijiensis originating from a deep-sea hydrothermal vent. *Carbohydr. Res.* 312(1–2), pp. 53–59.

Salehizadeh, H., Vossoughi, M., Alemzadeh, I., 2000. Some investigations on bioflocculant producing bacteria. *Biochem. Eng.* J. 5(1), pp. 39–44.

Samain, E., Miles, M., Bozzi, L., Dubreucq, G., Rinaudo, M., 1997. Simultaneous production of two different gel-forming exopolysaccharides by an *Alteromonas* strain originating from deep sea hydrothermal vents. *Carbohydr. Polym.* 34(4), pp. 235–241.

Sarmento, B., Ribeiro, A., Veiga, F., Ferreira, D., 2006. Development and characterization of new insulin containing polysaccharide nanoparticles. *Colloids Surf. B Biointerfaces*. 53(2), pp. 193–202.

Sathiyanarayanan, G., Kiran, G.S., Selvin, J., 2013. Synthesis of silver nanoparticles by polysaccharide bioflocculant produced from marine *Bacillus subtilis* MSBN17. *Colloids Surf. B Biointerfaces*. 102, pp. 13–20.

Schmid, J., Sieber, V., Rehm, B., 2015. Bacterial exopolysaccharides: Biosynthesis pathways and engineering strategies. *Front. Microbiol.* 6, pp. 496.

Schneiker, S., dos Santos, V.A.M., Bartels, D., Bekel, T., Brecht, M., Buhrmester, J., Goesmann, A., 2006. Genome sequence of the ubiquitous hydrocarbon-degrading marine bacterium *Alcanivorax borkumensis*. *Nat. Biotechnol.* 24(8), pp. 997–1004.

Shah, V., Ray, A., Garg, N., Madamwar, D., 2000. Characterization of the extracellular polysaccharide produced by a marine *Cyanobacterium Cyanothece* sp. ATCC 51142, and its exploitation toward metal removal from solutions. *Curr. Microbiol.* 40(4), pp. 274–278.

Shih, I.L., 2010. Microbial exo-polysaccharides for biomedical applications: Mini-rev. *Med. Chem.* 10(14), pp. 1345–1355.

Singh, R., Paul, D., Jain, R.K., 2006. Biofilms: Implications in bioremediation. *Trends Microbiol.* 14(9), pp. 389–397.
Sobhana, K.S., Geetha, P., Mohan, V., Surendran, A., 2015. Course manual-national workshop on effective management of e-resources in research libraries.
Spanò, A., Arena, A., 2016. Bacterial exopolysaccharide of shallow marine vent origin as agent in counteracting immune disorders induced by herpes virus. *J. Immunoassay Immunochem.* 37(3), pp. 251–260.
Su, Y.C., Liu, C., 2007. *Vibrio parahaemolyticus*: A concern of seafood safety. *Food Microbiol.* 24(6), pp. 549–558.
Sutherland, I.W., 1972. Bacterial exopolysaccharides. *Adv. Microbiol. Phy.* 8, pp. 143–213.
Sutherland, I.W., 1990. *Biotechnology of Microbial Exopolysaccharides*. Cambridge University Press, Cambridge, England, p. 9. DOI: 10.1002/pi.4990270216.
Ta-Chen, L., Chang, J.S., Young, C.C., 2008. Exopolysaccharides produced by *Gordonia alkanivorans* enhance bacterial degradation activity for diesel. *Biotechnol. Lett.* 30(7), pp. 1201–1206.
Venosa, A.D., Zhu, X., 2003. Biodegradation of crude oil contaminating marine shorelines and freshwater wetlands. *Spill Sci. Technol. Bull.* 8(2), pp. 163–178.
Vijayendra, S.V.N., Shamala, T.R., 2014. Film forming microbial biopolymers for commercial applications—A review. *Crit. Rev. Biotechnol.* 34(4), pp. 338–357.
Wai, S.N., Mizunoe, Y., Yoshida, S.I., 1999. How *Vibrio cholerae* survive during starvation. *FEMS Microbiol. Lett.* 180(2), pp. 123–131.
Wang, H., Jiang, X., Mu, H., Liang, X., Guan, H., 2007. Structure and protective effect of exopolysaccharide from *Pantoea agglomerans* strain KFS-9 against UV radiation. *Microbiol. Res.* 162(2), pp. 124–129.
Wang, L., Ma, F., Qu, Y., Sun, D., Li, A., Guo, J., Yu, B., 2011. Characterization of a compound bioflocculant produced by mixed culture of *Rhizobium radiobacter* F2 and *Bacillus sphaeicus* F6. *World J. Microbiol. Biotechnol.* 27(11), pp. 2559–2565.
Watanabe, M., Kawahara, K., Sasaki, K., Noparatnaraporn, N., 2003. Biosorption of cadmium ions using a photosynthetic bacterium, *Rhodobacter sphaeroides* S and a marine photosynthetic bacterium, *Rhodovulum* sp. and their biosorption kinetics. *J. Biosci. Bioeng.* 95(4), pp. 374–378.
Whitfield, C., 1988. Bacterial extracellular polysaccharides. *Can. J. Microbiol.* 34(4), pp. 415–420.
Williams, A., Wimpenny, J., 1978. Exopolysaccharide production by *Pseudomonas* NCIB 11264 grown in continuous culture. *J. Gen. Microbiol.* 104, pp. 47–57.
Williams, P.G., 2009. Panning for chemical gold: Marine bacteria as a source of new therapeutics. *Trends Biotechnol.* 27(1), pp. 45–52.
Wingender, J., Neu, T.R., Flemming, H.C., 1999. What are bacterial extracellular polymeric substances? In: Wingender, J., Neu, T.R., Flemming, H.C. (Eds.). *Microbial Extracellular Polymeric Substances*. Springer, Berlin, Heidelberg, pp. 1–19.
Wolfaardt, G.M., Lawrence, J.R., Korber, D.R., 1999. Function of EPS. In: Jost, W., Thomas, N., Hans-Curt, F. (Eds.). *Microbial Extracellular Polymeric Substances*. Springer, Berlin, Heidelberg, pp. 171–200.
Wrangstadh, M., Conway, P.L., Kjelleberg, S., 1986. The production and release of an extracellular polysaccharide during starvation of a marine *Pseudomonas* sp. and the effect thereof on adhesion. *Arch. Microbiol.* 145(3), pp. 220–227.
Yaskovich, G.A., 1998. The role of cell surface hydrophobicity in adsorption immobilization of bacterial strains. *Appl. Biochem. Biotechnol.* 34(4), pp. 373–376.
Yebra, D.M., Kiil, S., Dam-Johansen, K., 2004. Antifouling technology-past, present and future steps towards efficient and environmentally friendly antifouling coatings. *Prog. Org. Coat.* 50(2), pp. 75–104.

Zhang, D., Pan, X., Mostofa, K.M., Chen, X., Mu, G., Wu, F., Fu, Q., 2010. Complexation between Hg (II) and biofilm extracellular polymeric substances: An application of fluorescence spectroscopy. *J. Hazard. Mater*. 175(1–3), pp. 359–365.
Zhang, Z.Q., Bo, L., Xia, S.Q., Wang, X.J., Yang, A.M., 2007. Production and application of a novel bioflocculant by multiple-microorganism consortia using brewery wastewater as carbon source. *J. Environ. Sci*. 19(6), pp. 667–673.

19 An Overview of Protease Inhibitors

A New Wave of Drugs from Marine Actinobacteria

Veena Sreedharan and K.V. Bhaskara Rao

CONTENTS

19.1 INTRODUCTION

Proteases are without a doubt a focal need of the natural framework. Considered catalysts of scission prior, proteases are presently viewed as crucial proteins for the existing framework. All things considered, underneath their imperative capacities, lies the chance of being a risk. Upon interruption in the directed instrument of proteolytic handling, they can prompt certain oddities (Fitzpatrick, 2004). Nonetheless, they can be constrained by directed emission or articulation or potentially initiation of proteases by restraint of the proteolytic movement and debasement of developed chemicals. However, sicknesses can be constrained by hereditary adjustment, this thought is sketchy. The job of protease inhibitors becomes unmistakable here (Grant and Mackie, 1977). Protease inhibitors (PIs) are a universal class of catalysts that assume a

DOI: 10.1201/9781003303909-19

significant part in the life framework. They are significantly applicable attributable to their significance in directing physiological just as neurotic capacities in the living association. Normally accessible protease inhibitors, retrieved from plants and creatures, are bountiful in nature while little is known and concentrated about proteases inhibitors of microbial beginning. Microorganisms present a splendid wellspring of protease inhibitors as low-subatomic-weight peptidomimetic inhibitors. Having a place within eubacterial realm, actinobacteria are modernly and pharmacologically critical microorganisms esteemed for their compounds, antitoxins, antitumor specialists, and protein inhibitors (Dhanasekaran and Jiang, 2016). They produce exceptionally potential bioactive particles and subsequently are regularly evaluated for novel bioactive leads. The interest for protease inhibitors from actinobacteria traces all the way back to the 1960s. In contrast to the traditional protease inhibitors, protease inhibitors from actinobacteria are dominatingly little particles (Feinstein et al., 1967). The detachment and portrayal of leupeptin from actinobacteria by Aoyagi et al. was probably the earliest report. From that point forward, the creation and job of protease inhibitors have developed massively. The development of HIV protease inhibitor Norvir between 2000–2006 coming to 6.6 billion in worldwide deals is one such model (Aoyagi et al., 1969). Consequently, there is a promising future for protease inhibitors. In our survey, we present a course of events of protease inhibitors from actinobacteria with agents of a few striking protease inhibitors.

19.2 PROTEASE ENZYME

Proteases are seen in prokaryotes, fungi and animals and are very necessary for their survival. Proteases are enzymes that help to break down proteins in a method known as proteolysis. Such enzymes are present in a wide range of biological activities, from small protein digestion to extremely controlled cascades. Protease, like hormones, antibodies, and other enzymes, shows a vital physiological part in determining the life span of other proteins. In the physiology of organisms, this is one of the profligated "switching on" and "switching off" regulating systems. Proteases are secreted by a variety of bacteria to break the protein–peptide link into simple small monomers. As a result of multiple clinical trials suggesting their benefits in cancer studies include inflammations, immune regulations and blood flow control, their usage in medicine is garnering more and more attention. Many parasites are involved in pathogenesis, which includes parasite relocation through the host tissue barrier, hemoglobin and blood protein breakdown, immunological invasions, and inflammatory activation. Proteases thus show a decisive part in pathogenesis. Wild action, on the other hand, has negative consequences in the human body. These enzymes found within cancer cells have the ability to break other strong cell wall and membrane, allowing them to spread and grow into additional cell organ and part of the body, resulting in spread from one site to another (Figure 19.1).

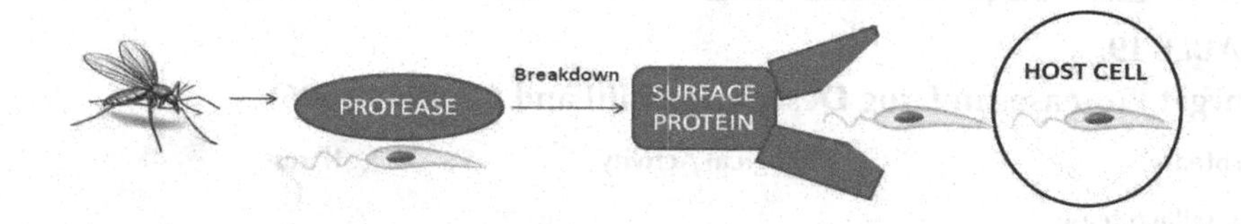

FIGURE 19.1 Role of protease.

19.2.1 Protease Classification

There are now six groups of proteases:

1. Serine
2. Threonine
3. Cysteine
4. Aspartate
5. Metallo
6. Glutamic acid.

Serine, aspartate, threonine, cysteine, glutamic acid groups, and metal ions all have a role in catalysis. All these enzymes can be found in bacteria. Glutamic proteases, for example, are only found in fungus. Many pathogenic microorganisms have serine, cysteine, and metalloproteases, which play important roles in evading host immune systems, acquiring nutrients for growth and proliferation, facilitating dissemination, and causing tissue damage during infection (Drag and Salvesen, 2010). Hence, proteases are the major targets in drug development studies for many infectious diseases as shown in Table 19.1 (Mittl and Grutter, 2006).

19.2.2 Enzymes Could Be a Possible Therapeutic Target

The first phase in the drug development process is to find and choose a pharmacological aim that is important in a specific biological pathway. To find a drug goal in a pathogen, there must be no doubt that the assumed target is either expressly preoccupied in the host or significantly vary from the host counterpart in order for it to be used as a medicine target. Ultimately, protein is critical for the microbe's existence or for directing a target toward certain pathway. Due to their involvement in the regulation of certain biochemical and metabolic pathways, enzymes are the most significant targets. For specific suppression of the aimed sites, the target enzymes should have considerable structural and functional variations from mammalian systems. As a result, specialized inhibitors that bind to the most potential sites of enzymes and led to the inhibition of enzymes and the viability of cell can be created or found in nature. Finally, it is critical that the chosen target be

TABLE 19.1
Target Proteases in Drug Designing (Mittl and Grutter, 2006)

Protease	Biological Activity	Illness
Metalloprotease		
Angiotensin altering enzyme	Brings changes in angiotensin	High blood pressure
Anthrax-endopeptidase	Splitting MAPKK	*Bacillus anthracis*
Carboxypeptidases-U	Tissue plasminogen cleavage	Blood coagulation
NAAG peptidase	Releases glutamate from Ac, Asp, and Glu present in the in brain tissue	Prostate cancer marker
Metallopeptidase-1	Connective tissue degradation	Helps in cleavage to tissue for tumor invasion
GP63	Involved in the leishmanial parasitic life cycle. From attachment to invasion inside human cells	Leishmania donovani
Serine protease		
Thrombase	fibrinogen breakdown	Blood clot in blood vessels
Xa-Factor	Formation of thrombin from prothrombin	Blood clot in blood vessels
Urokinase-type plasminogen activator	plasminogen stimulation	Cancerous
Flavivirus	Processing of diabetogenic protein gene	Viral disease
Dipeptidyl peptidase 4	Processing of hormone precursors	Diabetes
Proteasome	protein degradation dependent on ubiquitin	Cancerous
Aspartic protease		
HIV-protease	Viral-proprotein processing	AIDS
Angiotensinogenase	Angiotensinogen processing	High blood pressure
Beta-secretase 1	Activation of β-Secretase	Alzheimer's disease
Plm	Degradation of Hb	Malarial fever
Cysteine protease		
Cathepsin B	Processing of antigen	Cancerous
Cathepsin S	Proteolysis of lysosome	Septicity
FP	Degradation of Hb	Malarial fever
Caspase	Interleukin maturation	Endotoxic shocks
Caspase-7	Executioner in apoptosis	Cardiac and neural injuries
Calpain-2	Cytoskeletal protein degradation	Brain stroke

assayable, allowing compounds to be screened using a cost-effective particular assay technique (Shukla et al., 2010). Additionally, targeting multiple enzymes in a metabolic pathway is more beneficial and successful. Even if certain parasites are eukaryotes, their cell architecture differs significantly from that of mammalian cells, making it possible to identify parasite-specific targets.

19.3 PROTEASE INHIBITORS

PIs are proteins that go in or obstruct the putative site of a protease to limit substrate entree. When protease is activated, they change a sum of particular proteins, causing cellular renovation and cardiac failure. PI have been recognized from various sources, including animals, actinomycetes, plants, and fungi. Antibiotic research revealed the occurrence of PIs in bacteria, which function as an inhibitor of enzymes involved in microbial development and growth. In a biochemical investigation of biological activities and disorders, inhibitors of microbial derivation have been utilized as valuable tools (Fear et al., 2007). Table 19.2 summarizes the status of available PIs. PIs are being studied as a treatment for a number of disorders, including a variety of infectious and inflammatory disorders (Fear et al., 2007; Drag and Salvesen, 2010).

Proteases, a vast family of enzymes involved in a wide range of physiological processes, have been identified as prospective therapeutic targets (McKerrow et al., 2006). Increased proteolysis has been demonstrated to accentuate a variety of disease processes, making proteases chief therapeutic targets (Turk, 2006; Drag and Salvesen, 2010; Santos, 2009; Sabotic and Kos, 2012). Management of hypertension with angiotensin-converting-enzyme (ACE) inhibitors, AIDS treatment developed with HIV protease inhibitors, and multiple myeloma

TABLE 19.2
Protease Inhibitors and Clinical Trial Stages (Fear et al., 2007)

Diseases	Enzyme	Inhibitor and Stage
Cold	Human rhinovirus	AG7088 Clinical trail, Phase II
Ischemia	Serine-protease	Aprotinin Food and Drug Administration (FDA) approved
Liver inflammation	Serine-protease	Vertex, Phase II SCH 6, Phase I
	cysteine-aspartic proteases	IDN6556, Phase II
Heart failure	Metalloprotease	Captopril, benazepril, and quinapril—FDA approved.
HIV	Aspartyl-protease	Amprenavir and atazanavir—FDA approved
Cancer	Uro-kinase	WXUK1, Phase I
	Serine-proteases	BBI, Phase II
	MMP	AE941, Phase III
Diabetes	Dipeptidyl-peptidase	NVPLAF-237, Phase II MK0431, Phase II
Arthritis	Caspase	VX740, Phase II

treatment with protein complex inhibitors are all examples of effective PIs utilized in the therapeutic involvement of numerous illnesses. The development of chemicals that specifically inhibit enzymes that are critical for the survival of parasites inside the host and are part of parasites' metabolic processes is one prospective technique for treating parasitic disorders (Figure 19.2). Because of their functions in replication, metabolism, existence, and sickness, site proteases are interesting targets (Selzer et al., 1997).

19.3.1 Role of Protease Inhibitors

As natural inhibitors in all growth and metabolism are needed to control their complementary proteases, they are abundant in the live system. They are produced as small molecules or protein-sized large molecules by plants, animals, and microbes. PIs are classified as small molecules and high-quality PIs based on size. They can also be divided on the basis of their impeded enzymes, the protein structure of molecular weight, and disulfide bridges (Tamir et al., 1996). Notwithstanding, the order considered in this survey is the significant class of compounds they restrain, so they are partitioned as serine, cysteine, aspartic, and metalloprotease. Although they were confined to protein construction and response system contemplates, PIs have begun to be progressively utilized for pharmacological purposes (Rawlings et al., 2004). Weakened PIs are the foundation of various issues, including netherton condition, emphysema, innate angioneurotic edema, and epilepsy (Ritchie, 2003). Disorders in the creation and articulation of the PIs, for example, the disappointment of

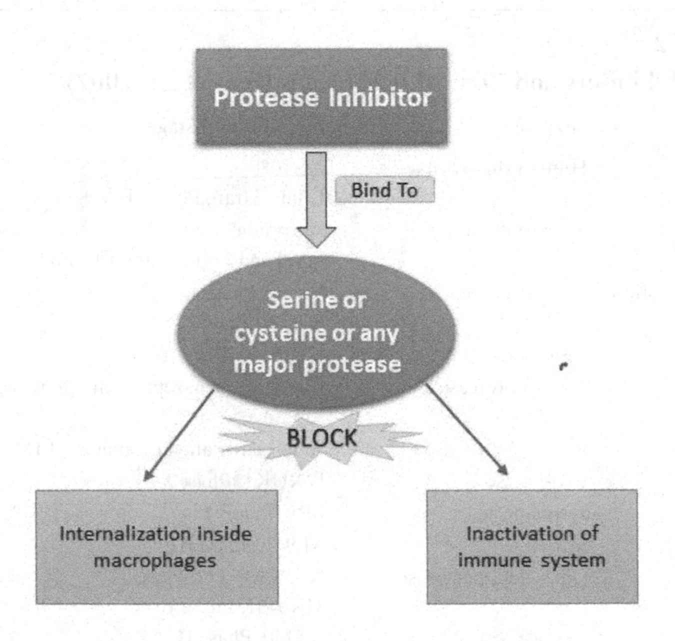

FIGURE 19.2 Effect of PIs on surface proteases of pathogens.

cystatin guidelines, might be connected to numerous obsessive conditions, including skin and neurological problems. Consequently, they can be proficient in finding markers. Consequently, they can be productive determination markers (Magister and Kos, 2013). As specialists for remedial use, PIs are significantly used to treat human immunodeficiency infection (HIV). This aspartyl protease inhibitor is controlled in combinational treatment to hold the viral titer under check; in any case, they likewise display hindrance when directed as a solitary prescription. Since the presentation of saquinavir, ritonavir and indinavir, in 1995, the range of endurance of people who are HIV-positive has been improved. The restoratively capacity to check hypertension is additionally being examined worldwide with angiotensin changing over compound inhibitors, captopril being the pioneer. These fruitful protease inhibitors expanded exploring their potential as restorative specialists against different clinically critical conditions (Gustafsson et al., 1998). Inhibitors like VX-950, rupintrivir, against various infectious viruses, including hepatitis C, human rhinovirus, herpes virus, human cytomegalovirus, severe acute respiratory syndrome, and picornavirus, are in various stages of clinical trials. Pathogenic proteases of several lethal diseases, including cancer, malaria and more, could also be inhibited by this class of bioactive leads (Zhang et al., 2010). These can be used for a plan of medication, crediting their exact restraint. Additionally, they present a decent decision for compelling restorative specialists against overall clinical difficulties, including malignancy, myocardial infections, and so forth, where restraint is needed (Bode et al., 1992). Likewise, they present a decent decision for successful remedial specialists to go against overall clinical difficulties including malignant growth, myocardial infections, and so forth (Hamilton et al., 2001). The system of activity of PIs can be portrayed in two ways: Possibly they act by restricting in the dynamic site of the protease, by re-creating the design of tetrahedral intermediates which are available in compound intervened responses, or they tie through an irreversible component by a conformational change of a peptide inside, in this way setting up a covalent bond. Consequently, concentrating on compound inhibitors tosses light to the component of catalysis by protein, catalyst substrate explicitness just as the useful gathering present at the dynamic site. PIs, for example, amastatin and bestatin, have discovered to be insusceptible modifiers (Umezawa, 1981).

19.3.2 Mechanism of Action of Protease Inhibitors

The activity of PIs on proteases can be reversible noncovalent holding or irreversible covalent connections. This degree of comprehension is significant in the improvement of protease inhibitors as restorative specialists. Reversible inhibitors or authoritative inhibitors are high proclivity inhibitors restricted to the dynamic site of the catalysts, mirroring a chemical substrate instrument. This is the most experienced system of activity. The inhibitor acts like a profoundly explicit substrate for their objective compounds. The inhibitor atoms have least one peptide bond called the responsive site which straightforwardly cooperates with the dynamic site of the protease framing the mind-boggling prompting restraint of the protease. Upon response, the inhibitor is changed from an unblemished structure to an altered inhibitor. Be that as it may, the affirmation of the two structures is generally saved, inferable from the disulfide linkage close to the responsive peptide securities that safeguards the peptide chains (Laskowski and Qasim, 2000). And on the whole, inhibitors shockingly share a typical protein restricting circle structure. Serine PIs are broadly concentrated on gathering sanctioned inhibitors. The irreversible or nonauthoritative protease inhibitors are ones that act through the N-end, accordingly shaping a short equal β sheet. The explicitness of site acknowledgment is additionally furnished by huge optional collaborations with the protease away from the dynamic site. This association likewise expands the speed and holding of the protease and the inhibitor. The best in the class is thrombin restraint by hirudin (Krowarsch et al., 2003).

19.4 SECONDARY METABOLITE PRODUCTION BY MARINE MICROORGANISMS

Pharmaceuticals and cosmetics products, fine chemicals and agrochemicals, all have benefited from the biological and chemical richness of the maritime environment (Ireland, 1993). The seas are a virtually unexplored resource for the discovery of new chemicals with practical applications. Although commercial achievement stories in biotechnology are common, there are considerably fewer such examples in marine biotechnology (Zilinskas et al., 1995). There has been a constant endeavor over the last two decades to understand more about the mostly uncharted area of marine products. Apart from actinobacteria, fungi, and bacteria, numerous additional marine creatures, such as shrimp, fish, crabs, algae, and plants, have been examined to tap into the collection of aquatic worlds (Debashish et al., 2005). In terms of the maritime environment, it's also worth noting that sessile marine invertebrates, such as ascidians, sponges, and cnidaria, account for the majority of the marine macrofauna's biomass. Sessile marine invertebrates' chemical defenses have diverse biological actions, including pharmacological, cosmeceutical, and nutraceutical (Blunt et al., 2012; Blunt et al., 2013; Blunt et al., 2015; Blunt et al., 2016). Scientists are particularly interested in properties like salt tolerance, cold adaptivity, barophilic, and ease of growth. Marine species flourish in settings like undersea caverns and certain regions with high pressure and no light; therefore, these traits are not expected in terrestrial sources (Ghosh, 2005). Organic molecules produced by both eukaryotes and prokaryotes are known as marine bioactive compounds. Such molecules aid the host organism in protecting itself and maintaining equilibrium in its surroundings. So yet, just about 1% of all marine species that produce bioactive metabolites have been discovered (Donia and Hamann, 2003). The seas (which encompass 70% of the globe) are a largely untapped and prospective source of new biologically energetic natural chemicals, as well as being connected to the marine environment's highest biodiversity when compared to the terrestrial microbes (Aneiros and Garateix, 2004; Blunt et al., 2012; Blunt et al., 2014; Mehbub et al., 2014; Reen et al., 2015).

19.4.1 Marine Bacteria

Another important source of active natural chemicals is marine bacteria, which were isolated from the marine ecosystem for the first time in the previous era (Faulkner, 2000; Amsler et al., 2001; Lindquist, 2002; Blunt et al., 2007). For their biological success in various maritime settings, they can store a variety of secondary metabolites (Figure 19.3). Because roughly 80% of marine bacteria are psychrophilic and require high Na+ concentrations for growth, they are quite different from terrestrial bacteria in terms of metabolic and physiological activities (MacLeod, 1965; Maeda and Taga, 1976). Maeda and Taga (1976) extracted and examined DNA from a marine *Vibrio sp.*, demonstrating that Mg2+ activated and Ca2+ stabilized the enzyme. Furthermore, an additional marine bacterium has been identified that can generate phosphatase with action at high hydrostatic pressures (1000 atm; Kobori and Taga, 1980), implying that marine enzymes work differently from terrestrial enzymes. Researchers are drawn to marine bacteria as they are rich in secondary metabolites with distinct biotic features (Fenical and Jensen, 1993). Until now, bioactive compounds were known to be produced by marine *Pseudomonas*, *Vibrio*, and *Bacillus* isolated from marine animals, seawater, algae, and sediments. Indole derivatives, peptides, alkaloids polyenes, terpenoids and macrolides are all produced by them. In 1947, the

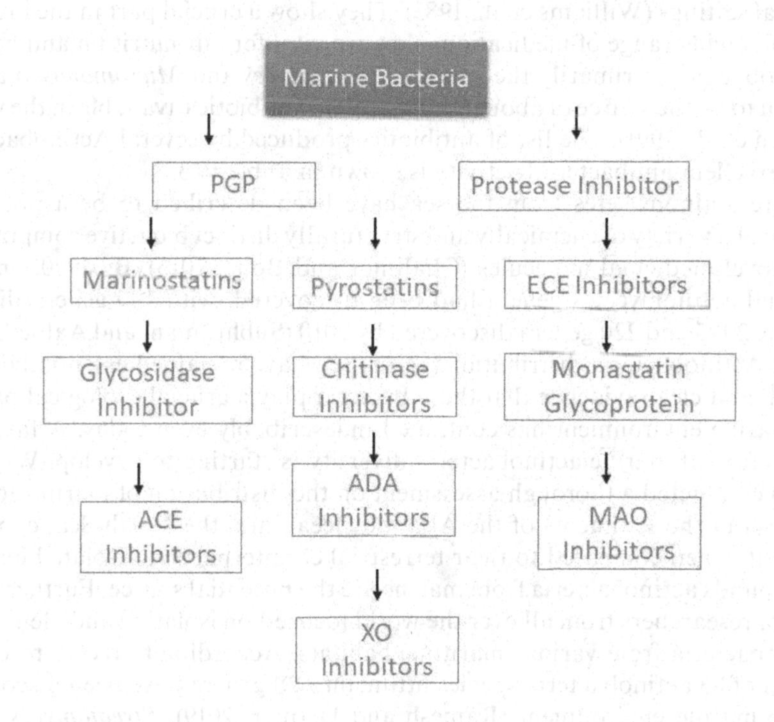

FIGURE 19.3 Secondary metabolites from marine bacteria.

first statement on secondary compounds produced by marine bacteria was published (Rosenfeld and Zobell, 1947). The first recognized marine-derived secondary metabolites were brominated pyrrole, discovered by Burkholder et al. (1966). Laatsch (2005) illustrated 260 marine secondary metabolites and identified from 170 isolated land bacteria between 2001 and 2004. At present, 15 metabolites derived from marine bacteria are in various stages of clinical trials, with many more in the preclinical stage (Mayer et al., 2010). As a result, researchers have focused their efforts on marine microbes and their metabolites in order to discover new medications.

19.4.2 Actinobacteria

Actinomycetes are bacteria that are nonmotile, aerobic, and gram-positive G+C (70 to 80%) and that have a relationship with bacteria based on 16s ribosomal arrangement studies. They are members of the Actinomycetales order, which produces substrate mycelium as well as aerial mycelia and spores. Aerial mycelium spore-bearing hyphae are slightly larger than substrate mycelium hyphae. Actinobacteria's aerial mycelium produces sporophores with a variety of structures. The spores are unaffected by drying and can live in soil for lengthy periods. This phase of the life cycle confers resilience to the soil's harsh environmental conditions, such as nutrient deficiency and water scarcity. Actinobacteria have a wide range of phenotypes and can be found in a variety of natural settings (Williams et al., 1983). They show a crucial part in the production of a wide range of medications that are vital for our nutrition and health. Actinobacteria, primarily the species *Streptomyces* and *Micromonospora*, are known to be the source of about 60–70% of the antibiotics available in the world (Jensen et al., 1991). The list of antibiotics produced by several Actinobacteria with excellent antibacterial activity is shown in Table 19.3.

Rare actinomycetes from the sea have been described to be a potential source of a variety of chemically and structurally distinct bioactive compounds, and novel medicinal molecules (Challinor and Bode, 2015). By 1970, only 11 unusual actinomycetes genera had been discovered, with 100 genera discovered by 2005 and 220 genera discovered by 2010 (Subhramani and Aalbersberg, 2012). Actinobacteria distributions in the sea have remained essentially unaffected, and clear evidence that these bacteria play a crucial ecological part in the marine environment has continued indescribably even today. A fascinating picture of marine actinobacteria diversity is starting to develop. Weyland (1969) conducted a thorough assessment on the distribution of marine actinobacteria in the sediments of the Atlantic Ocean and the North Sea, concluding that, when compared to their terrestrial counterparts, to isolate bioactive compounds actinobacteria from marine are the potential source. Furthermore, a lot of researchers from all over the world focused on isolating and identifying actinobacteria from various maritime habitats. According to recent research, a total of 83 actinobacteria species fitting into 28 genera have been discovered in the marine environment (Ramesh and Detmer, 2019). *Streptomyces* sp. is the most common among them. The majority of the genera that have been

TABLE 19.3
Antibiotics Extracted from Actinobacteria (Ranjani, 2016)

Antibiotic	Uses	Source
Anthracenedione	Used against tumor	*Streptomyces*
2-Allyloxyphenol	Antimicrobial	*Streptomyces sp*
Arenicolides A—C	Mild toxicity	*S. arenicola*
Avermectins	Used against parasites	*S. avermitilis*
Bafilomycin-A1	ATPase inhibitor	*S. halstedii*
Daryamides	Antioxidant	*Streptomyces*
Hygromycin	Immune suppression	*S. hygroscopicus*
Lincomycin	Protein biosynthesis inhibitor	*S. lincolnensis*
Valino-mycins	Transport particular ions across a lipid membrane in a cell.	*Streptomyces griseus*
ZHD	Used against cancer	*Actinomadura*
Chromomycin, A2	Used against tumors	*S. coelicolor*
Elaio-mycins, C	Used against tumors	*Streptomyces*- BK 190
Streptozotocin	Diabetogenic	*S. achromogenes*
Staurosporinone	Antitumor; phycotoxicity	*Streptomyces*
Salino-sporamide, C	Cell toxicity	*Streptomyces tropica*

identified are brand new to science. The diversity of marine actinobacteria could be expanded if more research is done.

The unfavorable features of the marine environment, such as salt, high pressure, salt, a lack of light, low temperatures, and so on, have evolved marine actinobacteria to thrive in this habitat. They need Na+ to proliferate since it's necessary to keep the osmotic environment stable for cellular integrity. Because of the limited number of nutrients available, oligotrophy is one more adaptation. Actinobacterial action, on the other hand, increases decomposition, deprivation and mineralization processes in soil and the covering water, as well as dissolved organic and inorganic material releases. Organic matter mineralization obtained from primary producers results in its recycling, making these substances available to primary producers once more. Any changes in water temperature, salinity, and other physicochemical features affect the distribution of actinobacteria. Actinobacteria are also vital food sources for a wide range of marine species. As a result, actinobacteria not only help to maintain the environment's pure state, but they also act as biological mediators in biogeochemical methods (Karthik et al., 2010). Actinobacterial spores can be transferred from land to water, where they can stay viable for a very long duration; a theory about the presence of indigenous populations of marine actinobacteria has arisen (Goodfellow and Williams, 1983; Bull et al., 2005; Mincer et al., 2002). As a result, actinobacteria isolated from marine materials are commonly thought to be of terrestrial origin. Although actinomycetes can be isolated from the deep sea (Weyland, 1969), and their derived products are highly active and can grow in marine water, this notion has remained (Jensen et al., 2005).

Despite the fact that dozens of antibiotics have been identified, their toxic nature has limited their use. To address this issue, researchers are looking for novel drugs that are both effective and not hazardous. Marine actinobacteria produce a huge number of natural compounds (60–70%), many of which are valuable in pharmacological, therapeutic, and agricultural applications as shown in Figure 19.4 (Baltz, 2005). Actinomycetes are found to create bioactive compounds in eight taxa, with 267 products stated from 98 marine actinomycetes (Subramani and Sipkema, 2019).

19.4.2.1 Enzyme Inhibitors from Marine Actinobacteria

Many enzyme inhibitors are also extracted from terrestrial organisms, even if they are structurally identical to marine microbial enzymes (Umezawa, 1972; Hamato et al., 1992). In contrast, the properties of enzyme inhibitors in marine creatures differ significantly from those in terrestrial organisms (Imada, 2004). One of the possible producers of enzyme inhibitors is *Streptomyces* isolated from the terrestrial environment (Umezawa, 1972). Actinomycin (Katz and Weissbach, 1962), candicidin (Liras et al., 1977), chloramphenicols (Jones and Westlake, 1974), and treptomycin (Liras et al., 1977) are examples of secondary metabolites extracted marine actinobacteria that act as enzyme inhibitors (Martin and Demain, 1980). The amount of research on enzyme inhibitors from marine microorganisms is limited

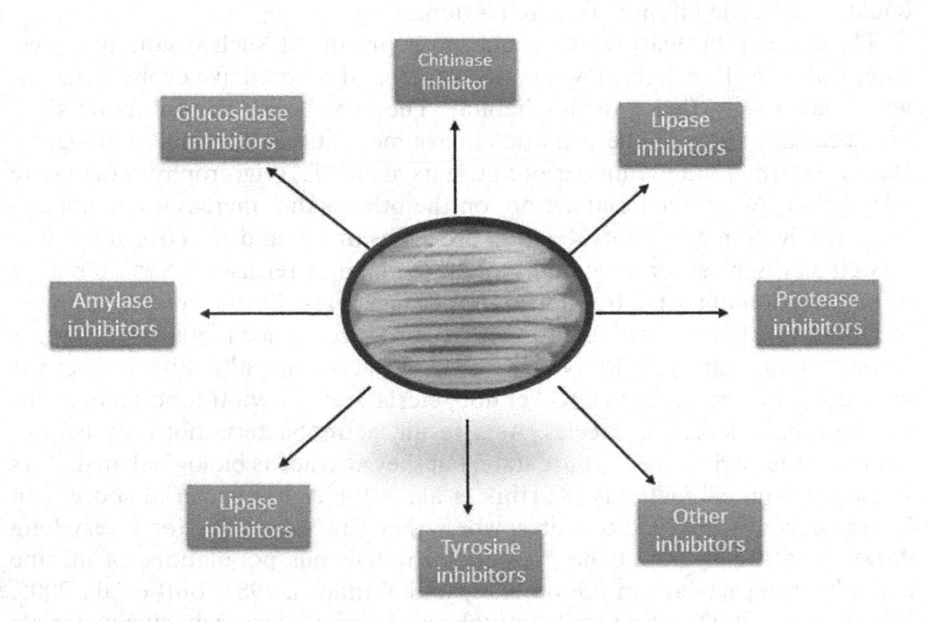

FIGURE 19.4 Natural products from marine actinobacteria.

(Newman and Cragg, 2007; Skropeta et al., 2011). Enzyme inhibitors have long been regarded as important tools in pharmacology and agriculture, owing to their biotechnological potential (Bode and Huber, 1992). (Terashita et al., 1980). The third-most important product of marine actinobacteria is enzyme inhibitors (Bode and Huber, 1992). Despite the fact that it is used to research enzyme shapes and reaction mechanisms, it was only recently introduced into pharmacology (Bode and Huber, 1992). These discriminating inhibitors can be employed to inactivate target proteases in disorders like high blood pressure, malaria, pancreatitis, emphysema, AIDS, and cancer (Demuth, 1990). Amylase inhibitors are effective tools for regulating carbohydrate-dependent illnesses, such as hyperlipemia, obesity diabetes (Puls and Keup, 1973; Bo-Linn et al., 1982). Karthik et al. (2014) extracted a PI (peptide) from deepsea Nicobar sand samples from marine actinobacterial that showed promising action against malarial parasites *P. falciparum*. The effectiveness of an inhibitor for α-glucosidase from sea actinobacteria for the treatment of hyperglycemia in *Streptozotocin*-induced diabetic male rats was recently reported by Sathish and Bhaskara Rao (2018).

19.5 WHY PROTEASE INHIBITORS FROM MARINE ACTINOBACTERIA?

Plants are the source of most naturally occurring protease inhibitors. Molecularly, they are clearly defined as serine protease inhibitors and most of them fall into that category. Thus, we can see that animals, microbes, and plants are not providing inhibitors with the same clarity of understanding as plants (Richardson, 1991). PIs from microorganisms, on the other hand, are small-molecular-weight biomolecules, as opposed to those from plants and animals. Usually found in microbes' culture filtrate, these molecules are produced when macromolecules are hydrolyzed (Umezawa, 1967). Since Umezawa's (1982) pioneering work, microorganisms have been used to produce protease inhibitors. Microbiology experts can easily prepare the organisms, extract them, and modify them. Microbial proteolytic enzyme inhibitors are molecule peptides that play virtually no role in microorganisms. The creation of a greater part of these effectual extracellular PIs is credited to actinobacteria, specifically, *Streptomyces* sp., enlarging to the exhibition of a many optional metabolites under use. They end up being biotechnologically the most assorted and effective metabolite makers among the prokaryotes, creating a lavish amount of antitoxins and making them the main antitoxin producers (Jensen et al., 2005). Although actinobacteria are generally subject to compounds like protease β glucosidase, amylase, cellullase, chitinase, and keratinase for nutraceuticals and in addition, they are an expected wellspring of chemical inhibitors (Manivasagan et al., 2014). It is astonishing to see that more than 10,000 of items were gotten from the 140 recorded genera of actinobacteria (Raja and Prabakarana, 2011). Hence, actinobacteria are been progressively investigated for novel mixtures, the emphasis being on marine actinomycetes recently. Marine climate has been known to create a colossal variety of mixtures. Notwithstanding the variety, the way that investigation of marine climate is as yet in the early stage adds to the significance of abuse of marine creatures.

19.6 CONCLUSION

Plants are known for producing PIs, but microbes, including actinobacteria, have increasingly been studied for their ability to produce PIs. As a therapeutic tool for treating many clinically challenging diseases, they are gaining more importance. A wide range of protease inhibitors is produced by terrestrial actinobacteria, a major exploitation group. *Streptomyces* sp., however, produces the most protease inhibitors of all actinobacteria, regardless of the condition. They have produced a wide range of protease inhibitors from the very beginning of their reports. Countable actinomycetes contribute only a small fraction of the total. Book chapter findings shed light on protease inhibitor potential in actinobacteria, which has been understudied, particularly in the rare group of actinomycetes. Considering the diversity and potency of actinobacteria, the less accessed the rare actinomycetes class can be regarded as an opulent source of novel metabolites. With the presence of numerous potential enzymes, actinobacteria will continue to produce a wide array of inhibitors directed toward many more ailments. The trend of protease inhibitors from actinobacteria will remain as this group of microbes is still less explored and exploited. This compilation depicts the current scenario and would open doors for future prospects. Actinomycetes are a less accessed class of actinobacteria that are capable of synthesizing a staggering number of compounds. A wide range of inhibitors directed toward many more diseases will be produced by actinobacteria as a result of their multiple enzymes. Actinobacteria will continue to be used as a source of PIs as they are still underexplored. Based on this compilation, we can anticipate future prospects. The chapter also provided an overview of the current situation.

REFERENCES

Amsler, C. D., Iken, K. B., McClintock, J. B. and Baker, B. J. 2001. Secondary Metabolites as Mediators of Trophic Interactions Among Antarctic Marine Organisms. *American Zoologist* 41:17–26.

Aneiros, A. and Garateix, A. 2004. Bioactive peptides from marine sources: Pharmacological properties and isolation procedures. *Journal of Chromatography B. Analytical Technology in Biomedical Life Sciences* 803:41–53.

Aoyagi, T., Takeuchi, T., Matsuzaki, A., Kawamura, K. and Kondo, S. 1969. Leupeptins, new protease inhibitors from actinomycetes. *Journal of Antibiotics* 22:283–286.

Baltz, R. 2005. Antibiotic discovery from actinomycetes: Will a renaissance follow the decline and fall? *SIM News* 55:186–196.

Blunt, J. W., Copp, B. R., Keyzers, R. A., Munro, M. H. G. and Prinsep, M. R. 2007. Marine natural products. *Natural Products Reports* 24:31–86.

Blunt, J. W., Copp, B. R., Keyzers, R. A., Munro, M. H. G. and Prinsep, M. R. 2012. Marine natural products. *Natural Products Reports* 29:144–222.

Blunt, J. W., Copp, B. R., Keyzers, R. A., Munro, M. H. G. and Prinsep, M. R. 2013. Marine natural products. *Natural Products Reports* 30:237–323.

Blunt, J. W., Copp, B. R., Keyzers, R. A., Munro, M. H. G. and Prinsep, M. R. 2014. Marine natural products. *Natural Products Reports* 30:237–323.

Blunt, J. W., Copp, B. R., Keyzers, R. A., Munro, M. H. G. and Prinsep, M. R. 2015. Marine natural products. *Natural Products Reports* 32:116–211.

Blunt, J. W., Copp, B. R., Keyzers, R. A., Munro, M. H. G. and Prinsep, M. R. 2016. Marine natural products. *Natural Products Reports* 33:382–431.

Bode, W. and Huber, R. 1992. Natural protein proteinase inhibitors and their interaction with proteinases, *European Journal of Biochemistry* 204:433–451.

Bo-Linn, G. W., Santa Ana, C. A., Morawski, S. G. and Fordtran, J. S. 1982. Starch blockers-their effect on calorie absorption from a high-starch meal. *The New England Journal of Medicine* 307:1413–1416.

Bull, A. T., Stach, J. E., Ward, A. C. and Goodfellow, M. 2005. Marine actinobacteria: Perspectives, challenges, future directions. *Antonie Van Leeuwenhoek* 87:65–79.

Burkholder, P. R., Pfister, R. M. and Leitz, F. H. 1966. Production of a Pyrrole Antibiotic by marine bacteria. *Applied and Environmental Microbiology* 14:649–653.

Challinor, V. L. and Bode, H. B. 2015. Bioactive natural products from novel microbial sources. *Annals of The New York Academy of Science* 1354:82–97.

Debashish, G., Malay, S., Barindra, S. and Joydeep, M. 2005. Marine enzymes. *Marine Biotechnology* 96:189–218.

Demuth, H. U. 1990. Recent developments in inhibiting cysteine and serine proteases. *Journal of Enzyme Inhibition and Medicinal Chemistry* 3: 249–278.

Dhanasekaran, D. and Jiang, Y. 2016. *Actinobacteria Basics and Biotechnological Applications*, 1st ed. In Tech Publishers, Croatia.

Donia, M. and Hamann, M. T. 2003. Marine natural products and their potential applications as anti-infective agents. *The Lancet Infectious Diseases* 3:338–348.

Drag, M. and Salvesen, G. S. 2010. Emerging principles in protease-based drug discovery. *Nature Review Drug Discovery* 9:690–701.

Faulkner, D. J. 2000. Marine pharmacology. *Antonie van Leeuwenhoek* 77:135–145.

Fear, G., Komarnytsky, S. and Raskin, I. 2007. Protease inhibitors and their peptidomimetic derivatives as potential drugs. *Pharmacology & Therapeutics* 113:354–368.

Feinstein, G., Malemud, C. G. and Janoff, A. 1967. The inhibition of human leucocyte elastase and chymotrypsin-like protease by elastatinal and Chymostatin. *BBA—Enzymology* 429:925–932.

Fenical, W. and Jensen, P. R. 1993. Marine microorganisms: A new biomedical resource. *Pharmaceutical and Bioactive Natural Products*. New York: Marine Biotechnology, pp. 419–457.

Fitzpatrick, F. 2004. Cyclooxygenase enzymes: Regulation and function. *Current Pharmacological Designs* 10:577–588.

Ghosh, D. 2005. Marine enzymes. *Advances in Biochemical Engineering and Biotechnology* 96:189–218.

Goodfellow, M and Williams, S. T. 1983. Ecology of actinomycetes. *Annual Review of Microbiology* 37:189–216.

Grant, P. and Mackie, A. 1977. Drugs from the sea-facts and fantasy. *Nature* 267:786–788.

Gustafsson, D., Antonsson, T., Bylund, R., Eriksson, U., Gyzander, E., Nilsson, I., Elg, M., Mattsson, C., Deinum, J., Pehrsson, S., Karlsson, O., Nilsson, A. and Sörensen, H. H. 1998. Effects of melagatran, a new low-molecular-weight thrombin inhibitor, on thrombin and fibrinolytic enzymes. *Thrombin Hemostatic* 79:110–118.

Hamato, N., Takano, R., Kamei-Hayashi, T., Imada, C. and Hara, S. 1992. Leupeptins produced by the marine Alteromonas sp. B-10-31. *Bioscience, Biotechnology, and Biochemistry* 56:1316–1318.

Hamilton, S. C., Farchaus, J. W. and Davis, M. C. 2001. DNA polymerases as engines for biotechnology. *Biotechniques* 31:370–383.
Imada, C. 2004. Enzyme inhibitors of marine microbial origin with pharmaceutical importance. *Marine Biotechnology* 6:193–198.
Ireland, C. M., Copp, B. R., Foster, M. D. and Donald, M. C. 1993. Biomedical potential of marine natural products. *Pharmaceutical and Bioactive Natural Products* 1:1–43.
Jensen, P. R., Dwight, R. Y. and Fenical, W. 1991. Distribution of actinomycetes in near-shore tropical marine sediments. *Applied Environmental Microbiology* 57:1102–1108.
Jensen, P. R., Mincer, T. J., Williams, P. G. and Fenical, W. 2005. Marine actinomycete diversity and natural product discovery. *Antonie van Leeuwenhoek* 87:43–48.
Jones, A. and Westlake, D.W. 1974. Regulation of chloramphenicol synthesis in Streptomyces sp. 3022a. Properties of arylamine synthetase, an enzyme involved in antibiotic biosynthesis. *Canadian Journal of Microbiology* 20:1599–1611.
Karthik, L., Gaurav, K. and Bhaskara Rao, K. V. 2010. Diversity of marine actinomycetes from nicobar marine sediments and its antifungal activity. *International Journal of Pharmacy and Pharmaceutical Sciences* 2:199–203.
Karthik, L., Gaurav, K., Tarun, K., Arindam, B., Sarath, C.S. and Bhaskara Rao, K. V. 2014. Protease inhibitors from marine actinobacteria as a potential source for antimalarial compound. *Plos One* 9:1–13.
Katz, E. and Weissbach, H. 1962. Biosynthesis of the actinomycin chromophore; enzymatic conversion of 4-methyl-3-hydroxyanthranilic acid to actinocin. *Journal of Biological Chemistry* 237: 882–886.
Kobori, H. and Taga, N. 1980. Extracellular alkaline phosphatase from marine bacteria: Purification and properties of extracellular phosphatase from a marine *Pseudomonas* sp. *Canadian Journal of Microbiology* 26:833–838.
Krowarsch, D., Cierpicki, T., Jelen, F. and Otlewski, J. 2003. Canonical protein inhibitors of serine proteases. *Cell Molecular Life Sciences* 60:2427–2444.
Laatsch, H. 2005. Marine bacterial metabolites. *Frontiers in Marine Biotechnology* 15:225–288.
Laskowski, M. and Qasim, M. A. 2000. What can the structures of enzyme-inhibitor complexes tell us about the structures of enzyme substrate complexes. *Biochemistry and Biophysics Acta* 1477:324–337.
Lindquist, N. 2002. Chemical defense of early stages of benthic marine invertebrates. *Journal of Chemical Ecology* 28:1987–2000.
Liras P., Villanueva J.R. and Martin J.F. 1977. Sequential expression of macromolecule biosynthesis and candicidin formation in Streptomyces griseus. *Journal of General Microbiology* 102:269–277.
MacLeod, R. A. 1965. The Question of the Existence of Specific Marine Bacteria. *Bacteriology Review* 29:9–23.
Maeda, M. and Taga, N. 1976. Extracellular nuclease produced by a marine bacterium, II: Purification and properties of extracellular nuclease from a marine *Vibrio* sp. *Canadian Journal of Microbiology* 22:1443–1452.
Magister, S. and Kos, J. 2013. Cystatins in immune system. *Journal of Cancer* 4:45–56.
Manivasagan, P., Venkatesan, J., Sivakumar, K. and Kim, S. 2014. Actinobacterial enzyme inhibitors- a review. *Critical Review on Microbiology* 41:261–272
Martin J.F. and Demain L.A. 1980. Control of antibiotic biosynthesis. *Microbiological Review* 44:230–251.
Mayer, A. M. S., Glaser, K. B. and Cuevas, C. 2010. The odyssey of marine pharmaceuticals: A current pipeline perspective. *Trends in Pharmacological Sciences* 31:255–265.
McKerrow, J. H., Caffrey, C., Kelly, B., Loke, P. and Sajid, M. 2006. Proteases in parasitic diseases. *Annual Review of Pathology* 1:497–536.

Mehbub, M. F., Lei, J., Franco, C. and Zhang, W. 2014. Marine sponge derived natural products between 2001 and 2010: Trends and opportunities for discovery of bioactives. *Marine Drugs* 12:4539–4577.

Mincer, T. J., Jensen., P. R., Kauffman, C. A. and Fenical, W. 2002. Widespread and persistent populations of a major new marine actinomycete taxon in ocean sediments. *Applied Environmental Microbiology* 68:5005–5011.

Mittl, P. R. and Grutter, M. G. 2006. Opportunities for structure-based design of protease-directed drugs. *Current Opinion in Structural Biology* 16:769–775.

Newman, D.J. and Cragg, G.M. 2007. Natural products as sources of new drugs over the last 25 years. *Journal of Natural Products* 70:461–477.

Puls, W. and Keup, U. 1973. Influence of an a-amylase inhibitor (BAY d 7791) on blood glucose, serum insulin and NEFA in starch loading tests in rats, dogs and man. *Diabetologia* 9:97–101.

Raja, A. and Prabakarana, P. 2011. Actinomycetes and drug-an overview. *American Journal of Drug Discovery and Development* 1:72–84

Ramesh, S. and Detmer, S. 2019. Marine rare actinomycetes: A promising source of structurally diverse and unique novel natural products. *Marine Drugs* 17:1–40.

Ranjani, A., Dhanasekaran, D. and Gopinath, P. M. 2016. *An Introduction to Actinobacteria.* Actinobacteria-Basics and Biotechnological Applications publication, InTech Publisher, India.

Rawlings, D. N., Tolle, P. D. and Barrett, A. J. 2004. Evolutionary families of peptidase inhibitors. *Biochemistry Journal* 378:705–716

Reen, F., Gutierrez-Barranquero, J., Dobson, A., Adams, C. and O'Gara, F. 2015. Emerging concepts promising new horizons for marine biodiscovery and synthetic biology. *Marine Drugs* 13:2924–2954.

Richardson, M. 1991. Seed storage proteins: The enzyme inhibitors. *Methods of Plant Biochem*istry 5:295–305.

Ritchie, B. C. 2003. Protease inhibitors in the treatment of hereditary angioedema. *Transfusion and Apheresis Science* 29:259–267.

Rosenfeld, W. D. and Zobell, C. E. 1947. Antibiotic production by marine microorganisms. *Journal of Bacteriology* 54:393–398.

Sabotic, J. and Kos, J. 2012. Microbial and fungal protease inhibitors—current and potential applications. *Applied Microbiology and Biotechnology* 93:1351–1375.

Santos, L. O. 2009. HIV aspartyl peptidase inhibitors interfere with cellular proliferation, ultrastructure and macrophage infection of *Leishmania amazonensis. Plos One* 4:4910–4918.

Sathish, K. and Bhaskara Rao, K.V. 2018. Efficacy of alpha glucosidase inhibitor from marine actinobacterium in the control of postprandial hyperglycaemia in streptozotocin (STZ) induced diabetic male albino wister rats, *Iranian Journal of Pharmaceutical Research* 17:202–214.

Selzer, P. M., Chen, X., Chan, V. J., Cheng, M., Kenyon, G. L., Kuntz, I. D., Sakanari, J. A., Cohen, F. E. and McKerrow, J. H. 1997. Leishmania major: Molecular modeling of cysteine proteases and prediction of new nonpeptide inhibitors. *Experimental Parasitology* 87:212–221.

Shukla, A. K., Singh, B. K., Patra, S. and Dubey, V. K. 2010. Rational approaches for drug designing against leishmaniasis. *Applied Biochemistry and Biotechnology* 160:2208–2218.

Skropeta, D., Pastro, N. and Zivanovic, A. 2011. Kinase inhibitors from marine sponges. *Marine Drugs* 9:2131–2154.

Subhramani, R. and Aalbersberg, W. 2012. Marine actinomycetes: An ongoing source of novel bioactive metabolites. Microbiology Research 167:571–580.

Tamir, S., Bell, J., Finlay, T. H., Sakal, E., Smirnoff, P., Gaur, S. and Birk, Y. 1996. Isolation, characterization and properties of a trypsin chymotrypsin inhibitor from amaranth seeds. *Journal of Protein Chemistry* 15:219–229.

Terashita, T., Kono, M. and Murao, S. 1980. Promoting effect of S-Pion fruiting of Lentinus edodes. *Transaction of the Mycology Society of Japan* 21:137–140.

Turk, B. 2006. Targeting proteases: Successes, failures and future prospects. *Nature Review Drug Discovery* 5:785–799.

Umezawa, H. 1967. Structures and activities of protease inhibitors of microbial origin. *Methods of Enzymology* 45:678–695:

Umezawa, H. 1972. Enzyme inhibitor of microbial origin, *Proceedings of the Sixth International Congress of Pharmacology, Mechanism of Toxicity and Metabolism*, Tokyo, pp. 17–31.

Umezawa, H. 1981. *Small Molecular Immunomodifiers of Microbial Origin: Fundamental and Clinical Studies of Bestatin*. Japan Scientific Society Press, Tokyo.

Umezawa, H. 1982. Low-·molecular-weight enzyme inhibitors of microbial origin. *Annual Review Microbiology* 36:75–99.

Weyland, H. 1969. Actinomycetes in North Sea and Atlantic Ocean sediments. *Nature* 223:850–858.

Williams, S. T., Goodfellow, M., Alderson, G., Wellington, E. M. H., Sneath, P. H. A. and Sackin, M. J. 1983. Numerical classification of Streptomyces and related genera. *Journal of General Microbiology* 129:1743–1813.

Zhang, X. N., Song, G. H., Jiang, T., Shi, B., Hu, Y. and Yuan, Z. 2010. Rupintrivir is a promising candidate for treating severe cases of Enterovirus-71 infection. *World Journal of Gastroenterology* 16:201–209.

Zilinskas, R. A., Colwell, R. R., Lipton, D. W. and Hill, R. T. 1995. *The Global Challenge of Marine Biotechnology*. A status report on the United States, Japan, Australia and Norway Maryland Sea Grant, Mayerland Sea College Publication, USA.

20 Nanoparticles from Marine Biomaterials for Cancer Treatment

Harika Atmaca and Suleyman Ilhan

CONTENTS

20.1 INTRODUCTION: BACKGROUND AND DRIVING FORCES

Cancer is a life-threatening disease defined by the uncontrolled multiplication of aberrant cells and their rapid spread to other organs. An abnormally dividing collection of cells suppresses the surrounding tissue or organ, preventing the tissue or organ from performing its function (Senga and Grose 2021). According to the World Health Organization (WHO), an estimated 9.6 million people died in 2018 due to this disease (Siegel et al. 2021).

The most important factors that complicate cancer treatment are that it is a heterogeneous disease, its course is different in each patient, and it shows different characteristics at each stage of the disease. Due to these difficulties in its treatment, it is the most important public health problem and the second-leading cause of death after cardiovascular diseases.

Radiotherapy, chemotherapy, and surgery are the main methods used in cancer treatment (Yildizhan et al. 2018). Surgical methods consist of resection of cancerous tissue. Organ loss, the risk of cancer recurrence, and the inability to apply to all cancer types are the disadvantages of these methods. In

DOI: 10.1201/9781003303909-20

radiotherapy, cancerous cells are burned with radiation in a specific frequency band and at a specific intensity. The disadvantages of this method are that healthy cells are damaged as well as cancerous cells, the radiation distribution is not equal to all cancer cells, and the loss of function in the tissue exposed to radiation. In chemotherapy, on the other hand, it is aimed to kill cancerous cells with drugs that have toxic effects and eliminate the mechanisms that allow cancerous cells to divide. Classical chemotherapy drugs do not act in a targeted manner in the body. The drugs used affect cancer cells as well as healthy cells. In addition, cancer cells do not reach the doses required for treatment. Chemotherapy weakens the patient's immune system and makes the patient more susceptible to other diseases. Another problem encountered is the multidrug resistance that develops against anticancer agents. All these important side effects are because chemotherapy drugs do not have a tissue-specific effect.

In cancer treatment, it is essential for the success of the treatment that chemotherapy drugs target tumors as much as possible and have a limited effect on healthy tissues. This issue is also important in terms of increasing the life span and quality of the patient. Advances in nano-oncology have brought important innovations in targeted drug delivery (Mu et al. 2020). In this way, while the intracellular concentrations of drugs in cancer cells can be increased, their toxic effects on healthy cells can be minimized.

While the diagnosis rate of cancer is constantly increasing with scientific studies, the cancer-related death rate remains almost constant (Clarke et al. 2020). This picture is promising that cancer is a winnable war. Various disadvantages of the methods used in cancer diagnosis and treatment reduce the effectiveness of these methods. Nanotechnology offers important advantages for developing more effective diagnostic and treatment methods.

Nanotechnology is one of the rapidly developing fields in science and technology and is widely used as a drug delivery system (DDS) for various therapeutic agents, such as small-molecule active substances, genes, proteins, and peptides. Nanosized systems are new-generation DDSs with particle sizes between 1–1000 nm (Bhardwaj and Nikkhah-Moshaie 2017; Martins et al. 2020). The use of nanoparticle (NP) DDSs in the treatment of cancer cells provides simultaneous diagnosis and treatment compared to traditional treatment methods.

From a biological perspective, nanotechnology deals with the interaction between engineered materials designed at nanoscales and cellular and biomolecular structures. The nanostructures used are of the same dimensions as the biological structures. For example, human cells are 10,000–20,000 nm in diameter. Hemoglobin, one of the bio-macromolecules, is 5 nm in diameter. Nanostructures smaller than 50 nm can easily enter most cells. Nanostructures smaller than 20 nm can be transported by blood circulation. Matter exhibits different physical, chemical, and biological properties at the nanoscale due to quantum mechanical reasons. It is possible to obtain nanostructures with desired properties (physical, chemical, etc.) by producing the substance in nano-dimensions without changing its chemical composition (Khan, Saeed, and Khan 2019).

Organic and inorganic NPs are the two major types of NPs. Organic NPs are made up of carbon NPs. Inorganic NPs include magnetic NPs, noble metal NPs (platinum, gold, silver), and semiconductor NPs (titanium dioxide; zinc oxide; Xu et al. 2006). Because of its unique characteristics such as ease of use, good functionality, biocompatibility, capacity to target particular cells, and controlled drug release, inorganic NPs are increasingly being employed in drug delivery. Nanotechnology has the potential to direct and selectively target chemotherapies to malignant cells and neoplasms, guide tumor surgical excision, and improve the therapeutic efficacy of radiation-based and other existing treatment methods. These possibilities provided by nanotechnology make it effective in cancer diagnosis and treatment.

Marine bio-nanotechnology is a fascinating and rapidly developing field of study. Nanoscience and nanotechnology have a lot of potential in the biologically diverse marine environment. The marine ecosystem is the largest on the planet, with more than 2 million different species living in it. Although only 9% have been discovered, they host a wide variety of creatures with different chemical and biological properties (Mora et al. 2011). Pharmaceutical and therapeutic research can benefit from a variety of marine-derived substances (Carroll et al. 2021). Nearly 30,000 natural resource mixes of marine origin have been identified from organisms such as various mollusks, delicate tunicates, robust seaweeds, large mangroves, porous sponges, powerful sea hares, microbes, advanced chordates, and evolutionarily conserved sharks (Hu et al. 2015). Antibacterial, antidiabetic, antiviral, antifungal, and anti-inflammatory properties of marine chemicals have been demonstrated. Many commercially available marine-derived chemicals have been shown to have anticancer properties.

The U.S. Food and Drug Administration (FDA) has approved a number of important anticancer drugs derived from marine organisms, such as cytarabine (Cytosar U®), a chemotherapeutic drug used to treat leukemia. Trabectedin (Yondelis®) is another antitumor agent that is used to treat advanced soft tissue sarcoma. Breast, prostate, and pediatric sarcomas are all in clinical studies. Another anticancer medication, eribulin mesylate (Halaven®), is used to treat breast cancer and liposarcoma. Brentuximab vedotin (Adcetris®) is an antibody-drug conjugate (ADC) medication used to treat HL and anaplastic large cell lymphoma that has relapsed or become resistant. Midostaurin (Rydapt®) is a protein kinase inhibitor used to treat acute myeloid leukemia and myelodysplastic syndrome (van Andel et al. 2018). Furthermore, 19 anticancer substances obtained from the sea are at various stages of clinical testing (Zuo and Kwok 2021). Numerous studies have also demonstrated the anticancer properties of marine-derived chemicals in vitro or in vivo.

The molecules might have a range of physicochemical properties, such as low toxicity and low manufacturing costs, that lead to fascinating biological characteristics in the biomedical field. Diverse biomaterials from marine creatures are also isolated for use in DDSs (Oliveira et al. 2020).

Within the nano-biotechnological sector, there is a rising interest in the use of nanomaterials in the development of cancer treatments. Marine-derived compounds might be a source of DDSs. Nano-formulations are linked to a wide range of NPs, and their polymerized structures are becoming a popular method for creating cancer medicines. It has also been shown to have a number of advantages for the encapsulation of biomaterials, genetic material, and other medicinal chemicals, including stability and protection, as well as improved solubility and encouraged sustained release, biocompatibility, and biodegradability (Brannon-Peppas and Blanchette 2012). They may be utilized to deliver a range of therapeutic compounds, and due to a leaky vascularization site and active cellular absorption, they tend to accumulate in tumor areas (Pérez-Herrero and Fernández-Medarde 2015). These systems also boost macromolecule circulation, shield medicines from enzymatic breakdown, regulate drug release, reduce cytotoxicity, and boost therapeutic index (Senapati et al. 2018). However, because NPs may be unstable throughout circulation and may be linked with unknown toxicity, these sorts of systems may have significant disadvantages, especially for long-term administration (Babu et al. 2014).

Marine organisms such as shellfish, macro-algae, fungi, micro-algae, and coral produce polysaccharides with variable structures and diverse biological activities that terrestrial organisms lack. This is because the marine environment is featured by high pressure, high salinity, low temperature, and oxygen deficiency.

In this chapter, we focus on the utilization of marine biomaterials, such as chitin, chitosan, alginate, carrageenan, fucoidan, and hyaluronan, to create NPs used in cancer therapy.

20.2 CHITIN (CH) AND CHITOSAN (CHT)

Chitin (β-(1–4)-poly-N-acetyl-d-glucosamine; CH), is the second-most abundant polysaccharide on the planet found in the cytoderm of fungus and green algae, as well as the shells of crustaceans like crabs and shrimp (Ehrlich et al. 2007). It was discovered in 1972 but did not receive the value it deserved until the 21st century (Peters 1972). Investigations identified that CH has low toxicity, is biocompatible, is biodegradable, and possesses hemostatic activity and antimicrobial capability (Synowiecki and Al-Khateeb 2003). Chitosan (CHT) is a natural polymer derived from CH that has biological features such as biodegradability and biocompatibility for specific applications and resembles glycosaminoglycans in some ways (Zhu, Liu, and Pang 2019). CHT is an amino-polysaccharide made up of (1–4)-linked amino acids. Due to CH's limited solubility, CHT, which is created by a deacetylation procedure, has gotten the most attention (Peptu et al. 2019). This procedure produces CHT with varying degrees of deacetylation and molecular weights, which affects its performance as a structural or bioactive component.

The CHT backbone is modified to change properties, such as solubility, mucoadhesion, and stability for use in a variety of applications (Li et al. 2018). The active sites for modification of CHT are both –NH2 and –OH groups. CHT polymers are made using techniques like blending, graft copolymerization, and curing.

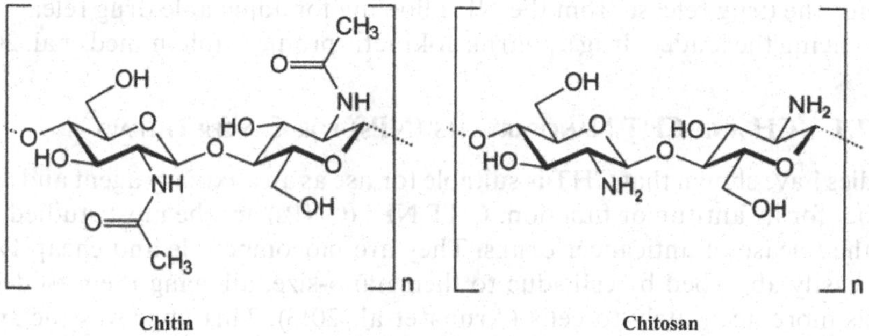

FIGURE 20.1 Structural formula of CH and CHT.

Blending is the physical process of combining two or more polymers. The quality and performance of the blend can be altered by changing the ratios of the polymers used. Blending is the most cost-effective method of tailoring polymer characteristics for specific uses.

Curing turns the polymers into a solidified mass by establishing three-dimensional linkages inside the polymer mass using thermal, electrochemical, or ultraviolet (UV) radiation treatment processes, whereas graft copolymerization includes covalent bonding of polymers (J. Li et al. 2018). Some hydrophilic polymers can be blended with poly (vinyl alcohol) (PVA), poly (vinyl pyrrolidone) (PVP) and poly (ethyl oxide) (PEO) for drug delivery. CHT-PVA blends improve its mechanical and barrier qualities. The intermolecular interactions between CHT and PVA produce PVA-CHT blends with better mechanical properties (tensile strength) for regulated drug administration. After blending, the characterization of CHT blends can be done via Fourier transform infrared spectroscopy (FTIR), differential scanning calorimetry (DSC), and X-ray diffraction (XRD; Mohammed et al. 2017).

Chemical modifications can be done via different methods such as chemical, radiation, photochemical, plasma-induced, and enzymatic grafting (Shukla et al. 2013). Several variants of CHT are formed as a result of chemical modification, including quaternized CHT, thiolated CHT, carboxylated CHT, amphiphilic CHT, CHT with chelating agents, PEGylated CHT, and lactose-modified CHT (Cui et al. 2009). The primary amine (–NH2) groups of CHT can react with sulfates, citrates, and phosphates, which can improve drug stability and encapsulation efficiency (Mohammed et al. 2017; Dambies et al. 2001). Grafting carboxylated chitosan with poly (methyl methacrylate) improves its pH-sensitive properties (Cui et al. 2009).

Swelling of the polymer, diffusion of the adsorbed drug, and drug diffusion across the polymer are some of the processes that control drug release from chitosan nanoparticles. The first burst release from CHT NPs is caused by either polymer swelling, which creates holes, or drug diffusion from the polymer's surface (Yuan et al. 2013). Because of the solubility of chitosan, drug release from CHT NPs is also pH-dependent (Siafaka et al. 2015). CHT compounds change the drug release from the NP, allowing for adjustable drug release and influencing the loaded drug's pharmacokinetic profile (Mohammed et al. 2017).

20.2.1 CH and CHT Nanoparticles (NPs) for Cancer Therapy

Studies have shown that CHT is suitable for use as an auxiliary agent and agent carrier for its antitumor function. CHT NPs (CNPs) are the most studied NPs for the release of anticancer drugs. They are biocompatible and cheap DDSs and easily absorbed by cells due to their nano-size, allowing them to deliver drugs more accurately to cells (Aruna et al. 2013). This improves the specificity of CNPs and explains why they have such a broad therapeutic use. As a result, they might be used as a vehicle for gene transfer. CNPs have proved to be quite effective in the treatment of a variety of cancers (Qi et al. 2005; Y. Xu et al. 2009). Its therapeutic value stems mostly from the alteration of CH-NP in the tumor microenvironment (Potdar D 2016). Increased temperature and the formation of acidic conditions are caused by a tumor microenvironment with inadequate vascular. When the CHT amino group is protonated in an acidic environment, CH-NPs expand, resulting in quicker drug release. Furthermore, the tumor microenvironment accumulates macromolecules due to inadequate vascular. Enhanced permeability and retention (EPR) were used to characterize this phenomenon (Maeda 2001). The major features for adopting the CNP medication system in the treatment of malignancies might include protonation under acid conditions and EPR. It's worth noting that chitosan is mostly used as a medication or nucleic acid carrier, with the therapeutic properties of the chitosan nanoparticle receiving less attention. Furthermore, research has revealed how CNPs are tailored for tumor cells. Various pharmaceutical applications, such as colon or cancer therapeutic foci, immunization transfer, mucosal processing, antioxidants, and genes, have all been captivated by CNPs (Divya and Jisha 2018). CHT is mixed with iron, graphene oxide, copper, silicon, and silver NPs and ionotropically produced into a variety of morphologies, including gelatinous, microemulsified, emulsified, polyelectrolyte, and micellar inversion (Frank et al. 2020).

Drug delivery with CH and CHT NPs for cancer therapy has been investigated broadly. CNPs modified with anticancer agent-loaded surfaces are developed to become more stable, porous, and bioactive over time. The marine biomaterials, anticancer agents, and cancer types used in these studies are summarized in Table 20.1. Overall, CHT is an essential chemical in cancer research that must be further investigated. This might soon lead to a breakthrough in cancer treatment that will benefit thousands of people suffering from various malignancies.

20.3 ALGINATE

Alginate (ALG) is a linear polysaccharide extracted from brown algae. It consist of 1–4-linked α-L-glucuronic acid (G) and β-d-mannuronic acid (M) (Severino et al. 2019).

TABLE 20.1
Overview of Chitin and Chitosan Nanoparticles Developed for Cancer Therapy

Tissue	Tested Cell Types/Organisms	Drug	Nanoparticle	Reference
Human breast adenocarcinoma cells	MCF-7	5-FU	Carboxymethyl chitosan	Mathew et al. (2010)
Human head and neck	FaDu	Doxorubicin	Chitosan oligosaccharide arachidic acid	Termsarasab et al. (2013)
Ovarian	OVK18 #2 tumor xenografts	HIF-1α siRNA	Folic acid poly (ethylene glycol)-chitosan oligosaccharide lactate (FA-PEG-COL) nanoparticles	Li et al. (2014)
Mouse osteosarcoma and mouse bone marrow mesenchymal cells	UMR-106 and mBMSCs	–	1D nanoparticles, chitin nanocrystals	Zhao et al. (2019)
Human breast adenocarcinoma cells	MCF-7	Curcumin	Chitin nanocrystals	Ou et al. (2018)
Human breast adenocarcinoma cells	MCF-7	–	Copper nanoparticles, Chitin nanoparticles, silver nanoparticles	Solairaj et al. (2017)
Human colorectal adenocarcinoma, Human breast adenocarcinoma cells	Caco-II and MCF-7	Doxorubicin, vinblastine, imatinib, pemetrexed	Chitosan-modified polymeric nanoparticles (chitosan diacetate, chitosan triacetate)	Khdair et al. (2016)
Human osteosarcoma	SaOs-2	Doxorubicin	Chitosan and *O*-HTCC (ammonium-quaternary derivative of chitosan) nanoparticles	Soares et al. (2016)
Human lung and liver carcinoma	A549 and HepG2	5-FU	Hyaluronic acid-coated chitosan nanoparticles	Wang et al. (2017)
Ovarian carcinoma cell	A2780	Cisplatin	Lipid-chitosan hybrid nanoparticles	Khan et al. (2019)
Human breast adenocarcinoma cells	MCF-7	Doxorubicin	Chitosan tripolyphosphate nanoparticles	Imran et al. (2018)
Human breast cancer cells	MDA-MB-231	Paclitaxel	K237-peptide-functionalized hybrid chitosan/poly (N-isopropyl acrylamide) nanoparticles	Qian et al. (2019)

(*Continued*)

TABLE 20.1
(Continued)

Tissue	Tested Cell Types/Organisms	Drug	Nanoparticle	Reference
Epidermoid carcinoma cell	A431	Resveratrol and ferulic acid	Chitosan nanoparticles	Balan et al. (2020)
Human glioblastoma cells	U-87 MG	Temozolomide	Chitosan nanoparticles	Irani et al. (2017)
Human gastric cancer cell	AGS	Taxanes	N-succinyl-chitosan and N-glutaryl-chitosan nanoparticles	Skorik et al. (2017)
Human hepatocarcinoma	HepG2	–	Chitin and silver nanoparticles	Vijayakumar et al. (2020)
Human liver cancer and lung cancer cell	A549 and HepG2	Honokiol	Epigallocatechin-3-gallate (EGCG) functionalized Chitin nanoparticles	Tang et al. (2018)
Mammary carcinoma	4T1 cells in vitro, 4T1-carrying BALB/c mice in vivo	DOX and α-tocopheryl succinate	ROS-fissile thioketal bonded silica nanoparticles by coating with carboxymethyl chitin	Ding et al. (2020)
Human colorectal adenocarcinoma, human colon carcinoma	HT-29, COLO-205	Paclitaxel	Amorphous chitosan nanoparticles	Smitha et al. (2013)
Human hepatocarcinoma	HepG2	Doxorubicin	Chitosan nanoparticles	Ye et al. (2018)
Bladder cell carcinoma	T24	5-FU	Chitosan-coated superparamagnetic iron oxide nanoparticles	Al-Musawi et al. (2019)
Human breast adenocarcinoma cells	MCF-7, MCF-7 xenograft	Paclitaxel	Estrone-modified glycol chitosan nanoparticles	Yang, Tang, and Yin (2018)
Human breast cancer cells	SK-BR-3	Trastuzumab	Docetaxel-loaded D-α-tocopherol polyethylene glycol 1000 succinate conjugated chitosan	Kumar Mehata et al. (2019)
Human breast adenocarcinoma cells	MCF-7	Methotrexate	Magnetite iron nanoparticles coated with chitosan	Ali et al. (2018)
Human osteosarcoma, Human breast adenocarcinoma cells	SaOs-2, MCF7 and T47D	Targeted delivery of adenosine 5′-triphosphate	Folic acid/chitosan-coated mesoporous hydroxyapatite nanoparticles	Feiz and Meshkini (2019)

Tissue	Tested Cell Types/Organisms	Drug	Nanoparticle	Reference
Human colorectal adenocarcinoma	HT-29	Docetaxel	poly(lactic-co-glycolic acid) (PLGA) and polycaprolactone (PCL) nanoparticles	Badran et al. (2018)
Human colorectal adenocarcinoma	HT-29	Andrographolide Analog (3A.1)	Polymeric nanoparticles made of naphthyl-grafted succinyl chitosan (NSC), octyl-grafted succinyl chitosan (OSC), and benzyl-grafted succinyl chitosan	Kansom et al. (2018)
Human hepatocarcinoma	HepG2	Ginsenoside compound K	Chitosan nanoparticles	Zhang et al. (2018)
Human cervical cancer cells, murine lung carcinoma cell subline	HeLa, M109-HiFR	Paclitaxel	Folate-grafted copolymer of polyethylene glycol (PEG) and chitosan solid lipid nanoparticles	Rosière et al. (2018)
Doxorubicin-resistant breast cancer cell line	MCF-7/DR	mRNA- cleaving (MDR1 gene) DNAzyme	Chitosan/Cyclodextrin Nanoparticle	Zokaei et al. (2019)
Human hepatocarcinoma	HepG2	Dunaliella Bardawil Biomass	N-succinyl chitosan nanoparticles	Kunjiappan et al. (2018)
Human lung cancer	A549	Methotrexate	Multi-functionalized chitosan nanoparticles	Guo et al. (2018)
Gastric cancer	MKN-45	(−)-epigallocatechin-3-gallate	fucose- carboxymethyl chitosan gold nanoparticles	Yuan et al. (2018)
Human breast cancer	4T1 in vitro, BALB/c nude mice in vivo	Gemcitabine	N-trimethyl chitosan nanoparticles and CSKSSDYQC peptide	Chen et al. (2018)
Human breast adenocarcinoma cells	MCF-7	Camptothecin	Chitosan/graphene oxide nanoparticle decorated with folic acid	Deb, Andrews, and Raghavan (2018)
Colorectal cancer	RKO	Metformin hydrochloride	Chitosan-based nanoparticles	Arafa et al. (2018)
Human lung cancer	L132, A549 in vitro, BALB/c-nu/nu athymic mice in vivo	Temozolomide	Folic acid decorated chitosan nanoparticles	Li et al. (2017)
Human colon adenocarcinoma	HCT15	Hesperetin	Chitosan folate hesperetin nanoparticle	Mary Lazer et al. (2018)
Human prostate cancer	LNCaP	Methotrexate	Chitosan nanoparticle	Gunel Nur et al. (2016)

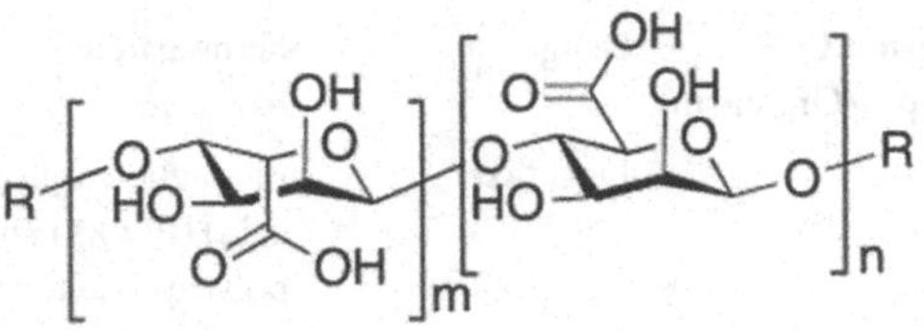

FIGURE 20.2 Structural formula of ALG.

To precipitate ALG, the extract is filtered, and sodium or calcium chloride is added to the filtrate (Qin 2008). This alginate salt may be converted to alginic acid by treating it with weak HCl. Water-soluble sodium ALG (NaAlg) powder is generated after purification and conversion.

After extraction, alginate is converted into NaAlg, which is utilized in medicinal dosage forms, via a chemical procedure. ALG's gelling characteristics, along with its lack of biological activity, make it a suitable polysaccharide for modifying drug release by crosslinking with calcium and other bivalent ions (Rinaudo 2008).

The physical and chemical properties of ALG are determined by the composition of the M and G units, as well as the length of their sequence and molecular weight (Hecht and Srebnik 2016).

The viscosity of the polymer changes proportionately with the amount of residues G and, more important, with the polymer's molecular weight. High G content is associated with bigger diameter and polydispersity particles, as well as high-porosity gels. A high M concentration, on the other hand, leads to weaker, more elastic, and more stable gels when frozen/thawed (Haug et al. 1959).

It is biocompatible and approved by the FDA for human use. It is widely used for the modified release of drugs since it has no toxicity and immunogenicity (D'Ayala et al. 2008).

20.3.1 ALG NPs for Cancer Therapy

There are two ways to make alginate NPs: The first one is complex formation in an aqueous solution, resulting in ALG nanoaggregates, or on the interface of an oil droplet, resulting in alginate nanocapsules. For ALG complexation, a crosslinker (e.g., calcium from calcium chloride) is employed, and the complexation can also be achieved by combining the ALG with an oppositely charged polyelectrolyte such as poly-L-lysine. The second one, ALG nanospheres, is formed by emulsifying alginate in oil and exterior or internal gelation of the ALG emulsion droplet (Severino et al. 2019).

Alginic acid is an ideal marine biomaterial for hydrogen-structured NP construction, as it has a high degree of water solubility, high porosity, biocompatibility, and a tendency to gel under the right conditions. When counter-ions are added to ALG, sequential crosslinking and polymer network creation results in hydrogel-structured drug delivery vehicles, including microparticles and

NPs (Tønnesen and Karlsen 2002). The gelification phenomenon may be controlled by the preparation techniques, resulting in the desired size ranges that are dependent on ALG concentration/viscosity, counter-ion concentration, and the speed with which the counter-ion solution is added to the alginate solution, among other parameters (Hamidi et al. 2008).

The biodegradability of ALG as a matrix for controlled release preparations is its most significant benefit since it is absorbed and destroyed by the body without causing any harmful effects during or after drug release (Shilpa et al. 2003). As a result, it may be an appropriate matrix for the long-term release of different medicines (Severino et al. 2019).

ALG has several characteristics that make it ideal for use as a drug delivery matrix: It is easily accessible, cheap, nontoxic and hemocompatible (Champagne et al. 2000). It is also a biodegradable material. Because it is water-soluble, it eliminates the need for noxious solvents during processing, reducing the stability, toxicological, and environmental issues associated with solvents (Ravi Kumar 2000). Because NPs without a coating may be picked up by the immune system, surface modifications such as hydrophilic polymers or surface-coating with cell-specific receptors help increase drug targeting and bioavailability.

The usage of cationic polyelectrolytes during the manufacture of NPs or between ALG and cationic pharmaceuticals to be loaded can crosslink alginate. With the improved long-term stability of NPs, the controlled release profile of the loaded pharmaceuticals can be achieved in both scenarios (Tønnesen and Karlsen 2002).

ALG is one of the most widely used marine biomaterials for the formation of (micro)particles in the literature for cancer treatment. Table 20.2 provides an overview of ALG NPs available in the literature.

20.4 CARRAGEENAN

Carrageenan is a linear sulfated polysaccharide isolated from the extracellular matrix of various red algae in the Florideophyceae class: *Agardhiella*, *Chondrus crispus*, *Eucheuma*, *Furcellaria*, *Gigartina*, *Hypnea*, *Iridaea*, *Sarconema*, and *Solieria* (Pacheco-Quito et al. 2020). It is composed of β-D-galactose and anhydrogalactose units, linked by glycosidic unions with ester sulfate groups (Silva et al. 2012). Its molecular weight is over 100 kDa and is classified into six basic forms, kappa (κ-), iota (ι-), lambda (λ-), mu (μ-), nu (ν-), beta (β-), and theta (θ-) based on the sulfate groups (Pacheco-Quito et al. 2020). κ is the most abundant form, and λ is the second-most abundant form of carrageenan. Due to its biocompatibility and unique properties, carrageenan is used to improve drug formulation properties, especially to prolong drug release (Khan et al. 2017).

It also aids in the development of medication delivery systems that are pH/temperature-sensitive. The required drug release characteristics such as zero-order release and pH-independent release cannot be attained when carrageenan is employed as the only matrix material to regulate drug release (Maderuelo et al. 2011). The interaction of carrageenan with other polymers is used to attain an ideal drug release profile. It has been studied that κ showed

TABLE 20.2
Overview of Alginate Nanoparticles for Cancer Therapy

Tissue	Tested Cell Types/Organisms	Drug	Nanoparticle	Reference
Human epithelial colorectal adenocarcinoma, hepatocellular carcinoma, breast cancer	CaCO-2, HepG2, MDA-MB-231	Curcumin	Diglutaric acid chitosan/alginate nanoparticles	Sorasitthiyanukarn et al. (2018)
Human breast adenocarcinoma cells	MCF-7	Doxorubicin-gelatin	Magnetic nanoparticles with a shell layer (Fe_3O_4-alginate)	Huang et al. (2020)
Human colorectal adenocarcinoma	HT-29	Docetaxel	pH- and redox-sensitive fluorescein-labeled agglutinin (fWGA) assembled disulfide crosslinked alginate nanoparticles	Chiu and Lim (2021)
Ehrlich carcinoma	balb/c mice	Doxorubicin	Alginate-coated caseinate nanoparticles	Elbialy and Mohamed (2020)
Human breast adenocarcinoma cells	MCF-7	Doxorubicin	Chitosan-alginate nanoparticles	Katuwavila et al. (2016)
Human hepatocarcinoma	HepG2	Paclitaxel	Folate-chitosan alginate nanoparticles	Wang et al. (2016)
Human breast cancer cells	MDA-MB-231	Curcumin diethyl disuccinate	Chitosan-alginate nanoparticles	Bhunchu et al. (2016)
Human cervical cancer cells	HeLa	Curcumin	Alginate-chitosan-pluronic composite nanoparticles	Das et al. (2010)
Human hepatocarcinoma, human cervical cancer cells	HepG2, HeLa	Doxorubicin	Novel disulfide crosslinked sodium alginate nanoparticles	Gao et al. (2017)
Keratin-forming tumor cell line HeLa	KB cells	Doxorubicin	Self-assembled folate-phytosterol-alginate nanoparticles	Wang et al. (2015)
Hepatocellular carcinoma cells	7703	Doxorubicin	Alginate nanoparticles	Zhang et al. (2009)

Tissue	Tested Cell Types/Organisms	Drug	Nanoparticle	Reference
Human hepatocarcinoma cells	HepG2	Doxorubicin	Glycyrrhetinic acid–modified alginate nanoparticles	Zhang et al. (2012)
Human hepatocarcinoma	HepG2, Kunming mice in vivo	Doxorubicin	Liver-targeted glycyrrhetinic acid–modified alginate nanoparticles	Guo et al. (2013)
Human cervical cancer cells	HeLa	Doxorubicin	Alginate/CaCO3/antitumor Gene hybrid nanoparticles	Zhao et al. (2012)
Human breast adenocarcinoma, Human cervical cancer cells	MCF-7, HeLa	Tamoxifen	Thiolated alginate-albumin nanoparticles	Martínez et al. (2012)

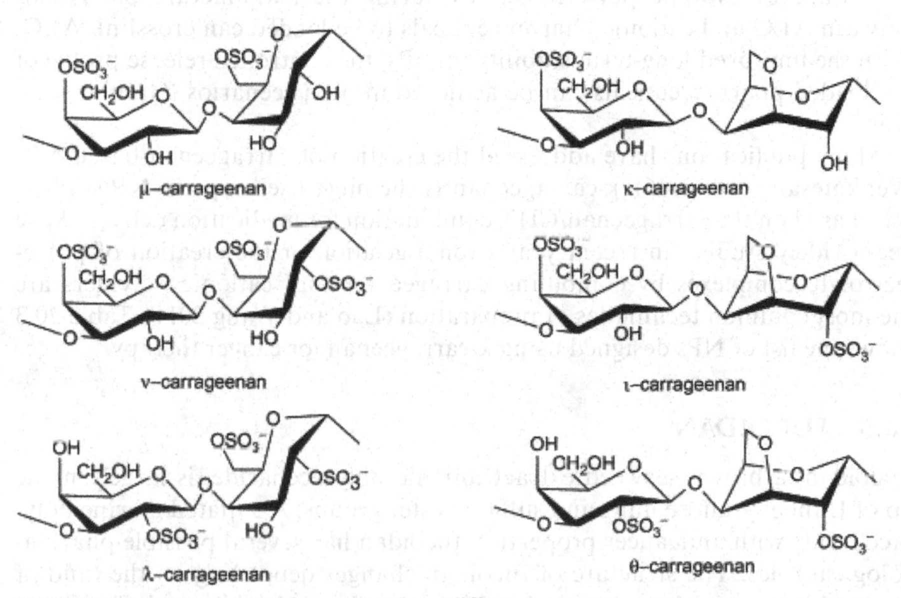

FIGURE 20.3 Structural formulas of six different carrageenans.

fast release because of fast degeneration or quick swelling (Ghanam and Kleinebudde 2011). Blending different types of carrageenan is also accepted as a good strategy for prolonging drug release (Nanaki et al. 2010). Another basic and important characteristic of carrageenan polymers is their ability to absorb water, which enhances drug solubility and hence raises the oral bioavailability of weakly water-soluble medicines (Peptu et al. 2014).

Chemical changes can be used to change the physicochemical properties of carrageenans. When subjected to mechanical deformation or aging, aqueous–carrageenan gels may display undesired syneresis. Gels made from hydroxyalkyl–carrageenan derivatives have less syneresis, allowing polysaccharides to be used more widely (Campo et al. 2009). Alkali modifications are one of the most well-known chemical reactions for carrageenans. At 80°C, lambda and kappa carrageenans undergo cyclization in the presence of 1 M sodium hydroxide. In carrageenans, the formation of 3,6-anhydro-D-galactose units from -D-galactose 6-sulfate occurs 20–60 times faster than in the family (Hosseinzadeh et al. 2005; Pourjavadi et al. 2004).

Because of their biocompatibility, high molecular weight, and gelling ability, carrageenans have lately gained popularity in medical, pharmaceutical, and biotechnological applications. In the pharmaceutical sector, carrageenan is commonly used as a raw material in the production of DDSs and cell capsules for cell treatments.

20.4.1 Carrageenan NPs for Cancer Treatment

The usage of cationic polyelectrolytes during the manufacture of NPs or between ALG and cationic pharmaceuticals to be loaded can crosslink ALG. With the improved long-term stability of NPs, the controlled release profile of the loaded pharmaceuticals can be achieved in both scenarios (Grenha et al. 2010).

Many publications have addressed the creation of carrageenan-based NPs (Venkatesan et al. 2016). κ-carrageenan is the most used type in NP studies. NPs based on the carrageenan/CHT combination for medication delivery have been widely studied in recent years. Ionic gelation or the creation of polyelectrolyte complexes by combining carrageenan with cationic polymers are the most common techniques of preparation (Luo and Wang 2014). Table 20.3 shows the list of NPs designed using κ-carrageenan for cancer therapy.

20.5 FUCOIDAN

Fucoidan, a brown seaweed extract anionic polysaccharide, is largely made up of L-fucopyranose units and sulfated ester groups. A sulfated marine polysaccharide with anticancer properties, fucoidan has several possible pharmacological roles. The structure of fucoidan changes depending on the kind of seaweed. However, there are two kinds of homofucose backbones in fucoidan. The first has repeated (1–3)-l-fucopyranose, whereas the second type has alternating and repeated (1–3)- and (1–4)-l-fucopyranose (Li et al. 2008).

To make modified fucoidans, scientists used chemical (e.g., hydrolysis and desulfation) and enzymatic (e.g., fucoidanases enzymes) methods. Chemical changes of fucoidan have been shown to allow the creation of novel agents with potential medical applications. Furthermore, the chemical conjugation of fucoidan with reactive functional groups, such as carboxylic, sulfate, and hydroxyl groups, has been shown to improve the absorption and solvation

TABLE 20.3
Overview of Nanoparticles with Carrageenan in Cancer Therapy

Tissue	Tested Cell Types/Organisms	Drug	Nanoparticle	Reference
Human colon, hepatocarcinoma	HCT-116, HepG2	Epirubicin	Carrageenan oligosaccharide gold nanoparticles	Chen et al. (2019)
Human colon carcinoma	HCT-116		Combination of iota carrageenan and a magnetic nanoparticle	Raman et al. (2015)
-	-	Sunitinib	Magnetic chitosan nanoparticles crosslinked with κ-carrageenan	Karimi et al. (2018)
Human breast adenocarcinoma cells	MCF-7	α-mangostin	Chitosan-κ-carrageenan-loaded nanoparticle	Wathoni et al. (2021)
Human colon, Human breast cancer cells	HCT-116, MDA-MB-231	–	Carrageenan oligosaccharide gold nanoparticles	Chen et al. (2018)
Mouse fibrosarcoma	WEHI 164	–	Cerium oxide nanoparticles using carrageenan	Nourmohammadi et al. (2018)

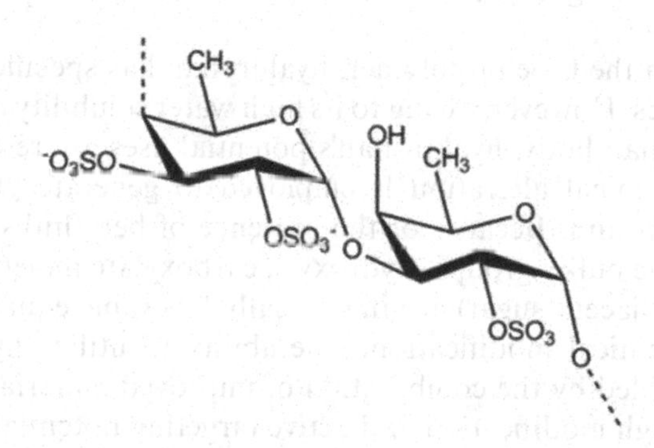

FIGURE 20.4 Structural formula of fucoidan.

characteristics of various medicines. According to recent research, natural polysaccharide-based materials can help with drug delivery, therapeutic methods, and tissue engineering. Sulfated polysaccharides including fucoidan, carrageenan, ALG, and CHT have better pharmacological qualities, such as biocompatibility, biodegradability, and bioactivity, and they have been studied extensively for their efficacy against cancer (Lu et al. 2017).

Fucoidans are also thought to be powerful cancer fighters. This is accomplished by controlling cancer cell apoptosis. The antitumor action of fucoidans is attributed to the ratio of monosaccharides, the number of sulfate groups, and the binding of sugar residues. Concentration, molecular weight, sulfate content, the amount of polymerization, the presence of substitution groups, the location and type of sugar, and glycosylation branching determine the antioxidant activity of fucoidans (Atashrazm et al. 2015).

20.5.1 Fucoidan NPs for Cancer Treatment

Natural products as antitumor medicines have attracted growing interest from patients due to their biological activity and lesser side effects when compared to synthetic drugs. As a natural compound, fucoidan has a potential for anticancer treatment that merits more research. Besides, fucoidan has been shown to have antitumoral properties in a range of human malignancies in numerous studies. Fucoidan can be used as an adjuvant in combination with chemotherapy to strengthen the immune system while minimizing side effects. As a result, nanosystems or NPs have been developed to enhance the bioavailability of fucoidan.

Table 20.4 shows studies using fucoidan nanoparticles in cancer treatment.

20.6 HYALURONAN

Hyaluronan is a linear polysaccharide consisting of repeating disaccharide units of N-acetyl-D-glucosamine and D-glucuronate (Tiwari and Bahadur 2019).

Depending on the type of polymer, hyaluronan has specific biotechnological characteristics. However, owing to its high water solubility and rapid breakdown in the human body, hyaluronan's potential uses are restricted (Schanté et al. 2011). Chemical alteration is employed to generate physico-chemical stabilities for this aim. Because of the presence of beta links in the hyaluronan structure, the bulky groups (hydroxyls, carboxylate moiety, and anomeric carbon on the adjacent sugar) are in sterically favorable equatorial locations, allowing for chemical modification. The ability to utilize hyaluronan more efficiently is enabled by the combination of improved material characteristics integrated through modification and active targeting potential of hyaluronan. Researchers have investigated the hydroxyl group (C-3) in GlcNAc as an alternate location for alteration, based on findings that carboxyl groups function as receptor recognition sites (Schanté et al. 2011; Tian et al. 2010). When the direct attachment of hydrophobic molecules in aqueous environments is not possible, modification can be used (Lee et al. 2008). Other methods include partial functional group protection using hydrophobic blocking agents and anionic group shielding by complexation with cationic moieties. Following conjugation,

TABLE 20.4
Nanoparticles Targeting Cancer Therapy with Fucoidan

Tissue	Tested Cell Types/ Organisms	Drug	Nanoparticle	Reference
Human breast cancer cells, 4T1 mouse breast cells	MDA-MB-231 in vitro, 4T1 induced tumor-bearing BALB/c mice in vivo	Doxorubicin	Fucoidan nanoparticles	Pawar et al. (2019)
Squamous carcinoma cells, Eye tumors	VX2 cells in vitro, xenograft tumors inoculation of VX2 cells to rabbits' eyes	Doxorubicin	Gold nanoparticles (AuNPs) coated with DOX-loaded fucoidan	Kim et al. (2017)
Human breast cancer cells	MDA-MB-231	Gemcitabine	Fucoidan/chitosan nanoparticles	Oliveira et al. (2018)
Colon carcinoma cells	HCT-8	Cisplatin	Fucoidan-cisplatin nanoparticles	Hwang et al. (2017)
Human cervical and human lung adenocarcinoma	HeLa, A549, and K562	-	Fucoidan-coated copper sulfide nanoparticles	Jang et al. (2018)
Human breast cancer cells	MDA-MB-231	Doxorubicin	Protamine/ fucoidan-based nanoparticles	Lu et al. (2017)
Human osteosarcoma cells, human colon adenocarcinoma cells, Mouse osteosarcoma cells	143B, Caco-2, LM8	-	Fucoidan lipid nanoparticles	Kimura et al. (2013)

FIGURE 20.5 Structure of hyaluronan.

blocking groups are cleaved in the first method. Carboxyl groups are used for amidation and esterification, whereas hydroxyl groups are responsible for the formation of ester and ether bonds. Dialdehyde groups formed during the periodate treatment of hyaluronan can be used for reductive amination and chemical modification (Collins and Birkinshaw 2013).

20.6.1 Hyaluronan NPs for Cancer Treatment

Biocompatible, anti-inflammatory, biodegradable, nonimmunogenic, and nontoxic characteristics of hyaluronan (HA) have boosted its use in a variety of sectors. Considerable research has been done to investigate the therapeutic and potential applications of hyaluronans, and they have been used in various biomedical and pharmaceutical applications. Because of the physicochemical and biological characteristics of HA, it can be used in drug delivery systems to treat cancer.

Another distinguishing feature of HA is its ability to bind with particular cancer cell receptors, such as CD44, and activate macrophages. It plays an essential role in cancer diagnostic and therapy research because of this characteristic. NPs can also be formulated into NPs for cancer treatment. As macromolecular prodrugs, HA-drug conjugates have been created for cancer treatment. For usage as a targeting moiety, HA has also been conjugated onto different drug-loaded NPs (Lapčík et al. 1998). When compared to free anticancer drugs, these HA conjugates containing anticancer agents such as paclitaxel, doxorubicin butyric acid, mitomycin C, and siRNA had better tumor targeting and therapeutic effectiveness (Choi et al. 2010). Studies with HA are more limited compared to other marine components. Park et al. created bioinspired monodisperse magnetite nanocrystals (HA-DN/MNC) for targeted cancer imaging with HA. They immobilized HA onto iron oxide nanocrystals to target cancer cells with CD44 overexpression. Via CD44-HA receptor–ligand interaction and ensure the detection of cancer cells through CD44-HA receptor–ligand interactions (Lee et al. 2008).

Xu et al. created a novel hyaluronidase-incorporated-hyaluronan-tyramine hydrogel for the delivery of trastuzumab. Adjustable antibody release, as well as trastuzumab-induced cell viability suppression and hydrogel disintegration regulated by hyaluronidase concentration, were discovered in release tests in vitro (Xu et al. 2015).

El-Dakdouki et al. designed HA-coated superparamagnetic iron oxide NPs (HA-SPION), with doxorubicin attached to the NPs' surface through an acid-sensitive hydrazone linkage. While HA-SPION was not toxic to cells, designed NPs were far more effective than free doxorubicin at killing drug-sensitive and multidrug-resistant cancer cells (El-Dakdouki et al. 2012).

20.7 PERSPECTIVES AND FUTURE DIRECTIONS

The immense potential of marine natural products is demonstrated by recent developments in the discovery, approval, and therapeutic use of marine pharmaceuticals. Because they are biocompatible and nontoxic, marine-derived components are frequently used to make DDSs, especially NPs. Chemical, enzymatic, and physical techniques can all be used to extract and modify these components. Chemical, enzymatic, and physical techniques may readily modify these natural components to create hydrogels, particles, and capsules that can store various bioactive chemicals. Researchers have developed several NPs that regulate the release of bioactive substances utilizing stimuli-sensitive polymers to generate a fast or sustained release. The utilization of marine biopolymers in the delivery of nanoparticle therapies in cancer was the subject of this chapter.

In-depth pharmacokinetic characterization can provide information that can help us better understand the molecular basis of pharmacological action, choose the right dosages and treatment plans, and use drugs more effectively. As a result, pharmacokinetics study on marine-derived chemicals is likely to grow in the near future.

REFERENCES

Ali, Ehab M. M., Aya A. Elashkar, Hala Y. El-Kassas, and Elsayed I. Salim. 2018. "Methotrexate Loaded on Magnetite Iron Nanoparticles Coated with Chitosan: Biosynthesis, Characterization, and Impact on Human Breast Cancer MCF-7 Cell Line." *International Journal of Biological Macromolecules* 120 (December): 1170–1180. doi:10.1016/j.ijbiomac.2018.08.118.

Al-Musawi, Sharafaldin, Ameer Jawad Hadi, Suhad Jameel Hadi, and Nada Khazal Kadhim Hindi. 2019. "Preparation and Characterization of Folated Chitosan/Magnetic Nanocarrier for 5-Fluorouracil Drug Delivery and Studying Its Effect in Bladder Cancer Therapy." *Journal of Global Pharma Technology* 11 (7): 628–637.

Andel, Lotte van, Hilde Rosing, Jan H. M. Schellens, and Jos H. Beijnen. 2018. "Review of Chromatographic Bioanalytical Assays for the Quantitative Determination of Marine-Derived Drugs for Cancer Treatment." *Marine Drugs*. doi:10.3390/md16070246.

Arafa, Kholoud, Rehab N. Shamma, Omaima N. El-Gazayerly, and Ibrahim M. El-Sherbiny. 2018. "Facile Development, Characterization, and Optimization of New Metformin-Loaded Nanocarrier System for Efficient Colon Cancer Adjunct Therapy." *Drug Development and Industrial Pharmacy*. doi:10.1080/03639045.2018.1438463.

Aruna, U., R. Rajalakshmi, Y. Indira Muzib, V. Vinesha, M. Sushma, K. R. Vandana, N. Vijay Kumar, and A. Rangampet. 2013. "Role of Chitosan Nanoparticles in Cancer Therapy." *International Journal of Innovative Pharmaceutical Research* 4 (3): 318–324.

Atashrazm, Farzaneh, Ray M. Lowenthal, Gregory M. Woods, Adele F. Holloway, and Joanne L. Dickinson. 2015. "Fucoidan and Cancer: A Multifunctional Molecule with Anti-Tumor Potential." *Marine Drugs*. doi:10.3390/md13042327.

Babu A., Templeton A.K., Munshi A., and Ramesh R. 2014. "Nanodrug Delivery Systems: A Promising Technology for Detection, Diagnosis, and Treatment of Cancer." *An Official Journal of the American Association of Pharmaceutical Scientists*. doi:10.1208/s12249-014-0089-8.

Badran, Mohammad M., Abdullah H. Alomrani, Gamaleldin I. Harisa, Abdelkader E. Ashour, Ashok Kumar, and Alaa Eldeen Yassin. 2018. "Novel Docetaxel

Chitosan-Coated PLGA/PCL Nanoparticles with Magnified Cytotoxicity and Bioavailability." *Biomedicine and Pharmacotherapy*. doi:10.1016/j.biopha.2018.07.102.

Balan, Poornima, Janani Indrakumar, Padmaja Murali, and Purna Sai Korrapati. 2020. "Bi-Faceted Delivery of Phytochemicals through Chitosan Nanoparticles Impregnated Nanofibers for Cancer Therapeutics." *International Journal of Biological Macromolecules* 142. Elsevier B.V.: 201–211. doi:10.1016/j.ijbiomac.2019.09.093.

Bhardwaj, Vinay, and Roozbeh Nikkhah-Moshaie. 2017. "Nanomedicine." In *Advances in Personalized Nanotherapeutics*, 1–10. Cham: Springer International Publishing. doi:10.1007/978-3-319-63633-7_1.

Bhunchu, S., C. Muangnoi, P. Rojsitthisak, and P. Rojsitthisak. 2016. "Curcumin Diethyl Disuccinate Encapsulated in Chitosan/Alginate Nanoparticles for Improvement of Its in Vitro Cytotoxicity against MDA-MB-231 Human Breast Cancer Cells." *Pharmazie*. doi:10.1691/ph.2016.6105.

Brannon-Peppas, Lisa, and James O. Blanchette. 2012. "Nanoparticle and Targeted Systems for Cancer Therapy." *Advanced Drug Delivery Reviews* 64 (December): 206–212. doi:10.1016/j.addr.2012.09.033.

Campo, Vanessa Leiria, Daniel Fábio Kawano, Dílson Braz da Silva, and Ivone Carvalho. 2009. "Carrageenans: Biological Properties, Chemical Modifications and Structural Analysis—A Review." *Carbohydrate Polymers* 77 (2). Elsevier Ltd: 167–180. doi:10.1016/j.carbpol.2009.01.020.

Carroll, Anthony R., Brent R. Copp, Rohan A. Davis, Robert A. Keyzers, and Michèle R. Prinsep. 2021. "Marine Natural Products." *Natural Product Reports*. doi:10.1039/d0np00089b.

Champagne, C. P., N. Blahuta, F. Brion, and C. Gagnon. 2000. "A Vortex-Bowl Disk Atomizer System for the Production of Alginate Beads in a 1500-Liter Fermentor." *Biotechnology and Bioengineering*. doi:10.1002/(SICI)1097-0290(20000620)68:6<681::AID-BIT12>3.0.CO;2-L.

Chen, Guanyu, Darren Svirskis, Weiyue Lu, Man Ying, Yuan Huang, and Jingyuan Wen. 2018. "N-Trimethyl Chitosan Nanoparticles and CSKSSDYQC Peptide: N-Trimethyl Chitosan Conjugates Enhance the Oral Bioavailability of Gemcitabine to Treat Breast Cancer." *Journal of Controlled Release*. doi:10.1016/j.jconrel.2018.03.013.

Chen, Xiangyan, Wenwei Han, Xia Zhao, Wei Tang, and Fahe Wang. 2019. "Epirubicin-Loaded Marine Carrageenan Oligosaccharide Capped Gold Nanoparticle System for PH-Triggered Anticancer Drug Release." *Scientific Reports* 9 (1). Springer US: 1–10. doi:10.1038/s41598-019-43106-9.

Chen, Xiangyan, Xia Zhao, Yanyun Gao, Jiaqi Yin, Mingyue Bai, and Fahe Wang. 2018. "Green Synthesis of Gold Nanoparticles Using Carrageenan Oligosaccharide and Their in Vitro Antitumor Activity." *Marine Drugs*. doi:10.3390/md16080277.

Chiu, Hock Ing, and Vuanghao Lim. 2021. "Wheat Germ Agglutinin-Conjugated Disulfide Cross-Linked Alginate Nanoparticles as a Docetaxel Carrier for Colon Cancer Therapy." *International Journal of Nanomedicine* 16: 2995–3020. doi:10.2147/IJN.S302238.

Choi, Ki Young, Hyunjin Chung, Kyung Hyun Min, Hong Yeol Yoon, Kwangmeyung Kim, Jae Hyung Park, Ick Chan Kwon, and Seo Young Jeong. 2010. "Self-Assembled Hyaluronic Acid Nanoparticles for Active Tumor Targeting." *Biomaterials* 31 (1): 106–114. doi:10.1016/j.biomaterials.2009.09.030.

Clarke, Christina A., Earl Hubbell, Allison W. Kurian, Graham A. Colditz, Anne-Renee Hartman, and Scarlett Lin Gomez. 2020. "Projected Reductions in Absolute Cancer—Related Deaths from Diagnosing Cancers Before Metastasis, 2006–2015." *Cancer Epidemiology Biomarkers & Prevention* 29 (5): 895–902. doi:10.1158/1055-9965.EPI-19-1366.

Collins, Maurice N., and Colin Birkinshaw. 2013. "Hyaluronic Acid Solutions—A Processing Method for Efficient Chemical Modification." *Journal of Applied Polymer Science*. doi:10.1002/app.39145.

Cui, Fuying, Feng Qian, Ziming Zhao, Lichen Yin, Cui Tang, and Chunhua Yin. 2009. "Preparation, Characterization, and Oral Delivery of Insulin Loaded Carboxylated Chitosan Grafted Poly(Methyl Methacrylate) Nanoparticles." *Biomacromolecules*. doi:10.1021/bm900035u.

Dambies, L., T. Vincent, A. Domard, and E. Guibal. 2001. "Preparation of Chitosan Gel Beads by Ionotropic Molybdate Gelation." *Biomacromolecules*. doi:10.1021/bm010083r.

Das, Ratul Kumar, Naresh Kasoju, and Utpal Bora. 2010. "Encapsulation of Curcumin in Alginate-Chitosan-Pluronic Composite Nanoparticles for Delivery to Cancer Cells." *Nanomedicine: Nanotechnology, Biology, and Medicine*. doi:10.1016/j.nano.2009.05.009.

D'Ayala, Giovanna Gomez, Mario Malinconico, and Paola Laurienzo. 2008. "Marine Derived Polysaccharides for Biomedical Applications: Chemical Modification Approaches." *Molecules* 13 (9): 2069–2106. doi:10.3390/molecules13092069.

Deb, Ananya, Nirmala Grace Andrews, and Vimala Raghavan. 2018. "Natural Polymer Functionalized Graphene Oxide for Co-Delivery of Anticancer Drugs: In-Vitro and in-Vivo." *International Journal of Biological Macromolecules*. doi:10.1016/j.ijbiomac.2018.02.153.

Ding, Xiao, Wenjie Yu, Yunfeng Wan, Mingyue Yang, Chenghuan Hua, Na Peng, and Yi Liu. 2020. "A PH/ROS-Responsive, Tumor-Targeted Drug Delivery System Based on Carboxymethyl Chitin Gated Hollow Mesoporous Silica Nanoparticles for Anti-Tumor Chemotherapy." *Carbohydrate Polymers* 245 (May). Elsevier: 116493. doi:10.1016/j.carbpol.2020.116493.

Divya, K., and M. S. Jisha. 2018. "Chitosan Nanoparticles Preparation and Applications." *Environmental Chemistry Letters*. doi:10.1007/s10311-017-0670-y.

Ehrlich, Hermann, Manuel Maldonado, Klaus-dieter Spindler, Carsten Eckert, Thomas Hanke, René Born, Caren Goebel, Paul Simon, Sascha Heinemann, and Hartmut Worch. 2007. "First Evidence of Chitin as a Component of the Skeletal Fibers of Marine Sponges. Part I. Verongidae (Demospongia: Porifera)." *Journal of Experimental Zoology Part B: Molecular and Developmental Evolution* 308B (4): 347–356. doi:10.1002/jez.b.21156.

Elbialy, Nihal Saad, and Noha Mohamed. 2020. "Alginate-Coated Caseinate Nanoparticles for Doxorubicin Delivery: Preparation, Characterisation, and in Vivo Assessment." *International Journal of Biological Macromolecules* 154. Elsevier B.V.: 114–122. doi:10.1016/j.ijbiomac.2020.03.027.

El-Dakdouki, Mohammad H., David C. Zhu, Kheireddine El-Boubbou, Medha Kamat, Jianjun Chen, Wei Li, and Xuefei Huang. 2012. "Development of Multifunctional Hyaluronan-Coated Nanoparticles for Imaging and Drug Delivery to Cancer Cells." *Biomacromolecules*. doi:10.1021/bm300046h.

Feiz, Mohadeseh S., and Azadeh Meshkini. 2019. "Targeted Delivery of Adenosine 5′-Triphosphate Using Chitosan-Coated Mesoporous Hydroxyapatite: A Theranostic PH-Sensitive Nanoplatform with Enhanced Anti-Cancer Effect." *International Journal of Biological Macromolecules*. doi:10.1016/j.ijbiomac.2018.08.158.

Frank, L. A., G. R. Onzi, A. S. Morawski, A. R. Pohlmann, S. S. Guterres, and R. V. Contri. 2020. "Chitosan as a Coating Material for Nanoparticles Intended for Biomedical Applications." *Reactive and Functional Polymers*. doi:10.1016/j.reactfunctpolym.2019.104459.

Gao, Cheng, Fan Tang, Jianxiang Zhang, Simon M. Y. Lee, and Ruibing Wang. 2017. "Glutathione-Responsive Nanoparticles Based on a Sodium Alginate Derivative for Selective Release of Doxorubicin in Tumor Cells." *Journal of Materials Chemistry B* 5 (12): 2337–2346. doi:10.1039/C6TB03032G.

Ghanam, Dima, and Peter Kleinebudde. 2011. "Suitability of κ-Carrageenan Pellets for the Formulation of Multiparticulate Tablets with Modified Release." *International Journal of Pharmaceutics*. doi:10.1016/j.ijpharm.2011.02.016.

Grenha, Ana, Manuela E. Gomes, Márcia Rodrigues, Vítor E. Santo, João F. Mano, Nuno M. Neves, and Rui L. Reis. 2010. "Development of New Chitosan/Carrageenan Nanoparticles for Drug Delivery Applications." *Journal of Biomedical Materials Research—Part A*. doi:10.1002/jbm.a.32466.

Gunel Nur, Selvi, Ozel Buket, Kipcak Sezgi, Aktan Cagdas, Akgun Cansu, Ak Guliz, Yilmaz Habibe, Biray Avci Cigir, Dodurga Yavuz, and Hamarat Sanlier Senay. 2016. "Synthesis of Methotrexate Loaded Chitosan Nanoparticles and in Vitro Evaluation of the Potential in Treatment of Prostate Cancer." *Anti-Cancer Agents in Medicinal Chemistry*. doi:10.2174/1871520616666160101120040.

Guo, Hua, Quanyong Lai, Wei Wang, Yukun Wu, Chuangnian Zhang, Yuan Liu, and Zhi Yuan. 2013. "Functional Alginate Nanoparticles for Efficient Intracellular Release of Doxorubicin and Hepatoma Carcinoma Cell Targeting Therapy." *International Journal of Pharmaceutics*. doi:10.1016/j.ijpharm.2013.04.025.

Guo, Xueling, Qianfen Zhuang, Tianjiao Ji, Yinlong Zhang, Changjian Li, Yueqi Wang, Hong Li, Hongying Jia, Yang Liu, and Libo Du. 2018. "Multi-Functionalized Chitosan Nanoparticles for Enhanced Chemotherapy in Lung Cancer." *Carbohydrate Polymers*. doi:10.1016/j.carbpol.2018.04.087.

Hamidi, Mehrdad, Amir Azadi, and Pedram Rafiei. 2008. "Hydrogel Nanoparticles in Drug Delivery." *Advanced Drug Delivery Reviews*. doi:10.1016/j.addr.2008.08.002.

Haug, Arne, Kerstin Claeson, Svend E. Hansen, R. Sömme, Einar Stenhagen, and H. Palmstierna. 1959. "Fractionation of Alginic Acid." *Acta Chemica Scandinavica* 13: 601–603. doi:10.3891/acta.chem.scand.13-0601.

Hecht, Hadas, and Simcha Srebnik. 2016. "Structural Characterization of Sodium Alginate and Calcium Alginate." *Biomacromolecules*. doi:10.1021/acs.biomac.6b00378.

Hosseinzadeh, H., A. Pourjavadi, G. R. Mahdavinia, and M. J. Zohuriaan-Mehr. 2005. "Modified Carrageenan. 1. H-CarragPAM, a Novel Biopolymer-Based Superabsorbent Hydrogel." *Journal of Bioactive and Compatible Polymers* 20 (5): 475–490. doi:10.1177/0883911505055164.

Hu, Yiwen, Jiahui Chen, Guping Hu, Jianchen Yu, Xun Zhu, Yongcheng Lin, Shengping Chen, and Jie Yuan. 2015. "Statistical Research on the Bioactivity of New Marine Natural Products Discovered during the 28 Years from 1985 to 2012." *Marine Drugs*. doi:10.3390/md13010202.

Huang, C-H., T-J. Chuang, C-J. Ke, and C-H. Yao. 2020. "Doxorubicin—Gelatin/Fe 3 O 4—Alginate Dual-Layer Magnetic Nanoparticles as Targeted Anticancer Drug Delivery Vehicles." *Polymers* 12 (8):1747: 1–15.

Hwang, Pai An, Xiao Zhen Lin, Ko Liang Kuo, and Fu Yin Hsu. 2017. "Fabrication and Cytotoxicity of Fucoidan-Cisplatin Nanoparticles for Macrophage and Tumor Cells." *Materials*. doi:10.3390/ma10030291.

Imran, Muhammad, Abdur Rauf, Imtiaz Ali Khan, Muhammad Shahbaz, Tahira Batool Qaisrani, Sri Fatmawati, Tareq Abu-Izneid, Ali Imran, Khaliq Ur Rahman, and Tanweer Aslam Gondal. 2018. "Thymoquinone: A Novel Strategy to Combat Cancer: A Review." *Biomedicine and Pharmacotherapy* 106 (April): 390–402. doi:10.1016/j.biopha.2018.06.159.

Irani, Mohammad, Gity Mir Mohamad Sadeghi, and Ismaeil Haririan. 2017. "A Novel Biocompatible Drug Delivery System of Chitosan/Temozolomide Nanoparticles Loaded PCL-PU Nanofibers for Sustained Delivery of Temozolomide." *International Journal of Biological Macromolecules* 97. Elsevier B.V.: 744–751. doi:10.1016/j.ijbiomac.2017.01.073.

Jang, Bian, Madhappan Santha Moorthy, Panchanathan Manivasagan, Li Xu, Kyeongeun Song, Kang Dae Lee, Minseok Kwak, Junghwan Oh, and Jun O. Jin. 2018. "Fucoidan-Coated CuS Nanoparticles for Chemo-and Photothermal Therapy against Cancer." *Oncotarget*. doi:10.18632/oncotarget.23898.

Kansom, Teeratas, Warayuth Sajomsang, Rungnapha Saeeng, Purin Charoensuksai, Praneet Opanasopit, and Prasopchai Tonglairoum. 2018. "Apoptosis Induction and Antimigratory Activity of Andrographolide Analog (3A.1)-Incorporated Self-Assembled Nanoparticles in Cancer Cells." *AAPS PharmSciTech* 19 (7): 3123–3133. doi:10.1208/s12249-018-1139-4.

Karimi, Mohammad Hasan, Gholam Reza Mahdavinia, and Bakhshali Massoumi. 2018. "PH-Controlled Sunitinib Anticancer Release from Magnetic Chitosan Nanoparticles Crosslinked with κ-Carrageenan." *Materials Science and Engineering C* 91 (May). Elsevier: 705–714. doi:10.1016/j.msec.2018.06.019.

Katuwavila, Nuwanthi P., A. D. L. Chandani Perera, Sameera R. Samarakoon, Preethi Soysa, Veranja Karunaratne, Gehan A. J. Amaratunga, and D. Nedra Karunaratne. 2016. "Chitosan-Alginate Nanoparticle System Efficiently Delivers Doxorubicin to MCF-7 Cells." *Journal of Nanomaterials*. doi:10.1155/2016/3178904.

Khan, Abida Kalsoom, Ain Us Saba, Shamyla Nawazish, Fahad Akhtar, Rehana Rashid, Sadullah Mir, Bushra Nasir, et al. 2017. "Carrageenan Based Bionanocomposites as Drug Delivery Tool with Special Emphasis on the Influence of Ferromagnetic Nanoparticles." *Oxidative Medicine and Cellular Longevity*. doi:10.1155/2017/8158315.

Khan, Ibrahim, Khalid Saeed, and Idrees Khan. 2019. "Nanoparticles: Properties, Applications and Toxicities." *Arabian Journal of Chemistry*. doi:10.1016/j.arabjc.2017.05.011.

Khan, Muhammad Muzamil, Asadullah Madni, Vladimir Torchilin, Nina Filipczak, Jiayi Pan, Nayab Tahir, and Hassan Shah. 2019. "Lipid-Chitosan Hybrid Nanoparticles for Controlled Delivery of Cisplatin." *Drug Delivery* 26 (1). Taylor & Francis: 765–772. doi:10.1080/10717544.2019.1642420.

Khdair, Ayman, Islam Hamad, Hatim Alkhatib, Yasser Bustanji, Mohammad Mohammad, Rabab Tayem, and Khaled Aiedeh. 2016. "Modified-Chitosan Nanoparticles: Novel Drug Delivery Systems Improve Oral Bioavailability of Doxorubicin." *European Journal of Pharmaceutical Sciences* 93. Elsevier B.V.: 38–44. doi:10.1016/j.ejps.2016.07.012.

Kim, Hyejin, Van Phuc Nguyen, Panchanathan Manivasagan, Min Jung Jung, Sung Won Kim, Junghwan Oh, and Hyun Wook Kang. 2017. "Doxorubicin-Fucoidan-Gold Nanoparticles Composite for Dual-Chemo-Photothermal Treatment on Eye Tumors." *Oncotarget* 8 (69): 113719–113733. doi:10.18632/oncotarget.23092.

Kimura, Ryuichiro, Takayoshi Rokkaku, Shinji Takeda, Masachika Senba, and Naoki Mori. 2013. "Cytotoxic Effects of Fucoidan Nanoparticles against Osteosarcoma." *Marine Drugs*. doi:10.3390/md11114267.

Kumar Mehata, Abhishesh, Shreekant Bharti, Priya Singh, Matte Kasi Viswanadh, Lakshmi Kumari, Poornima Agrawal, Sanjay Singh, Biplob Koch, and Madaswamy S. Muthu. 2019. "Trastuzumab Decorated TPGS-g-Chitosan Nanoparticles for Targeted Breast Cancer Therapy." *Colloids and Surfaces B: Biointerfaces*. doi:10.1016/j.colsurfb.2018.10.007.

Kunjiappan, Selvaraj, Theivendren Panneerselvam, Balasubramanian Somasundaram, Murugesan Sankaranarayanan, Pavadai Parasuraman, Shrinivas D. Joshi, Sankarganesh Arunachalam, and Indhumathy Murugan. 2018. "Design Graph Theoretical Analysis and In Silico Modeling of Dunaliella Bardawil Biomass Encapsulated N-Succinyl Chitosan Nanoparticles for Enhanced Anticancer Activity." *Anti-Cancer Agents in Medicinal Chemistry*. doi:10.2174/1871520618666180628155223.

Lapčík, Lubomír, Lubomír Lapčík, Stefaan De Smedt, Joseph Demeester, and Peter Chabreček. 1998. "Hyaluronan: Preparation, Structure, Properties, and Applications." *Chemical Reviews*. doi:10.1021/cr941199z.

Lee, Hyukjin, Kyuri Lee, and Gwan Park Tae. 2008. "Hyaluronic Acid-Paclitaxel Conjugate Micelles: Synthesis, Characterization, and Antitumor Activity." *Bioconjugate Chemistry*. doi:10.1021/bc8000485.

Lee, Yuhan, Haeshin Lee, Young Beom Kim, Jaeyoon Kim, Taeghwan Hyeon, Hyun-Wook Park, Phillip B. Messersmith, and Tae Gwan Park. 2008. "Bioinspired Surface Immobilization of Hyaluronic Acid on Monodisperse Magnetite Nanocrystals for Targeted Cancer Imaging." *Advanced Materials*, October, NA-NA. doi:10.1002/adma.200800756.

Li, Bo, Fei Lu, Xinjun Wei, and Ruixiang Zhao. 2008. "Fucoidan: Structure and Bioactivity." *Molecules* 13 (8): 1671–1695. doi:10.3390/molecules13081671.

Li, Jianghua, Chao Cai, Jiarui Li, Jun Li, Jia Li, Tiantian Sun, Lihao Wang, Haotian Wu, and Guangli Yu. 2018. "Chitosan-Based Nanomaterials for Drug Delivery." *Molecules* 23 (10): 2661. doi:10.3390/molecules23102661.

Li, Kaidi, Naixin Liang, Huaxia Yang, Hongsheng Liu, and Shanqing Li. 2017. "Temozolomide Encapsulated and Folic Acid Decorated Chitosan Nanoparticles for Lung Tumor Targeting: Improving Therapeutic Efficacy Both in Vitro and in Vivo." *Oncotarget*. doi:10.18632/oncotarget.22791.

Li, Tony Shing Chau, Toshio Yawata, and Koichi Honke. 2014. "Efficient SiRNA Delivery and Tumor Accumulation Mediated by Ionically Cross-Linked Folic Acid-Poly(Ethylene Glycol)-Chitosan Oligosaccharide Lactate Nanoparticles: For the Potential Targeted Ovarian Cancer Gene Therapy." *European Journal of Pharmaceutical Sciences* 52 (1). Elsevier B.V.: 48–61. doi:10.1016/j.ejps.2013.10.011.

Lu, Kun Ying, Rou Li, Chun Hua Hsu, Cheng Wei Lin, Shen Chieh Chou, Min Lang Tsai, and Fwu Long Mi. 2017. "Development of a New Type of Multifunctional Fucoidan-Based Nanoparticles for Anticancer Drug Delivery." *Carbohydrate Polymers*. doi:10.1016/j.carbpol.2017.02.065.

Luo, Yangchao, and Qin Wang. 2014. "Recent Development of Chitosan-Based Polyelectrolyte Complexes with Natural Polysaccharides for Drug Delivery." *International Journal of Biological Macromolecules*. doi:10.1016/j.ijbiomac.2013.12.017.

Maderuelo, Cristina, Aránzazu Zarzuelo, and José M. Lanao. 2011. "Critical Factors in the Release of Drugs from Sustained Release Hydrophilic Matrices." *Journal of Controlled Release*. doi:10.1016/j.jconrel.2011.04.002.

Maeda, Hiroshi. 2001. "The Enhanced Permeability and Retention (EPR) Effect in Tumor Vasculature: The Key Role of Tumor-Selective Macromolecular Drug Targeting." *Advances in Enzyme Regulation* 41 (1): 189–207. doi:10.1016/S0065-2571(00)00013-3.

Martínez, A., M. Benito-Miguel, I. Iglesias, J. M. Teijón, and M. D. Blanco. 2012. "Tamoxifen-Loaded Thiolated Alginate-Albumin Nanoparticles as Antitumoral Drug Delivery Systems." *Journal of Biomedical Materials Research Part A* 100A (6): 1467–1476. doi:10.1002/jbm.a.34051.

Martins, João Pedro, José das Neves, María de la Fuente, Christian Celia, Helena Florindo, Nazende Günday-Türeli, Amirali Popat, et al. 2020. "The Solid Progress of Nanomedicine." *Drug Delivery and Translational Research*. doi:10.1007/s13346-020-00743-2.

Mary Lazer, Lizha, Balaji Sadhasivam, Kanagaraj Palaniyandi, Thangavel Muthuswamy, Ilangovan Ramachandran, Anandan Balakrishnan, Surajit Pathak, Shoba Narayan, and Satish Ramalingam. 2018. "Chitosan-Based Nano-Formulation Enhances the Anticancer Efficacy of Hesperetin." *International Journal of Biological Macromolecules*. doi:10.1016/j.ijbiomac.2017.10.064.

Mathew, Manjusha Elizabeth, Jithin C. Mohan, K. Manzoor, S. V. Nair, H. Tamura, and R. Jayakumar. 2010. "Folate Conjugated Carboxymethyl Chitosan-Manganese Doped Zinc Sulphide Nanoparticles for Targeted Drug Delivery and Imaging of Cancer Cells." *Carbohydrate Polymers* 80 (2). Elsevier Ltd: 442–448. doi:10.1016/j.carbpol.2009.11.047.

Mohammed, Munawar A., Jaweria T. M. Syeda, Kishor M. Wasan, and Ellen K. Wasan. 2017. "An Overview of Chitosan Nanoparticles and Its Application in Non-Parenteral Drug Delivery." *Pharmaceutics*. doi:10.3390/pharmaceutics9040053.

Mora, Camilo, Derek P. Tittensor, Sina Adl, Alastair G. B. Simpson, and Boris Worm. 2011. "How Many Species Are There on Earth and in the Ocean?" *PLoS Biology*. doi:10.1371/journal.pbio.1001127.

Mu, Weiwei, Qihui Chu, Yongjun Liu, and Na Zhang. 2020. "A Review on Nano-Based Drug Delivery System for Cancer Chemoimmunotherapy." *Nano-Micro Letters*. doi:10.1007/s40820-020-00482-6.

Nanaki, Stavroula, Evangelos Karavas, Lida Kalantzi, and Dimitrios Bikiaris. 2010. "Miscibility Study of Carrageenan Blends and Evaluation of Their Effectiveness as Sustained Release Carriers." *Carbohydrate Polymers*. doi:10.1016/j.carbpol.2009.10.067.

Nourmohammadi, Esmail, Reza Kazemi Oskuee, Leila Hasanzadeh, Mohammad Mohajeri, Alireza Hashemzadeh, Majid Rezayi, and Majid Darroudi. 2018. "Cytotoxic Activity of Greener Synthesis of Cerium Oxide Nanoparticles Using Carrageenan towards a WEHI 164 Cancer Cell Line." *Ceramics International*. doi:10.1016/j.ceramint.2018.07.201.

Oliveira, Catarina, Ana C. Carvalho, Rui L. Reis, Nuno N. Neves, Albino Martins, and Tiago H. Silva. 2020. "Marine-Derived Biomaterials for Cancer Treatment." In *Biomaterials for 3D Tumor Modeling*, 551–576. Elsevier. doi:10.1016/B978-0-12-818128-7.00023-X.

Oliveira, Catarina, Nuno M. Neves, Rui L. Reis, Albino Martins, and Tiago H. Silva. 2018. "Gemcitabine Delivered by Fucoidan/Chitosan Nanoparticles Presents Increased Toxicity over Human Breast Cancer Cells." *Nanomedicine*. doi:10.2217/nnm-2018-0004.

Ou, Xianfeng, Jingqi Zheng, Xiujuan Zhao, and Mingxian Liu. 2018. "Chemically Cross-Linked Chitin Nanocrystal Scaffolds for Drug Delivery." *ACS Applied Nano Materials*. doi:10.1021/acsanm.8b01585.

Pacheco-Quito, Edisson Mauricio, Roberto Ruiz-Caro, and María Dolores Veiga. 2020. "Carrageenan: Drug Delivery Systems and Other Biomedical Applications." *Marine Drugs* 18 (11). doi:10.3390/md18110583.

Pawar, Vivek K., Yuvraj Singh, Komal Sharma, Arpita Shrivastav, Abhisheak Sharma, Akhilesh Singh, Jaya Gopal Meher, et al. 2019. "Improved Chemotherapy against Breast Cancer through Immunotherapeutic Activity of Fucoidan Decorated Electrostatically Assembled Nanoparticles Bearing Doxorubicin." *International Journal of Biological Macromolecules*. doi:10.1016/j.ijbiomac.2018.09.059.

Peptu, Catalina Anisoara, Lacramioara Ochiuz, Liana Alupei, Cristian Peptu, and Marcel Popa. 2014. "Carbohydrate Based Nanoparticles for Drug Delivery across Biological Barriers." *Journal of Biomedical Nanotechnology*. doi:10.1166/jbn.2014.1950.

Peptu, Cristian, Andra Cristina Humelnicu, Razvan Rotaru, Maria Emiliana Fortuna, Xenia Patras, Mirela Teodorescu, Bogdan Ionel Tamba, and Valeria Harabagiu.

2019. "Chitosan-Based Drug Delivery Systems." In *Chitin and Chitosan: Properties and Applications*. doi:10.1002/9781119450467.ch11.
Pérez-Herrero E., and Fernández-Medarde A. 2015. "Advanced Targeted Therapies in Cancer: Drug Nanocarriers, the Future of Chemotherapy." *European Journal of Pharmaceutics and Biopharmaceutics*. doi:10.1016/j.ejpb.2015.03.018.
Peters, W. 1972. "Occurrence of Chitin in Mollusca." *Comparative Biochemistry and Physiology—Part B: Biochemistry*. doi:10.1016/0305-0491(72)90117-4.
Potdar D., Pravin. 2016. "Chitosan Nanoparticles: An Emerging Weapon against the Cancer." *MOJ Cell Science & Report*. doi:10.15406/mojcsr.2016.03.00049.
Pourjavadi, A., A. M. Harzandi, and H. Hosseinzadeh. 2004. "Modified Carrageenan 3. Synthesis of a Novel Polysaccharide-Based Superabsorbent Hydrogel via Graft Copolymerization of Acrylic Acid onto Kappa-Carrageenan in Air." *European Polymer Journal* 40 (7): 1363–1370. doi:10.1016/j.eurpolymj.2004.02.016.
Qi, Li Feng, Zi Rong Xu, Yan Li, Xia Jiang, and Xin Yan Han. 2005. "In Vitro Effects of Chitosan Nanoparticles on Proliferation of Human Gastric Carcinoma Cell Line MGC803 Cells." *World Journal of Gastroenterology*. doi:10.3748/wjg.v11.i33.5136.
Qian, Qianqian, Shiwei Niu, Gareth R. Williams, Jianrong Wu, Xueyi Zhang, and Li Min Zhu. 2019. "Peptide Functionalized Dual-Responsive Chitosan Nanoparticles for Controlled Drug Delivery to Breast Cancer Cells." *Colloids and Surfaces A: Physicochemical and Engineering Aspects* 564 (October 2018). Elsevier: 122–130. doi:10.1016/j.colsurfa.2018.12.026.
Qin, Yimin. 2008. "Alginate Fibres: An Overview of the Production Processes and Applications in Wound Management." *Polymer International* 57 (2): 171–180. doi:10.1002/pi.2296.
Raman, Maya, Viswambari Devi, and Mukesh Doble. 2015. "Biocompatible I-Carrageenan-γ-Maghemite Nanocomposite for Biomedical Applications—Synthesis, Characterization and in Vitro Anticancer Efficacy." *Journal of Nanobiotechnology* 13 (1): 1–13. doi:10.1186/s12951-015-0079-3.
Ravi Kumar, M. N. 2000. "Nano and Microparticles as Controlled Drug Delivery Devices." *Journal of Pharmacy & Pharmaceutical Sciences : A Publication of the Canadian Society for Pharmaceutical Sciences, Societe Canadienne Des Sciences Pharmaceutiques* 3 (2): 234–258. http://www.ncbi.nlm.nih.gov/pubmed/10994037.
Rinaudo, Marguerite. 2008. "Main Properties and Current Applications of Some Polysaccharides as Biomaterials." *Polymer International*. doi:10.1002/pi.2378.
Rosière, Rémi, Matthias Van Woensel, Michel Gelbcke, Véronique Mathieu, Julien Hecq, Thomas Mathivet, Marjorie Vermeersch, Pierre Van Antwerpen, Karim Amighi, and Nathalie Wauthoz. 2018. "New Folate-Grafted Chitosan Derivative to Improve Delivery of Paclitaxel-Loaded Solid Lipid Nanoparticles for Lung Tumor Therapy by Inhalation." *Molecular Pharmaceutics*. doi:10.1021/acs.molpharmaceut.7b00846.
Schanté, Carole E., Guy Zuber, Corinne Herlin, and Thierry F. Vandamme. 2011. "Chemical Modifications of Hyaluronic Acid for the Synthesis of Derivatives for a Broad Range of Biomedical Applications." *Carbohydrate Polymers*. doi:10.1016/j.carbpol.2011.03.019.
Senapati S., Mahanta A.K., Kumar S., and Maiti P. 2018. "Controlled Drug Delivery Vehicles for Cancer Treatment and Their Performance." *Signal Transduction and Targeted Therapy*. doi:10.1038/s41392-017-0004-3.
Senga, Sasi S., and Richard P. Grose. 2021. "Hallmarks of Cancer—the New Testament." *Open Biology*. doi:10.1098/rsob.200358.
Severino, Patricia, Classius F. da Silva, Luciana N. Andrade, Daniele de Lima Oliveira, Joana Campos, and Eliana B. Souto. 2019. "Alginate Nanoparticles for Drug Delivery and Targeting." *Current Pharmaceutical Design*. doi:10.2174/1381612825666190425163424.

Shilpa, Anu, S. S. Agrawal, and Alok R. Ray. 2003. "Controlled Delivery of Drugs from Alginate Matrix." *Journal of Macromolecular Science—Polymer Reviews*. doi:10.1081/MC-120020160.

Shukla, Sudheesh K., Ajay K. Mishra, Omotayo A. Arotiba, and Bhekie B. Mamba. 2013. "Chitosan-Based Nanomaterials: A State-of-the-Art Review." *International Journal of Biological Macromolecules* 59 (August): 46–58. doi:10.1016/j.ijbiomac.2013.04.043.

Siafaka, Panoraia I., Alexandra Titopoulou, Emmanuel N. Koukaras, Margaritis Kostoglou, Efthimios Koutris, Evangelos Karavas, and Dimitrios N. Bikiaris. 2015. "Chitosan Derivatives as Effective Nanocarriers for Ocular Release of Timolol Drug." *International Journal of Pharmaceutics*. doi:10.1016/j.ijpharm.2015.08.100.

Siegel, Rebecca L., Kimberly D. Miller, Hannah E. Fuchs, and Ahmedin Jemal. 2021. "Cancer Statistics, 2021." *CA: A Cancer Journal for Clinicians*. doi:10.3322/caac.21654.

Silva, T. H., A. Alves, B. M. Ferreira, J. M. Oliveira, L. L. Reys, R. J. F. Ferreira, R. A. Sousa, S. S. Silva, J. F. Mano, and R. L. Reis. 2012. "Materials of Marine Origin: A Review on Polymers and Ceramics of Biomedical Interest." *International Materials Reviews*. doi:10.1179/1743280412Y.0000000002.

Skorik, Yury A., Anton A. Golyshev, Andreii S. Kritchenkov, Ekaterina R. Gasilova, Daria N. Poshina, Amal J. Sivaram, and Rangasamy Jayakumar. 2017. "Development of Drug Delivery Systems for Taxanes Using Ionic Gelation of Carboxyacyl Derivatives of Chitosan." *Carbohydrate Polymers* 162 (17). Elsevier Ltd.: 49–55. doi:10.1016/j.carbpol.2017.01.025.

Smitha, K. T., A. Anitha, T. Furuike, H. Tamura, Shantikumar V. Nair, and R. Jayakumar. 2013. "In Vitro Evaluation of Paclitaxel Loaded Amorphous Chitin Nanoparticles for Colon Cancer Drug Delivery." *Colloids and Surfaces B: Biointerfaces* 104. Elsevier B.V.: 245–253. doi:10.1016/j.colsurfb.2012.11.031.

Soares, Paula I. P., Ana Isabel Sousa, Jorge Carvalho Silva, Isabel M. M. Ferreira, Carlos M. M. Novo, and João Paulo Borges. 2016. "Chitosan-Based Nanoparticles as Drug Delivery Systems for Doxorubicin: Optimization and Modelling." *Carbohydrate Polymers* 147. Elsevier Ltd.: 304–312. doi:10.1016/j.carbpol.2016.03.028.

Solairaj, Dhanasekaran, Palanivel Rameshthangam, and Gnanapragasam Arunachalam. 2017. "Anticancer Activity of Silver and Copper Embedded Chitin Nanocomposites against Human Breast Cancer (MCF-7) Cells." *International Journal of Biological Macromolecules* 105. Elsevier B.V.: 608–619. doi:10.1016/j.ijbiomac.2017.07.078.

Sorasitthiyanukarn, Feuangthit Niyamissara, Chawanphat Muangnoi, Pahweenvaj Ratnatilaka Na Bhuket, Pornchai Rojsitthisak, and Pranee Rojsitthisak. 2018. "Chitosan/Alginate Nanoparticles as a Promising Approach for Oral Delivery of Curcumin Diglutaric Acid for Cancer Treatment." *Materials Science and Engineering C* 93. Elsevier B.V: 178–190. doi:10.1016/j.msec.2018.07.069.

Synowiecki, Józef, and Nadia Ali Al-Khateeb. 2003. "Production, Properties, and Some New Applications of Chitin and Its Derivatives." *Critical Reviews in Food Science and Nutrition*. doi:10.1080/10408690390826473.

Tang, Peixiao, Qiaomei Sun, Hongqin Yang, Bin Tang, Hongyu Pu, and Hui Li. 2018. "Honokiol Nanoparticles Based on Epigallocatechin Gallate Functionalized Chitin to Enhance Therapeutic Effects against Liver Cancer." *International Journal of Pharmaceutics* 545 (1–2): 74–83. doi:10.1016/j.ijpharm.2018.04.060.

Termsarasab, Ubonvan, Hyun Jong Cho, Dong Hwan Kim, Saeho Chong, Suk Jae Chung, Chang Koo Shim, Hyun Tae Moon, and Dae Duk Kim. 2013. "Chitosan Oligosaccharide-Arachidic Acid-Based Nanoparticles for Anti-Cancer Drug Delivery." *International Journal of Pharmaceutics* 441 (1–2). Elsevier B.V.: 373–380. doi:10.1016/j.ijpharm.2012.11.018.

Tian, Qin, Xiuhua Wang, Wei Wang, Chuangnian Zhang, Yuan Liu, and Zhi Yuan. 2010. "Insight into Glycyrrhetinic Acid: The Role of the Hydroxyl Group on Liver Targeting." *International Journal of Pharmaceutics*. doi:10.1016/j.ijpharm.2010.08.032.

Tiwari, Sanjay, and Pratap Bahadur. 2019. "Modified Hyaluronic Acid Based Materials for Biomedical Applications." *International Journal of Biological Macromolecules*. doi:10.1016/j.ijbiomac.2018.10.049.

Tønnesen, Hanne Hjorth, and Jan Karlsen. 2002. "Alginate in Drug Delivery Systems." *Drug Development and Industrial Pharmacy*. doi:10.1081/DDC-120003853.

Venkatesan, Jayachandran, Sukumaran Anil, Se Kwon Kim, and Min Suk Shim. 2016. "Seaweed Polysaccharide-Based Nanoparticles: Preparation and Applications for Drug Delivery." *Polymers*. doi:10.3390/polym8020030.

Vijayakumar, Mayakrishnan, Kannappan Priya, Soundharrajan Ilavenil, Balakarthikeyan Janani, Vadanasundari Vedarethinam, Thiyagarajan Ramesh, Mariadhas Valan Arasu, Naif Abdullah Al-Dhabi, Young Ock Kim, and Hak Jae Kim. 2020. "Shrimp Shells Extracted Chitin in Silver Nanoparticle Synthesis: Expanding Its Prophecy towards Anticancer Activity in Human Hepatocellular Carcinoma HepG2 Cells." *International Journal of Biological Macromolecules* 165. Elsevier B.V: 1402–1409. doi:10.1016/j.ijbiomac.2020.10.032.

Wang, Fang, Siqian Yang, Jian Yuan, Qinwei Gao, and Chaobo Huang. 2016. "Effective Method of Chitosan-Coated Alginate Nanoparticles for Target Drug Delivery Applications." *Journal of Biomaterials Applications* 31 (1): 3–12. doi:10.1177/0885328216648478.

Wang, Jianting, Ming Wang, Mingming Zheng, Qiong Guo, Yafan Wang, Heqing Wang, Xiangrong Xie, Fenghong Huang, and Renmin Gong. 2015. "Folate Mediated Self-Assembled Phytosterol-Alginate Nanoparticles for Targeted Intracellular Anticancer Drug Delivery." *Colloids and Surfaces B: Biointerfaces* 129 (May): 63–70. doi:10.1016/j.colsurfb.2015.03.028.

Wang, Tao, Jiahui Hou, Chang Su, Liang Zhao, and Yijie Shi. 2017. "Hyaluronic Acid-Coated Chitosan Nanoparticles Induce ROS-Mediated Tumor Cell Apoptosis and Enhance Antitumor Efficiency by Targeted Drug Delivery via CD44." *Journal of Nanobiotechnology* 15 (1). BioMed Central: 1–12. doi:10.1186/s12951-016-0245-2.

Wathoni, Nasrul, Lisna Meylina, Agus Rusdin, Ahmed Fouad Abdelwahab Mohammed, Dorandani Tirtamie, Yedi Herdiana, Keiichi Motoyama, et al. 2021. "The Potential Cytotoxic Activity Enhancement of α-Mangostin in Chitosan-Kappa Carrageenan-Loaded Nanoparticle against Mcf-7 Cell Line." *Polymers*. doi:10.3390/polym13111681.

Xu, Keming, Fan Lee, Shujun Gao, Min Han Tan, and Motoichi Kurisawa. 2015. "Hyaluronidase-Incorporated Hyaluronic Acid-Tyramine Hydrogels for the Sustained Release of Trastuzumab." *Journal of Controlled Release*. doi:10.1016/j.jconrel.2015.08.015.

Xu, Yinglei, Zhengshun Wen, and Zirong Xu. 2009. "Chitosan Nanoparticles Inhibit the Growth of Human Hepatocellular Carcinoma Xenografts through an Antiangiogenic Mechanism." *Anticancer Research* 29:5103–5109.

Xu, Zhi Ping, Qing Hua Zeng, Gao Qing Lu, and Ai Bing Yu. 2006. "Inorganic Nanoparticles as Carriers for Efficient Cellular Delivery." *Chemical Engineering Science*. doi:10.1016/j.ces.2005.06.019.

Yang, Huan, Cui Tang, and Chunhua Yin. 2018. "Estrone-Modified PH-Sensitive Glycol Chitosan Nanoparticles for Drug Delivery in Breast Cancer." *Acta Biomaterialia* 73 (April): 400–411. doi:10.1016/j.actbio.2018.04.020.

Ye, Bai liang, Ru Zheng, Xiao jiao Ruan, Zhi hai Zheng, and Hua jie Cai. 2018. "Chitosan-Coated Doxorubicin Nano-Particles Drug Delivery System Inhibits its Cell Growth of Liver Cancer via P53/PRC1 Pathway." *Biochemical and*

Biophysical Research Communications 495 (1). Elsevier Inc.: 414–420. doi:10.1016/j.bbrc.2017.10.156.

Yildizhan, Hatice, Nezehat Pınar Barkan, Seçil Karahisar Turan, Özerk Demiralp, Fatma Duygu Özel Demiralp, Bengi Uslu, and Sibel A. Özkan. 2018. "Treatment Strategies in Cancer from Past to Present." In *Drug Targeting and Stimuli Sensitive Drug Delivery Systems*, 1–37. Elsevier. doi:10.1016/B978-0-12-813689-8.00001-X.

Yuan, Xiaoming, Yan He, Guangrong Zhou, Xiangwei Li, Aiwen Feng, and Weiwei Zheng. 2018. "Target Challenging-Cancer Drug Delivery to Gastric Cancer Tissues with a Fucose Graft Epigallocatechin-3-Gallate-Gold Particles Nanocomposite Approach." *Journal of Photochemistry and Photobiology B: Biology*. doi:10.1016/j.jphotobiol.2018.04.026.

Yuan, Zeting, Yajing Ye, Feng Gao, Huihui Yuan, Minbo Lan, Kaiyan Lou, and Wei Wang. 2013. "Chitosan-Graft-β-Cyclodextrin Nanoparticles as a Carrier for Controlled Drug Release." *International Journal of Pharmaceutics*. doi:10.1016/j.ijpharm.2013.02.024.

Zhang, Chuangnian, Wei Wang, Tong Liu, Yukun Wu, Hua Guo, Ping Wang, Qin Tian, Yongming Wang, and Zhi Yuan. 2012. "Doxorubicin-Loaded Glycyrrhetinic Acid-Modified Alginate Nanoparticles for Liver Tumor Chemotherapy." *Biomaterials*. doi:10.1016/j.biomaterials.2011.11.045.

Zhang, Chuangnian, Wei Wang, Chunhong Wang, Qin Tian, Wei Huang, Zhi Yuan, and Xuesi Chen. 2009. "Cytotoxicity of Liver Targeted Drug-Loaded Alginate Nanoparticles." *Science in China, Series B: Chemistry*. doi:10.1007/s11426-009-0178-9.

Zhang, Jianmei, Yijun Wang, Yunyao Jiang, Tingwu Liu, Yanyan Luo, Enjie Diao, Yufeng Cao, et al. 2018. "Enhanced Cytotoxic and Apoptotic Potential in Hepatic Carcinoma Cells of Chitosan Nanoparticles Loaded with Ginsenoside Compound K." *Carbohydrate Polymers*. doi:10.1016/j.carbpol.2018.06.121.

Zhao, Dong, Chuan Jun Liu, Ren Xi Zhuo, and Si Xue Cheng. 2012. "Alginate/CaCO3 Hybrid Nanoparticles for Efficient Codelivery of Antitumor Gene and Drug." *Molecular Pharmaceutics*. doi:10.1021/mp3002123.

Zhao, Xiujuan, Qinli Wan, Xiaodie Fu, Xiao Meng, Xianfeng Ou, Ruowei Zhong, Qinghua Zhou, and Mingxian Liu. 2019. "Toxicity Evaluation of One-Dimensional Nanoparticles Using Caenorhabditis Elegans: A Comparative Study of Halloysite Nanotubes and Chitin Nanocrystals." *ACS Sustainable Chemistry and Engineering* 7 (23): 18965–18975. doi:10.1021/acssuschemeng.9b04365.

Zhu, Yuefei, Yiyang Liu, and Zhiqing Pang. 2019. "Chitosan in Drug Delivery Applications." In *Natural Polysaccharides in Drug Delivery and Biomedical Applications*. doi:10.1016/B978-0-12-817055-7.00004-2.

Zokaei, Elham, Arastoo Badoei-dalfrad, Mehdi Ansari, Zahra Karami, Touba Eslaminejad, and Seyed Noureddin Nematollahi-Mahani. 2019. "Therapeutic Potential of DNAzyme Loaded on Chitosan/Cyclodextrin Nanoparticle to Recovery of Chemosensitivity in the MCF-7 Cell Line." *Applied Biochemistry and Biotechnology*. doi:10.1007/s12010-018-2836-x.

Zuo, Weimin, and Hang Fai Kwok. 2021. "Development of Marine-Derived Compounds for Cancer Therapy." *Marine Drugs*. doi:10.3390/md19060342.

21 Sponge Enzyme's Role in Biomineralization and Human Applications

Moin Merchant and Maushmi S. Kumar

CONTENTS

21.1 INTRODUCTION

The phylum Porifera (colloquially known as sponges) belong to Animalia kingdom and is not having a major biodiversity weightage, such as arthropods, mollusks, or chordates, with an estimated record of around 15,000 extinct and 5000 living species (Hooper and Van Soest, 2002). Sponges are sessile animals that scarcely move during their adult lives and can live for thousands of years. They strain food particles from ambient water bodies and grow continuously. They contribute greatly to global ecology and understanding of animal evolution as sponge evolution holds a distinct role in understanding early animal evolution because they have diverged from the base of the animal family tree. The phylum Porifera is classified into three groups, among which the Hexactinellida and Demospongia have siliceous skeletons, while Calcarea possess calcium

DOI: 10.1201/9781003303909-21

carbonate skeleton (Müller et al., 2004). There are fundamental differences in morphology and development of the Calcareous from the Hexactinellida/Demospongiae (Silicispongea). Unlike Silicispongea, calcareans lack distinct microscleres, and their spicules are secreted intercellularly within an organic sheath. Figure 21.1 details the classification of the phylum Porifera.

Biomineralization is the process by which organic molecules combine with inorganic molecules by a living organism (Estroff, 2008). As a result of the living organism's metabolism, biominerals can be deposited within the organism as well as in its immediate surroundings or environment. The organisms' communication to one another and the formation of skeletal minerals, as well as their forms and functions, are affected by natural selection over geologic time. They are evolving in a controlled manner, resulting in monophyletic groups also known as clades. Biomineralization helps maintain the hardness and stiffness in the body form and shape of multicellular creatures (Weiner and Wagner, 1998). These activities take place both extracellularly and intracellularly, in the skeletons of animals and plants as well as in the creation of Fe_3O_4-based nanocrystals in magnetotactic bacteria, respectively. Biomineral production is governed by the same mechanisms that govern the formation of other biological macromolecules in living organisms (Faivre and Schüler, 2008). It is carried out by enzymes and operates at ambient and physiological conditions with non-saturating concentrations. The morphogenetic biomineralization processes are genetically controlled and integrated with the anabolic and catabolic metabolic pathways (Uriz, 2006). The genetically controlled biomineralization processes differ from the physiologically induced biomineralization reactions. Bio-seeds nucleate the subsequent formation of mineral deposits of various forms and sizes during the biological induction of the mineralization process in deepsea nodules or crusts. The organisms have good control over the nucleation process and the shape and size of the mineral deposits in genetically driven mineralization processes (Wang and Müller, 2009). As the functional groups of organic molecules can influence the deposition of ions, salts, and inorganic monomeric and polymeric units, they can greatly enhance biomineralization process with an interplay between organic chemical macromolecules and inorganic components (Westbroek, 1983; Lowenstam and Weiner, 1983). These nuances are fundamental for the formation of the skeleton, even for the stabilization of bones in vertebrates. The organic reactions in complex biological systems were elucidated only after the discovery of enzymes, which was accelerated by the decoding of the genetic code and subsequent application of recombinant techniques (Weiner and Wagner, 1998). The production and deposition of silica and calcium carbonate ($CaCO_3$) are driven enzymatically in sponges, as supported by the evolutionary studies (Müller et al., 2004).

21.2 ROLE OF SILICATEIN IN BIOSILICIFICATION

Biosilicification is the biological formation of opal-like amorphous hydrated silica in a wide variety of organisms, including protists, radiolarian, foraminifera, sponges, mollusks, brachiopods, copepods, ascidians, diatoms, and

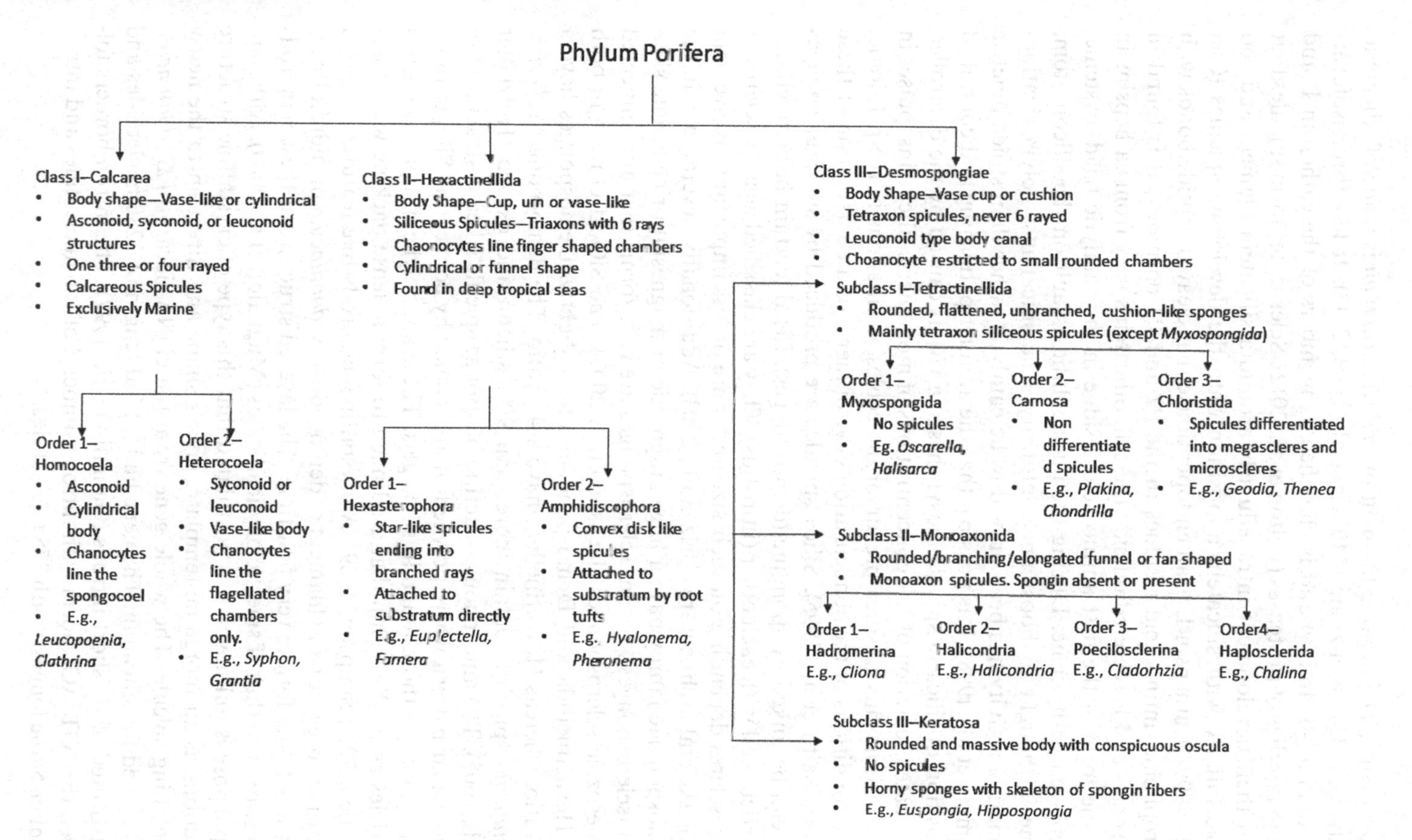

FIGURE 21.1 Classification of the phylum Porifera.

higher plants. The organic filament of *Tethya aurantium* is made of silicatein, a cathepsin L-like enzyme (Müller et al., 2003, 2013). It has three isoforms (silicatein-α, -β, and -γ) and is classified as members of the cathepsin L and papain families of proteases (Belton et al., 2012). Sclerocytes form spicules that begin their development intracellularly around a silicatein filament and end extracellularly with silicatein apposition onto the developing spicules (Cha et al., 1999). Cathepsin L is an endopeptidase that cleaves peptide bonds with hydrophobic amino acid residues in the P2 and P3 positions. It is found in lysosomes and is secreted outside the cell. Silicateins vary from cathepsins in that the form at the first amino acid residue in the catalytic triad, cysteine, is replaced with serine. During the nucleophilic assault on the silicon atom, serine is thought to boost nucleophilicity. The silicateins' polymerization-promoting activity has been proved to be catalytic rather than stoichiometric (Shimizu et al., 1998). It is possible that the annular patterning and continued deposition of silica in spicule biosynthesis *in vivo* is caused by the controlled punctuated secretion of low concentrations of monomeric silicateins, pulses in the transitory flux of silica-precursor molecules, oscillations in the pH, ionic, other conditions of the condensation environment, or a combination of these factors (Saito et al., 1995). Silica spicules are produced by siliceous sponges and can be linked by spongine (demosponges), fused (certain hexactinellids), or articulated with each other (lithistids). They are classified as megascleres or microscleres depending on their size and more or less important involvement in the skeletal architecture (Uriz et al., 2003). Additionally, several taxa in the Demospongiae family have a huge aragonitic or magnesium-rich calcite skeleton (sclerosponges). Extant sclerosponges are relic forms that are connected to mesozoic sclerosponges (Jackson et al. 2007). The SiO_2:H_2O ratio in both the Hexactinellida and Demospongiae differs slightly. Demosponges have a 5:1 ratio, whereas Hexactinellida have a 4:1 ratio. The demosponge *Suberites domuncula* spicules contain more than 90% silicon and oxygen (Holzhüter et al., 2005). The amorphous structure of sponge spicule silica is structured in a molecular pattern where polyhedral are formed by connecting SiO_4 tetrahedral, similar to those found in silica gels. These are polyhedral produce silica particles as small as 3nm that are linked to form a dense packing with pores (Ehrlich, 2011; Simpson, 1989). A crystalline nanorods-like features localized within the organic axial filament of demosponge *S. domuncula* primordial spicules was found for the very first time. The layered structure of these nanorods was similar to that of smectic phyllosilicates (Mugnaioli et al. 2009). Although its purpose is unknown, it is postulated that this type of crystalline structure functions as an inorganic template, giving shape and direction to the newly developing spicule. The whole gene from the cDNA library of *S. domuncula* was cloned for silicatein. The gene had a total length of 2280 nucleotides and constituted of six short introns (Gamulin et al., 1997). The intron borders followed the "GT–AG" norm with intron number 2 of 166 nucleotides and intron number 5 (the longest) with 375 nucleotides.

21.2.1 Mechanisms of Biosilicification

21.2.1.1 Cellular Events

There are two possible mechanisms for enzyme catalysis: (1) to stabilize one molecule of deprotonated silicic acid (the nucleophile) at the active site, which will then react with another molecule of silicic acid, or (2) stabilize a protonated silicic acid (the electrophile) at the active site, which will then react with another molecule of silicic acid (Fairhead et al., 2008). Sclerocytes release axial filaments that are primarily made up of silicateins but also include other organic compounds. The highly structured mesoporous axial filament is formed using silicatein as a structural template. Silicate is utilized or actively taken up into the sclerocyte and complexed with certain proteins to generate an organic-silica substrate that has yet to be characterized. The organo-silica compound is delivered to the "silica deposition space" by silica lemma. The complexing protein is released and may be recycled to the sclerocyte cytoplasm after silica from the organo-silica substrate polymerizes as nanospheres on the outer surface of the mesoporous axial filaments, possibly at serine and histidine catalytic center sites. The axial filament continues to grow at spicule tips, giving primarily a spicule pattern, while silica transport and deposition proceed on lateral spicule surfaces between the tips. The axial filament's silicateins can no longer exhibit enzymatic activity once the first few layers of silica nanospheres have encased it. Silicatein at active areas on the silica lemma facilitates and controls the continued deposition of silica on the outer surface of spicules, resulting in the creation of distinct patterned morphologies of spicules. Several aspects of this approach are still unproven. The roles of minor non-silicatein organic compounds incorporated in axial filaments, whether silicateins associated with the silica lemma are the same forms as those in the axial filaments, and how the final high-fidelity submicroscopic external patterning on spicule surfaces is genetically controlled are some of the other unknown factors. However, in the nonspicule-forming marine sponge *Acanthodendrilla* spp. silicatein genes were discovered (AcSilA and AcSilB; Veremeichik et al., 2011).

Enzymatically driven silica polycondensation, or biosilica production, adheres to thermodynamic principles, including chemical equilibrium kinetics during sol-gel silica polymerization (Bastos-Neto et al., 2020). Condensation of the silicic acid monomers to oligomers or polymers with the release of water is observed during the creation of the siloxane linkages before reaching the isoelectric point (Hench and Wang, 2006). The polycondensation reaction is pH-dependent; H^+ acts as a catalyst at a pH of about 2, whereas OH catalysis the reaction in the alkaline pH. Cyclic silica oligomers containing three to four silicic acid units develop initially during the silica formation and continue to expand. The silica species are weakly ionized in pH range of 2–7, when the primary particles develop slowly. Above pH 7, the particles gain more negative charges, allowing faster particle development. Because repulsive electrostatic charges are mild in the absence of cations at pH 7, particles form fibrillary aggregates.

During chain elongation, silica monomers can be added to both the main and side chains, forming a three-dimensional open gel network filled with water. The electrostatic interactions between charged particles are significant at pH > 7, which inhibits aggregation. As a result, the size of the primary particles generated increases, but the quantity decreases (Coradin and Lopez, 2003). The smaller particles dissolve, hydrolyze, and redeposit onto the surface of the larger ones as their solubility reduces as their size increases. Finally, a sol made of monodisperse particles is created, with increased stability at higher pH values. A coarsening of the pore structure can occur during the dissolution and re-polycondensation processes, which has a significant impact on the later phases of drying and sintering, as well as the reactions involved in silica aging. The casting of silica gels for structure-building applications is possible during this phase (Satoh et al., 2002).

This method, which allows monolithic silica gels to be formed, necessitates the removal of water from the wet gel and a further drying procedure at a higher temperature. The monolithic silica gels are heated to >800°C to generate dense silica rods for the gel–silica–glass chemical manufacturing reaction, with optical transmission properties at wavelengths of around 170nm (Hench and Vasconcelos, 1990).

21.2.1.2 Chemical Events

The proposed reaction mechanism is based on the assumption that serine (Ser) and histidine (His) residues in the enzyme's catalytic center are required for the catalytic mechanism, particularly for the formation of hydrogen bonds between the Ser OH and, later, the enzyme-bound silicic acid OH and the imidazole nitrogen of the His residue (Figure 21.2). The nucleophilic attack (SN2 type) of the partial electronegative oxygen atom of the OH group of the

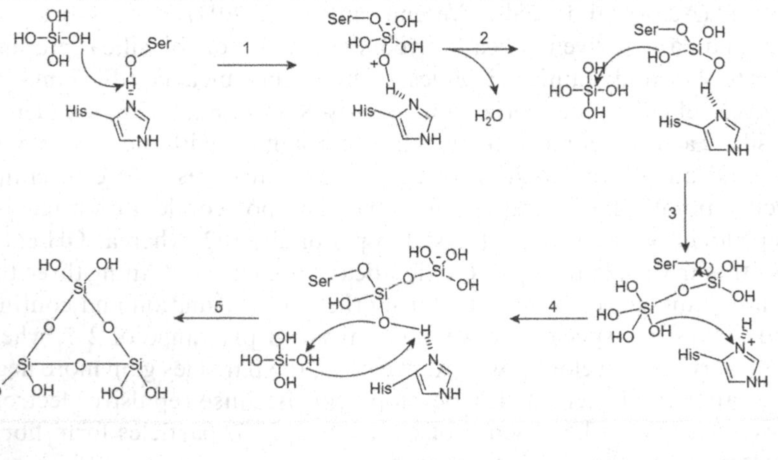

FIGURE 21.2 Reaction mechanism of biosilicification via silicatein re-created with partial modifications from Schröder et al. (2012).

Ser at the (electropositive) silicon atom of the silicic acid molecule is the first catalytic step (Figure 21.2, step 1). The hydrogen bond created between the Ser-OH and the His imidazole facilitates this interaction, increasing the nucleophilicity of the Ser-OH group. A water molecule is released when a proton is transferred from the His N (Ser-His hydrogen bond) to one of the silicon OH ligands of the pentavalent intermediate produced (transition stage; Figure 21.2, step 2). A silicic acid species is produced, which is covalently linked to the Ser in the catalytic center in a transitional state. The Ser-bound silicic acid undertakes a nucleophilic attack at the silicon atom of a second orthosilicate in step 3 (Figure 21.2). The nucleophilicity of the attacking oxygen atom of the OH ligand from the first silicic acid is raised by the formation of a hydrogen bond with the nitrogen imidazole of the His residue in the enzyme's catalytic center. The loss of a second water molecule, aided by proton transfer from His N imidazole, produces a disilicic acid molecule that is linked to silicatein via an ester-like link. A second OH ligand of the enzyme-bound silicic acid unit interacts with the nitrogen imidazole of the catalytic center His residue after the ester bond is rotated (Figure 21.1, step 4). This process allows the disilicic acid to expand even more by nucleophilic attack on a third orthosilicate. The formation of higher membered silicic acid oligomers (tetrasilicic acid, etc.) can be achieved by repeating this reaction cycle (nucleophilic attack, proton transfer, loss of water, rotation). The cyclization of the oligomer is postulated to trigger the release of silicatein-bound trisicilic acid—the scheme indicated (Figure 21.2, step 5). The mechanisms of siloxane polymer cyclization have been extensively dealt (Schröder et al., 2012; Jacobson and Stockmayer, 1950; Flory and Semlyen, 2002).

21.2.2 Silicatein in the Biomineralization of Other Substrates

Silicatein is capable of synthesizing a wide range of industrially significant compounds at low temperatures around a neutral pH. Few of them are polysilsesquioxanes (silicones) such as polyphenylsilsesquioxane (from phenyltriethoxysilane) and polymethylsilsesquioxane (from methyltriethoxysilane) and a large number of metal oxides, including the nanocrystalline anatase (metastable metal) polymorph TiO_2 (from the water-stable and -soluble precursor, titanium bis-ammonium lactato; Sumerel et al., 2003). Others include the orthorhombic perovskite-like (perovskite = a calcium titanium oxide mineral) $BaTiOF_4$ from $BaTiF_6$ (Brutchey et al., 2006) and the polymorph of γ-Ga_2O_3 from $Ga(NO_3)_3$ (Kisailus et al., 2006). In addition, the catalytic ability of silicatein is demonstrated in the synthesis and surface deposition of a biodegradable polymer, poly(L)-lactide (Curnow et al., 2006), $TiPO_4$ (Curnow et al., 2005), and polymerized NCN (two nitrogens bound to one carbon)- and PCP (two phosphine bound to one carbon)-pincer metal materials (pincer complexes are formed by the binding of a chemical structure to a metal atom with at least one carbon–metal bond) useful for gas-sensing applications (O'Leary et al., 2009). The enzymatic synthesis of metal oxides is particularly intriguing, as it occurs when silicatein reacts with a variety of simple inorganic metal salts, such as $Ga(NO_3)_3$ in water. This synthesis is carried out by the hydrolysis of metal salts, a mechanism that actually refers to the hydrolysis of the aquo-complexes (stereospecific hydrates) of incompletely dissociated metal salts, for which Livage's Partial Charge Model can be used to understand and quantitatively predict the structure- and pH-specific stabilities (Livage et al., 1988; Kisailus et al., 2006). The extremely wide range of nonnatural substrates is due to evolutionary relaxation of the structural constraints in the active site cleft of silicateins—presumably necessary to accommodate the sugar, sugar alcohol, or polyol adduct of silicic acid suggested to be the natural substrate for biosilicification (Kinrade et al., 2001). This is in contrast to the otherwise highly similar catalytic triad hydrolases, which have very strict structural requirements.

21.3 ROLE OF CARBONIC ANHYDRASES IN BIOCALCIFICATION IN SPONGES

Biocalcification is the most common type of biomineralization among animal phyla (Murdock and Donoghue, 2011). It evolved independently, resulting in several recruitments of the same genes for biomineralization in various lineages (Murdock, 2020). Members of the carbonic anhydrase (CAs) gene family are required for biomineralization (Roy et al., 2014). CAs are zinc-binding enzymes that catalyze the reversible conversion of carbon dioxide and water to one proton and bicarbonate (Tripp et al., 2001). Three His residues are required for the protein's catalytic activity and mediate Zn binding (Aspatwar et al., 2014; Kim et al., 2020). CAs role in a variety of physiological processes requires ion management and carbon transport for the regulated precipitation of carbonate biominerals (Supuran, 2016). Sixteen distinct CAs are expressed in specific organs in defined subcellular compartments in mammals (Hassan et al., 2013). They can be classified in cytosolic, mitochondrial, membrane-bound, and secretory forms. These categories have grown and shrunk in different animal groups (Roy et al., 2014). In different metazoan lineages, including sponges, certain CAs are engaged in carbonate biomineralization (Jackson et al., 2007; Germer et al., 2015). RNA-Seq data analysis revealed the role of apically overexpressed genes in biomineralization in *Sycon ciliatum*, particularly on bicarbonate transporters. Solute carrier 4 (SLC4) family bicarbonate transporters play a role in carbon transport and pH regulation (Romero et al., 2013), and a specific version has been identified as a major biomineralization gene in scleractinian corals (Zoccola et al., 2015). According to investigations of amino acids content from proteins isolated from the spicules of diverse species, Aspartate (Asp) or Asparagine (Asn), i.e., Asx-rich proteins (ARPs) appear to represent a prominent component of the spicule matrix proteome of calcareous sponges (Aizenberg, 1996; Aizenberg et al., 1996). Two SLC4 proteins and three ARPs with signal peptides among the apically overexpressed transcripts (ARP1–3) were found. One of the detected, presumably sclerocyte-specific SLC4-proteins of *S. ciliatum* belonged to Na^+/HCO_3^- group of co-transport proteins (NCBT-like), while the other belonged to HCO_3^-/Cl^- anion exchange proteins (AE-like).

SciNCBT-like1 and SciAE-like1 are the names given to these *Sycon ciliatum* SLC4 proteins. SciNCBT-like2, SciAE-like2, and SciAE-like3 are additional SLC4 proteins found in *S. ciliatum* that belonging to two SLC4 groups. The expression pattern of SciNCBT-like1 and SciAE-like1 was identical (Voigt et al., 2014). Although SLC4 proteins can be classified as AE-like, NCBT-like, or BOR-like based on their evolutionary affinities, only a few members of each group have known functions and stoichiometry of transport, excluding sponge proteins (Parker and Boron, 2013). As a result, for the two sclerocyte-specific SLC4 proteins in *S. ciliatum*, neither the direction nor the mechanism of transport (Na^+ independent Cl^- cotransport for SciAE-like1, or Na^+ coupled for SciNCBT-like1) can be deduced. Even so, it is fair to assume that the proteins are involved in the assisted transport of bicarbonate to the calcification site through the sclerocyte, that is, trafficking bicarbonate from the mesohyl into the sclerocyte and/or trafficking bicarbonate formed within the sclerocyte through SciCA1 activity to the intercellular space at the site of calcification (Voigt et al., 2014).

Many organisms, notably stony corals and coralline demosponges, contain acidic proteins with a high aspartic acid content in the organic matrix of carbonate skeletons (Puverel et al., 2005; Drake et al., 2013). These proteins are thought to have important roles in the biomineralization process. Aspartic acid residues in these proteins, for example, can bind to Ca^{2+} ions, and some of them can interact with certain crystal faces of growing biominerals, altering crystal form. Acidic skeletal organic matrix (SOM) proteins have been shown to hinder or promote crystallization. (Marin and Luquet, 2007).

21.4 NOVEL APPLICATIONS OF BIOMINERALIZATION ENZYMES

The biomineralization process has been carefully investigated and replicated *in vitro*. It has permeated many fields in our lives, including mechanical, electrical, environmental, and biomedical engineering, based on the aforementioned forming process by biomineralization enzymes combined with other emerging technologies to synthesize new nanomaterials. Biomedical engineering, as the forerunner of human health and life, has gained a lot of attention and has grown into a thriving sector.

21.4.1 Formation of Silicones Using Silicatein

Silicateins, which are only present in sponges, catalyze the production of silica from monomeric silicic acid esters *in vivo*. Recombinant silicatein converts alkoxy silanes to silicones (siloxanes) in a phase transfer reaction at neutral pH under physiological circumstances *in vitro*. Mass spectrometry, ultraviolet/visible light (UV/Vis), and nuclear magnetic resonance (NMR) methods revealed a significant increase in the production rate of oligomers and the chain length of silicones. The first enzymatically driven organometallic condensation step is the production of silicones in the presence of silicatein. The enzymatic activity of Silicatein was determined using UV/VIS measurements and ^{29}Si DEPT

NMR as an analytical probe, which demonstrated the hydrolytic activity of silicatein (Wolf et al., 2010). Silicatein has a remarkable ability to synthesize silica bio-catalytically under physiological conditions of neutral pH, ambient temperature, and pressure and can be used for the production of silicones. This bio-catalytical process could have an impact on the long-term production of silicones (Bains ànd Tacke, 2003).

21.4.2 Biomedical Application to Bone Defects

Biosilica is a naturally occurring polymer found in sponges, diatoms, and mammals. Silicon has been found to build up in mammals, particularly in tissues near bone-forming sites (Pastero et al., 2004). It is shown that silica, specifically biosilica produced with silicatein, had anabolic effects on bone-forming cells (osteoblasts; Liu et al., 2018) and antagonistic effects on bone-resorbing cells (osteoclasts; Lei et al., 2005). Biosilica has a good stimulating effect on osteoblasts in terms of hydroxyapetite (HA) crystallite production, as evidenced by the fact that it generated differential gene expression in Saos-2 cells cultivated *in vitro* and encouraged osteoblast differentiation (Liu et al., 2015). Biosilica increased the steady-state level of osteoprotegerin (OPG) transcripts while keeping the amount of the osteoclastogenic ligand, receptor activator of nuclear factor-κB (NFκB) ligand (RANKL) unchanged (Liu et al., 2018). Osteoclast differentiation is inhibited with the rise in the ratio of OPG to RANKL. As an anabolically acting polymer, biosilica was postulated to offer a significant biomedical potential for the treatment and prophylaxis of osteoporotic illnesses. Biosilica also has osteogenic potential *in vitro*, as seen by enhanced [3H] thymidine ([3H] dThd) incorporation into DNA as well as increased HA production. The wingless-related integration site (WNT) and sonic hedgehog (SHH) pathways, which both become activated by biosilica, have recently been narrowed down as the molecular mechanism by which silica and biosilica function in morphogenetic ways (Pacella and Gray, 2018).

21.4.3 Cancer/Tumor Target Engineering

Various nanomaterial-based therapeutic platforms have been developed for the diagnosis and treatment of cancer/tumor, since the emergence of nanotechnology (Chen et al., 2013). Aside from their high biocompatibility, synthetic nanomaterials inspired by biomineralization have unique optical, magnetic, and thermal properties that make it easier to build out-of-body control systems for better cancer/tumor location, imaging, and diagnostics. An existing kind of energy, adenosine triphosphate (ATP), was discovered to mediate the drug-liberating process during cancer therapy and operate as an antineoplastic agent *in vitro* (Jiang et al., 2015). An ATP-loaded nanocomposite was created that induced death in tumor cells (Srivastava et al., 2018). The surface of mesoporous silica nanoparticles (MSNPs), a biocompatible delivery system was created using tetraethyl orthosilicate (TEOS) and silicatein and was first coated with ATP. Following that, the cancer medication doxorubicin (DOX) was put into the poriferous nanoparticles. The pores on the surface of MSNPs were closed, and DOX was encapsulated after biomineralization of $CaCO_3$ using CAs and ATP. $CaCO_3$ dissolved in an acidic environment around the tumor while ATP was exposed to bind with antigen on the tumor's surface, which opened the pores, and DOX got released to impact the tumor (Wang et al., 2014, 2016).

21.4.4 Bionic and Three-Dimensional Printing Engineering

The molecularly controlled processes of biomineralization using CAs and their application to bionic manufacturing is recently summarized (Arakaki et al., 2015). Three critical elements, including catalysts, surface recognitions, and template effects, influence the biomineralization formation process. Controlling the molecular composition of these three components could result in biomineralization with diverse structures, morphologies, functions, and performance and that is exactly how CAs bring about biocalcification. Biomineralization-based bionic natural molluscan shell design and three-dimensional (3D) printing were used and described using a layer-by-layer process (Yadav et al., 2018). When compared to nacre and foliated materials, the crossed lamellar microstructure was found to have superior impact properties and a faster wear rate. This microstructural-dependent site-specific printing approach could be used for bionic and biomedical engineering on advanced functionalized materials. Biomineralization-based cell-micro-reactors were proposed to be used like 3D-printing assemblies that would imitate natural bone healing (Itel et al. 2018)

21.4.5 Deposition of Silicatein on Culture Plates

Simple physisorption was used to deposit silicatein onto culture plates, followed by the addition of TEOS to deposit biosilica, resulting in a significant increase in the production of calcium phosphate nodules by human osteosarcoma Saos-2 cells (Schröder et al., 2005). An even more innovative technique is to add an 8-Glu tag to the protein's N-terminus, which confers hydroxyapatite binding and promotes biosilica formation on synthetic hydroxyapatite nanofibrils and dental hydroxyapatite when biosilica precursor is added (Natalio et al., 2010). Surface flaws and dentinal tubules could be sealed with this "smart glue," reducing the risk of tooth decay and dental hypersensitivity. It would be interesting to see if limiting silicatein expression on these cells and subsequent deposition of biosilica directly on the cell surface (cf. the bacterial display system in *E. coli*) could enhance osteoblast mineralization even more, although there is a clear disadvantage in continuous silicatein expression and the potential for undesirable biosilica formation.

21.4.6 Drug and Cell-Therapy Engineering

Biomineralization, as an organic-controlled inorganic structure formation technique, solves the problem of biocompatibility by providing a method to protect curative viral vectors from molecular recognition (Rebek, 2009) in the context of medication and cell-therapy engineering. A biomimetic method to silica-encapsulated yeast cells with surface functionalization was proposed, in which first thiol functionalization into silicatein directed silica encapsulation, as well as fluorescein and streptavidin onto the silica shells for better observation was introduced (Yang et al., 2011). Biomineralization-based silica encapsulation of *Saccharomyces cerevisiae* cells achieved the control of silica shell thickness (Lee et al., 2014). The manipulation in exact thickness might improve the physicochemical stability of the nanomaterial while also controlling the functional activities of internal medicines and metabolic activities of encapsulated cells. When biomineralization was used in the synthesis, it improved the drug biocompatibility while overcoming drug resistance and allowed better and longer cancer chemotherapy (Xiao et al., 2018).

21.5 CONCLUSION

The majority of biomineralization occurs in soft tissues, near to cells, allowing cells and minerals to interact and utilize one another. Temperature, pH, and organics are the three main parameters that control the biomineralization process, and specifically in sponges, it is brought about by silicatein and CA enzymes. Biomineralization provides excellent support for synthesizing various nanomaterials to protect drugs/cells/viruses from physiological clearance during transportation to the target tumor/cancer, and improves the nanomaterial's thermal, magnetic, and optical properties for better real-time imaging diagnosis and treatment of diseases. A recent paradigm shift demonstrating that not only organic but also inorganic polymers are generated enzymatically in at least some biological systems opens up new avenues for biotechnology in general and new human medicinal interventions in particular. We believe that biosilica and biocalcite will play a significant role in future regenerative medicine.

ACKNOWLEDGMENT

The authors would like to acknowledge SVKM's NMIMS for all the support and required facility to carry out this work.

REFERENCES

Aizenberg, J., M. Ilan, S. Weiner, and L. Addadi. 1996. "Intracrystalline Macromolecules Are Involved in the Morphogenesis of Calcitic Sponge Spicules." In *Connective Tissue Research*, 34:255–261. Connect Tissue Res. doi:10.3109/03008209609005269.

Aizenberg, Joanna. 1996. "Stabilization of Amorphous Calcium Carbonate by Specialized Macromolecules in Biological and Synthetic Precipitates." *Advanced Materials* 8 (3). John Wiley & Sons, Ltd: 222–226. doi:10.1002/adma.19960080307.

Arakaki, Atsushi, Katsuhiko Shimizu, Mayumi Oda, Takeshi Sakamoto, Tatsuya Nishimura, and Takashi Kato. 2015. "Biomineralization-Inspired Synthesis of Functional Organic/Inorganic Hybrid Materials: Organic Molecular Control of

Self-Organization of Hybrids." *Organic & Biomolecular Chemistry* 13 (4). The Royal Society of Chemistry: 974–989. doi:10.1039/C4OB01796J.

Aspatwar, Ashok, Martti E. E. Tolvanen, Csaba Ortutay, and Seppo Parkkila. 2014. "Carbonic Anhydrase Related Proteins: Molecular Biology and Evolution." *Sub-Cellular Biochemistry* 75. Springer, Dordrecht: 135–156. doi:10.1007/978-94-007-7359-2_8.

Bains, William, and Reinhold Tacke. 2003. "Silicon Chemistry as a Novel Source of Chemical Diversity in Drug Design." *Current Opinion in Drug Discovery and Development* 6 (4): 526–543. https://europepmc.org/article/med/12951816.

Bastos-Neto, Moises, Diana Cristina Silva de Azevedo, and Sebastião Mardônio Pereira de Lucena. 2020. "Adsorption." *Kirk-Othmer Encyclopedia of Chemical Technology*, March. American Cancer Society, 1–59. doi:10.1002/0471238961.0104191518212008.A01.PUB3.

Belton, David J., Olivier Deschaume, and Carole C. Perry. 2012. "An Overview of the Fundamentals of the Chemistry of Silica with Relevance to Biosilicification and Technological Advances." *FEBS Journal*. doi:10.1111/j.1742-4658.2012.08531.x.

Brutchey, Richard L., Edward S. Yoo, and Daniel E. Morse. 2006. "Biocatalytic Synthesis of a Nanostructured and Crystalline Bimetallic Perovskite-like Barium Oxofluorotitanate at Low Temperature." *Journal of the American Chemical Society* 128 (31): 10288–10294. doi:10.1021/ja063107g.

Cha, Jennifer N., Katsuhiko Shimizu, Yan Zhou, Sean C. Christiansen, Bradley F. Chmelka, Galen D. Stucky, and Daniel E. Morse. 1999. "Silicatein Filaments and Subunits from a Marine Sponge Direct the Polymerization of Silica and Silicones in Vitro." *Proceedings of the National Academy of Sciences of the United States of America* 96 (2): 361–365. doi:10.1073/pnas.96.2.361.

Chen, Wei, Yun Xiao, Xueyao Liu, Yanhong Chen, Jiaojiao Zhang, Xurong Xu, and Ruikang Tang. 2013. "Overcoming Cisplatin Resistance in Chemotherapy by Biomineralization." *Chemical Communications* 49 (43). The Royal Society of Chemistry: 4932–4934. doi:10.1039/c3cc41872c.

Coradin, Thibaud, and Pascal Jean Lopez. 2003. "Biogenic Silica Patterning: Simple Chemistry or Subtle Biology?" *ChemBioChem*. John Wiley & Sons, Ltd. doi:10.1002/cbic.200390044.

Curnow, Paul, Paul H. Bessette, David Kisailus, Meredith M. Murr, Patrick S. Daugherty, and Daniel E. Morse. 2005. "Enzymatic Synthesis of Layered Titanium Phosphates at Low Temperature and Neutral PH by Cell-Surface Display of Silicatein-α." *Journal of the American Chemical Society* 127 (45): 15749–15755. doi:10.1021/ja054307f.

Curnow, Paul, David Kisailus, and Daniel E. Morse. 2006. "Biocatalytic Synthesis of Poly(L-Lactide) by Native and Recombinant Forms of the Silicatein Enzymes." *Angewandte Chemie—International Edition* 45 (4). John Wiley & Sons, Ltd: 613–616. doi:10.1002/anie.200502738.

Drake, Jeana L., Tali Mass, Liti Haramaty, Ehud Zelzion, Debashish Bhattacharya, and Paul G. Falkowski. 2013. "Proteomic Analysis of Skeletal Organic Matrix from the Stony Coral Stylophora Pistillata." *Proceedings of the National Academy of Sciences of the United States of America* 110 (10): 3788–3793. doi:10.1073/pnas.1301419110.

Ehrlich, Hermann. 2011. "Silica Biomineralization, Sponges." *Encyclopedia of Earth Sciences Series*, no. 9781402092114. Springer, Dordrecht: 796–808. doi:10.1007/978-1-4020-9212-1_31.

Estroff, Lara A. 2008. "Introduction: Biomineralization." *Chemical Reviews* 108 (11). American Chemical Society: 4329–4331. doi:10.1021/CR8004789.

Fairhead, Michael, Kenneth A. Johnson, Thomas Kowatz, Stephen A. McMahon, Lester G. Carter, Muse Oke, Huanting Liu, James H. Naismith, and Christopher F. Van Der Walle. 2008. "Crystal Structure and Silica Condensing Activities of Silicatein α-Cathepsin L Chimeras." *Chemical Communications*, no. 15 (April). The Royal Society of Chemistry: 1765–1767. doi:10.1039/b718264c.

Faivre, Damien, and Dirk Schüler. 2008. "Magnetotactic Bacteria and Magnetosomes." *Chemical Reviews* 108 (11). American Chemical Society: 4875–4898. doi:10.1021/CR078258W.

Flory, P. J., and J. A. Semlyen. 2002. "Macrocyclization Equilibrium Constants and the Statistical Configuration of Poly(Dimethylsiloxane) Chains." *Journal of the American Chemical Society* 88 (14): 3209–3912. doi:10.1021/JA00966A006.

Gamulin, Vera, Alexander Skorokhod, Vadim Kavsan, Isabel M. Müller, and Werner E. G. Müller. 1997. "Experimental Indication in Favor of the Introns-Late Theory: The Receptor Tyrosine Kinase Gene from the Sponge Geodia Cydonium." *Journal of Molecular Evolution* 44 (3): 242–252. doi:10.1007/PL00006141.

Germer, Juliane, Karlheinz Mann, Gert Wörheide, and Daniel John Jackson. 2015. "The Skeleton Forming Proteome of an Early Branching Metazoan: A Molecular Survey of the Biomineralization Components Employed by the Coralline Sponge Vaceletia Sp." *PLoS ONE* 10 (11): e0140100. doi:10.1371/journal.pone.0140100.

Hassan, Imtaiyaz, Md, Bushra Shajee, Abdul Waheed, Faizan Ahmad, and William S. Sly. 2013. "Structure, Function and Applications of Carbonic Anhydrase Isozymes." *Bioorganic and Medicinal Chemistry* 21 (6): 1570–1582. doi:10.1016/j.bmc.2012.04.044.

Hench, L. L., and Wander Vasconcelos. 1990. "Gel-Silica Science." *Annual Review of Materials Science* 20 (1): 269–298. doi:10.1146/annurev.ms.20.080190.001413.

Hench, L. L., and S. H. Wang. 2006. "The Sol-Gel Glass Transformation of Silica." 24–26 (2). Taylor & Francis Group: 785–834. doi:10.1080/01411599008210251.

Holzhüter, Gerd, Kamatchi Lakshminarayanan, and Thomas Gerber. 2005. "Silica Structure in the Spicules of the Sponge Suberites Domuncula." *Analytical and Bioanalytical Chemistry* 382 (4): 1121–1126. doi:10.1007/s00216-005-3220-6.

Hooper, John N. A., and Rob W. M. Van Soest. 2002. "Systema Porifera. A Guide to the Classification of Sponges." *Systema Porifera*. Springer US, 1–7. doi:10.1007/978-1-4615-0747-5_1.

Itel, Fabian, Jesper Skovhus Thomsen, and Brigitte Städler. 2018. "Matrix Vesicles-Containing Microreactors as Support for Bonelike Osteoblasts to Enhance Biomineralization." *ACS Applied Materials & Interfaces* 10 (36): 30180–30190. doi:10.1021/ACSAMI.8B10886.

Jackson, Daniel J., Luciana Macis, Joachim Reitner, Bernard M. Degnan, and Gert Wörheide. 2007. "Sponge Paleogenomics Reveals an Ancient Role for Carbonic Anhydrase in Skeletogenesis." *Science* 316 (5833): 1893–1895. doi:10.1126/science.1141560.

Jacobson, Homer, and Walter H. Stockmayer. 1950. "Intramolecular Reaction in Polycondensations. I. The Theory of Linear Systems." *The Journal of Chemical Physics* 18 (12). American Institute of PhysicsAIP: 1600–1606. doi:10.1063/1.1747547.

Jiang, Jean X., Manuel A. Riquelme, and Jade Z. Zhou. 2015. "ATP, a Double-Edged Sword in Cancer." *Oncoscience*. doi:10.18632/oncoscience.230.

Kim, Jin Kyun, Cheol Lee, Seon Woo Lim, Aniruddha Adhikari, Jacob T. Andring, Robert McKenna, Cheol Min Ghim, and Chae Un Kim. 2020. "Elucidating the Role of Metal Ions in Carbonic Anhydrase Catalysis." *Nature Communications* 11 (1). Nature Publishing Group: 1–10. doi:10.1038/s41467-020-18425-5.

Kinrade, Stephen, Robin J. Hamilton, Andrew S. Schach, and Christopher T. G. Knight. 2001. "Aqueous Hypervalent Silicon Complexes with Aliphatic Sugar Acids." *Journal of the Chemical Society, Dalton Transactions*, no. 7 (January). The Royal Society of Chemistry: 961–963. doi:10.1039/B010111G.

Kisailus, David, Quyen Truong, Yosuke Amemiya, James C. Weaver, and Daniel E. Morse. 2006. "Self-Assembled Bifunctional Surface Mimics an Enzymatic and Templating Protein for the Synthesis of a Metal Oxide Semiconductor." *Proceedings of the National Academy of Sciences of the United States of America* 103 (15): 5652–5657. doi:10.1073/pnas.0508488103.

Lee, Hojae, Daewha Hong, Ji Yu Choi, Ji Yup Kim, Sang Hee Lee, Ho Min Kim, Sung Ho Yang, and Insung S. Choi. 2015. "Layer-by-Layer-Based Silica Encapsulation of Individual Yeast with Thickness Control." *Chemistry—An Asian Journal* 10 (1). John Wiley & Sons, Ltd: 129–132. doi:10.1002/ASIA.201402993.

Lei, M., P. G. Li, Z. B. Sun, and W. H. Tang. 2006. "Effects of Organic Additives on the Morphology of Calcium Carbonate Particles in the Presence of CTAB." *Materials Letters* 60 (9–10): 1261–1264. doi:10.1016/j.matlet.2005.11.023.

Liu, Xiaojun, Shimei Zeng, Shaojian Dong, Can Jin, and Jiale Li. 2015. "A Novel Matrix Protein Hic31 from the Prismatic Layer of Hyriopsis Cumingii Displays a Collagen-Like Structure." *Plos One* 10 (8): e0135123. doi:10.1371/JOURNAL.PONE.0135123.

Liu, Yujing, Ying Tan, Jie Ren, Hongzheng Chen, and Hanying Li. 2018. "Assessing the Synergy Effect of Additive and Matrix on Single-Crystal Growth: Morphological Revolution Resulted from Gel-Mediated Enhancement on CIT-Calcite Interaction." *Chinese Chemical Letters* 29 (8). Elsevier B.V.: 1296–1300. doi:10.1016/J.CCLET.2018.07.005.

Livage, J., M. Henry, and C. Sanchez. 1988. "Sol-Gel Chemistry of Transition Metal Oxides." *Progress in Solid State Chemistry* 18 (4). Pergamon: 259–341. doi:10.1016/0079-6786(88)90005-2.

Lowenstam, H. A., and S. Weiner. 1983. "Mineralization by Organisms and the Evolution of Biomineralization." *Biomineralization and Biological Metal Accumulation*. Springer Netherlands, 191–203. doi:10.1007/978-94-009-7944-4_17.

Marin, F., and G. Luquet. 2007. "Unusually acidic proteins in biomineralization." In *Handbook of Biomineralization, Vol 1: The Biology of Biominerals Structure Formation*, E. Baeuerlein ed., Wiley-VCH Weinheim, Chapter 16, pp. 273 290.

Mugnaioli, Enrico, Filipe Natalio, Ute Schloßmacher, Xiaohong Wang, Werner E. G. Müller, and Ute Kolb. 2009."Crystalline Nanorods as Possible Templates for the Synthesis of Amorphous Biosilica during Spicule Formation in Demospongiae." *ChemBioChem* 10 (4). John Wiley & Sons, Ltd: 683–689. doi:10.1002/cbic.200800623.

Müller, Werner E. G., Anatoli Krasko, Gaël Le Pennec, and Heinz C Schröder. 2003. "Biochemistry and Cell Biology of Silica Formation in Sponges." *Microscopy Research and Technique* 62: 368–377. doi:10.1002/jemt.10402.

Müller, Werner E. G., Heinz C. Schröder, Zaklina Burghard, Dario Pisignano, and Xiaohong Wang. 2013. "Silicateins—A Novel Paradigm in Bioinorganic Chemistry: Enzymatic Synthesis of Inorganic Polymeric Silica." *Chemistry—A European Journal* 19 (19): 5790–5804. doi:10.1002/chem.201204412.

Müller, Werner E. G., Matthias Wiens, Teresa Adell, Vera Gamulin, Heinz C. Schröder, and Isabel M. Müller. 2004. "Bauplan of Urmetazoa: Basis for Genetic Complexity of Metazoa." *International Review of Cytology* 235: 53–92. doi:10.1016/S0074-7696(04)35002-3.

Murdock, Duncan J. E. 2020. "The 'Biomineralization Toolkit' and the Origin of Animal Skeletons." *Biological Reviews* 95 (5). Biol Rev Camb Philos Soc: 1372–1392. doi:10.1111/brv.12614.

Murdock, Duncan J. E., and Philip C. J. Donoghue. 2011. "Evolutionary Origins of Animal Skeletal Biomineralization." In *Cells Tissues Organs*, 194:98–102. doi:10.1159/000324245.

Natalio, Filipe, Thorben Link, Werner E. G. Müller, Heinz C. Schröder, Fu Zhai Cui, Xiaohong Wang, and Matthias Wiens. 2010. "Bioengineering of the Silica-Polymerizing Enzyme Silicatein-α for a Targeted Application to Hydroxyapatite." *Acta Biomaterialia* 6 (9): 3720–3728. doi:10.1016/j.actbio.2010.03.010.

O'Leary, Patrick, Cornelis A. Van Walree, Nilesh C. Mehendale, Jan Sumerel, Daniel E. Morse, William C. Kaska, Gerard Van Koten, and Robertus J. M. Klein Gebbink. 2009. "Enzymatic Immobilization of Organometallic Species: Biosilification of NCN- and PCP-Pincer Metal Species Using Demosponge Axial Filaments." *Dalton Transactions*, no. 22 (May). The Royal Society of Chemistry: 4289–4291. doi:10.1039/b900341j.

Pacella, Michael S., and Jeffrey J. Gray. 2018. "A Benchmarking Study of Peptide—Biomineral Interactions." *Crystal Growth and Design* 18 (2). American Chemical Society: 607–616. doi:10.1021/ACS.CGD.7B00109.

Parker, Mark D., and Walter F. Boron. 2013. "The Divergence, Actions, Roles, and Relatives of Sodium-Coupled Bicarbonate Transporters." *Physiological Reviews* 93 (2). American Physiological Society: 803. doi:10.1152/PHYSREV.00023.2012.

Pastero, Linda, Emanuele Costa, Marco Bruno, Marco Rubbo, Giulio Sgualdino, and Dino Aquilano. 2004. "Morphology of Calcite (CaCO3) Crystals Growing from Aqueous Solutions in the Presence of Li+ Ions. Surface Behavior of the {0001} Form." *Crystal Growth and Design* 4 (3). American Chemical Society: 485–490. doi:10.1021/cg034217r.

Puverel, S., E. Tambutté, L. Pereira-Mouriès, D. Zoccola, D. Allemand, and S. Tambutté. 2005. "Soluble Organic Matrix of Two Scleractinian Corals: Partial and Comparative Analysis." *Comparative Biochemistry and Physiology—B Biochemistry and Molecular Biology* 141 (4): 480–487. doi:10.1016/j.cbpc.2005.05.013.

Rebek, Julius. 2009. "Introduction to the Molecular Recognition and Self-Assembly Special Feature." *Proceedings of the National Academy of Sciences* 106 (26): 10423–10424. doi:10.1073/PNAS.0905341106.

Romero, Michael F., An Ping Chen, Mark D. Parker, and Walter F. Boron. 2013. "The SLC4 Family of Bicarbonate (HCO3-) Transporters." *Molecular Aspects of Medicine*. doi:10.1016/j.mam.2012.10.008.

Roy, Nathalie Le, Daniel J. Jackson, Benjamin Marie, Paula Ramos-Silva, and Frédéric Marin. 2014. "The Evolution of Metazoan Aα-Carbonic Anhydrases and Their Roles in Calcium Carbonate Biomineralization." *Frontiers in Zoology*. BioMed Central. doi:10.1186/s12983-014-0075-8.

Saito, Yuriko, Tetsuhiko Isobe, and Mamoru Senna. 1995. "Incipient Chemical Reaction on the Scratched Silicon {111} Surface with Ethoxy and Hydroxy Groups." *Journal of Solid State Chemistry* 120 (1). Academic Press Inc.: 96–100. doi:10.1006/JSSC.1995.1382.

Satoh, Shin, Kenzo Susa, and Iwao Matsuyama. 2002. "Sol-Gel Preparation and Optical Properties of SiO2-Ta2O5 Glass." *Journal of Non-Crystalline Solids* 306 (3): 300–308. doi:10.1016/S0022-3093(02)01192-4.

Schröder, Heinz C., Oleksandra Boreiko, Anatoli Krasko, Andreas Reiber, Heiko Schwertner, and Werner E. G. Müller. 2005. "Mineralization of SaOS-2 Cells on Enzymatically (Silicatein) Modified Bioactive Osteoblast-Stimulating Surfaces." *Journal of Biomedical Materials Research Part B: Applied Biomaterials* 75B (2). John Wiley & Sons, Ltd: 387–392. doi:10.1002/JBM.B.30322.

Schröder, Heinz C., Matthias Wiens, Ute Schloßmacher, David Brandt, and Werner E. G. G. Müller. 2012. "Silicatein-Mediated Polycondensation of Orthosilicic Acid: Modeling of a Catalytic Mechanism Involving Ring Formation." 4 (1). Springer: 33–38. https://link.springer.com/article/10.1007/s12633-010-9057-4.

Shimizu, Katsuhiko, Jennifer Cha, Galen D. Stucky, and Daniel E. Morse. 1998. "Silicatein α: Cathepsin L-like Protein in Sponge Biosilica." *Proceedings of the National Academy of Sciences of the United States of America* 95 (11): 6234–6238. doi:10.1073/pnas.95.11.6234.

Simpson, Tracy L. 1989. "Silicification Processes in Sponges: Geodia Asters and the Problem of Morphogenesis of Spicule Shape." In *Origin, Evolution, and Modern Aspects of Biomineralization in Plants and Animals*, edited by Rex E. Crick, 125–136. Boston, MA: Springer US. doi:10.1007/978-1-4757-6114-6_9.

Srivastava, Prateek, Sumit Kumar Hira, Divesh Narayan Srivastava, Vivek Kumar Singh, Uttam Gupta, Ranjeet Singh, Ram Adhar Singh, and Partha Pratim Manna. 2018. "ATP-Decorated Mesoporous Silica for Biomineralization of Calcium Carbonate and P2 Purinergic Receptor-Mediated Antitumor Activity against Aggressive Lymphoma." *ACS Applied Materials and Interfaces* 10 (8). American Chemical Society: 6917–6929. doi:10.1021/acsami.7b18729.

Sumerel, Jan L., Wenjun Yang, David Kisailus, James C. Weaver, Joon Hwan Choi, and Daniel E. Morse. 2003. "Biocatalytically Templated Synthesis of Titanium Dioxide." *Chemistry of Materials* 15 (25). American Chemical Society: 4804–4809. doi:10.1021/cm030254u.

Supuran, Claudiu T. 2016. "Structure and Function of Carbonic Anhydrases." *Biochemical Journal*. doi:10.1042/BCJ20160115.

Tripp, Brian C., Kerry Smith, and James G. Ferry. 2001. "Carbonic Anhydrase: New Insights for an Ancient Enzyme." *Journal of Biological Chemistry*. doi:10.1074/jbc.R100045200.

Uriz, María J. 2006. "Mineral Skeletogenesis in Sponges." *Canadian Journal of Zoology* 84 (2): 322–356. doi:10.1139/Z06-032.

Uriz, Maria J., X. Turon, M. A. Becerro, and G. Agell. 2003. "Siliceous Spicules and Skeleton Frameworks in Sponges: Origin, Diversity, Ultrastructural Patterns, and Biological Functions." *Microscopy Research and Technique* 62 (4). 279–299. doi:10.1002/JEMT.10395.

Veremeichik, Galina N., Yuri N. Shkryl, Victor P. Bulgakov, Sergey V. Shedko, Valery B. Kozhemyako, Svetlana N. Kovalchuk, Vladimir B. Krasokhin, Yuri N. Zhuravlev, and Yuri N. Kulchin. 2011. "Occurrence of a Silicatein Gene in Glass Sponges (Hexactinellida: Porifera)." *Marine Biotechnology* 13 (4): 810–819. doi:10.1007/s10126-010-9343-6.

Voigt, Oliver, Marcin Adamski, Kasia Sluzek, and Maja Adamska. 2014. "Calcareous Sponge Genomes Reveal Complex Evolution of α-Carbonic Anhydrases and Two Key Biomineralization Enzymes." *BMC Evolutionary Biology* 14 (1). doi:10.1186/s12862-014-0230-z.

Wang, Xiaohong, and Werner E. G. Müller. 2009. "Marine Biominerals: Perspectives and Challenges for Polymetallic Nodules and Crusts." *Trends in Biotechnology*. doi:10.1016/j.tibtech.2009.03.004.

Wang, Xiaohong, Heinz C. Schröder, and Werner E. G. Müller. 2014. "Enzyme-Based Biosilica and Biocalcite: Biomaterials for the Future in Regenerative Medicine." *Trends in Biotechnology* 32 (9): 441–447. doi:10.1016/j.tibtech.2014.05.004.

Wang, Zhantong, Peng Huang, Orit Jacobson, Zhe Wang, Yijing Liu, Lisen Lin, Jing Lin, et al. 2016. "Biomineralization-Inspired Synthesis of Copper Sulfide-Ferritin Nanocages as Cancer Theranostics." *ACS Nano* 10 (3): 3453–3460. doi:10.1021/acsnano.5b07521.

Weiner, S., and H. D. Wagner. 1998. "The Material Bone: Structure-Mechanical Function Relations." *Annual Review of Materials Science* 28 (1): 271–298. doi:10.1146/annurev.matsci.28.1.271.

Westbroek, P. 1983. "Introduction Biological Metal Accumulation and Biomineralization in a Geological Perspective." In *Biomineralization and Biological Metal Accumulation*, 1–11. Springer Netherlands. doi:10.1007/978-94-009-7944-4_1.

Wolf, Stephan E., Ute Schlossmacher, Anna Pietuch, Bernd Mathiasch, Heinz C. Schröder, Werner E. G. Müller, and Wolfgang Tremel. 2010. "Formation of Silicones Mediated by the Sponge Enzyme Silicatein-α." *Dalton Transactions* 39 (39). The Royal Society of Chemistry: 9245–9249. doi:10.1039/b921640e.

Xiao, Bing, Xiaoxuan Zhou, Hongxia Xu, Bihan Wu, Ding Hu, Hongjie Hu, Kanyi Pu, et al. 2018. "Integration of Polymerization and Biomineralization as a Strategy to Facilely Synthesize Nanotheranostic Agents." *ACS Nano* 12 (12): 12682–12691. doi:10.1021/acsnano.8b07584.

Yadav, Ramdayal, Rajendra Goud, Abhishek Dutta, Xungai Wang, Minoo Naebe, and Balasubramanian Kandasubramanian. 2018. "Biomimicking of Hierarchal Molluscan Shell Structure Via Layer by Layer 3D Printing." *Industrial & Engineering Chemistry Research* 57 (32). American Chemical Society: 10832–10840. doi:10.1021/ACS.IECR.8B01738.

Yang, Sung Ho, Eun Hyea Ko, Young Hwan Jung, and Insung S. Choi. 2011. "Bioinspired Functionalization of Silica-Encapsulated Yeast Cells." *Angewandte Chemie International Edition* 50 (27). John Wiley & Sons, Ltd: 6115–6118. doi:10.1002/ANIE.201102030.

Zoccola, Didier, Philippe Ganot, Anthony Bertucci, Natacha Caminiti-Segonds, Nathalie Techer, Christian R Voolstra, Manuel Aranda, et al. 2015. "Bicarbonate Transporters in Corals Point towards a Key Step in the Evolution of Cnidarian Calcification." *Scientific Reports* 5 (1). Nature Publishing Group: 1–11. doi:10.1038/srep09983.

22 Nutritional and Health Benefits of Marine Mollusks

Wafa Boulajfene

CONTENTS

22.1 INTRODUCTION

The marine ecosystems have always been Earth's most precious source of food owing to their great specific diversity (Simat et al. 2020). Today they are also considered an important and versatile source of bioactive compounds, and they have earned the label "Natural Medicine Chest of the New Millennium," constituting a supreme reservoir of new molecules identification and marine nutraceuticals development (Simat et al. 2020). Marine organisms, namely, sponges, bryozoans, tunicates, mollusks, algae, fishes and crustaceans, among others, are continuously exploited as valuable sources of natural components in comparison to other terrestrial creatures (Simat et al. 2020). They provide bioactive components such as protein and peptides (collagen, gelatin, and albumins), polysaccharides (carrageenan, agar-agar, fucans, fucanoids, chitin, chitosan, derivatives), ω-3 polyunsaturated fatty acids (PUFA), polyphenolic compounds,

DOI: 10.1201/9781003303909-22

pigments (phlorotannins, β-carotene, chlorophylls, lutein), enzymes (gastric proteases, pepsins, gastricsins, chymosins, serine, cysteine, lipases, transglutaminase) and fat- and water-soluble vitamins (Simat et al. 2020). The health-promoting repercussions of numerous products extracted from marine organisms were admitted following successful clinical trials (Coulson et al. 2013; Kean et al. 2013). It turned out that these components present various properties, such as antiviral, antibacterial, anticoagulant, antioxidant, antihypertensive, antiarthritic, radioprotective, antiparasitic, anti-inflammatory, anti-obesity and anticarcinogenic, among others (Sharanagat et al. 2020; Simat et al. 2020). In fact, the power of those natural nutraceuticals exists in their ability to improve the quality of life, prevent or even treat some conditions without any adverse side effects (Simat et al. 2020). Certainly the safe character of marine nutraceuticals and pharmaceuticals makes them more preferable and may be beneficial to develop nontoxic but efficient natural agents as substitutes to the chemical compounds (Nalini et al. 2018). Thus, more than 20,000 bioactive elements were isolated from marine organisms, but just a minor proportion of them have been meticulously studied and exploited (Simat et al. 2020). In 2018, the international market for marine derived products was over US$10 billion, and it is predicted to rise to US$22 billion by 2025 at a composite annual growth rate of 11.3% from 2019 to 2025 (Infinium Global Research 2019).

Marine mollusks constitute the most interesting group in this regard since they are among the largest and economically important marine organisms. This taxon, determined by Cuvier in 1797, comprises more than 130,000 species with more than 52,000 identified and characterized species (Benkendorff 2010). It gathers together eight classes: Solenogastres, Caudofoveates, Polyplacophores, Monoplacophores, Cephalopods, Bivalves, Scaphopods and Gastropods (Le Cointre and Le Guyader, 2001; Khan and Liu 2019). Out of these eight classes, bivalves (mussels, oysters, clams and scallops), cephalopods (squid, cuttlefish and octopus) and gastropods (whelks, sea snail, cockle and abalone) represent the economically considerable mollusks (Venugopal and Gopakumar 2017). These groups come under edible shellfish that are used worldwide in huge quantities as a traditional functional food, believing consumption provides health benefits. Marine mollusks also get supplementary attention since they are used in other ways such as for making clothes, as a craft, yarn, dying cotton, and the like (Flores-Garza et al. 2012). This extremely significant biodiversity of marine mollusks, as well as their immense efficacy, as a source of food and their beneficial and nutritional values made them greatly considered by the scientific communities. Furthermore, they now play an important role as practical ingredients for the food industries, and they convey various benefits for the health of humans, directly or after processing (Kumari et al. 2020). Indeed, the species belonging to this phylum represent rich sources of chemical diversity and health products, allowing the evolution of nutritional supplements, drug candidates, cosmetics and molecular probes (Kumari et al. 2020). Moreover, they are activators or inhibitors of significant enzymes and transcription factors, competitors of sequestrants and transporters that adjust different physiological pathways. In addition, consuming shellfish as an element of the daily diet can be useful in preventing many ailments given their richness in vital nutrients and active secondary metabolites and their ability to enhance the immune response (Kumari et al. 2020).

This chapter collects information from the existing literature relating to the general availability of marine mollusks, their consumption, pretreatment and manipulating procedures. It also summarizes the use of those organisms in the food industry and the contemporary research carried out for the segregation, classification and characterization of bioactive compounds and their efficiency for health promotion by highlighting their therapeutic features.

22.2 WORLDWIDE MOLLUSKS CONSUMPTION

Contemporary dietary habits and behaviors in developed countries have spawned an increasing number of diseases such as obesity, type 2 diabetes mellitus, cancer, metabolic syndrome or neurodegenerative illnesses (Mateos et al. 2020). Nevertheless, consuming marine products as part of a daily diet has exhibited health advantages since they are full in vital nutrients (Benkendorff 2010). Among marine animals, mollusks are particularly important as a source of bioactive compounds. Diverse species of this phylum, namely, cephalopods, clams and snails, constitute common seafood items in human consumption (Zhukova 2019). In fact, this group is characterized by a wide chemodiversity given that the mode of communication and defense in mollusks are basically through secondary metabolites (Benkendorff 2014). For example, the nudibranchs are a group of gastropods, among others, known to utilize small molecules for many targets, including reproduction, communication and defense against predators (Wayan Mudianta et al. 2014). Thus, the edible mollusks are continuously harvested and cultured, and their consumption is increasing every year (Surm et al. 2015; Khan and Liu 2019). Note that the availability of seafood in 2014 counted up to 167.2 MMT (million metric tons), among which cephalopods counted for 4.3 MMT. Out of the quickly growing cephalopods, ***Illex argentines*** (Argentine short-fin squid; Castellanos, 1960) and ***Dosidicus gigas*** (jumbo flying squid; Orbigny, 1835) accounted for the most marketed squids, while cuttlefish (300,000 tons) and octopus (350,000 tons) were fished in quite unwavering quantities since 2008 (FAO 2016a, 2016b). The constant rise in demand for seafood has paved the way for their ever-increasing aquaculture, culminating in 73.8 MMT overall production in 2014 with an appraised market value of US$160.2 billion (FAO 2016a, 2016b).

The share of molluscan aqua cultural species (16.1 MMT) was evaluated at US$19 billion in 2014 (FAO 2016a). The aquaculture of mollusks in Asia generally comprises farmed squids (***Loligo duvauceli***, Orbigny, 1848, ***Doryteuthis sibogae***; Adam, 1954; and ***Sepioteuthis sp.***), cuttlefish (***Sepia pharaonic***, Ehrenberg, 1831; ***Sepia aculeate***, Van Hasselt, 1835; ***Sepia officinalis***, Linnaeus, 1758; and ***Sepia elliptica***, Hoyle, 1885) and scallop (***Pecten yesoensis***, Jay, 1857) (FAO 2016a, 2016b), while abalone (***Haliotis spp.***) aquacultures are mostly found in China (Lou et al. 2013). The amplified attention in abalone aquaculture resulted in a fivefold increase in global abalone supply throughout the past four decades

(Suleria et al. 2017). In addition, aquaculture is also in tendency for other mollusks like ***Ruditapes philippinarum*** (Manila clam; Adams and Reeve, 1850), ***Patinopecten yessoensis*** (Yesso scallop, Jay, 1857), ***Mytilus edulis*** (blue mussel, Linnaeus, 1758), ***Perna viridis*** (Asian green mussel, Linnaeus, 1758), ***Perna canaliculus*** (green-lipped mussel; Gmelin, 1791) and ***Anadara granosa*** (Linnaeus, 1758; Khan and Liu 2019). Furthermore, a yearly rise of 39% is anticipated for aquaculture, with the resultant annually production of approximately 102 MMT in 2025 (FAO 2016a).

22.3 EFFECTS OF PRETREATMENTS, CONSERVATION, AND COOKING

The evolution of optimized nutraceuticals and pharmaceuticals from marine natural products requires advanced extraction and purification procedures in order to intensify the output and the quality of the bioactive components. Conventional liquid solvent extraction systems provoke a number of perturbations for downstream applications. Disadvantages consist of inherent toxicity from organic volatile impurities in pharmaceutical preparations, variability in income, stabilization of the active compound and the relative expense of analytical class solvents (Reverchon et al. 2006; Wakimoto et al. 2011). As for the conservation process, extended storage of mollusks exhibits some troubles due to their delicate nature, thereby requiring some particular preservation techniques (Khan and Liu 2019). The accumulation of microorganisms and particulate matters is mainstream in mollusks, namely, mussels, oysters and clams duly to continuous filtration of immense quantities of water. In addition, prolonged freeze storage may generate a metal-catalyzed oxidation of polyunsaturated fatty acids (PUFAs; Huss 2003; Venugopal 2006). In parallel, an increased moisture rate, lowered nitrogen rates and altered viscoelastic properties are related to storage in frost (Binsi et al. 2007). Such mollusks should instantly be kept in water for a couple of days in order to depurate them (Venugopal and Gopakumar 2017). Oysters are exposed, after washing, to steam action or infrared heating to open up their shells (Martin and Hall 2006). Cephalopods ought to be cautiously cleaned and immediately iced up. Their exposure to direct sunshine or wind must be avoided due to their extremely delicate nature (Kreuzer 1984). Note that commercially important cephalopods are frozen by passing them through chilled air or by adding cryogens (liquid N2) for the purpose of transnational trade (Venugopal 2006; Gokoglu and Yerlikaya 2015). Textural modifications, lipid oxidation, altered protein functionality and modified flavor during defrosting generate diminished quality next to extended freeze storage (Venugopal 2006). Squid rings or fillets and scallops can be individually quick frozen (IQF) in expedient "ready to-cook" quantities (Venugopal and Gopakumar 2017). A breadcrumb coating before freezing may also be made for common bivalve IQF products, such as cockles, vacuum-packaged half-shell oysters, scallops and clams (Venugopal 2006). Some supplementary methods may also be carried out to conserve seafoods other than freezing, such as heat treatment, breading, altered atmosphere packaging (MAP), dehydration and high hydrostatic pressure (HHP) (Kreuzer 1984; Venugopal 2006).

Actually, numerous Asian communities consume molluscan species entirely for the associated health-promoting effects (Kim and Pallela 2012). It is, therefore, obvious that some molluscan bioactivities are retained even after cooking and are not lost during digestion in the gastrointestinal tract. Moreover, no important modifications in molluscan contiguous compositions are correlated to thermal treatments such as boiling or steaming (Kreuzer 1984; Su and Liu 2013; Venugopal 2006). Conversely, standard cooking normally rises digestibility of proteins by denaturing them (Venugopal 2006) and can lead to the production of bioactive peptides from the corresponding proteins (Suleria et al. 2015). Indeed, protein-reacting highly reactive peroxides may be formed in correlation with PUFA oxidation as a result of oxygen-supplemented heating or even sun-drying (Huss 2003). Furthermore, Kalogeropoulos et al. (2004) reported an enhancement in the levels of squalene (skin protectant) and monounsaturated fatty acids in mussel and squid when fried in olive oil, while Ozogul et al. (2015) suggested a substantial decline in the contents of sitosterol and cholesterol related to frying. Likewise, riboflavin (bioactive vitamin B2) remains quite unchanged during cooking (Venugopal 2006). However, prolonged cooking destroys vitamin B6, while both storage and cooking result in fractional degradation of vitamin B12, resulting loss of its bioactivity. Similarly, lowered pH, heat and ionization affect thiamin (vitamin B1).

22.4 FUNCTIONAL FOODS, NUTRACEUTICALS, AND HEALTH-PROMOTING EFFICACIES OF MARINE MOLLUSKS

It has been well known for a long time that the choice of consumed foods has a major outcome on our well-being, environment, and society (Khan and Liu 2019). Historically, marine mollusks (soft tissue, basal parts, mucilage, even shells) constituted an essential ingredient in traditional Chinese medicine recipes, based on data carried throughout generations and on repeated trials and errors, for curing a huge number of medical conditions including gastrointestinal pains, dotage, cancer, menstrual disarrays, asthma, influenza, tuberculosis, pneumonia, skin abscesses, eczema, burns, osteoporosis and osteoarthritis (Kehinde et al. 2015; Ahmad et al. 2018). Mollusks were also employed against many ailments by the medieval Eastern Mediterranean and European societies, primeval Greco-Romans and in India, Latin America and Zimbabwe (Lev and Amar 2007; Ahmad et al. 2018). Such prevalent efficiency of mollusks in several civilizations throughout history has prompted the scientists to induct in vitro investigations of this diverse phylum (Khan and Liu 2019). The results from such efforts are absolutely promising and overall validating the therapeutic efficacy of these organisms against a wide array of infections and medical disorders. Thus, it is actually acknowledged that the medicinal value of these invertebrates comes from the presence of several active ingredients in the form of carbohydrates, proteins, lipids, minerals, nucleosides, sterols among others (Grienke et al. 2014; Kehinde et al. 2015; Cheong et al. 2017). In fact, mollusks' abundance and simplicity to be collected and bred make of them suitable candidates for a number of industrial processing for functional foods and nutraceuticals (Khan and Liu 2019). Functional foods are defined as food products that generate extra beneficial health values beyond their normal nutritive effect, such as additional calcium in orange juice, omega-3 fatty acids in eggs or lycopene in tomatoes (Khan and Liu 2019). On the other hand, nutraceuticals are food-derived products referring to health-promoting and/or disease-preventing bioactive components found in foods, dietary supplements and herbal products such as polyphenols, terpenoids, glucosinolates and sulfur-containing elements of the Alliaceae family (Asparagales; GRS 2018). They are marketed as medicines rather than foods, in the form of powder, tablets, drops, syrup and capsules (Khan and Liu 2019).

Note that the researches on marine mollusks bioactive peptides, proteins and amino acids are continuous with intent to also determine their applications (Chi Fai Cheung et al. 2015). In fact, the security and eligibility of functional foods and food-derived products for human consumption is a point demanding serious considerations in the food industry (Khan and Liu 2019). That is why producers are increasingly tending to use safe biological isolates that substitute their eyebrow-raising synthetic counterparts and to valid analytical instrumentation that fully analyze the ingredients during manufacturing and the finished food products before they are distributed. This assessment will not only ensure the effects expected from such products but also guarantee their safety and estimate their shelf life (Khan and Liu 2019). Marine mollusks provide bioactive peptides with properties such as antimicrobial, cardioprotective (anticoagulant, antihypertensive, anti-atherosclerotic), antioxidant, radioprotective, antiparasitic, anti-inflammatory and anticancerous activities (Thakur 2020). The large bioactivity spectrum of those peptides has high potential medicinal values that attract the attention of the pharmaceutical and nutraceutical industry (Khan and Liu 2019).

22.4.1 Antimicrobial Potency

Yet mollusks are devoid of a well-defined adaptive immune system; they utilize various chemical antimicrobial agents (proteins and glycoproteins) as a defense against infections (viral, fungal and bacterial) in addition to their corporal barriers represented by the epithelium, the shell and the operculum in some (Hooper et al. 2007; Liu et al. 2009). Thus, several commercial antimicrobial medicines owe their existence either to their containment of secondary metabolites isolated from living organisms or to in vitro synthesis inspired from physiological and biochemical defense mechanisms of these organisms (Khan and Liu 2019). Until now, only about 0.3% of known molluscan species were exploited for the extraction of more than 1120 natural products, of which less than 50% are being tested for their pharmacological value (Benkendorff 2010; 2014). Moreover, just 6% of the isolated metabolites were studied regarding antimicrobial efficacy (Benkendorff 2014; Dang et al. 2015).

The number of viruses in oceans is enormous (about 107 particles/milliliter) and the number of marine viral infections is estimated at 1023 infections/second (Suttle 2007). Some metabolites isolated from eight gastropods (***Rapana venosa***, Valenciennes, 1846; ***Haliotis rubra***, Leach, 1814; ***Haliotis laevigata***, Donovan, 1808; ***Haliotis rufescens***, Swainson, 1822; *Buccinulum corneum*, Linnaeus, 1758; *Buccinum undatum*, Linnaeus, 1758; *Tegula gallina*, Forbes, 1850; and *Littorina littorea*, Linnaeus, 1758) and nine bivalves (*Ruditapes philippinarum*, Adams & Reeve, 1850; *Mercenaria mercenaria*, Linnaeus, 1758; *Mytilus galloprovincialis*, Lamarck, 1819; *Mya arenaria*, Linnaeus, 1758; *Cerastoderma edule*, Linnaeus, 1758; *Crenomytilus grayanus*, Dunker, 1853; *Crassostrea gigas*, Thunberg, 1793; *Crassostrea virginica*, Gmelin, 1791; and *Ostrea edulis*, Linnaeus, 1758) were described as powerful against many human viruses (Khan and Liu 2019). For examples, hemocyanin (from *Rapana venosa*, Valenciennes, 1846) was perceived active against respiratory syncytial virus, herpes simplex virus (HSV)-1 and -2 and Epstein–Barr virus (EBV; Genova-Kalou et al. 2008; Dolashka-Angelova et al. 2009; Dolashka et al. 2010); kelletinin A (from *Buccinulum corneum*, Linnaeus, 1758) against human T-cell leukemia virus type 1 (Orlando et al. 1996), mytilin (from *Mytilus galloprovincialis*, Lamarck, 1819) against white spot syndrome virus (Dupuy et al. 2004), defensin (also from *Mytilus galloprovincialis*) against human immunodeficiency virus type 1 (HIV-1; Roch et al. 2004), while lectin (from *Crenomytilus grayanus*, Dunker, 1853) against HIV (Luk'yanov et al. 2007). Moreover, some representatives of the families Podoviridae (T3 coliphage; Bachère et al. 1990), Adenoviridae (human adenovirus type 5; Carriel-Gomes et al. 2007), Herpesviridae (HSV type 1) and Birnaviridae (infectious pancreatic necrosis virus; Olicard et al. 2005) turned out vulnerable to active ingredients present in oyster hemolymph. These active elements apparently neutralize invading viruses either by directly inactivating them and preventing their binding to or access into target cells or by inhibiting their replication and transcription (Dang et al. 2015). Thereby, several mollusks constitute widely available and economically viable sources of antiviral compounds.

Furthermore, marine mollusks provide an extended collection of other antimicrobial compounds including chlorinated acetylenes, indole alkaloids, peptides and glycol proteins (Khan and Liu 2019). For instance, the presence of antimicrobial peptides was reported in the mucus of *Achatina fulica* (giant snail, Férussac, 1821; (Kubota et al. 1985), in the egg mass and purple fluid of *Aplysia kurodai* (Baba, 1937) and in the mantel of *Dolabella auricularia* Lightfoot, 1786; (Khan and Liu 2019). Furthermore, human infectious agents (*Klebsiella pneumoniae*, Trevisan, 1887, and *Proteus mirabilis*, Hauser, 1885), fish pathogen (*Aeromonas hydrophila*, Stanier, 1943), pathogenic fungi (*Aspergillus niger*, van Tieghem, 1867) and *Candida albicans* (Berkhout, 1923) and the biofilm-forming *Micrococus* sp. were found to be extremely vulnerable to active substances isolated from two marine molluscan species: *Thais tissoti* (Petit de la Saussaye, 1852) and *Babylonia spirata* (Linnaeus, 1758; Kumaran et al. 2011). Similarly, compounds isolated from "the giant tun" (*Tonna galea*, Linnaeus, 1758) turned out efficient against the gram-negative *Vibrio cholerae*

(Pacini, 1854) and *Aeromonas hydrophila* (Stanier, 1943; Santhi et al. 2011), while extracts from *Babylonia zeylanica* (Bruguière, 1789) seem to inhibit the growth of *Escherichia coli* (Escherich, 1885), *Aeromonas hydrophila* (Stanier, 1943) and *Salmonella typhi* (Lignières, 1900; Santhi et al. 2013). Correspondingly, antibacterial and antifungal peptides were purified from the hemolymph of *Mytilus edulis*, while an antimicrobial peptide from the mRNA and plasma of *Mytilus galloprovincialis* (Khan and Liu 2019). Moreover, four proline rich antimicrobial peptides were isolated from the hemolymph of *Rapana venosa* belonging to Muricidae family (Dolashka et al. 2011). In fact, these identified peptides owe their high potential nutraceutical and pharmaceutical values to their wide spectra of bioactivities (Chi Fai Cheung et al. 2015).

In addition, as part of a screening research on the antimicrobial characteristics of molluscan egg masses, *Dicathais orbita* (Gmelin, 1791) was acknowledged as a species of particular interest, providing the lipophylic extracts that show strong activity against a range of human and marine bacterial pathogens (Benkendorff 1999). Bio-guided fractionation reported that the brominated indole precursors of Tyrian purple is responsible for this activity (Benkendorff et al. 2000). Based on this action, Benkendorff et al. (2000) suggested that resistance of the developing muricidae embryos against marine pathogens could be the natural role of these brominated indoles in the evolution occurrence. In agreement with this, the surface of the egg capsules of the gastropod *D. orbita* was found to shelter low levels of bacterial biofilm, with high percentages of dead bacteria estimated by live/dead bacterial coloration (Lim et al. 2007). The same egg capsules were also shown to hold no protists on their surfaces and were relatively free of algal fouling compared to other species egg masses (Przeslawski and Benkendorff 2005). This low surface fouling seems to be due to a combination of chemical, physical and mechanical protection mechanisms preventing bacterial attachment on the eggs surface (Lim et al. 2007). Nutracuticals examples comprise the Paolin which is a drug made from abalones (gastropods) juice constituting an effective inhibitor of penicillin-resistant strains of bacteria (Kim and Pallela 2012).

22.4.2 Anti-Inflammation, Immune Modulation, and Wound Healing

Steroidal and nonsteroidal anti-inflammatory agents are the common drugs for inflammations managing (Borquaye et al. 2017). However, reports of disturbing side effects necessitated research for new anti-inflammatory agents that provoke minimal complications. Marine-derived natural products contribute significantly in the pharmaceutical, nutraceutical and cosmeceutical industries, and a number of extracts and compounds from marine origin have shown potential as anti-inflammatory agents (Borquaye et al. 2017). In this context, numerous marine molluscan products and derivatives were used in customary medicines for the treatment of inflammatory conditions since mollusks are known to provide several extracts of lipids containing fatty acids, beneficial PUFAs, vitamin E and sterols. Examples of mollusks generating lipid extracts include the New Zealand green-lipped mussel (*Perna canaliculus*, Gmelin, 1791), the Indian green mussel (*Perna viridis*, Linnaeus, 1758), *Filopaludina bengalensis* (Lamarck, 1822), *Mytilus unguiculatus* (hard-shelled mussel, Valenciennes, 1858), sea hares (*Aplysia fasciata*, Poiret, 1789 and *Aplysia punctata*, Cuvier, 1803), among others (Ahmad et al. 2019). Reports from clinical trials advocated the effectiveness of those lipid extracts against familiar anti-arthritis agents, rheumatism, moderate asthma in kids, knee osteoarthritis, hypertension, skin burns, injuries and gastrointestinal dysfunctions, producing nonsignificant modifications in the intestine microbiota (Coulson et al. 2013; Lyprinol 2018). For instance, the lipid extracts derived from the muricid *Rapana venosa* (Valenciennes, 1846) have very good anti-inflammatory and healing effects in skin burns in Wistar rats. From a histological point of view it was revealed that mice treated with *Rapana venosa* lipid extracts have a reduced healing time of no more than 13–15 days instead of 20–22 days in control animals (Kumari et al. 2020). Complete regeneration of the skin (dermis, epidermis, hypodermis) was observed in the presence of newly formed blood vessels and new epithelium in the provisional fibrin matrix collagen fibers and basal membrane. Other than lipids, amino acid extracts from the same gastropod were also noticed to accelerate the skin healing by intensifying the epidermal and dermal neoformation in Wistar rats (Kumari et al. 2020).

In a similar way, some supplementary essential amino acids (EAAs) are recently extracted from marine mollusks and employed in vitro evaluations of anti-inflammatory and immune-modulatory efficacies (Ahmad et al. 2018). For example, brominated indole precursors of the dye Tyrian purple are amid the notable biologically active natural elements derived from muricidae species (Khan and Liu 2019). A fusion of brominated and nonbrominated indirubin with indigo was disclosed in the purple secretion of some muricidae species. In fact, indirubin helps prove the increase in reactive oxygen species (ROS) from macrophage cells (Khan and Liu 2019). Correspondingly, anti-inflammatory effect is also linked to a further indirubin precursor called isatin, discovered to inhibit formation of prostaglandin E2 (PGE2), nitric oxide (NO), cyclooxgenase 2 (COX-2), tumor necrosis factor-alpha (TNF-α) and inducible nitric oxide synthase (iNOS) in a lipopolysaccharide (LPS; Khan and Liu 2019). Similarly, the potency of 6-bromoisatin (oxidation product), a biologically available multifunctional chemical agent of supreme importance, was effectively and safely demonstrated in animal models regarding the prevention of acute lung inflammation (Ahmad et al. 2018; Khan and Liu 2019). Additionally, homogenized-tissue supernatant isolated from *Filopaludina bengalensis* (Lamarck, 1822) expressively enhanced the levels of hydroxyproline, calcium, phosphate, glucosamine and creatinine in urine of osteoarthritic and osteoporotic rats. In addition, the levels of serum acid phosphatase (ACP), alkaline phosphatase (ALP), tartrate-resistant acid phosphatase (TRAP), creatinine, calcium, TNF- α, IL-1β and cytokine-induced neutrophil chemoattractant-1 also explained obvious increase (Sarkar et al. 2013). On the other hand, another natural bioactive EAA called taurine, derived from mollusks and other marine phyla, showed several physiological and biological properties in humans, namely the stabilization of cell membranes, helping the development of the retina (the innermost light-sensitive layer of the eye) and the central nervous system and immune-modulatory effects (Simat et al. 2020). As well, ground pearls from oysters were proven to be an anti-inflammatory agent in a painful condition called conjunctivitis, during which the conjunctiva becomes inflamed and painful (Kim and Pallela 2012). As for cephalopods, their extracts were also verified to have anti-inflammation properties in vivo experimental works. In this context, writhing provoked in mice by acetic acid was inhibited while mice latency period was extended by application of *Sepia officinalis* (Linnaeus, 1758) ink extracts (Khan and Liu 2019). Correspondingly, paw edema induced by either carrageenan or formalin reacted positively to liver oil from *Sepia pharaonis* (Ehrenberg, 1831; Khan and Liu 2019). Also, cuttlefish bones are employed as cure for rachitis, as healing agent in the remedy of gastrointestinal troubles, as local antihemorrhagic and as an antiseptic in the case of inflammation of the middle ear (Kim and Pallela 2012). Additionally, an increase in capillary permeability, as well as inhibition of leukocytes movement and protein exudation into pouch fluid in mouse granuloma-pouch model was reported for *Ommastrephes bartramii* (Lesueur, 1821; Khan and Liu 2019).

In spite of the promising anti-inflammatory action of lipid extracts from mollusks, their marketing remains restricted and the only natural anti-inflammatory nutraceuticals available over the counter are Lyprinol and Biolane derived from the lipid extracts of the New Zealand green-lipped mussel *Perna canaliculus* (Gmelin, 1791) and Cadalmin ™, the lipid extract isolated from the related species of bivalve *Perna viridis* (Linnaeus, 1758) in India (Ahmad et al. 2019).

22.4.3 Antitumor Efficacy

Marine mollusks supply wide range of anticancer peptides. For instance, the marine hare *Dolabella auricularia* (Lightfoot, 1786) generate a group of cyclic and linear peptides that significantly suppress cell growth. Examples of marine-derived peptides include dolastatin 10, which greatly contributed to clinical trials of phase I state patients suffering from solid tumor (Shukla 2016). Dolastatin 10 is a pentapeptide providing several exclusive amino acid subunits (Pangestuti and Kim 2017). Cytotoxic activity of this peptide against mouse lymphocytic leukemia (L1210), human promyelocytic leukemia (HL-60), human acute myelomonocytic leukemia (ML-2), human monocytic (THP-1), multiple lymphoma, small-cell lung cancer (NCI-H69, -H82, -H446, and -H510) and PC-3 cells were reported (Pangestuti and Kim 2017). Previously, marine cyclic depsipeptides, belonging to Kahalalides family, were isolated from the Hawaiian marine mollusk *Elysia rufescens* (Pease, 1871; Shukla 2016). Among the seven isolated Kahalalides (A—F), Kahalalide F showed significant cytotoxic activity against cell lines and tumor specimens derived from various human solid tumors including melanoma and prostate, breast, lung, ovarian and colon carcinomas. It probably induces cell death via oncosis preferentially in tumor cells (Shukla 2016; Pangestuti and Kim 2017). Another cytotoxic peptide isolated from *Pleurobranchus forskalii* is the Keenamide A. This hexapeptide revealed considerable activity against the P-388, MEL-20, A-549 and HT-29 tumor cell lines (Pangestuti and Kim 2017). Moreover, Liu et al. (2012) discovered a linier peptides (Mere15) derived from the bivalve *Meretrix meretrix* (Linnaeus, 1758). Mere15 inhibited the multiplication of leukemia (K562) cells, and its cytotoxicity was correlated to the apoptosis induction, cell-cycle interruption and microtubule disassembly. A further antitumor agent ES-285 was obtained in the North Atlantic clam *Mactromeris polynyma* (Stimpson, 1860; known also as *Spisula polynyma*, Stimpson, 1860; Rudd and Benkendorff 2014). Furthermore, species belonging to the family of Chromodorididae (order Nudibranchia) commonly provide oxygenated diterpene metabolites (Wayan Mudianta et al. 2014). Terpenes 1 and 2 both showed bioactivity against P388 leukemia cells in mice and were later isolated from the Chinese nudibranch *Chromodoris sinensis* (Rudman, 1985; now known as *Goniobranchus sinensis*, Rudman, 1985; Wayan Mudianta et al. 2014).

As muricidae represent a conventional element of African, European, Mediterranean and Asian diets, there is excellent ability for the development of the Australian gastropod *Dicathais orbita* (Gmelin, 1791) as a novel medicinal food, especially for colorectal cancer prevention (Benkendorff 2013). Organic extracts from *D. orbita* hypobranchial glands, egg masses and mucus secretion efficiently inhibit the proliferation of some cancer cell lines. Bioassay guided fractionations reported that this activity is mainly related to the brominated indoles tyrindoleninone, tyrindolinone and 6-bromoisatin. These derivatives contained in a crude chloroform extract were found to prevent primary-stage tumor formation by stimulating the critical apoptotic response to DNA-damage in the distal colon of mice (Benkendorff 2013). In addition, the muricid *D. orbita* is characterized by the biosynthesis of various biologically active secondary metabolites. Early research concentrated on the identification of the precursors to the Tyrian purple dye and disclosed a fascinating interrelation between these brominated indole precursors and choline esters (Benkendorff 2013). The muscle-relaxant, neurotoxic and anticancer properties of choline esters derived from muricidae were well reported in the literature. Thus, the fusion of biologically active compounds provided by *D. orbita* generates interesting capability for nutraceutical progress, but the separation of choline esters, from the brominated indole fraction, is crucial since the choline esters are toxic at higher concentrations (Rudd and Benkendorff 2014). Until now, the bioactive indole compounds are only extracted with chlorinated solvents, which are improper for human use (Benkendorff 2013; Rudd and Benkendorff 2014).

Additionally, it was confirmed that n-3 PUFAs, mostly isolated from gastropods, possess a potential for prevention and remedy of several types of cancers and can also improve the efficacy and acceptability of chemotherapy (Zhukova 2019). This promising effect of n-3 PUFAs on certain types of cancer is explained by their ability to modulate membrane-associated signal transductions and gene expression involved in cancer pathogenesis, as well as to suppress systemic inflammation (Zhukova 2019). Dietary intake of these essential components, as substances with therapeutic action, may maintain health, prevent the development of many diseases and mitigate a number of pathological conditions. Nevertheless, n-6 PUFAs was reported, according to other studies, to induce progression of certain types of cancer (Huerta-Yépez et al. 2016). Epidemiological studies suggested a correlation between a polyunsaturated fatty acid supplemented diet and the proliferation of some types of cancer, namely colon and colorectal carcinoma, neuroblastoma and breast, prostate and lung cancers (Huerta-Yépez et al. 2016). Indeed, the diet of gastropods, represented by the greatest number of species, differs according to their trophic group. Their trophic specialization influence the composition of their fatty acids, particularly PUFAs only biosynthesized by microalgae and protozoa, which can differ fundamentally with diverse diets and become an essential dietary component for higher trophic levels (Zhukova 2019).

22.4.4 Cardioprotective and Antioxidative Efficacies

Many beneficial cardiovascular effects were attributed to PUFAs, isolated from marine mollusks, including hypolipidemic, antithrombotic, antihypertensive, anti-inflammatory and antiarrhythmic properties, as well as the mitigation of coronary heart disease (Zhukova 2019). The effectiveness of n-3 polyunsaturated fatty acids for the prevention of cardiovascular diseases (CVDs) is based on multiple molecular operations, including membrane change whereby n-3 PUFAs are incorporated into the lipid bilayer and affect membrane fluidity, of lipid micro-domains formation and other mechanisms, such as the attenuation of ion channels, the regulation of pro-inflammatory gene expression and the production of lipid mediators (Zhukova 2019). Thus, the use of n-3 PUFAs is recommended for mitigating the CVD risk factors.

On the other hand, EAA supplements of cysteine, leucine, histidine, methionine, proline, hydroxyproline, tyrosine, threonine, trans-4-hydroxy-proline and valine, derived from marine mollusks, showed an antioxidant activity due to drastic scavenging activity and lipid peroxidation inhibition (Simat et al. 2020). They help with human homeostasis, principally due to their function, in the regulation of various cellular mechanisms and act as precursors of other molecules (e.g., nitrogenous bases and hormones) and as protein building blocks (Simat et al. 2020).

22.4.5 Neuroprotective and Antidiabetic Activities

The docosahexaenoic acid (DHA) was described as the chief n-3 fatty acid within the brain and the retina since it ensures a central role in neural functions (Zhukova 2019). This fatty acid displays neuroprotective properties and constitutes a potential treatment against a variety of neurological and neurodegenerative disorders (Zhukova 2019). Its completely beneficial effect in preventing or mitigating age-related cognitive decline was proved by a clinical study. The DHA, as well as all the n-3 LC-PUFAs, deploys positive impacts on brain structure and memory functioning in healthy seniors and supports the neurological progress of the infant brain by improving the cognitive performance relating to knowledge, memory and level of self-performance of cognitive tasks

(Zhukova 2019). Ethyl eicosapentaenoic acid (EPA) and DHA play an essential role in neurotransmission and neuronal cell functions, as well as in inflammatory and immune activities that are involved in neuropsychiatric disease states (Zhukova 2019). Moreover, it was reported that low intake of dietary EPA and DHA is associated with increased risk of the development of Alzheimer's disease, as well as with poor fetal development, including neuronal, retinal and immune function (Zhukova 2019). Low maternal supplementation of DHA could also increase the risk of early preterm birth and asthma in children. Notice that an intake of 1 g/day of PUFAs, either in capsules or by marine mollusks products, confirmed a preventive action against metabolic syndrome as hyper-triglyceridemia, hyperlipidemia, or type 2 diabetes (Zhukova 2019).

As a part of the ongoing research program to explore novel sources of bioactivities from the marine organisms, different species of deepsea cephalopods namely *Cystopus indicus*, *Sepiella inermis* and *Amphioctopus marginatus* were identified as potential sources of bioactive leads with respect to antidiabetic properties (Chakraborty and Joy 2017). The utilities of 1H-NMR to assess the abundance of the bioactive functional groups present in the EtOAc-MeOH extracts of the cephalopod species and to illustrate the principles regarding the presence of these functional groups vis-à-vis antidiabetic activities were demonstrated. Thus, some cephalopod species can potentially be considered as new sources of important health foods and can be used in formulating various nutraceuticals and functional food ingredients in combating diabetes and metabolic disorders (Chakraborty and Joy 2017).

22.4.6 Ailments Suppressing

Some marine mollusks could be particularly useful as nutraceticals or medicinal foods since they provide extracts containing muscle-relaxing and analgesic properties (Benkendorff 2013). For instance, a Food and Drug Administration ratified commercial drug in 2004, Prialt®, is an analgesic agent administered for approximately one decade for the cure of chronic pains (Chi Fai Cheung et al. 2015; Khan and Liu 2019). It is essentially based on ziconotide extracted from the venom of the gastropod *Conus magus* Linnaeus, 1758 (predatory Indo-Pacific cone snail) (Khan and Liu 2019). Ziconotide showed interestingly hundred times higher analgesic property than standard morphine in animal models. It turned out effective in blocking N-type calcium channels (Shukla 2016; Khan and Liu 2019), thereby potent in providing relief during enduring pains. Thus, as part of everyday diet, this snail can be helpful in avoiding many ailments (Benkendorff 2010). Similarly, the pain-relieving efficacy of a commercial extract from the mollusk *Perna canaliculus* (BioLex) in osteoarthritis was investigated in a clinical trial, which advocated its efficiency as a long-term treatment based on observations in the postintervention period (Khan and Liu 2019).

On the other hand, the mixture of a range of elements with various bioactivities contained in the muscle-sedative extracts derived from the murcid *Dicathais orbita* (Gmelin, 1791) is also of particular significance for nutraceuticals progress. In fact, the 5-bromoisatin isolated from this gastropod, was patented as an analgesic with relaxing activity that reduce bleeding time in mice (Benkendorff 2013).

22.4.7 Other Benefits

Many groups of mollusks widely participate in the world's consumption of marine food since they provide diverse bioactive peptides that play a major role in human well-being. For example, abalones are key marine animals possessing significant medical importance (Kim and Pallela 2012). Besides the sale of soft bodies in food markets, even dried abalones are sold in medical shops in Hong Kong, Singapore and Southeast Asia. Refined and handled abalone shells are particularly regarded as cures for some eye diseases and are used as a calcium supplements (Kim and Pallela 2012). Furthermore, the powdered flat shell of the abalone can be taken orally to ameliorate vision, mitigate keratoses (cataracts) and control some conditions such as hemeralopia. Likewise, oyster shells have a wide variety of uses in Vietnam: Powdered oyster shell is orally administered to treat acid indigestion and fatigue and to stop hemorrhage (Kim and Pallela 2012).

22.5 CONCLUSION

Nowadays, the food industry becomes among the most important manufacturing sectors and its significance in provisioning the dietary needs and economy is further improved with the advent of contemporary functional food and nutraceutical markets. The abundance and diversity of marine organisms make of them important elements in the supply chain of this growing manufacturing sector. Apart from being an important source of food, marine organisms also attract the interest of scientific communities given their role in the functional foods and nutraceuticals industries. Mollusks, representing the second-largest group of marine animals with confirmed medicinal values, are relatively easy to collect and constitute the most interesting group of organisms in this regard. Their consumption as part of everyday diet seems to provide various health benefits as they are rich in vital nutrients and active secondary metabolites. Their several properties, namely, antimicrobial, antioxidative, immunomodulatory, antitumor, analgesic, cardioprotective, antidiabetic and neuroprotective, deserve the consideration of the pharmaceutical industry, which contemplates using them in the treatment or prevention of various diseases. Some marine molluscan peptides or their derivatives have reached the pharmaceutical and nutraceutical markets owing to their high commercial values.

Marine mollusks constitute, therefore, a profuse source of natural products that can potentially be used as flagship components in drug discovery or inspire organic chemists as synthetic targets. Nevertheless, despite the great scientific attention this phylum has received and despite the fact that a number of commercial nutraceuticals of molluscan origin are widely marketed, the number of compounds isolated is not proportionate with the number of species therein. In fact, chemical investigations have only been performed on about 0.4% of molluscan species, as many studies report analyses of the same species in various ecological habitats. The promising potential of shellfish and products needs more assessment and validation in order to fully exploit this huge source of food and medicine. The nutraceuticals and functional foods industry has obviously profited from the work available in this regard and these gains can be multiplied by many efforts in this way.

REFERENCES

Ahmad, T. B., Liu, L., Kotiw, M., and Benkendorff, K. 2018. Review of anti-inflammatory, immune-modulatory and wound healing properties of molluscs. *Journal of Ethnopharmacology* 210: 156–178.

Ahmad, T. B., Rudd, D., Kotiw, M., Liu, L., and Benkendorff, K. 2019. Correlation between fatty acid profile and anti-inflammatory activity in common australian seafood by-products. *Marine Drugs* 17(3): 155.

Bachère, E., Hervio, D., Mialhe, E., and Grizel, H. 1990. Evidence of neutralizing activity against T3 coliphage in oyster *Crassostrea gigas* hemolymph. *Developmental & Comparative Immunology* 14(3): 261–268.

Benkendorff, K. 1999. Molluscan resources: The past present and future value of molluscs. In *The Other 99%. The Conservation and Biodiversity of Invertebrates*, eds. W. Ponder and D. Lunney, p. 454. Sydney, Australia: Mosman.

Benkendorff, K. 2010. Molluscan biological and chemical diversity: Secondary metabolites and medicinal resources produced by marine molluscs. *Biological Reviews* 85(4): 757–775.

Benkendorff, K. 2013. Natural Product Research in the Australian Marine Invertebrate *Dicathais orbita*. *Marine Drugs* 11: 1370–1398.

Benkendorff, K. 2014. Chemical diversity in molluscan communities: From natural products to chemical ecology. In *Neuroecology and Neuroethology in Molluscs: The Interface between Behaviour and Environment*, eds. A. Cosmo and W. Winlow, pp. 13–41. New York, NY: Nova Science

Benkendorff, K., Bremner, J. B., and Davis, A. R. 2000. Tyrian purple precursors in the egg masses of the Australian muricid, *Dicathais orbita*: A possible defensive role. *Journal of Chemical Ecology* 26: 1037–1050.

Binsi, P., Shamasundar, B., and Dileep, A. 2007. Physico-chemical and functional properties of proteins from green mussel (Perna viridis) during ice storage. *Journal of the Science of Food and Agriculture* 87(2): 245–254.

Borquaye, L. S., Darko, G., Laryea, M. K., Roberts, V., Boateng, R., and Gasu, E. N. 2017. Anti-inflammatory activities of extracts from *Oliva sp.*, *Patella rustica*, and *Littorina littorea* collected from Ghana's coastal shorelines. *Cogent Biology* 3: 1.

Carriel-Gomes, M. C., Kratz, J. M., Barracco, M. A., Bachère, E., Barardi, C. R. M., and Sim˜oes, C. M. O. 2007. In vitro antiviral activity of antimicrobial peptides against herpes simplex virus 1, adenovirus, and rotavirus. *Memorias do Instituto Oswaldo Cruz* 102(4): 469–472.

Chakraborty, K., and Joy, M. 2017. Anti-diabetic and anti-inflammatory activities of commonly available cephalopods. *International Journal of Food Properties* 20(7): 1655–1665.

Cheong, K. L., Xia, L. X., and Liu, Y. 2017. Isolation and characterization of polysaccharides from oysters (*Crassostrea gigas*) with anti-tumor activities using an aqueous two-phase system. *Marine Drugs* 15(11): 338.

Chi Fai Cheung, R., Ng, T. B., and Wong, J. H. 2015. Marine peptides: Bioactivities and applications. *Marine Drugs* 13: 4006–4043.

Coulson, S., Butt, H., Vecchio, P., Gramotnev, H., and Vitetta, L. 2013. Green-lipped mussel extract (*Perna canaliculus*) and glucosamine sulphate in patients with knee osteoarthritis: Therapeutic efficacy and effects on gastrointestinal microbiota profiles. *Inflammopharmacology* 21(1): 79–90.

Dang, V. T., Benkendorff, K., Green, T., and Speck, P. 2015. Marine snails and slugs: A great place to look for antiviral drugs. *Journal of Virology* 89: 8114–8118.

Dolashka, P., Moshtanska, V., Borisova, V. et al. 2011. Antimicrobial proline-rich peptides from the hemolymph of marine snail *Rapana venosa*. *Peptides* 32: 1477–1483.

Dolashka, P., Velkova, L., Shishkov, S., et al. 2010. Glycan structures and antiviral effect of the structural subunit RvH2 of Rapana hemocyanin. *Carbohydrate Research* 345(16): 2361–2367.

Dolashka-Angelova, P., Lieb, B., Velkova, L., et al. 2009. Identification of glycosylated sites in Rapana hemocyanin by mass spectrometry and gene sequence, and their antiviral effect. *Bioconjugate Chemistry* 20(7): 1315–1322.

Dupuy, J. W., Bonami, J. R., and Roch, P. 2004. A synthetic antibacterial peptide from *Mytilus galloprovincialis* reduces mortality due to white spot syndrome virus in palaemonid shrimp. *Journal of Fish Diseases* 27(1): 57–64.

Flores-Garza, R. et al. 2012. Commercially important marine mollusks for human consumption in acapulco, México. *Natural Resources* 3: 11–17.

Food and Agriculture Organization (FAO). 2016a. *The State of World Fisheries and Aquaculture: Contributing to Food Security and Nutrition for all*. Rome. 200 pp.

Food and Agriculture Organization (FAO)/INFOODS. 2016b. *Global Food Composition Database for Fish and Shellfish*. Version 10-uFiSh10. Rome, Italy.

Genova-Kalou, P., Dundarova, D., Idakieva, K., Mohmmed, A., Dundarov, S., and Argirova, R. 2008. Anti-herpes effect of hemocyanin derived from the mollusk *Rapana thomasiana*. r *Zeitschrift für Naturforschung C* 63(5–6): 429–434.

Global Regulatory Services (GRS). 2018. *Nutraceuticals & Functional Foods*. Global House, Cambridgeshire, UK. https://globalregulatoryservices.com/industry-sectors/nutraceuticalsfunctional-foods

Gokoglu, N., and Yerlikaya, P. 2015. *Seafood Chilling, Refrigeration and Freezing: Science and Technology*. Chichester, UK: Wiley Blackwell.

Grienke, U., Silke, J., and Tasdemir, D. 2014. Bioactive compounds from marine mussels and their effects on human health. *Food Chemistry* 142: 48–60.

Hooper, C., Day, R., Slocombe, R., Handlinger, J., and Benkendorff, K. 2007. Stress and immune responses in abalone: Limitations in current knowledge and investigative methods based on other models. *Fish & Shellfish Immunology* 22(4): 363–379.

Huerta-Yépez, S., Tirado-Rodriguez, A. B., and Hankinson, O. 2016. Role of diets rich in omega-3 and omega-6 in the development of cancer. *Boletín Médico del Hospital Infantil de México* 73: 446–456.

Huss, H. 2003. *Assessment and Management of Seafood Safety and Quality*. Technical Paper No. 444. Rome, Italy: Food and Agriculture Organization.

Infinium Global Research. 2019. *Marine-Derived Drugs Market Growing at a CAGR of 11.20% and Expected to Reach $21,955.6 Million by 2025*. https://www.medgadget.com/2019/07/marine-derived-drugs-market-growing-at-a-cagr-of-11-20-and-expected-to-reach-21955–6-million-by-2025-exclusive-report-by-infinium-global-research.html (accessed October, 15 2020)

Kalogeropoulos, N., Andrikopoulos, N. K., and Hassapidou, M. 2004. Dietary evaluation of Mediterranean fish and molluscs pan-fried in virgin olive oil. *Journal of the Science of Food and Agriculture* 84(13): 1750–1758.

Kean, J. D., Camfield, D., Sarris, J., et al. 2013. A randomized controlled trial investigating the effects of PCSO-524, a patented oil extract of the New Zealand green lipped mussel (*Perna canaliculus*), on the behaviour, mood, cognition and neurophysiology of children and adolescents (aged 6–14 years) experiencing clinical and sub-clinical levels of hyperactivity and inattention: Study protocol ACTRN12610000978066. *Nutrition Journal* 12(1): 100.

Kehinde, O., Mariam, Y., Adebimpe, O., and Blessing, A. 2015. Traditional utilization and biochemical composition of six mollusc shells in Nigeria. *Revista de Biologia Tropical* 63(2): 459–464.

Khan, B. M., and Liu, Y. 2019. Marine Mollusks: Food with Benefits. *Comprehensive Reviews in Food Science and Food Safety* 18: 548–564.

Kim, S. K., and Pallela, R. 2012. Medicinal uses of marine animals: Current state and prospects. In *Advances in Food and Nutrition Research,* ed. S. K. Kim, 65, pp. 1–9. New York, NY: Academic Press.

Kreuzer, R. 1984. *Cephalopods: Handling Processing and Products* (p. 108). Fisheries Technical Paper, 254. Rome, Italy: Food and Agriculture Organization.

Kubota, Y., Watanabe, Y., Otsuka, H., Tamiya, T., Tsuchiya, T., and Matsumoto, J. J. 1985. Purification and characterization of an antibacterial factor from snail mucus. *Comparative Biochemistry and Physiology. Part C, Comparative Pharmacology* 82(2): 345–348.

Kumaran, N. S., Bragadeeswaran, S., and Thangaraj, S. 2011. Screening for antimicrobial activities of marine molluscs *Thais tissoti* (Petit, 1852) and *Babylonia spirata* (Linnaeus, 1758) against human, fish and biofilm pathogenic microorganisms. *African Journal of Microbiology Research* 5(24): 4155–4161.

Kumari, M., Priyanka, P., and Malik, S. 2020. Molluscans: Applications from Basic Life Science to Biotechnology. *Bulletin of Environment, Pharmacology and Life Science* 9(5): 134–141.

Le Cointre, G., and Le Guyader H. 2001. *Classification phylogénique du vivant*. Paris: Belin, p. 542.

Lev, E., and Amar, Z. 2007. *Practical Material Medica of the Medieval Eastern Mediterranean according to the Cairo Genizah*. Leiden, Netherlands: Brill.

Lim, N. S. H., Everuss, K. J., Goodman, A. E., and Benkendorff, K. 2007. Comparison of surface microfouling and bacterial attachment on the egg capsules of two molluscan species representing Cephalopoda and Neogastropoda. *Aquatic Microbial Ecology* 47: 275–287.

Liu, H., S¨oderh¨all, K., and Jiravanichpaisal, P. 2009. Antiviral immunity in crustaceans. *Fish & Shellfish Immunology* 27(2): 79–88.

Liu, M., Zhao, X., Zhao, J., et al. 2012. Induction of apoptosis, G0/G1 phase arrest and microtubule disassembly in K562 leukemia cells by Mere15, a novel polypeptide from Meretrix meretrix Linnaeus. *Marine Drugs* 10: 2596–2607.

Lou, Q. M., Wang, Y. M., and Xue, C. H. 2013. Lipid and fatty acid composition of two species of abalone, *Haliotis discus hannai. Journal of Food Biochemistry* 37(3): 296–301.

Luk'yanov, P., Chernikov, O., Kobelev, S., Chikalovets, I., Molchanova, V., and Li, W. 2007. Carbohydrate-binding proteins of marine invertebrates. *Russian Journal of Bioorganic Chemistry* 33(1): 161–169.

Lyprinol. 2018. *Lyprinol.* http://www.lyprinol.com/

Martin, D. E., and Hall, S. G. 2006. Oyster shucking technologies: Past and present. *International Journal of Food Science & Technology* 41(3): 223–232.

Mateos, R., Pérez-Correa, J. R., and Domínguez, H. 2020. Bioactive properties of marine phenolics. *Marine Drugs* 18: 501.

Nalini, S., Richard, D., Riyaz, S. U., Kavitha, G., and Inbakandan, D. 2018. Antibacterial macro molecules from marine organisms. *International Journal of Biological Macromolecules* 115: 696–710.

Olicard, C., Didier, Y., Marty, C., Bourgougnon, N., and Renault, T. 2005. In vitro research of anti-HSV-1 activity in different extracts from Pacific oysters *Crassostrea gigas*. *Diseases of Aquatic Organisms* 67(1–2): 141–147.

Orlando, P., Strazzullo, G., Carretta, F., De Falco, M., and Grippo, P. 1996. Inhibition mechanisms of HIV-1, Mo-MuLV and AMV reverse transcriptases by Kelletinin A from *Buccinulum corneum*. *Experientia* 52(8): 812–817.

Ozogul, F., Kuley, E., and Ozogul, Y. 2015. Sterol content of fish, crustacean and mollusk: Effects of cooking methods. *International Journal of Food Properties* 18(9): 2026–2041.

Pangestuti, R., and Kim, S. K. 2017. Bioactive peptide of marine origin for the prevention and treatment of non-communicable diseases. *Marine Drugs* 15(3): 67.

Przeslawski, R., and Benkendorff, K. 2005. The role of surface fouling in the development of encapsulated gastropod embryos. *Journal Molluscan Studies* 71: 75–83.

Reverchon, E., and De Marco, I. 2006. Supercritical fluid extraction and fractionation of natural matter. *Supercritical Fluids* 38: 146–166.

Roch, P., Beschin, A., and Bernard, E. 2004. Antiprotozoan and antiviral activities of non-cytotoxic truncated and variant analogues of mussel defensin. *Evidence-Based Complementary and Alternative Medicine* 1(2): 167–174.

Rudd, D., and Benkendorff, K. 2014. Supercritical CO2 extraction of bioactive Tyrian purple precursors from the hypobranchial gland of a marine gastropod. *The Journal of Supercritical Fluids* 94: 1–7.

Santhi, V., Sivakumar, V., Thangathirupathi, A., and Thilaga, R. 2011. Analgesic, antipyretic and anti-inflammatory activities of chloroform extract of prosobranch mollusc *Purpura persica*. *International Journal of Pharmacology & Biological Sciences* 5(2): 9–15.

Santhi, V., Sivakumar, V., Mukilarasi, M., and Kannagi, A. 2013. Antimicrobial substances of potential biomedical importance from *Babylonia zeylanica*. *Journal of Chemical and Pharmaceutical Research* 5(9): 108–115.

Sarkar, A., Datta, P., Gomes, A., Das Gupta, S., and Gomes, A. 2013. Anti-osteoporosis and anti-osteoarthritis activity of fresh water snail (*Viviparous bengalensis*) flesh extract in experimental animal model. *Open Journal of Rheumatology and Autoimmune Diseases* 3(1): 10–17.

Sharanagat, V. S., Singla, V., and Singh, L. 2020. Bioactive compounds from marine sources. In *Technological Processes for Marine Foods-from Water to Fork: Bioactive Compounds, Industrial Applications and Genomics,* eds. M. R. Goyal, H. A. Rasul Suleria, and S. Kirubanandan. Oakville, ON, Canada: Apple Academic Press, Inc.

Shukla, S. 2016. Therapeutic importance of peptides from marine source: A mini review. *Indian Journal of Geo-Marine Sciences* 45(11): 1422–1431.

Simat, V., Elabed, N., Kulawik, P., et al. 2020. Recent advances in marine-based nutraceuticals and their health benefits. *Marine Drugs* 18(627): 1–40.

Su, Y. C., and Liu, C. 2013. Shellfish handling and primary processing. In *Seafood Processing: Technology, Quality and Safety*, ed. I. S. Boziaris, pp. 9–32. Chichester, UK: John Wiley & Sons.

Suleria, H. A. R., Addepalli, R., Masci, P., Gobe, G., and Osborne, S. A. 2017. In vitro anti-inflammatory activities of blacklip abalone (Haliotis rubra) in RAW 264.7 macrophages. *Food and Agricultural Immunology* 28(4): 711–724.

Suleria, H. A. R., Osborne, S., Masci, P., and Gobe, G. 2015. Marine-based nutraceuticals: An innovative trend in the food and supplement industries. *Marine Drugs* 13(10): 6336–6351.

Surm, J. M., Prentis, P. J., and Pavasovic, A. 2015. Comparative analysis and distribution of omega-3 LCPUFA biosynthesis genes in marine molluscs. *PLoS ONE 10*: e0136301.

Suttle, C. A. 2007. Marine viruses—Major players in the global ecosystem. *Nature Reviews Microbiology* 5(10): 801–812.

Thakur, M. 2020. Marine bioactive components: Sources, health benefits, and future prospects. In *Technological Processes for Marine Foods-from Water to Fork: Bioactive Compounds, Industrial Applications and Genomics*, pp. 61–72. New York, NY: Apple Academic Press.

Venugopal, V. 2006. *Seafood Processing: Adding Value Through Quick Freezing, Retortable Packaging and Cook-chilling*. Boca Raton, FL: CRC Press/Taylor & Francis.

Venugopal, V., and Gopakumar, K. 2017. Shellfish: Nutritive value, health benefits, and consumer safety. *Comprehensive Reviews in Food Science and Food Safety* 16(6): 1219–1242.

Wakimoto, T., Kondo, H., Nii, H. et al. 2011. Furan fatty acid as an anti-inflammatory component from the green-lipped mussel *Perna canaliculus*. *Proceedings of the National Academy of Sciences United States of America* 108(42): 17533–17537.

Wayan Mudianta, I., White, A. M., Suciati, et al. 2014. Chemoecological studies on marine natural products: Terpene chemistry from marine mollusks. *Pure and Applied Chemistry* 86(6): 995–1002.

Zhukova, N. V. 2019. Fatty acids of marine mollusks: Impact of diet, bacterial symbiosis and biosynthetic potential. *Biomolecules* 9: 857.

Index

E

F

G

H

I

K

L

M

N

O

P

R

S

T

U

V

X

Z

Printed in the United States
by Baker & Taylor Publisher Services